Optical Communication Systems

Optical Communication Systems

System Analysis, Design and Optimization

J. Franz/V.K. Jain

(450 pages and 148 figures)

John Wiley & Sons (Asia) Pte Ltd

Narosa Publishing House

Prof. Dr.-Ing. Jürgen Franz
Fachhochschule Düsseldorf
Optical Communications Laboratory, Faculty of Electrical Engineering,
Josef-Gockeln-Straβe 9, 40474 Düsseldorf, Germany

Dr. Virander K. Jain
Associate Professor
Department of Electrical Engineering, Indian Institute of Technology, Delhi
New Delhi 110016, India

Text : Camera Ready by Authors

ISBN 0-471-19005-5   John Wiley & Sons

Cataloging-in-Publication Data:

Franz, J., Jain, V. K.
     Optical Communication Systems
     p. cm.
     ISBN: 0-471-19005-5

Printed in India

PREFACE

For several years now, optical fiber communication systems are extensively being used all over the world for telecommunication and data transmission purposes. Optical communications offer advantage of ultrahigh speed and highly reliable data transmission. In telecommunication engineering, optical fiber and free-space communications have become more and more important. Optical communication links have now been preferred over the other high bit rate point-to-point communication links. Further, light intensity modulation and direct detection optical transmission links are also most economical.

With coherent detection, optical communications have reached a new generation, characterized by a number of new applications. Coherent optical communication systems offer significant advantages based on increased receiver sensitivity and improved selectivity. First advantage is partially reduced by optical preamplifier direct detection system, but the second advantage still remains. It is quite important for coherent multichannel communication systems. These systems represent key component of highly sophisticated multiple-access networks. Moreover, optical multichannel systems with coherent detection offer possibility of using fully the large optical bandwidth available with the fibers. Besides microelectronic and software, coherent optical communication represents a key technology of modern telecommunication systems.

This book has been written with the aim of providing basic material required for advanced study in theory and application of optical fiber and space communication systems with and without optical amplifiers. The background required to study the book is only that of typical senior level engineering students. Specifically, it is presumed that the reader has been introduced to the principles of electromagnetic theory and digital and analog communications. It would be helpful if the reader has some exposure to spectral analysis and statistics. Some of the major topics are briefly reviewed in this book also.

The book is recommended to those, who have interest in optical communications. It is written specifically as a textbook for senior level engineering students. The practicing engineers and physicists will also find it useful to update their knowledge in the field. In addition, this book will also be useful as a working reference in the selection and design of optical fiber and free-space communication systems.

We are pleased to thank our colleagues, students and friends who made many valuable suggestions and skillful services in the preparation of the manuscript. Last but not least, we would like to thank our wives and children for their patience during the time we devoted to writing this book.

The author wishes all readers a successful study.

J. Franz
V. K. Jain

CONTENTS

Physical Constants and Conversion Formulas

Physical Constants

Constant	Physical meaning	Quantity
c	Velocity of light	$2.998 \cdot 10^8$ m/s
e	Electron charge	$1.601 \cdot 10^{-19}$ As
h	Planck's constant	$6.624 \cdot 10^{-34}$ Ws^2
k_B	Boltzmann's constant	$1.379 \cdot 10^{-23}$ Ws/K
ϵ_0	Dielectricity constant	$8.854 \cdot 10^{-12}$ As/(Vm)
μ_0	Permeability constant	$1.256 \cdot 10^{-6}$ Vs/(Am)

Conversion Formulas

$1 \ \mu$m $= 10^{-6}$ m $= 10^{-4}$ cm

$1 \ \text{Å} = 10^{-4} \ \mu$m $= 10^{-10}$ m

1 Np $= 8.686$ dB

Frequency f in Hz $\approx 3 \cdot 10^{14}$/wavelength λ in μm

Bandwidth Δf in Hz (at centre wavelength λ) $\approx (c/\lambda^2) \cdot$ bandwidth $\Delta\lambda$ in wavelength

1   INTRODUCTION

In conventional fiber optic communication systems, information is carried only by modulating the optical *power* or *intensity* of a laser or a light emitting diode (LED). In receiver, demodulation is simply performed by converting the modulated light to an equivalent electrical current by means of a photodiode. This detection scheme is termed as *direct detection*. Optical communication systems with *intensity modulation* and direct detection (IM/DD) normally have poor sensitivity. In *coherent optical communication system*s, information is contained either in amplitude, frequency or phase of an optical carrier signal. The receiver analyzes the received *electric field* with respect to modulation scheme applied to the optical carrier. This is done by mixing the electric fields of incoming optical signal and local laser, which operates as a local oscillator (LO). Thereby, an intermediate frequency (IF), equal to the difference between LO and received carrier frequencies is generated. The IF signal is simply a frequency-translated replica of the original incoming optical signal. As the IF is usually in the GHz range, well-established microwave signal processing techniques can be employed to process this signal.

A direct detection receiver is the optical analogue of an old-fashioned crystal radio, whereas a coherent receiver is that of a modern superheterodyne receiver used in radio and television systems. Just as heterodyning has revolutionized radio communications, coherent optical detection opens the door to more sophisticated advanced optical communications.

Coherent optical communication systems have two fundamental advantages over the IM/DD systems. The first advantage is a high gain in receiver sensitivity up to 20 dB and more. This leads to an enlargement of repeater spacing in long-haul point-to-point applications. With a single-mode fiber of attenuation 0.2 dB/km, increase in repeater spacing of more than 100 km can be achieved. The second advantage is a significant improvement in receiver selectivity, which offers the possibility to realize a real *optical frequency division multiplex* (OFDM) system with a large number of optical channels at different closely-spaced optical carriers. This transmission scheme is often termed as *coherent multichannel communication* (CMC). In the receiver, optical channels are separated in the electrical domain by means of a tunable local laser and sharp microwave filters unlike direct detection receiver. Theoretically, CMC systems with more than hundred optical high-bandwidth channels are possible. Thus, CMC systems with coherent detection can make use fully the large optical bandwidth available with the optical fiber. In combination with optical time division multiplexing (OTDM), which offers a transmission speed of up to 100 Gbit/s at a single optical carrier, CMC systems represent the basic for multigigabit transmission. Even fiber optic Tbit/s-transmission links becomes realistic, offering various attractive new applications.

All advantages of coherent optical communications are at the cost of an increased complexity. Thus, additional components such as the local laser, optical coupler and additional control and tracking circuits like the polarization control unit are required. Moreover, some advantages mentioned above have partially been reduced by the advent of optical amplifiers and dispersion equalization in the optical domain.

Since many years the telecommunication market is world wide growing very rapidly and the pace of growth is still increasing. The enormous progress in communication technology during the last decades was primarily based on various milestones such as digitalization, microprocessor-controlled switching, optical fiber transmission, local area network (LAN), integrated services digital network (ISDN), cellular mobile radio and ATM (asynchronous transmission mode). Nowadays, modern communications is fundamentally characterized by three key technologies: microelectronics, software-controlling and last but not least, optical communications. The next generation of telecommunication will primarily be characterized by highly sophisticated, high bit rate information superhighways, interactive broadband multimedia services and intelligent multiple-access networks combining ISDN with broadband services (which results in BISDN) and combining the flexible ATM transmission scheme with the standardization of the synchronous digital hierarchy (SDH). Furthermore, photonics may more and more replace electronics. In the frame of advanced optical communications, coherent systems will surely represent a key technology of significant importance.

First experiment on coherent optical communication was performed in 1967 and second in 1974 by using gas lasers and free-space medium [84, 187]. However, atmospheric distortion and small transmission range limited interest in such systems at that time. Similar experiments were restarted in 1980, but now by employing optical fibers and semiconductor laser diodes [228]. Within a short time, the operation principle of fiber-based coherent optical communication was demonstrated. However, much improvement in receiver sensitivity was not achieved. Subsequently, research activities increased rapidly in three different areas. The first area includes all system experiments and the measurement technique required to determine and to assess the system performance. Practical research in system design is primarily characterized by performing long-period field experiments, realizing high-quality, fully-engineered single and multicarrier systems and developing commercial low-cost systems of high-reliability. In the development of high bit rate single carrier coherent systems, modulation speed has progressed from some Mbit/s to more than 10 Gbit/s. The second field of research is focused on components since the performance of coherent optical communication systems fundamentally depends on components quality. Linewidth narrowing is one of the key points in realizing coherent optical systems. Depending on modulation and demodulation schemes used, linewidth requirement ranges from comparable to the bit rate to a small fraction of the bit rate (see Chapters five and seven). For the realization of CMC systems, a tuneable local laser with broad tuning range at a fixed linewidth and optical output power are most important. The last research field is dedicated on theory. This area especially includes system analysis such as the calculation of bit error rate (BER) and system optimization. Due to complexity of coherent optical communication systems, this task must frequently be done by means of powerful approximations or by computer simulations. The latter approach represents a very useful tool offering a fast and low-cost analysis of coherent optical communication systems. Furthermore, computer simulation gives a clear insight into the complex interactions of the signals and components in a coherent system. For instant, effects of a change in system parameter can quickly be assessed.

This book is mainly focused on system analysis, design and optimization of optical communication systems. To understand this, some knowledge of basic material on optical communications is unavoidable. Therefore, a great part of the book is earmarked for basic material.

Chapter two provides the background required for advanced study of coherent optical communication systems. This Chapter describes the principle of coherent optical receivers and gives an introduction to the basic laws. In order to assess the performance of coherent receivers, it is useful to compare its features with the direct detection receiver, which is well-developed and established in commercial optical communication networks since many years. Therefore, this Chapter gives a brief review of conventional direct detection receiver also. Performance gain of coherent over direct detection receiver is then analyzed. It also gives an overview of the various components and signals in coherent optical link. Finally, it reviews some important aspects of digital communication systems which are required to analyze coherent optical communication systems. The eye pattern technique and probability of error, also referred as BER, are refreshed as two important criteria for systems comparison.

The performance of coherent optical transmission systems can seriously be deteriorated by the intensity/amplitude and phase noise of both transmitter and local lasers. In particular, the laser phase noise is a dominant source of noise in coherent optical communication systems. The analysis of coherent systems requires a basic knowledge of laser physics, in particular a knowledge of the statistical properties of laser noise. *Chapter three* first gives a brief review of the basic principle of laser operation and discusses the physical reasons of laser noise. Subsequently, statistical properties of laser noise by means of a simple model are derived which especially allow to determine its degrading effect on the communication system. This Chapter also briefly examines the effects of laser relaxation oscillations and presents some interesting and generally valid aspects on filtering signals perturbed by phase noise. Here, a remarkable difference between phase noise and additive Gaussian receiver noise (i.e., shot noise and thermal noise) has been illustrated. Finally, some important technical solutions to reduce the effects of laser phase noise are suggested.

Besides laser noise, polarization fluctuations in the fiber also seriously deteriorate the performance of coherent optical communication systems. *Chapter four* is focused on the reasons, effects and technologies of handling the problems arising due to polarization fluctuations.

Based on the fundamentals of Chapter two, *Chapter five* gives the analyses of various coherent optical communication systems. Modulation and demodulation schemes are used as a characteristic feature to distinguish and to group the different systems. In the analysis, all relevant sources of noise and system imperfections have been considered. In this Chapter, system optimization is always accomplished with a goal to minimize the probability of bit error. For this purpose, various methods of calculation and approximation are explained and compared. In addition, this Chapter also presents some interesting results obtained by computer simulation.

One of the most exciting development in the field of optics has been the discovery of optical amplifier. It is probably the most significant development since the discovery of single-mode fiber. *Chapter six* describes different types of optical amplifiers and presents their comparative study. Further, various applications of optical amplifiers in the optical communication systems are discussed.

Main objective of *Chapter seven* is to compare the optical communication systems based on the analysis given in Chapters five and six. For comparison, different criteria are taken into account: bit rate, laser linewidth requirements, eye pattern, possible applications etc.

Over the past several years, optical fiber communication systems are extensively used all over the world for telecommunication and data transmission purposes. Therefore, design of optical links to meet a given requirements has become an important issue. In *Chapter eight*, design procedure of a digital point-to-point link has been given. Various types of the optical networks have been described and design issues involved in their implementation have been discussed.

In *Chapter nine*, a very promising application of optical communications for space links i.e., optical free-space communications has been discussed. This last Chapter serves to provide an overall view to applications and advantages of optical space communications relative to other transmission media such as fiber and microwave communications.

2 FUNDAMENTALS

This Chapter provides the fundamental background required for advanced study of coherent optical communication systems.

Section 2.1 describes the principle function of coherent optical receivers and gives an introduction to the basic laws. In order to assess the performance of coherent receivers, it is useful to compare their features with a reference receiver. For this purpose, direct detection receiver is well-suited since this receiver is well-developed and established in commercial optical communication networks [87, 134, 275]. Therefore, Section 2.1 begins with a brief review of conventional direct detection receiver.

In coherent detection communication systems, there are two different receiver configurations: heterodyne and homodyne. Both concepts are briefly described in Section 2.2. We then analyze the performance gain of coherent receivers over direct detection receiver. Sections 2.3 and 2.4 give an overview of various components and signals in coherent optical fiber systems. Finally, Section 2.5 reviews some important aspects of digital communication systems, which are required to analyze coherent optical communication systems. The eye pattern technique and probability of error, also referred as BER, are refreshed as two important and powerful measures for systems comparison.

2.1 PRINCIPLE OF DIRECT AND COHERENT DETECTION RECEIVERS

In a *direct detection receiver,* incoming optical light wave electric field (E_r) at the receiver input is directly converted to an electrical current by means of a PIN-photodiode. When E_r is quite weak, an avalanche photodiode (APD) is often used. The photodiode current in both the cases is directly proportional to the square of E_r and, therefore, proportional to optical power (P_r) incident upon the photodiode. Let us consider the current in an APD receiver:

$$I_{APD} \sim P_r \sim E_r^2$$

$$I_{APD} = M \, R_0 \, P_r \tag{2.1}$$

Here, R_0 [A/W] represents the responsivity and M the average gain of photodiode. The responsivity R_0 is given by the well-known expression [275]

$$R_0 = \frac{e \, \eta}{h \, f} \tag{2.2}$$

where e is the electron charge, η the quantum efficiency of photodiode, h the Planck's constant and f the frequency of light.

(a) Direct detection receiver

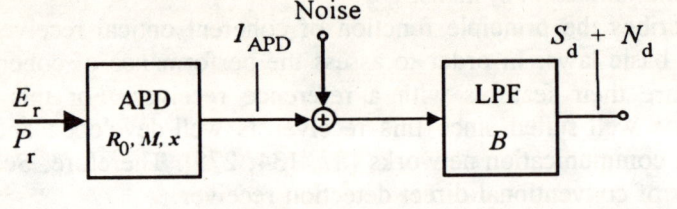

$$I_{APD} \sim E_r^2 \sim P_r$$

$$S_d = R_0^2 M^2 P_r^2 \qquad \text{signal power}$$

$$N_d = (G_{ds} + G_{dark} + G_{th})\,B = e\,M^{2+x}\,(R_0 P_r + I_{dark})\,B + G_{th}\,B \qquad \text{noise power}$$

(b) Coherent detection receiver

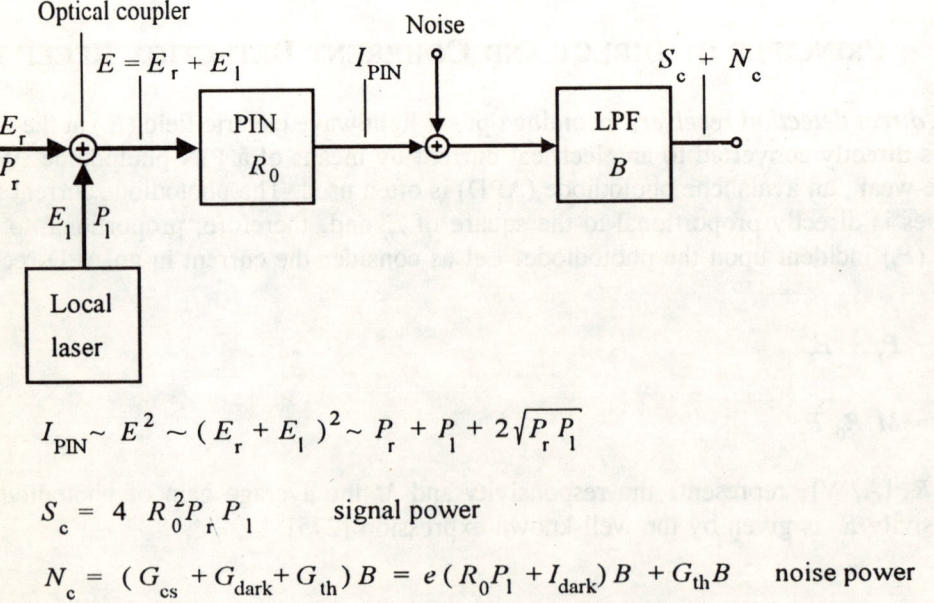

$$I_{PIN} \sim E^2 \sim (E_r + E_l)^2 \sim P_r + P_l + 2\sqrt{P_r P_l}$$

$$S_c = 4\,R_0^2 P_r P_l \qquad \text{signal power}$$

$$N_c = (G_{cs} + G_{dark} + G_{th})\,B = e\,(R_0 P_l + I_{dark})\,B + G_{th}\,B \qquad \text{noise power}$$

Fig. 2.1: Principle block diagram of optical (a) direct and (b) coherent detection receivers

It can be seen from Fig. 2.1a that the photodiode current I_{APD} is disturbed by noise with constant power spectral density (psd) $G_d(f) = G_d$. In an optical receiver, this additive noise arises from shot noise (G_{ds}) of the photodiode and electronic or thermal noise of the amplifiers and resistors (G_{th}) frequently referred as circuit noise. For simplicity, amplifiers are not added to the block diagram in Fig. 2.1. The shot noise have two components: one component depends on the received optical power level (G_{dp}), while the other component is independent of it (G_{dark}). The latter one is caused by the dark current I_{dark} of photodiode. The overall double-sided psd is given by

$$G_d(f) = G_d = G_{ds} + G_{th} = G_{dp} + G_{dark} + G_{th}$$

$$= e \, M^2 \, F(M) \left[R_0 \, P_r + I_{dark} \right] + G_{th} \tag{2.3}$$

Here, suffix d represents direct detection and will only be used if there is a difference to coherent detection. In most practical applications, excess noise factor $F(M)$ can be approximated by

$$F(M) \approx M^x \tag{2.4}$$

where x is the excess noise exponent which depends on the photodiode material (0.2-0.5 for Silicon and 0.9-1.0 for Germanium). If B is the double-sided noise equivalent bandwidth of the low-pass filter, then signal power, noise power and signal-to-noise ratio (it is abbreviated as SNR or S/N ratio also) will be

$$S_d = (R_0 \, M \, P_r)^2 \tag{2.5}$$

$$N_d = e \, M^{2+x} \left(R_0 \, P_r + I_{dark} \right) B + G_{th} \, B \tag{2.6}$$

and

$$SNR_d = \left(\frac{S}{N} \right)_d = \frac{(R_0 \, M \, P_r)^2}{e \, M^{2+x} \left(R_0 \, P_r + I_{dark} \right) B + G_{th} \, B} \tag{2.7}$$

Sometimes, N_d and SNR_d are expressed in terms of single-sided psd. In that case, double-sided noise equivalent bandwidth will be replaced by single-sided bandwidth. It can be seen from Eqs. (2.5) and (2.6) that signal and noise power are related to a reference resistor of 1 Ω. Hence, S_d and N_d are expressed in Amperes squared (A^2) instead of Watts (W). In order to simplify calculation and focus our interest on the principle in this Section, effect of low-pass filter on signal power S_d is not considered here. For this reason, we assume a constant, time-invariant optical power P_r at the input to the receiver and, consequently, a constant shot noise psd G_{ds}. The received optical power P_r is actually a function of time since intensity modulation is normally applied to the transmitter laser source. Therefore, optical systems with intensity modulation and direct detection (IM/DD) are characterized by *signal-dependent noise*.

Next we consider the coherent receiver shown in Fig. 2.1b. Here, incoming optical signal and signal from a local laser are superimposed on the surface of photodiode. This is in contrast to direct detection receiver wherein received signal is directly made to incident on the photodiode. A coherent receiver is called *homodyne receiver* if the frequencies of received and local laser light waves are equal and *heterodyne receiver* if these are different. In Sections 2.3 and 2.4, heterodyne and homodyne receivers have been considered in more detail. The superposition of light waves is simply described by

$$E = E_r + E_l \tag{2.8}$$

where E_l is the electric field of local laser light wave and P_l the corresponding power level. Like in direct detection receiver, photodiode current is again proportional to optical power at the input of photodiode. Assuming a PIN-photodiode (APD is generally not used in coherent receivers), current at the output of the photodiode is given by

$$I_{PIN} \sim (E_r + E_l)^2 \sim P_r + P_l + 2\sqrt{P_r P_l}$$

$$I_{PIN} = R_0 \left(P_r + P_l + 2\sqrt{P_r P_l} \right) \tag{2.9}$$

The photodiode current I_{PIN} has three components, but only the third component

$$I = 2 R_0 \sqrt{P_r P_l} \tag{2.10}$$

is of interest in coherent receivers. With respect to frequency, this mixing produces a baseband signal if homodyne receiver is used (this case is assumed here), whereas it is an intermediate frequency signal in case of heterodyne receiver. The intermediate frequency (IF) itself is equal to the difference in local laser and received optical light wave frequencies. The first and also the second components in Eq. (2.9) can practically be eliminated by means of filtering or by using a balanced receiver with two photodiodes instead of one (Section 2.4). Similar to the constant noise psd G_d of a direct detection receiver, noise psd in coherent heterodyne or homodyne receiver is given by

$$G_c(f) = G_c = G_{cs} + G_{th} = G_{cp} + G_{dark} + G_{th}$$
$$= e \left[R_0 P_1 + I_{dark} \right] + G_{th} \tag{2.11}$$

In the strict sense, P_r must also be considered in addition to P_1 to determine noise psd from Eq. (2.11). In most practical applications, however, this is not required because P_r is always some order of magnitudes lower than P_1 i.e.,

$$P_r < P_1 \tag{2.12}$$

This important relationship is generally valid due to attenuation of the optical link (for example, fiber attenuation). Let us consider now signal power, noise power and SNR:

$$S_c = 4 R_0^2 P_r P_1 \tag{2.13}$$

$$N_c = e (R_0 P_1 + I_{dark}) B + G_{th} B \tag{2.14}$$

and

$$SNR_c = \left(\frac{S}{N} \right)_c = \frac{4 R_0^2 P_r P_1}{e (R_0 P_1 + I_{dark}) B + G_{th} B} \tag{2.15}$$

It may be mentioned that this equation is valid only for the simplified coherent receiver configuration shown in Fig. 2.1b. From Eq. (2.14), it can be seen that noise power N_c at the output of coherent receiver is independent of the received power level P_r and applied modulation. Hence, coherent optical systems are characterized by *signal-independent noise* which is an advantage in comparison to direct detection system. However, in presence of laser phase noise this advantage no longer exists (Chapter three).

2.2   PERFORMANCE GAIN

Comparing the signal power S_d at the output of direct detection receiver (Eq. 2.5) and signal power S_c at the output of coherent receiver (Eq. 2.13), it becomes clear that the avalanche gain M of an APD is merely replaced by the capability of the local laser light power P_1 to amplify the signal in the same way: $S_d \sim M^2$ and $S_c \sim P_1$.

Difference in noise power N_d and N_c is, however, much more predominant. In direct detection receiver, signal-dependent shot noise (G_{ds}) is an important part of the receiver noise and it increases with M^{2+x}. As x is always greater than zero, M^{2+x} is always more than M^2. For this reason, shot noise in direct detection receiver is always amplified more than the signal power: $G_{dp} \sim M^{2+x}$ and $S_d \sim M^2$. Hence, there is an optimum avalanche gain M which maximizes SNR_d. It is given by

$$M_{opt} = \left(\frac{2 \, G_{th}}{x \, e \, (R_0 \, P_r + I_{dark})} \right)^{\frac{1}{2+x}} \tag{2.16}$$

Unlike direct detection, signal power S_c and signal-independent shot noise (G_{cs}) are amplified in like manner if heterodyne or homodyne detection is used. Therefore, a rise in P_l always improves SNR_c (an exception is explained in [60]). If P_l is sufficiently large, then all sources of noise which are independent of P_l can be neglected. As system performance is now limited by the shot noise of photodiode only, we call this most important boundary case the *shot noise limit*. Here, SNR_c exhibits a saturation and reaches its maximum value which is much higher than SNR_d of direct detection receiver: $SNR_c \gg SNR_d$. It represents a significant advantage of coherent detection over direct detection.

Emergence of performance gain is shown in Fig. 2.2 which gives a first qualitative comparison of both types of optical receivers, coherent and direct detection. Fig. 2.2a and 2.2b compare the signal power S_c and S_d and noise power N_c and N_d respectively by defining the ratios

$$V_S = \frac{S_c}{S_d} = 4 \, M^{-2} \, \frac{P_l}{P_r} \tag{2.17}$$

and

$$V_N = \frac{N_c}{N_d} = \frac{R_0 \, P_l + I_{dark} + \dfrac{G_{th}}{e}}{M^{2+x}(R_0 \, P_r + I_{dark}) + \dfrac{G_{th}}{e}} \tag{2.18}$$

as a function of P_l. It is seen from Eqs. (2.17) and (2.18) that V_S and V_N increase proportional to P_l. However, increase in V_N is less than the increase in V_S. This highlights the reason why coherent receivers offer a performance gain over a direct detection receiver.

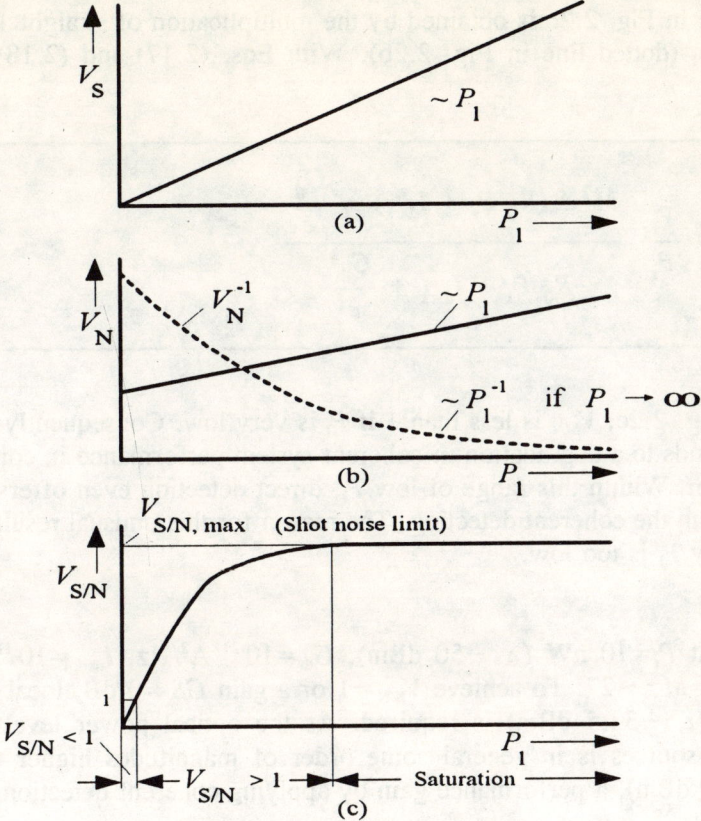

Fig. 2.2: Variations of relative (a) signal power, (b) noise power and (c) signal-to-noise ratio for coherent and direct detection receivers

A powerful measure to assess the performance gain of coherent receivers is defined by the gain

$$GS = 10 \log(V_{S/N}) \tag{2.19}$$

in SNR, where

$$V_{S/N} = \frac{\left(\dfrac{S}{N}\right)_c}{\left(\dfrac{S}{N}\right)_d} = V_S \, V_N^{-1} \tag{2.20}$$

The $V_{S/N}$ shown in Fig. 2.2c is obtained by the multiplication of straight V_s (Fig. 2.2a) and hyperbolic $1/V_N$ (dotted line in Fig. 2.2b). With Eqs. (2.17) and (2.18), above equation becomes

$$V_{S/N} = 4\,M^{-2}\,\frac{P_l}{P_r}\;\frac{M^{2+x}\,(R_0\,P_r + I_{dark}) + \dfrac{G_{th}}{e}}{R_0\,P_l + I_{dark} + \dfrac{G_{th}}{e}} \qquad (2.21)$$

As shown in Fig. 2.2c, $V_{S/N}$ is less than 1 if P_l is very low. Consequently, GS is negative, which corresponds to a degradation in coherent system performance in comparison to direct detection system. Within this range of low P_l, direct detection even offers a better receiver performance than the coherent detection. The reason for this unusual result is that the signal amplification by P_l is too low.

Example 2.1

Consider that $P_r = 10$ nW ($\triangleq -50$ dBm), $G_{th} = 10^{-23}$ A²/Hz, $I_{dark} = 10^{-11}$ A, $R_0 = 1$ A/W, $x = 0.9$ and $M = M_{opt} \approx 27$. To achieve $V_{S/N} = 1$ or a gain $GS = 0$ dB, local laser light power $P_l \approx 0.56\ \mu$W ($\triangleq -32.5$ dBm) is required. As the optical power level of commercially available laser sources is in general some order of magnitudes higher than -32.5 dBm (-10 dBm to 0 dBm), a performance gain by applying coherent detection is always obtainable in practice.

As seen from Fig. 2.2c, performance gain of coherent detection rises first very steeply with the increase in P_l. Finally, shot noise limited receiver operation is achieved when P_l becomes sufficiently large. The maximum gain GS_{max} in SNR at shot noise limit is

$$GS_{max} = 10\,\log\left(4\,M^x\left[1 + \frac{M^{2+x}\,I_{dark} + \dfrac{G_{th}}{e}}{M^{2+x}\,R_0\,P_r}\right]\right) \qquad (2.22)$$

Example 2.2

With the same system parameters as given in the previous example 2.1, it can be deduced from above equation that the performance gain due to coherent detection at high P_l is $GS_{max} = 20.5$ dB implying $V_{S/N} = V_{S/N,max} \approx 112$. Even for a local laser light power of only $P_l \approx 0.1$ mW ($\triangleq -10$ dBm), GS from Eq. (2.21) is 18.4 dB implying $V_{S/N} = 69$.

Besides the gain GS in SNR, gain in receiver sensitivity (GR) is another very important and most efficient measure to compare systems performance. GR is defined as the logarithm of the ratio of required optical power at the input of both types of receivers (direct and coherent) for $SNR_c = SNR_d$ (i.e., $V_{S/N} = 1$ and $GS = 0$ dB). It is given by

$$GR = 10 \log(V_r) = 10 \log\left(\frac{P_{r,d}}{P_{r,c}}\right) \tag{2.23}$$

The required optical input power $P_{r,d}$ and $P_{r,c}$ can be determined from Eqs. (2.7) and (2.15). These are given by

$$P_{r,d} = \frac{M^x e B}{2 R_0}\left(\frac{S}{N}\right)_d \left[1 + \sqrt{1 + \frac{4}{M}\frac{M^{2+x} I_{dark} + \dfrac{G_{th}}{e}}{M^{2+x} e B \left(\dfrac{S}{N}\right)_d}}\right] \tag{2.24}$$

and

$$P_{r,c} = \frac{e B}{4 R_0}\left(\frac{S}{N}\right)_c \frac{R_0 P_l + I_{dark} + \dfrac{G_{th}}{e}}{R_0 P_l} \tag{2.25}$$

For the same SNR i.e., $(S/N)_d = (S/N)_c = (S/N)$, gain in receiver sensitivity is given by

$$V_r = 2 M^x \frac{R_0 P_l}{R_0 P_l + I_{dark} + \dfrac{G_{th}}{e}}\left[1 + \sqrt{1 + \frac{4}{M}\frac{M^{2+x} I_{dark} + \dfrac{G_{th}}{e}}{M^{2+x} e B \dfrac{S}{N}}}\right] \tag{2.26}$$

In a coherent receiver, excess noise exponent x, I_{dark} and G_{th}/e do not affect the system performance provided P_l is high enough. Use of a coherent receiver is more advantageous when the influence of x, I_{dark} and G_{th}/e on the receiver performance is higher. Hence, a rise in x, I_{dark} or G_{th}/e offers an improvement in the receiver sensitivity. In order to obtain high SNR, large optical power $P_{r,d}$ or $P_{r,c}$ is required at the receiver input. Higher the desired SNR, larger the difference in $P_{r,d}$ and $P_{r,c}$ and, therefore, higher the GR. Under shot noise limited receiver operation (i.e., $P_1 \to \infty$), maximum gain in receiver sensitivity GR_{max} is achieved. It is given by

$$GR_{max} = 10 \log \left(2 M^x \left[1 + \sqrt{1 + \frac{4}{M} \frac{M^{2+x} I_{dark} + \frac{G_{th}}{e}}{M^{2+x} e B \frac{S}{N}}} \right] \right) \qquad (2.27)$$

In the following, both gain definitions i.e., gain in signal-to-noise ratio and gain in receiver sensitivity are summarized.

Gain in signal-to-noise ratio, GS

With equal optical power levels $P_{r,d}$ and $P_{r,c}$ at the input of direct and coherent detection receivers, GS represents the performance gain with respect to the signal-to-noise ratios $(S/N)_d$ and $(S/N)_c$ at the receivers output. Therefore, GS is a direct measure to assess the improvement in bit error rate obtained by using a coherent instead of a direct detection receiver (Section 2.5).

Gain in receiver sensitivity, GR

With equal signal-to-noise ratios $(S/N)_d$ and $(S/N)_c$ at the output of direct and coherent detection receivers, GR represents the performance gain with respect to the optical power $P_{r,d}$ and $P_{r,c}$ required at the receivers input. Therefore, GR is a direct measure to determine the improvement in repeater spacing (in km)

$$\Delta l = GR \cdot \alpha^{-1} \qquad (2.28)$$

provided the fiber attenuation α [dB/km] is known.

Example 2.3

Consider that equal signal-to-noise ratio $(S/N)_d = (S/N)_c = S/N = 40.48$ is required. Receiver parameters are the same as given in the examples 2.1 and 2.2. The coherent receiver is shot noise limited and the double-sided bandwidth of the low-pass filter is $B = 1120$ MHz. From Eqs. (2.27) and (2.28), it can be deduced that GR_{max} is 19.5 dB. With $\alpha = 0.2$ dB/km, Δl from Eq. (2.28) will be 97.5 km. It corresponds to an increase in repeater spacing by about 97 km.

Coherent communication systems are capable to offer a remarkable system performance gain of up to 20 dB in contrast to conventional direct detection system. However, such a high improvement in system performance can only be achieved under ideal conditions. A more realistic system analysis requires to consider also the following important aspects:

- *vector* and not scaler superposition of electric fields,

- polarization fluctuation,

- laser noise,

- intersymbol interference,

- influence of filters on laser phase noise,

- different degrading effects on binary symbols due to laser phase noise and

- system optimization.

It may be mentioned that a high gain in system performance can also be achieved by means of optical preamplifier in front of direct detection receiver. Assuming ideal conditions, direct detection with optical preamplification may also be able to reach shot noise limited receiver operation. In that case, no gain in system performance due to coherent detection remains ($GR \rightarrow 0$ dB). Since optical amplifiers are commercially available, the advantage of coherent detection in terms of a high gain in receiver sensitivity is no longer evident. The second advantage of coherent receivers i.e., its excellent selectivity still remains. Therefore, coherent optical receivers will especially find extensive use in high-capacity and highly sophisticated broadband communication networks offering services such as interactive multimedia or video-on-demand. These networks will primarily be based on high-density optical frequency division multiplex (OFDM) systems which are referred as coherent multichannel (CMC) systems (see Chapter seven).

2.3   COMPONENTS OF COHERENT OPTICAL COMMUNICATION SYSTEM

Block diagram of digital optical communication system with coherent detection is shown in Fig. 2.3. The first component of this system is the *digital source*. The signal $s(t)$ at the output of this source is characterized by discrete values, for example 0V and 1V. This signal represents the binary symbol sequence $<q_\nu>$ and, therefore, the information to be transmitted. A typical pattern $<q_\nu>$ is shown in Fig. 2.4. In the *electrical transmitter*, input signal $q(t)$ is converted to an electrical signal $s_e(t)$ which modulates the optical carrier wave $\vec{E}_C(t)$ of the transmitter laser in amplitude, frequency or phase. In Fig. 2.3, modulation is performed by *external optical modulator* which allows to apply all modulation schemes well-known and well-established in the radio frequency range such as amplitude shift keying (ASK), frequency shift keying (FSK), phase shift keying (PSK) or differential phase shift keying (DPSK).

In addition to the binary modulation schemes mentioned above, multilevel modulation schemes such as QPSK or 4-FSK can also be applied [44]. The basic function of external optical modulators is based either on electrooptical or acoustooptic phenomenon [268]. Electrooptical waveguide modulators based on $LiNbO_3$ allow a modulation frequency of more than 10 GHz.

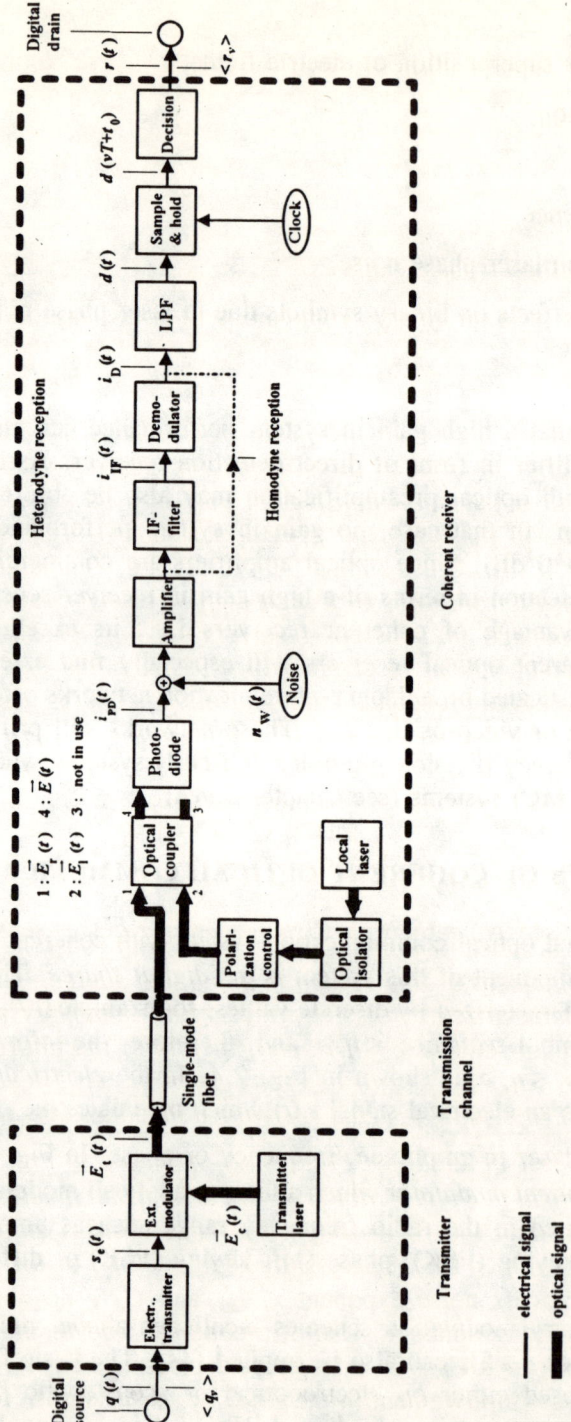

Fig. 2.3: Block diagram of fiber-based coherent optical communication system

In order to achieve high bit rate and long range data transmission with low bit error rate, a *single-mode transmitter laser* with an excellent emission spectrum is required. In particular, linewidth of the spectrum should be as small as possible. An ideal laser oscillates at one single frequency only which appears as a Dirac delta function at the frequency of the carrier in the emission spectrum. The reasons for undesired line broadening of real lasers and technical steps for its reduction will be explained in Chapter three. The optical carrier wave can mathematically be expressed by its complex electric field vector $\vec{E}_c(t)$ where vector representation describes the polarization of laser light wave.

There can be undesired reflections of the carrier light wave between transmitter laser and external modulator. These reflections impair the laser emission spectrum and, consequently, deteriorate the quality of transmission and reduce the overall system performance. To avoid or at least to reduce these impairments, an *optical isolator* (not shown in Fig. 2.3) is placed between transmitter laser and external modulator. The operation of optical isolator is based on the magnetooptical Faraday effect [52, 268].

The external modulator can be eliminated by using direct modulation technique (also referred as internal modulation) which is most efficient when semiconductor diode lasers are used. Such lasers can easily be modulated via the injection current. Direct modulation of diode laser allows to modulate the optical carrier wave in amplitude as well as in frequency. Further, DPSK modulation is also possible [186]. The primary advantage of using direct modulation is to avoid the external modulator as a component which exhibits a dominant insertion loss and increases the entire system cost.

As shown in Fig. 2.3, modulated optical wave $\vec{E}_t(t)$ is applied to *single-mode fiber* which is often called monomode fiber. The primary disturbances to the signal caused by the fiber arise from polarization fluctuations due to thermal and mechanical effects (Chapter four) and dispersion.

In the receiver front end, received light wave $\vec{E}_r(t)$ and local laser light wave $\vec{E}_l(t)$ are superimposed. To achieve better system performance, quality of the emission spectrum of the *local laser* must be similar to that of transmitter laser. The superposition of light waves $\vec{E}_r(t)$ and $\vec{E}_l(t)$ is performed by an *optical coupler*, for example optical waveguide coupler. The coupler normally has two fiber inputs and two fiber outputs called the fiber pigtails. The power splitting ratio with respect to input and output is defined by the coupling efficiency or coupling ratio k (Section 2.4).

The superimposed optical wave $\vec{E}(t)$ at the output of optical coupler is applied to a *photodiode* which generates an electric current $i_{PD}(t)$ proportional to optical input power. For this purpose, PIN or APD can be used in principle. As the gain of APD improves the performance of coherent optical communication systems only insignificantly [60], PIN-photodiode is often used in practice. As explained in Section 2.1, optical power at the input of photodiode is proportional to the square of input electric field. Owing to the fact that the superposition of fields $\vec{E}_r(t)$ and $\vec{E}_l(t)$ is linear, photodiode current $i_{PD}(t)$ also contains an intermediate frequency signal $i_{IF}(t)$ referred as IF signal. In coherent heterodyne receivers, we are primarily interested in this IF signal. The IF itself is given by the difference of local laser frequency (f_l) and the frequency (f_r) of received light wave.

Signal and noise are amplified by the *electrical amplifier(s)* in the same way. Therefore, we can assume identical amplification without the loss of generality.

When the frequencies f_l and f_r are equal (i.e., $f_{IF}=0$), coherent receiver is called *homodyne receiver*. Consequently, the modulated optical signal at the input to receiver is directly converted to baseband. To achieve homodyne detection, great care and accuracy have to be undertaken for the phase and frequency stabilization and tracking of received and local laser light waves. In *heterodyne receiver* ($f_{IF} \neq 0$), IF signal is selected by an *IF filter* and then demodulated with respect to amplitude, frequency or phase in the demodulator or detector circuit. All types of well-known radio frequency demodulators can be used: synchronous demodulator (coherent demodulator), envelope detector or frequency detector. The output signal $i_d(t)$ of the demodulator also contains undesired components such as double-frequency component which can be eliminated usually by a low-pass filter (baseband filter). In homodyne detection receiver, low-pass filter will reduce additionally the influence of noise by decreasing the noise equivalent bandwidth. In a heterodyne receiver, noise reduction is primarily performed by the IF filter.

The white Gaussian noise $n_w(t)$ is caused by the shot noise of photodiode and thermal noise of resistors and amplifiers, mainly the first amplifier stage next to photodiode. In *optical free-space communication system*, background noise is also a source of noise (Chapter nine) in addition to the above two noise sources.

The last components in our coherent system shown in Fig. 2.3 are the sample and hold and the decision circuits which have to recover the digital information $r(t)$. The noisy and disturbed signal at the input of sample and hold circuit is referred as *detected signal $d(t)$* in this book. A data transmission is without any error if transmitted and received sequence of symbols are same (i.e., $<r_\nu> = <q_\nu>$). Due to various imperfections and disturbances such as laser noise, shot noise, thermal noise, background noise, dispersion and polarization fluctuations, symbol errors are unavoidable. The most efficient measure to assess the quality of digital communication system is the probability of symbol error, also called the probability of error. This probability is quite important to analyze, optimize and compare digital optical as well as electrical communication systems. Section 2.5 discusses the fundamentals of error probability calculation in more detail.

The coherent optical communication systems require great efforts in the design of high-quality *phase- and frequency tracking* circuits. For simplicity, these circuits have not been shown in the block diagram in Fig. 2.3. In order to design and analyze coherent optical communication systems, basic knowledge of the principle of phase- and frequency tracking is unavoidable. Section 5.1.3 reviews the prime features of tracking as far as required in this book.

2.4   SIGNALS IN COHERENT OPTICAL COMMUNICATION SYSTEM

In this Section, signals in the transmitter, channel and receiver of coherent communication system will be discussed in more detail. In order to obtain a simplified description and straight forward calculation, complex signal representation is preferred. Transfer to the real signals, which can actually be measured, is easily accomplished by taking the real part of complex signals.

2.4.1 TRANSMITTER

(i) DIGITAL SOURCE

The digital source in the transmitter generates symbols q_v at equal temporal intervals of duration T. The digital information to be transmitted is represented by the sequence $<q_v>$. We call T the symbol length or symbol period and $1/T$ the symbol rate. The symbols q_v are taken out of a store of M symbols. On the basis of M, we have to distinguish between binary source ($M=2$) and multilevel source ($M>2$). In case of binary source with

$$q_v \in \{L, H\} \text{ or } q_v \in \{"0","1"\} \text{ or } q_v \in \{\text{zero, one}\} \text{ or } q_v \in \{\text{space, mark}\} \qquad (2.29)$$

symbol period T and symbol rate $1/T$ are frequently called a bit period or duration and bit rate respectively. In this Chapter, we assume that all the symbols q_v are binary and statistically independent. The probabilities of occurrence $p(q_v=H)$ and $p(q_v=L)$ are equal. Hence,

$$p(q_v=H) = p(q_v=L) = 0.5 \qquad (2.30)$$

The physical representation of the binary symbol sequence $<q_v>$ is given by

$$q(t) = \hat{q} \sum_{v=-\infty}^{\infty} a_v \, \text{rect}\left(\frac{t - vT}{T}\right) \qquad (2.31)$$

The peak value of this signal is $\hat{q}$, for example $\hat{q}=1$ V. In this book, we always provide a rectangular source signal $q(t)$ such as shown in Fig. 2.4. For this reason, we use

$$\text{rect}(x) = \begin{cases} 1 & \text{for } |x| \le 0.5 \\ 0 & \text{otherwise} \end{cases} \qquad (2.32)$$

to describe rectangular pulses mathematically. The interrelation between physical source signal $q(t)$ and symbols q_v is given by the amplitude coefficients

$$a_v=1 \text{ for } q_v=H \quad \text{and} \quad a_v=0 \text{ for } q_v=L \qquad (2.33)$$

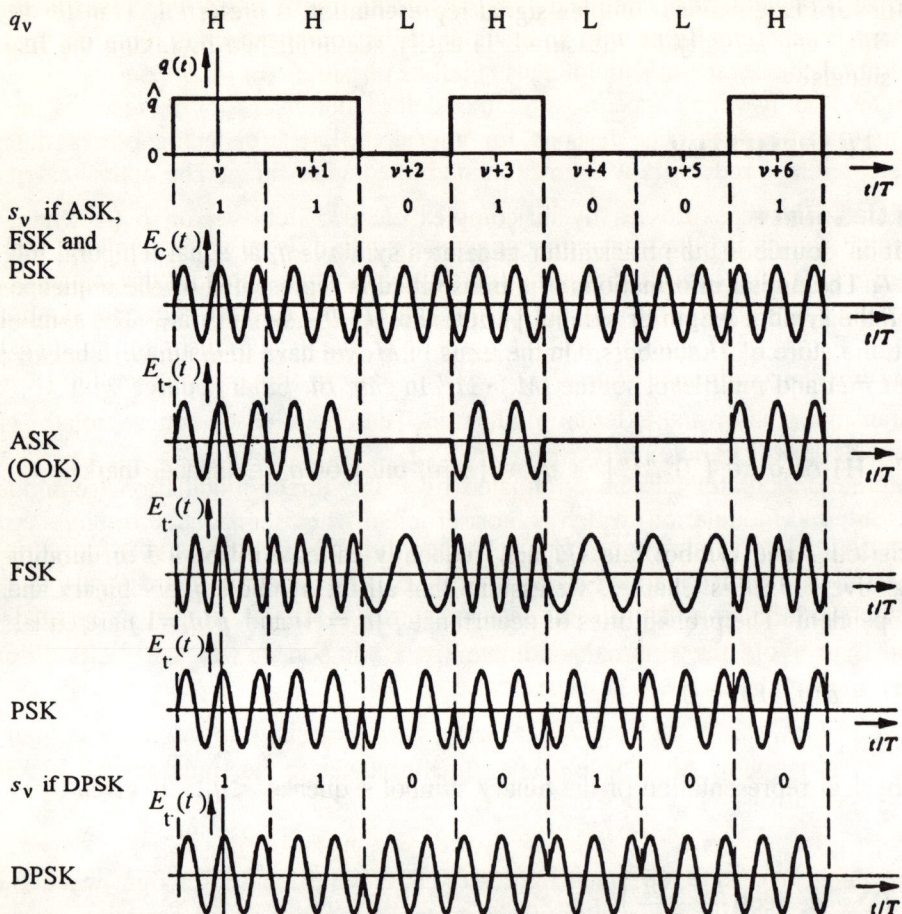

Fig. 2.4: Typical signals in coherent optical transmitter in absence of noise (OOK: on-off keying)

(ii) ELECTRICAL TRANSMITTER

In the electrical transmitter, source signal $q(t)$ at the input is converted to an electrical signal $s_e(t)$ at the output. For simplicity, we assume that this signal already includes the modulation, although this is actually not true (see Subsection "Optical Modulator"). By using complex representation, $\underline{s}_e(t)$ for different modulation schemes can be written as follows:

Amplitude Shift Keying (ASK)

$$\underline{s}_e(t) = \hat{s}_e \sum_{v=-\infty}^{\infty} s_v \, \text{rect}\left(\frac{t - vT}{T}\right) = \hat{s}_e \, s(t) = s_e(t) \tag{2.34}$$

Phase and Differential Phase Shift Keying (PSK and DPSK)

$$\underline{s}_e(t) = \hat{s}_e \exp\left(j \sum_{v=-\infty}^{\infty} \pi (1-s_v) \, \mathrm{rect}\left(\frac{t - vT}{T}\right) \right) = \hat{s}_e \, \underline{s}(t) \tag{2.35}$$

Frequency Shift Keying (FSK)

$$\underline{s}_e(t) = \hat{s}_e \exp\left(j \sum_{v=-\infty}^{\infty} 2\pi (2s_v-1) f_d t \, \mathrm{rect}\left(\frac{t - vT}{T}\right) \right) = \hat{s}_e \, \underline{s}(t) \tag{2.36a}$$

Continuous Phase Frequency Shift Keying (CPFSK)

$$\underline{s}_e(t) = \hat{s}_e \exp\left(j \sum_{v=-\infty}^{\infty} \int_{\tau=-\infty}^{t} 2\pi f_d (2s_v-1) \, \mathrm{rect}\left(\frac{t - vT}{T}\right) d\tau \right) = \hat{s}_e \, \underline{s}(t) \tag{2.36b}$$

The peak value $\hat{s}_e$ of electrical transmitter output signal $\underline{s}_e(t)$ is independent of modulation scheme. In case of ASK, $\underline{s}_e(t)$ is a real signal (Eq. 2.34). Normalizing $\underline{s}_e(t)$ with respect to $\hat{s}_e$, we obtain $\underline{s}(t)$ which is dimensionless. The quantity f_d in Eqs. (2.36a) and (2.36b) is called the frequency shift or frequency deviation.

Modulation coefficients s_v and amplitude coefficients a_v are related by

$$s_v = a_v \tag{2.37}$$

for ASK, PSK, FSK and CPFSK, whereas

$$\begin{aligned} s_v &= s_{v-1} \quad \text{if} \quad a_v = 1 \\ s_v &= \bar{s}_{v-1} \quad \text{if} \quad a_v = 0 \end{aligned} \tag{2.38}$$

in case of DPSK. Here, $\bar{s}_v$ represents the inversion of s_v, i.e., $\bar{s}_v = 1-s_v$, where $s_v \in \{0, 1\}$. An example of the relationship between s_v and a_v is given in Fig. 2.4.

(iii) TRANSMITTER LASER

With a single-mode laser in the transmitter, optical carrier wave can be written as

$$\vec{E}_c(t) = \begin{pmatrix} E_{cx}(t) \\ E_{cy}(t) \end{pmatrix} = E_c(t) \, e^{2\pi f_c t} \, \vec{e}_c \tag{2.39}$$

Here, normalized unit vector

$$\vec{e}_c = \begin{pmatrix} e_{cx} \\ e_{cy} \end{pmatrix} = \begin{pmatrix} |e_{cx}| \; e^{j\psi_{cx}} \\ |e_{cy}| \; e^{j\psi_{cy}} \end{pmatrix} \quad \text{with} \quad \vec{e}_c \, \vec{e}_c^* = 1 \tag{2.40}$$

describes the *state of polarization* (SOP) of the optical carrier wave (Chapter four). It is often called the polarization unit vector. Laser phase and amplitude noise are included in the complex carrier envelope

$$E_c(t) = |E_c(t)| \; e^{j\Phi_c(t)} \approx \hat{E}_c \; e^{j\Phi_c(t)} \tag{2.41}$$

As an example, Fig. 2.4 shows the real part $\text{Re}\{E_c(t)\}$ of the periodical carrier wave in absence of any noise. In contrast to phase noise $\phi_c(t)$, envelope noise $|E_c(t)|$ of the laser belongs to the subordinate source of noise in coherent optical communication systems and will not be considered in detail in this Chapter ($|E_c(t)| \approx \hat{E}_c$). Optical communication systems with direct detection receiver are not influenced by phase noise, whereas coherent systems are very sensitive to this type of noise. Laser phase noise has been discussed in details in Chapter three. It includes physical reasons, formation and characteristic features.

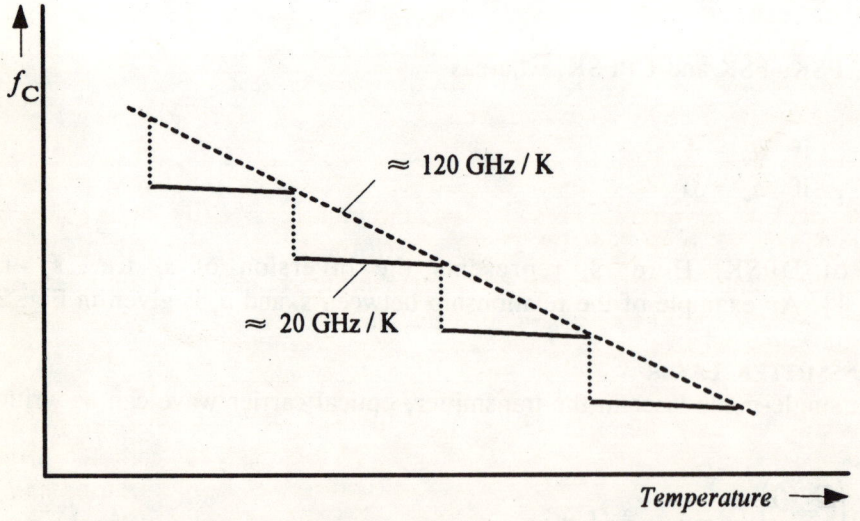

Fig. 2.5: Frequency temperature characteristic of typical semiconductor laser

In a real semiconductor laser, carrier frequency f_c is not constant; rather it changes continuously with time. Fluctuations in temperature are the prime reason for these undesired changes. Even high frequency jumps called mode hopping can occur if temperature is changing continuously (Fig. 2.5). Hence, temperature stabilization and automatic frequency control circuits are unavoidable in coherent optical communication systems.

(iv) OPTICAL MODULATOR

Depending on modulation scheme and information to be transmitted, carrier wave $\underline{E}_c(t)$ of the transmitter laser is changed in amplitude, phase or frequency by the modulator. Thereby, modulated electric baseband signal $\underline{s}_e(t)$ as shown in Fig. 2.4 is shifted to optical frequency. The modulator output signal is represented by $\vec{E}_t(t)$. To describe the modulation process mathematically, an ideal modulator (i.e., a simple multiplier) is formally considered [65]. Taking into account modulator coefficient K_m, modulated optical wave at the output of transmitter is given by

$$\vec{E}_t(t) = K_m \, \hat{s}_o \, \underline{s}(t) \, \underline{E}_c(t) \, \vec{e}_c = \underline{E}_t(t) \, \vec{e}_t \quad \text{with} \quad \vec{e}_c = \vec{e}_t \qquad (2.42)$$

The right side of above equation is due to simple formal separation of electric field vector $\vec{E}_t(t)$ in two components: one is the polarization unit vector $\vec{e}_t$ and the other modulated complex envelope $\underline{E}_t(t) = K_m \hat{s}_o \underline{s}(t) \underline{E}_c(t)$. To simplify our discussions, we assume that $\vec{e}_t$ and $\vec{e}_c$ are constant and equal. It has been made more clear by Fig. 2.4 which shows typical signals obtained by taking the real part of $\underline{E}_t(t)$.

In order to obtain a simple mathematical representation of the transmitted signals, we have formally carried out the modulation process in two steps: electrical modulation in the virtual electrical transmitter as considered above (Fig. 2.4) and optical modulation in the real optical modulator. In this simple model, optical modulation process is described by a simple multiplication of the electrical transmitter output signal $\underline{s}_e(t)$ and the transmitter laser signal $\underline{E}_c(t)$. However, in a real coherent optical transmitter the electric transmitter subsystem is usually a simple booster amplifier. It provides the required electric input power to optical modulator, wherein the modulation is actually performed. Therefore, signals given in Eqs. (2.34) to (2.36) cannot be measured in a real transmitter as they exist only in our model.

2.4.2 TRANSMISSION CHANNEL

In this Chapter, single-mode fiber is always used as transmission channel. Free-space channel will be considered in Chapter nine. Due to non-ideal transmission properties of single-mode fibers, transmitted optical signal $\underline{E}_t(t)$ is disturbed during transmission. Especially the polarization is influenced by various thermal and mechanical strains in the fiber. As a result, state of polarization at the input to the receiver is changing randomly with time. Direct detection receivers are independent of state of polarization, whereas coherent optical receivers are extremely dependent. Thus, polarization fluctuations are a fundamental source of noise in coherent optical communication systems and will be considered in more detail in Chapter four.

A second important imperfection of single-mode fibers is dispersion. Three different types of dispersion have to be considered: waveguide, chromatic (frequently called material) and polarization dispersion. The first two types of dispersion are important in every digital optical communication system and the last one, in addition, becomes important in long-range multigigabit systems. Since the influence of dispersion is considered in various publications, we desist from a detailed discussion on it in this book [87, 134, 274]. In absence of dispersion, fiber output signal or the receiver input signal is given by

$$\vec{E}_r(t) = s(t) \, E_r(t) \, e^{j2\pi f_r t} \, \vec{e}_r(t) \tag{2.43}$$

where $f_r = f_t$. Unlike the transmitted signal, polarization unit vector

$$\vec{e}_r(t) = \begin{pmatrix} e_{rx}(t) \\ e_{ry}(t) \end{pmatrix} = \begin{pmatrix} |e_{rx}(t)| \; e^{j\Psi_{rx}(t)} \\ |e_{ry}(t)| \; e^{j\Psi_{ry}(t)} \end{pmatrix} \quad \text{with} \quad \vec{e}_r(t) \, \vec{e}_r(t)^* = 1 \tag{2.44}$$

of received signal is now a function of time and describes the polarization fluctuations at the input of coherent optical receiver. Due to various imperfections and disturbances from the fiber (Chapter four), polarization at the input to the receiver can be regarded as a random process. Owing to the inherent losses of fiber, optical signal will get attenuated during transmission. At the output end of fiber of length L (in km), attenuated amplitude of electric field is given by

$$E_r(t) = \hat{E}_r \, e^{j\phi_r(t)} = \hat{E}_t \, e^{-\alpha_{Np} L} \, e^{j\phi_t(t)} \tag{2.45a}$$

where α_{Np} represents the fiber attenuation in Np/km. In terms of optical power levels, above equation yields the well-known relation

$$20 \, \log\left(\frac{\hat{E}_r}{\hat{E}_t}\right) = 10 \, \log\left(\frac{P_r}{P_t}\right) = -\alpha_{dB} \, L \tag{2.45b}$$

where P_r and P_t represent the received and transmitted optical power levels respectively. The relationship between α_{Np} (in Np/km) and α_{dB} (in dB/km) is : 1 Np = 8.686 dB.

Without loss of generality, we assume that the phase noise at the transmitter output and receiver input are same that is, $\phi_t(t) = \phi_r(t)$. As we have neglected fiber dispersion, optical signal $\underline{E}_t(t)$ at the transmitter output and $\underline{E}_r(t)$ at the receiver input only differ in amplitude ($\hat{E}_r \ll \hat{E}_t$) and polarization ($\vec{e}_t$ is usually constant and $\vec{e}_r(t)$ is random).

2.4.3 COHERENT RECEIVER

(i) LOCAL LASER

For a local laser with same spectral properties as the transmitter laser, output wave can be expressed as

$$
\vec{E}_1(t) = \begin{pmatrix} E_{1x}(t) \\ E_{1y}(t) \end{pmatrix} = E_1(t)\ e^{2\pi f_1 t}\ \vec{e}_1
\tag{2.46}
$$

Similar to the transmitter laser (Eqs. 2.39 and 2.40), we again assume a constant and linear polarization for local laser source. Hence, the local laser polarization unit vector is given by

$$
\vec{e}_1 = \begin{pmatrix} e_{1x} \\ e_{1y} \end{pmatrix} = \begin{pmatrix} |e_{1x}|\ e^{j\psi_{1x}} \\ |e_{1y}|\ e^{j\psi_{1y}} \end{pmatrix} \quad \text{with} \quad \vec{e}_1\ \vec{e}_1^{\,*} = 1
\tag{2.47}
$$

The complex envelope

$$
E_1(t) = |E_1(t)|\ e^{j\Phi_1(t)} \approx \hat{E}_1\ e^{j\Phi_1(t)}
\tag{2.48}
$$

contains laser phase and amplitude noise. As the phase noise $\phi_1(t)$ is a dominant source of noise, we are allowed again to neglect amplitude noise ($|\underline{E}_1(t)| \approx \hat{E}_1$).

In comparison to the received optical wave $\underline{E}_r(t)$, local laser wave $\underline{E}_1(t)$ is not attenuated by the transmission channel i.e., by the fiber. Therefore, the following relations are always valid in coherent optical communication systems:

$$
\hat{E}_1 > \hat{E}_r = \hat{E}_t e^{-\frac{\alpha_{NP} L}{1\ Np}} \quad \text{and} \quad P_1 > P_r = P_t 10^{-\frac{\alpha_{dB} L}{10\ dB}}
\tag{2.49}
$$

Here, P_1 is the power of local laser.

(ii) OPTICAL COUPLER

The modulated optical wave $\underline{E}_r(t)$ at the input of coherent receiver and local laser wave $\underline{E}_l(t)$ are linearly superimposed in an optical coupler. Usually, this coupler is a four-port coupler with two fiber inputs and two fiber outputs called the pigtails. As shown in Fig. 2.3, input ports are taken as 1 and 2 and output ports as 3 and 4. An optical coupler is primarily characterized by its coupling efficiency k ($0 < k < 1$) that defines the ratio of the input to the output power. Considering a symmetrical coupler without any internal loss, the coupler can be described by the following matrix [103].

$$
\begin{pmatrix} \vec{E}_4(t) \\ \vec{E}_3(t) \end{pmatrix} = \begin{pmatrix} \sqrt{1-k} & j\sqrt{k} \\ j\sqrt{k} & \sqrt{1-k} \end{pmatrix} \begin{pmatrix} \vec{E}_1(t) \\ \vec{E}_2(t) \end{pmatrix}
\tag{2.50}
$$

This coupler is exactly same as given in block diagram of Fig. 2.3 except that $\underline{E}_1(t)$ and $\underline{E}_2(t)$ are replaced by $\underline{E}_r(t)$ and $\underline{E}_l(t)$ respectively. Considering the optical power $P_4(t)$ and $P_3(t)$ at the optical coupler output, following relationships can be given:

$$
\begin{aligned}
P_4(t) \sim \vec{E}_4(t)\vec{E}_4^*(t) &= \left|\vec{E}_4(t)\right|^2 \\
&= (1-k)\left|\vec{E}_r(t)\right|^2 + k\left|\vec{E}_l(t)\right|^2 + 2\sqrt{k(1-k)}\,\mathrm{Im}\left\{\vec{E}_r(t)\,\vec{E}_l^*(t)\right\}
\end{aligned}
\tag{2.51}
$$

and

$$
\begin{aligned}
P_3(t) \sim \vec{E}_3(t)\vec{E}_3^*(t) &= \left|\vec{E}_3(t)\right|^2 \\
&= (1-k)\left|\vec{E}_l(t)\right|^2 + k\left|\vec{E}_r(t)\right|^2 - 2\sqrt{k(1-k)}\,\mathrm{Im}\left\{\vec{E}_r(t)\,\vec{E}_l^*(t)\right\}
\end{aligned}
\tag{2.52}
$$

In order to obtain a better signal representation, we now take the advantage of complex description of optical power $P_4(t) = \mathrm{Re}\{\underline{P}_4(t)\}$ and $P_3(t) = \mathrm{Re}\{\underline{P}_3(t)\}$. If we consider that the proportionality between optical power and electric field is same for all the participating electric fields $\underline{E}_r(t)$, $\underline{E}_l(t)$, $\underline{E}_3(t)$ and $\underline{E}_4(t)$, we obtain

$$
\begin{aligned}
\underline{P}_4(t) = k\,P_l &+ (1-k)\,|\underline{s}(t)|^2\,P_r \\
&+ 2\sqrt{k(1-k)}\,\sqrt{P_r\,P_l}\,\underline{s}(t)\,a_p(t)\,e^{j\phi(t)}\,e^{j\phi_p(t)}\,e^{j2\pi f_{IF}t}
\end{aligned}
\tag{2.53}
$$

and

$$
\begin{aligned}
P_3(t) = {} & (1-k) \, P_1 + k \, |s(t)|^2 \, P_r \\
& - 2 \, \sqrt{k(1-k)} \, \sqrt{P_r \, P_1} \, s(t) \, a_p(t) \, e^{j\phi(t)} \, e^{j\phi_p(t)} \, e^{j2\pi f_{IF} t}
\end{aligned}
\tag{2.54}
$$

In the above Eqs. (2.53) and (2.54), P_1 and P_r represent the mean optical power level of the local laser light wave and unmodulated received light wave respectively. The intermediate frequency is given by

$$
f_{IF} = f_1 - f_r \quad \text{with} \quad f_r = f_t
\tag{2.55}
$$

and the overall resulting phase noise follows the relation

$$
\phi(t) = \phi_1(t) - \phi_r(t) \quad \text{with} \quad \phi_r(t) = \phi_t(t)
\tag{2.56}
$$

It becomes evident from Eqs. (2.53) and (2.54) that polarization fluctuations caused by fiber imperfections firstly, result in amplitude fluctuations $a_p(t)$ and secondly, phase fluctuations $\phi_p(t)$. These fluctuations are sources of noise and may seriously deteriorate the performance of coherent communication systems. In the worst-case, detected signal can even become zero as shown in Chapter four. Both sources of noise can be deduced from the following simple expression:

$$
\vec{\varepsilon}_r(t) \, \vec{\varepsilon}_1^*(t) = a_p(t) \, e^{j\phi_p(t)}
\tag{2.57}
$$

Using Eqs. (2.44) and (2.47), we obtain the following equations for amplitude and phase fluctuations:

$$
\begin{aligned}
a_p(t) = {} & \left| \vec{\varepsilon}_r(t) \, \vec{\varepsilon}_1^*(t) \right| = \sqrt{ \left[\mathrm{Re}\{\vec{\varepsilon}_r(t) \, \vec{\varepsilon}_1^*(t)\} \right]^2 + \left[\mathrm{Im}\{\vec{\varepsilon}_r(t) \, \vec{\varepsilon}_1^*(t)\} \right]^2 } \\
= {} & \left[|\varepsilon_{1x}|^2 \, |\varepsilon_{rx}(t)|^2 + |\varepsilon_{1y}|^2 \, |\varepsilon_{ry}(t)|^2 \right. \\
& + \left. 2 \, |\varepsilon_{1x}| \, |\varepsilon_{rx}(t)| \, |\varepsilon_{1y}| \, |\varepsilon_{ry}(t)| \cos(\psi_{rx}(t) - \psi_{1x} - \psi_{ry}(t) + \psi_{1y}) \right]^{1/2}
\end{aligned}
\tag{2.58}
$$

and

$$\phi_p(t) = \arctan\left(\frac{\text{Im}\{\vec{\varepsilon}_r(t)\ \vec{\varepsilon}_l^*\}}{\text{Re}\{\vec{\varepsilon}_r(t)\ \vec{\varepsilon}_l^*\}}\right)$$

(2.59)

$$= \arctan\left(\frac{|\varepsilon_{lx}|\ |\varepsilon_{rx}(t)|\ \sin(\psi_{rx}(t) - \psi_{lx}) + |\varepsilon_{ly}|\ |\varepsilon_{ry}(t)|\ \sin(\psi_{ry}(t) - \psi_{ly})}{|\varepsilon_{lx}|\ |\varepsilon_{rx}(t)|\ \cos(\psi_{rx}(t) - \psi_{lx}) + |\varepsilon_{ly}|\ |\varepsilon_{ry}(t)|\ \cos(\psi_{ry}(t) - \psi_{ly})}\right)$$

The output of coupler is converted to an electric signal by means of PIN or APD. Since an optical coupler normally contains two fiber outputs, either of the outputs can be used and the other output remains unused. However, instead of using only one output of the coupler, it is much more efficient to use both the outputs. Both possibilities are discussed below.

(iii) PHOTODIODE

The photodiode generates a current $i_{PD}(t)$ proportional to the absorbed optical power. Consider first a coherent receiver with only one single photodiode. We call it a *single-diode receiver*. Assuming, for example, that port 4 is used, the photodiode current is given by

$$i_{PD4}(t) = k\ R_0\ P_1 + (1-k)\ |\underline{s}(t)|^2\ R_0\ P_r$$

$$+ 2\ R_0\ \sqrt{k(1-k)}\ \sqrt{P_1\ P_r}\ \underline{s}(t)\ a_p(t)\ e^{j\phi(t)}\ e^{j\phi_p(t)}\ e^{j2\pi f_{IF}t}$$

(2.60)

This current consists of three components: first component kR_0P_1 is a DC without any information. Second component $(1-k)\ |\underline{s}(t)|^2R_0P_r$ truly contains information, but is negligibly small as compared to third component since $P_r \ll P_1$. Therefore, only the third component is of practical interest in coherent receiver. This component contains transmitted information $\underline{s}(t)$ which has to be detected and recovered in the demodulator. However, it is severely corrupted by polarization fluctuations in the fiber and laser phase noise.

Next we consider a coherent receiver with two photodiodes, which is termed as a *balanced receiver*. In this type of receiver, optical power available at both the output ports is fully used. To get maximum signal power, photodiode currents $i_{PD4}(t)$ and $i_{PD3}(t)$ have to be subtracted to obtain the output current $i_{PD}(t) = i_{PD4}(t) - i_{PD3}(t)$. It is given by

$$i_{PD}(t) = (2k-1)\ R_0\ P_1 + (1-2k)\ |\underline{s}(t)|^2\ R_0\ P_r$$

$$+ 4\ R_0\ \sqrt{k(1-k)}\ \sqrt{P_1\ P_r}\ \underline{s}(t)\ a_p(t)\ e^{j\phi(t)}\ e^{j\phi_p(t)}\ e^{j2\pi f_{IF}t}$$

(2.61)

By comparing Eqs. (2.60) and (2.61), it becomes clear that the amplitude of third component is twice now. It implies a gain of 3 dB in signal power. This gain is an important feature of balanced receiver over single-diode receiver. A second advantage of significant practical interest is obtained when coupling ratio is chosen to be $k = 0.5$. In that case, DC component $(2k-1)R_0 P_1$ as well as the baseband component $(1-2k)\,|\,\underline{s}(t)\,|\,^2 R_0 P_r$ in Eq. (2.61) become zero. Therefore, a balanced receiver configuration becomes more useful particularly when homodyne receivers are used since third component of $i_{PD}(t)$ is already at a baseband frequency ($f_{IF}=0$).

The phase noise $\phi(t)$ caused by local and transmitter lasers is a fundamental source of noise in coherent optical communication systems. This noise shows a strong influence on BER and, therefore, on the overall transmission quality. Polarization fluctuations due to fiber imperfections also degrade the system performance. Several techniques exist to suppress the effect of polarization fluctuations almost completely (Chapter four).

With a fixed and stable state of polarization, negligible DC and baseband components, photodiode current can now be expressed as

$$i_{PD}(t) = \hat{i}_{PD}\ \underline{s}(t)\ e^{j\phi(t)}\ e^{2\pi f_{IF}t} \tag{2.62}$$

with

$$\hat{i}_{PD} = 2K_R\ \sqrt{k(1-k)}\ R_0\ \sqrt{P_1\,P_r} \tag{2.63}$$

Here, K_R is a new constant which will be 2 for balanced receiver and 1 for single-diode receiver. For a symmetrical balanced receiver with $k = 0.5$, above equation becomes same as Eq. (2.10) derived in Section 2.1 under ideal conditions.

(iv) RECEIVER NOISE

As shown in Fig. 2.3, photodiode current $i_{PD}(t)$ is disturbed by an additive and nearly white Gaussian noise $n_w(t)$. As stated earlier, main sources of this noise are the shot noise of photodiode(s) and the thermal noise of resistors and amplifiers, especially of the first amplifier stage. As the frequency components of both sources of noise are uniformly distributed, this noise is called *white noise*. Its double-sided psd measured in Amperes squared per Hertz (A^2/Hz) is given by

$$G_c(f) = G_c = e\,(R_0\,k\,P_1 + I_{dark}) + G_{th} \tag{2.64}$$

Except the coupling ratio k, above equation is same as Eq. (2.11). In the strict sense, both P_r and P_1 must be considered to determine the shot noise psd i.e., $eR_0 k(P_1 + P_r)$. As $P_r \ll P_1$, this is practically not required.

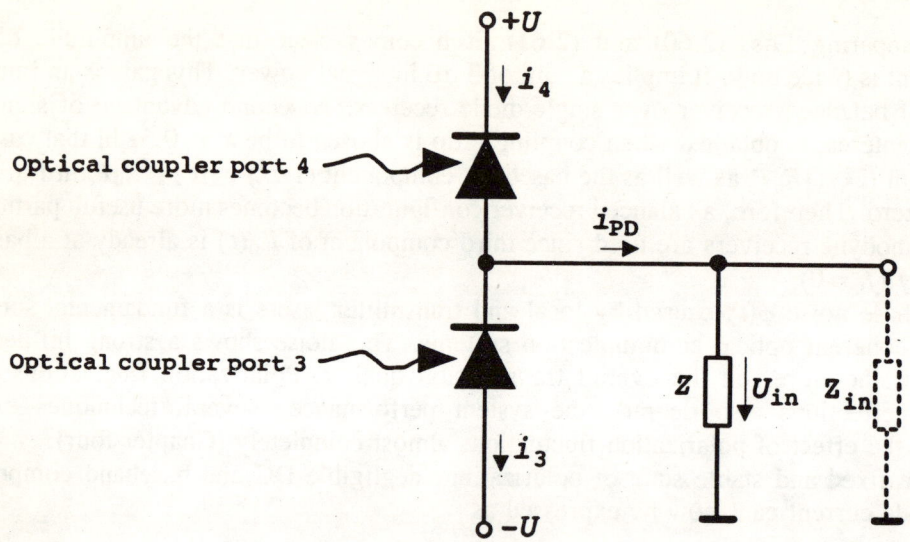

Fig. 2.6: Balanced receiver configuration

Let us now consider the balanced receiver shown in Fig. 2.6, which is exclusively used in practice. The shot noise of both photodiodes is absolutely uncorrelated. Therefore, in a symmetrical balanced receiver ($k=0.5$), noise psd is always twice the psd in a single-diode receiver:

$$G_c(f) = G_c = 2\ e\ (R_0\ k\ P_1 + I_{dark}) + G_{th} \qquad (2.65)$$

With K_R and k as variables, psd of noise in both balanced and single-diode receivers can be expressed by the following common formula:

$$G_c(f) = G_c = e\ K_R\ (R_0\ k\ P_1 + I_{dark}) + G_{th} \qquad (2.66)$$

It is infer from this equation that first amplifier stage and, hence, the source of the thermal noise is to be placed after the subtraction of photodiode currents. If amplifier is placed immediately after the photodiodes, two amplifiers would be required. In that case, thermal noise G_{th} will get doubled.

In order to reduce the effect of noise, amplified photodiode current must be filtered. It is an IF filter if a heterodyne receiver is used and a low-pass filter for a homodyne receiver. Thereby, the white noise $n_w(t)$ at the filter input will be converted to *band-limited coloured noise* $n(t)$ at the filter output. An important measure of $n(t)$ is its variance. In heterodyne receiver, this variance is given by

$$\sigma_{het}^2 = G_c \int\limits_{-\infty}^{+\infty} |H_{IF}(f)|^2 df = 2G_c \int\limits_{-\infty}^{+\infty} |H_B(f)|^2 df \tag{2.67}$$

where $H_{IF}(f)$ and $H_B(f)$ represent the frequency response (system function) of the IF filter and equivalent virtual baseband filter respectively. In heterodyne receiver, baseband filter function $H_B(f)$ is used to describe $H_{IF}(f)$ analytically. The IF filter and its baseband representation are related by the following relationships:

$$H_{IF}(f) = H_B(f-f_{IF}) + H_B(f+f_{IF}) \tag{2.68}$$

and

$$h_{IF}(t) = h_B(t)e^{j2\pi f_{IF}t} + h_B(t)e^{-j2\pi f_{IF}t} \tag{2.69}$$

Here, $h_{IF}(t)$ and $h_B(t)$ are the impulse response of IF filter and equivalent baseband filter respectively. As mentioned earlier, low-pass filter $H_{LP}(f)$ in heterodyne receiver (Fig. 2.3) has to reject the double-frequency components caused by the demodulation mechanism. Hence, signal and noise are not influenced by this filter. In homodyne receiver, $H_{LP}(f)$ is responsible for noise reduction also. At the output of this filter, noise variance is given by

$$\sigma_{hom}^2 = G_c \int\limits_{-\infty}^{+\infty} |H_{LP}(f)|^2 df \tag{2.70}$$

Under the condition that heterodyne IF filter and homodyne low-pass filter are of same type that is $H_{LP}(f) = H_B(f)$, following important relationship is obtained:

$$\sigma_{het}^2 = 2\sigma_{hom}^2 = 2G_c \int\limits_{-\infty}^{+\infty} |H_B(f)|^2 df \tag{2.71}$$

It becomes clear from the above equation that a homodyne receiver always offers a 3 dB higher signal-to-noise ratio or a 3 dB better receiver sensitivity than a heterodyne receiver provided signal power is same. However, it is only valid if ideal laser sources without any laser noise have been used. Further, modulation scheme (ASK or PSK) must be same in both types of systems. When the effect of phase noise is considered, heterodyne detection may even offer a better performance than homodyne detection (Chapters five and seven).

In order to analyze coherent communication systems, receiver noise $n(t)$ is considered to be a zero mean Gaussian random process. Its probability density function (pdf) is

$$f_{\underline{n}}(n) = \frac{1}{\sqrt{2\pi}\sigma_{\underline{n}}} e^{-(n^2/2\sigma_{\underline{n}}^2)} \quad \text{with} \quad \sigma_n = \begin{cases} \sigma_{het} = \sqrt{2}\sigma_{hom} & \text{for heterodyning} \\ \sigma_{hom} & \text{for homodyning} \end{cases} \quad (2.72)$$

In the analysis of heterodyne receivers, use of the narrow-band representation

$$n(t) = x(t)\cos(2\pi f_{IF}t) + y(t)\sin(2\pi f_{IF}t) \qquad (2.73)$$

is very advantageous. The in-phase and quadrature components $x(t)$ and $y(t)$ are statistically independent, zero mean Gaussian random processes with

$$\sigma_n = \sigma_x = \sigma_y \qquad (2.74)$$

Narrow-band representation is valid if

$$\frac{B_{IF}}{f_{IF}} < 1 \qquad (2.75)$$

where B_{IF} is the noise equivalent bandwidth of IF filter. The complex representation of narrow-band noise is

$$\underline{n}(t) = x(t)\, e^{j2\pi f_{IF}t} - jy(t)\, e^{j2\pi f_{IF}t}$$

$$= (x(t) - jy(t))\, e^{j2\pi f_{IF}t} \quad \text{with} \quad n(t) = \text{Re}\{\underline{n}(t)\} \qquad (2.76)$$

In this book, we have primarily considered Gaussian filters. It is characterized by

$$H_B(f) = e^{-\pi\left(\frac{f}{2f_g}\right)^2} \quad\bullet\!\!-\!\!\circ\quad h_B(t) = 2f_g\, e^{-\pi(2f_g t)^2} \qquad (2.77)$$

where f_g is the cut-off frequency. With this Gaussian filter the double-sided noise equivalent bandwidth is $B = \sqrt{2}\, f_g$. Thus, the noise variance at the output of the filter yields

$$\sigma_{het}^2 = 2\sigma_{hom}^2 = 2\sqrt{2}\, G_c f_g \qquad (2.78)$$

(v) DEMODULATION

In heterodyne receiver, demodulator has to recover the transmitted information which is included in the IF signal

$$i_{IF}(t) = \int_{-\infty}^{+\infty} i_{PD}(\tau) \, h_{IF}(t-\tau) \, d\tau + \underline{n}(t)$$

$$= \hat{i}_{PD} \int_{-\infty}^{+\infty} \underline{s}(\tau) \, e^{j\phi(\tau)} \, h_B(t-\tau) \, d\tau \; e^{j2\pi f_{IF} t} \tag{2.79a}$$

$$+ \; \hat{i}_{PD} \int_{-\infty}^{+\infty} \underline{s}(\tau) \, e^{j\phi(\tau)} \, h_B(t-\tau) \, e^{j4\pi f_{IF}\tau} d\tau \; e^{-j2\pi f_{IF}t} + \underline{n}(t)$$

When the narrow-band condition (2.75) is satisfied, second integral in the above equation containing the double-frequency term $\exp(j\,2\pi\,2f_{IF}\,\tau)$ is approximately zero. In that case, Eq. (2.79a) can be approximated as

$$i_{IF}(t) \approx \left[\hat{i}_{PD} \int_{-\infty}^{+\infty} \underline{s}(\tau) \, e^{j\phi(\tau)} \, h_B(t-\tau) \, d\tau \; + \; x(t) \; - \; jy(t) \right] e^{j2\pi f_{IF}t} \tag{2.79b}$$

Depending on modulation scheme (ASK, FSK, PSK or DPSK), different types of demodulators can be used, such as synchronous demodulator (which is often called coherent demodulator), envelope detector, frequency discriminator, one- and two-filter demodulator, autocorrelation demodulator etc. In this book, both heterodyne and homodyne are called coherent receivers. With respect to the demodulation, however, we have to distinguish between noncoherent detection (for example, envelope detection) and coherent detection (for example, electrical or optical synchronous detection). In heterodyne receivers, either noncoherent or coherent demodulation can be used, whereas homodyne receivers always use coherent demodulation.

As mentioned earlier, low-pass filter in heterodyne receiver is responsible only for suppressing the undesired products of demodulation such as signal component of twice the intermediate frequency. Noise reduction and signal shaping are usually not performed by this filter.

In contrast to heterodyne receiver, demodulation process in homodyne receiver is already performed by transferring the optical signal to baseband. Homodyne detection and demodulation are actually one single and joint process which cannot be separated. Thus, homodyne receiver can be regarded either as an optical synchronous receiver or as a coherent receiver with coherent detection (Chapter seven). The low-pass filter in homodyne receiver is responsible for the noise reduction. In addition, this filter is also responsible for undesired signal shaping and, therefore, intersymbol interference due to a restricted low-pass filter bandwidth B. To maximize system performance, bandwidth and frequency response of this filter must be optimized. In this book, output signal of the low-pass filter is referred as *detected signal* $d(t)$ irrespective of type of receiver (heterodyne or homodyne), modulation scheme and demodulation process (coherent or noncoherent) used.

(vi) SAMPLE AND HOLD CIRCUIT

In the sample and hold circuit, detected signal $d(t)$ is sampled in equal temporal intervals of duration T (i.e., bit period). Finally, the sampled signal $d(\nu T+t_0)$ is applied to the decision circuit (Fig. 2.3). The time t_0 exactly defines the sampling point. Thus, t_0 represents an important system parameter which must be optimized to maximize system performance. In case of symmetrical filters, for example a Gaussian filter, optimum decision time is always at the centre of each symbol (bit) i.e., $t_0=0$.

(vii) DECISION CIRCUIT

Depending on whether $d(\nu T+t_0)$ is above or below a fixed threshold, symbol $r_\nu=$"0" or $r_\nu=$"1" is decided by the decision circuit and sent to the data output of digital receiver. In addition to filter transfer function $H_B(f)$ and sampling time t_0, threshold voltage E is a third optimizable system parameter of primary importance (Chapter five). For the calculation of bit error probability or BER, signal shape of the detected signal $d(t)$ and statistical features of its sampled values $d(\nu T+t_0)$ play a major role.

2.5   EYE PATTERN AND PROBABILITY OF ERROR

This Section presents a brief review of two important and powerful measures to determine the transmission quality of digital communication systems: eye pattern and probability of error. The general results obtained in this Section will be applied to coherent optical communication systems in Chapter five. The readers who are already familiar with this topic may escape this Section and move over to the next Chapter.

2.5.1   INFLUENCE OF NOISE AND INTERSYMBOL INTERFERENCE

The signals in digital communication system are significantly perturbed by noise and intersymbol interference (ISI). ISI is caused by the restricted bandwidth of various components such as filters. The most important signal which directly determines the transmission quality, is the detected signal $d(t)$ at the input of sample and hold circuit. It is given by

$$d(t) = d_0(t) + n(t) \tag{2.80}$$

This can be regarded as a linear superposition of signal $d_0(t)$ and noise $n(t)$. The signal $d_0(t)$ in presence of ISI will look like as in Fig. 2.7.

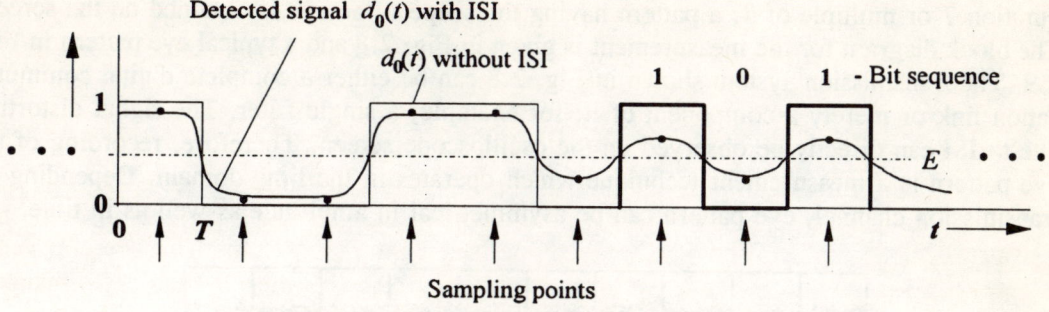

Fig.2.7: Detected signal $d_0(t)$ with and without ISI

At a given bit rate $R = 1/T$, influence of ISI can be reduced by increasing the available transmission bandwidth of all responsible system components and filters. In case of an infinite bandwidth, signal $d_0(t)$ would be rectangular in shape (i.e., without any ISI). However, large bandwidth will increase the noise. Thus, an obvious trade-off exists between bandwidth and transmission quality. It is clear that the bandwidth is an optimizable system parameter of prime importance.

In the development and realization of digital communication systems, some measure to determine the degree of ISI is required. For this reason, eye pattern recording represents a very simple and frequently used measurement technique. The eye pattern is briefly described below.

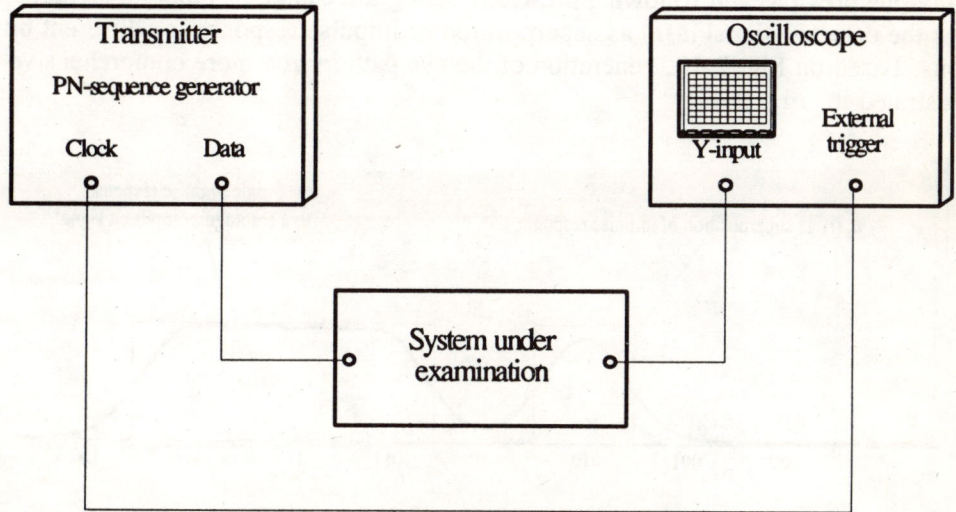

Fig. 2.8: System configuration for recording eye pattern

2.5.2   EYE PATTERN

The detected signal $d_0(t)$ when fed to the input of an oscilloscope triggered on bit duration T or multiple of T, a pattern having the shape of an eye is obtained on the screen. The block diagram for the measurement is given in Fig. 2.8 and a typical eye pattern in Fig. 2.9. The transmission system shown in Fig. 2.8 can be either a complete digital communication link or merely a component of it; for example, a single filter. The signal distortion due to ISI can directly be observed on the oscilloscope screen. Therefore, recording of the eye pattern is a measurement technique which operates in the time domain. Depending on transmission channel, eye pattern can be asymmetrical in amplitude as well as in time.

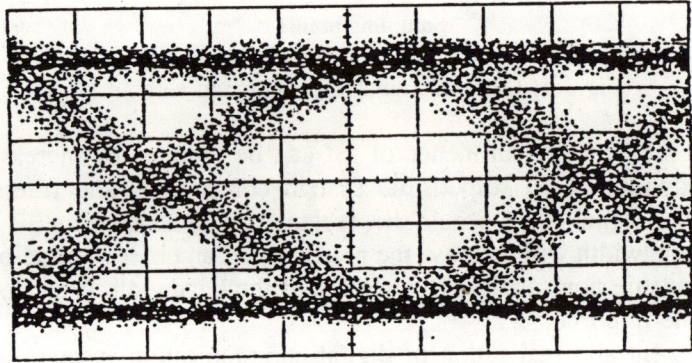

Fig. 2.9: Typical eye pattern

Generation of eye pattern is explained in Figs. 2.10 and 2.11. Here, each bit is disturbed by only one previous and following bit. Considering a rectangular input bit signal, Fig. 2.10 shows the detected signal $d_0(t)$ as superposition of impulse response for different bit combinations. Based on Fig. 2.10, generation of the eye pattern in a more comprehensive manner is illustrated in Fig. 2.11.

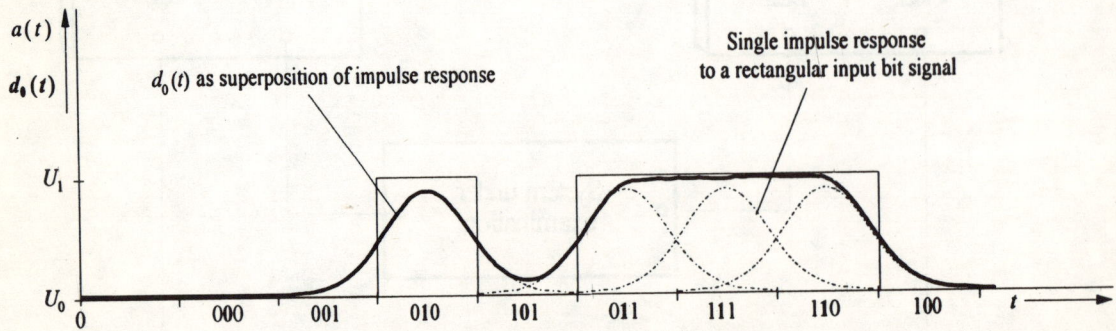

Fig. 2.10: Principle of eye pattern recording (explanation 1)

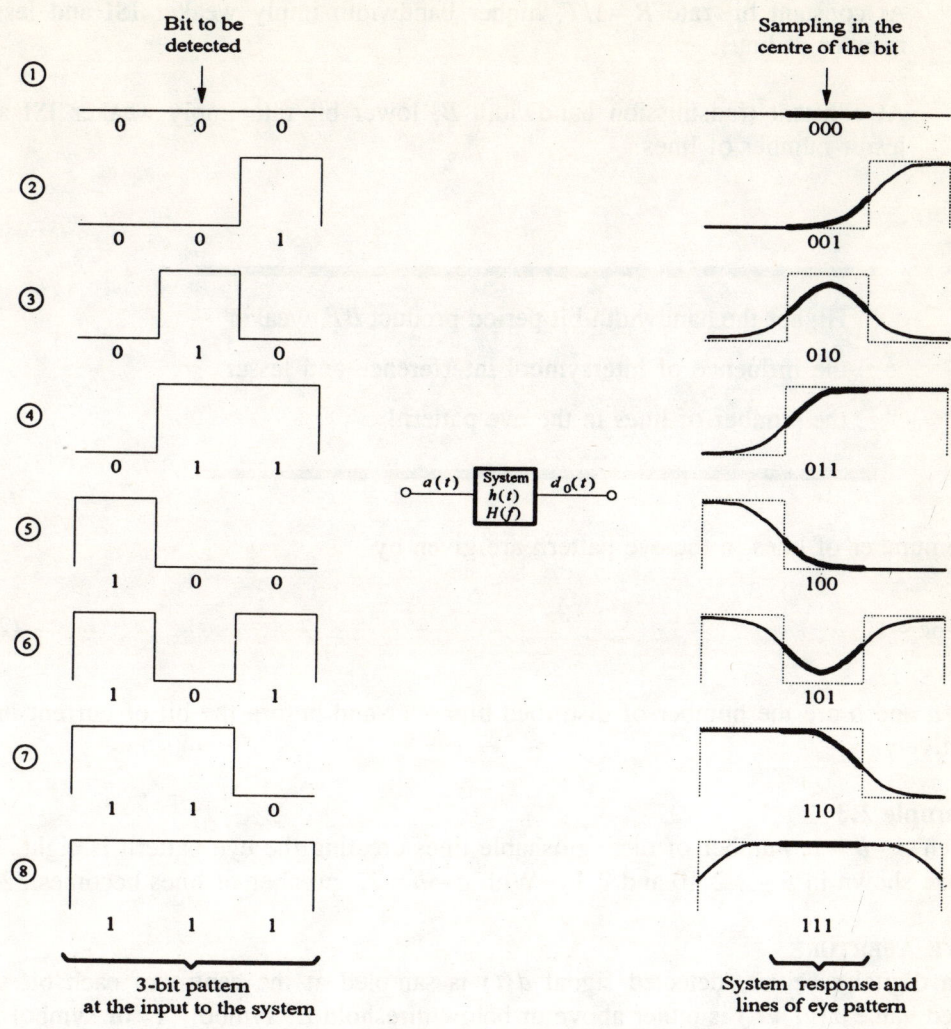

Fig. 2.11: Principle of eye pattern recording (explanation 2)

(i) NUMBER OF EYE LINES

The fundamental reason of signal distortion and, therefore, the primary reason of the different lines in the eye pattern is the pulse dispersion or pulse broadening arising from restricted transmission bandwidth (for example, filtering). Pulse dispersion yields interference of neighbouring pulses or bits and, therefore, a mutual distortion. System performance degradation due to ISI is higher when more neighbouring bits have mutually interfered. Simultaneously, the number of lines in the eye pattern will increase. It leads to the following conclusion:

(a)   At constant bit rate $R = 1/T$, higher bandwidth imply weaker ISI and lesser number of lines.

(b)   At constant transmission bandwidth B, lower bit rate imply weaker ISI and lesser number of lines.

Therefore,

> Higher the bandwidth-bit period product BT, weaker the influence of intersymbol interference and lesser the number of lines in the eye pattern!

The number of lines in the eye pattern are given by

$$K = 2^{a+b+1}$$

(2.81)

where a and b are the number of disturbed bits after and before the bit of current interest respectively.

Example 2.3

When $a = b = 1$, number of distinguishable lines creating the eye pattern is eight. These lines are shown in Figs. 2.10 and 2.11. With $a = b = 2$, number of lines becomes 32.

(II) EYE APERTURE

In a digital receiver, detected signal $d(t)$ is sampled at the centre of each bit. If the sampled value $d(\nu T + t_0)$ is either above or below threshold E, symbol "1" or symbol "0" is detected respectively i.e.,

$$d(\nu T + t_0) \geq E \rightarrow 1,$$

$$d(\nu T + t_0) < E \rightarrow 0.$$

When the sampled value $d_0(\nu T + t_0)$ is very close to threshold, even a small noise level may be high enough to cause an error. The probability of error becomes maximum for those bit sequences which are responsible for the inner lines in the eye pattern. We call these special sequences the *worst-case pattern*. In practice, the worst-case pattern are most widely given by

··· 0 0 0 1 0 0 0 ··· (single one, upper worst-case line) and

··· 1 1 1 0 1 1 1 ··· (single zero, lower worst-case line).

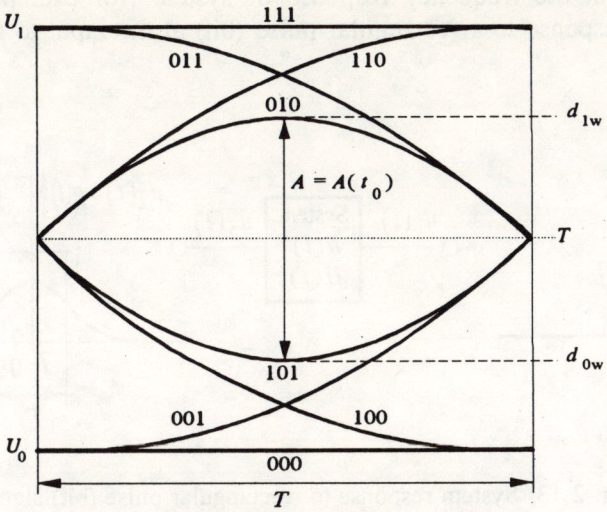

Fig. 2.12: Eye aperture in the eye pattern

As shown in Fig. 2.12, worst-case pattern define the inner open part of the eye. The absolute and normalized eye aperture A and a are

$$A = d_{1w} - d_{0w} = (U_1 - U_0) - 2d_{0w} = 2d_{1w} - (U_1 - U_0) \qquad (2.82)$$

and

$$a = \frac{A}{U_1 - U_0} \, 100\% \qquad (2.83)$$

The eye aperture A is a well-suited and powerful measure to assess system performance and to determine the probability of error (discussed in Section 2.5.3). In general:

> Smaller the eye aperture A, higher the probability of error!

Normally, the eye pattern and eye aperture are measured by the set up given in Fig. 2.8. To analyze and optimize digital communication systems, theoretical calculation or at least an estimation of the eye aperture is frequently required. A method to calculate the eye aperture A has been briefly discussed below.

Let us assume that the frequency response of system (for example, single filter) be $H(f)$. The system response to a rectangular pulse (bit) at the input to the system is shown in Fig. 2.13.

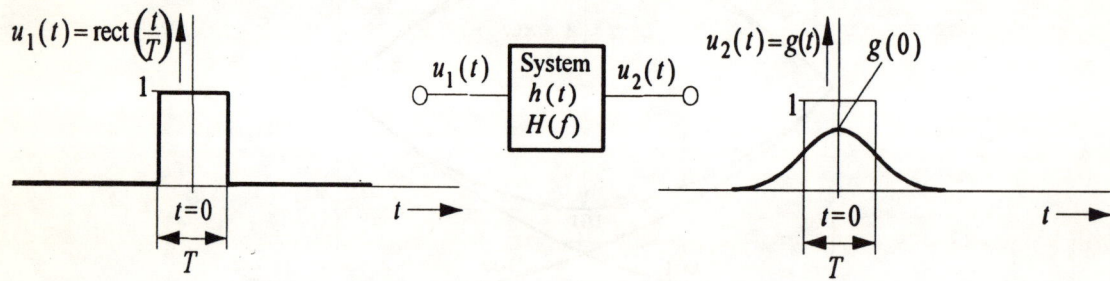

Fig. 2.13: System response to rectangular pulse (bit) signal

The eye aperture A can be determined by the general expression

$$A = 2\left(U_1 - U_0\right)\left[g(t_0) - \sum_{v=0}^{v=+\infty} |g(t_0-vT)| - \sum_{\mu=0}^{\mu=+\infty} |g(t_0+\mu T)|\right] \qquad (2.84)$$

where $g(t)$ represents the system response to a rectangular pulse as shown in Fig. 2.13 and U_0 and U_1 are the input signal levels corresponding to bit "0" and "1" respectively.

When the worst-case pattern of the system are known, calculation of the eye aperture is simplified. With worst-case pattern of "single one" and "single zero", we obtain

$$A = \left(U_1 - U_0\right)\left[2g(t_0) - H(0)\right] \qquad (2.85)$$

where $H(0)$ represents the DC frequency response of the system. In passive systems, $H(0)$ is normally unity i.e., $H(0)=1$:

2.5.3   PROBABILITY OF ERROR AND BIT ERROR RATE

Due to ISI, noise and other inherent system imperfections, bit sequence at the output of a digital receiver is usually different from the transmitted bit sequence i.e., $<q_v> \neq <r_v>$.

Each error in detection is called a *bit error*. Higher the number of errors within a certain time, lower is the quality of a digital transmission system. Therefore, probability of bit error or simply probability of error represents the most important measure to quantify system performance and assess system quality. In practice, probability of error is frequently called the *Bit Error Rate* (BER). Actually, this is a misnomer, because the term "rate" is always related to certain interval of time; for example, the bit *rate* of digital systems is in unit of bit/s. Therefore, the unit of bit error *rate* must also be bit/s or error/s, whereas the probability of error p is a dimensionless quantity. As an example, consider a bit rate $R = 10$ Gbit/s (i.e., $T = 0.1$ ns) and a probability of error $p = 10^{-9}$. This means that only one single bit out of 10^9 bits (1 Gbit) is detected incorrectly. In other words, ten bit errors occur during a time interval of 1 second. Hence, the rate of bit errors or *BER* is 10 bit/s.

Generally, the relationship between bit rate R, probability of error p and *BER* is given by the simple expression

$$BER = p \, R \qquad\qquad (2.86)$$

In practice, both p and *BER* are used interchangeably. In this book, we also follow the same trend.

As discussed in the previous Section, probability of error depends upon the transmitted bit sequence. Therefore, each transmitted bit pattern has its own characteristic probability of error. For this reason, calculation of the average BER (p_a) is very comprehensive. In practice, it is much more convenient to determine the worst-case BER (p_w) which is given by

$$p_w = \frac{1}{2}(p_1 + p_0) \geq p_a \qquad\qquad (2.87)$$

Here, p_1 and p_0 are the probabilities of error for equiprobable symbols "1" and "0" respectively in the worst-case pattern. Eq. (2.87) can be taken as a very good and well-suited approximation ($p_w \approx p_a$). It can also be taken as an upper limit on the probability of error ($p_a \leq p_w$). As discussed in Section 2.5.2, worst-case pattern are frequently given by "single one" and "single zero". Therefore,

$$p_1 = p(1 \Rightarrow 0) \text{ in case of pattern } \cdots 0001000 \cdots \text{ and}$$

$$p_0 = p(0 \Rightarrow 1) \text{ in case of pattern } \cdots 1110111 \cdots.$$

The worst-case pattern correspond to the worst-case sample values d_{1w} and d_{0w} which are much closer to threshold than those of all other bit pattern (Section 2.5.2). As a consequence, worst-case pattern are most sensitive to noise and yield much higher probability of error than the other pattern. Hence, worst-case pattern and worst-case BER are of great interest in practical communication engineering.

(i) CALCULATION OF WORST-CASE BER

As shown in Fig. 2.14, bit error occurs with a probability determined by the probability density function (pdf) $f_d(d)$ of the sampled detected signal

$$d(t_0) = d_0(t_0) + n(t_0) \tag{2.88}$$

at the input to the decision circuit. Here $d_0(t_0)$ represents the signal component. Under worst conditions, it will be

$$d_0(t_0) = \begin{cases} d_{1w} & \text{in case of pattern} \quad \cdots 0001000 \cdots \\ d_{0w} & \text{in case of pattern} \quad \cdots 1110111 \cdots \end{cases} \tag{2.89}$$

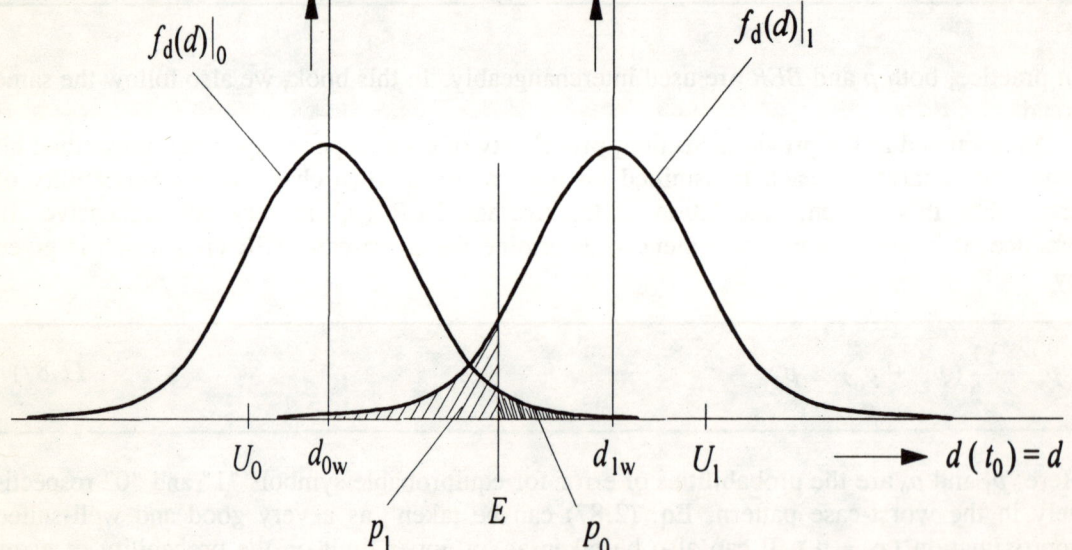

Fig. 2.14: Probability density functions of detected signals and error regions for bit "0" and "1"

In practice, fast estimation of BER is often required. For this purpose, d_{1w} and d_{0w} can be approximated by

$$d_{1w} \approx U_1 \quad \text{and} \quad d_{0w} \approx U_0 \tag{2.90}$$

which means that ISI has been neglected.

In the absence of phase noise, noise in coherent optical communication systems can be regarded as *Additive White Gaussian Noise* (AWGN). Hence, the sampled noise $n(t_0)$ will be Gaussian distributed with zero mean and variance σ_n^2 as given by Eq. (2.72). In that case, BER can be calculated as follows:

$$p = \frac{1}{2}(p_1 + p_0) = \frac{1}{2}\int_E^\infty f_d(d)\big|_0 \, dd + \frac{1}{2}\int_{-\infty}^E f_d(d)\big|_1 \, dd$$

(2.91)

$$= \frac{1}{2}\frac{1}{\sqrt{2\pi}\sigma_n}\int_E^\infty e^{-(d-d_{0w})^2/2\sigma_n^2}\,dd + \frac{1}{2}\frac{1}{\sqrt{2\pi}\sigma_n}\int_{-\infty}^E e^{-(d-d_{1w})^2/2\sigma_n^2}\,dd$$

Here, $f_d(d)\big|_0$ and $f_d(d)\big|_1$ represent the pdf of detected signal for the transmitted symbol "0" and symbol "1" respectively. Substitution of $x=(d-d_{0w})/\sigma_n$ and $y=(d-d_{1w})/\sigma_n$ in the above equation yields

$$p = \frac{1}{2}\frac{1}{\sqrt{2\pi}}\int_{\frac{E-d_{0w}}{\sigma_n}}^\infty e^{-x^2/2}\,dx + \frac{1}{2}\frac{1}{\sqrt{2\pi}}\int_{-\infty}^{\frac{E-d_{1w}}{\sigma_n}} e^{-y^2/2}\,dy$$

(2.92)

$$= \frac{1}{2}Q\left(\frac{E-d_{0w}}{\sigma_n}\right) + \frac{1}{2}Q\left(\frac{d_{1w}-E}{\sigma_n}\right)$$

where

$$Q(x) = \frac{1}{\sqrt{2\pi}}\int_x^{+\infty} e^{-u^2/2}\,du = \frac{1}{2}\,\text{erfc}\left(\frac{x}{\sqrt{2}}\right)$$

(2.93)

is the complementary Gaussian error function and erfc the complementary error function [1, 256]. It may be mentioned that $f_d(d)\big|_0$ and $f_d(d)\big|_1$ are Gaussian distributed only in the absence of phase noise. It becomes clear from Fig. 2.14 that BER reaches minimum if threshold level E is located at the cross point where pdf $f_d(d)\big|_0$ and pdf $f_d(d)\big|_1$ meet i.e.,

$$E_{opt} = \frac{d_{1w} + d_{0w}}{2}$$

(2.94)

With the above optimum threshold E_{opt} and eye aperture $A = d_{1w}-d_{0w}$ (Section 2.5.2), we finally get the simple and well-known expression.

$$p = Q\left(\frac{A}{2\sigma_n}\right) = Q\left(\frac{1}{2}\sqrt{\frac{S}{N}}\right) \tag{2.95}$$

It must be mentioned that the above equation is valid only for a binary digital baseband system. For multilevel systems or carrier systems with modulation schemes such as FSK, PSK or DPSK , it has to be modified appropriately (Chapters five and seven).

In most digital systems with AWGN, there is a steep fall in p with the increase in S/N ratio (Fig. 2.15). When phase noise disturb the system in addition, this behaviour is no longer valid. In such cases, error rate curve exhibits a characteristic error rate floor (Chapter five) and is, therefore, different from the curve shown in Fig. 2.15.

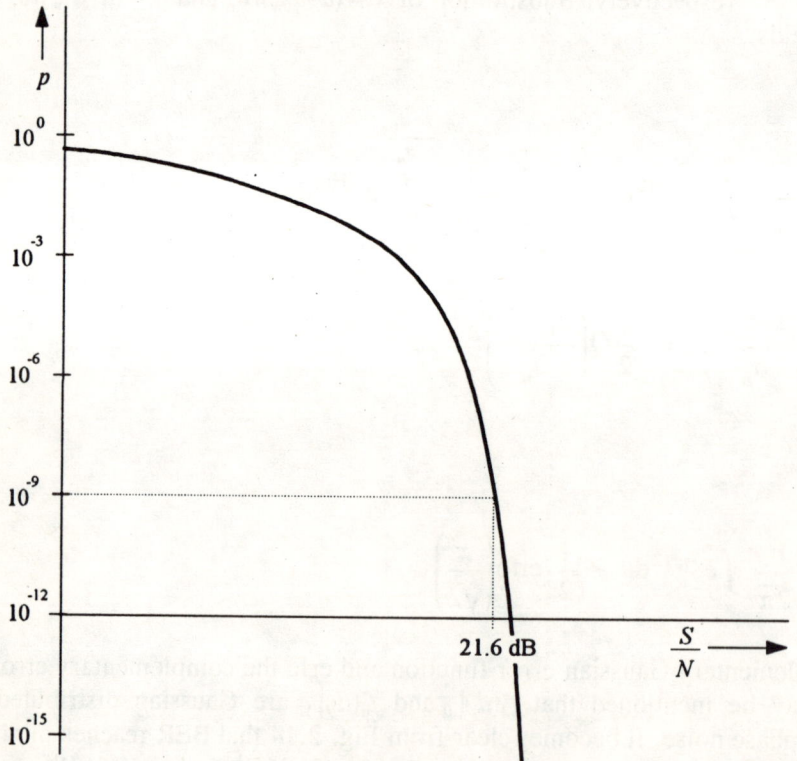

Fig. 2.15: Variations of probability of error with signal-to-noise ratio in digital system with AWGN

In practice, $p = 10^{-9}$ is a frequently demanded BER. For this BER, required S/N ratio can be determined with a high accuracy using the following approximation:

$$Q(6) \approx 10^{-9} \hspace{8cm} (2.86)$$

Finally, some typical eye pattern and their characteristic BERs are shown in Fig. 2.16. All eye pattern shown in this figure have been measured in a realized digital optical communication system [280]. It becomes evident from this figure that there is a strong relationship between eye pattern, eye aperture and bit error rate. With some training and practical experience, it is possible to estimate the bit error rate only by considering the eye pattern. Since a BER measuring equipment is usually more expensive than the set up required for the recordings of eye pattern particularly when the bit rate is high, this method of estimation represents a powerful as well as a low-cost technique.

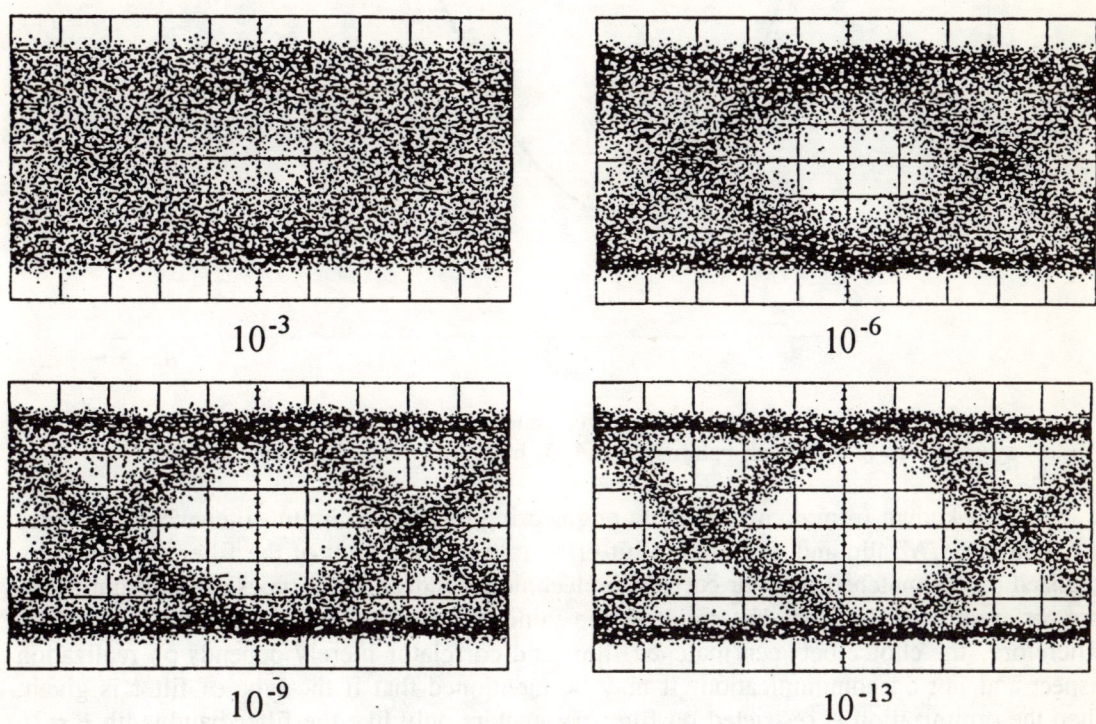

Fig. 2.16: Correlation between eye pattern and BER

(ii) INFLUENCE OF FILTERING

Eye pattern and bit error rate are strongly influenced by filtering. As discussed in Section 2.5.2, eye pattern and, consequently, the eye aperture A are functions of the receiver bandwidth. Noise power (N) increases linearly with bandwidth. It means

(a)   large bandwidth B imply large eye aperture A and low BER and

(b)   large bandwidth B imply large noise power and high BER.

Both the statements are contradictory. For this reason, there exists an optimum filter bandwidth B (actually an optimum filter frequency response $H(f)$) for which S/N ratio is maximum and BER is minimum (Fig. 2.17).

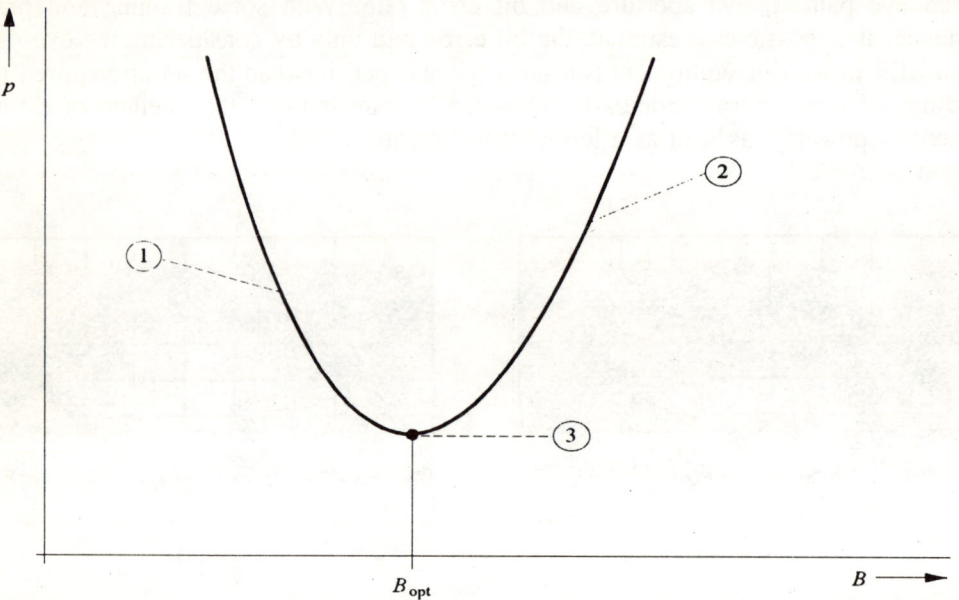

Fig. 2.17: BER p as function of receiver bandwidth B for (1) low noise, but small eye aperture (2) optimum bandwidth and (3) high noise, but large eye aperture

It is well-known from communication engineering that the optimum filter which provides a maximum S/N ratio and a minimum bit error rate at the output of the filter can either be realized by the matched filter or correlation technique. Indeed, both types of optimum filters are different in realization, but result in the same S/N ratio and bit error rate [256, 269]. Therefore, the choice between matched filter and correlator merely depends on realization aspect and not on communication. It may be mentioned that if the type of filter is given, then the optimization is restricted on filter parameters only like the filter bandwidth $B = 2f_g$ for a Gaussian filter.

In digital systems with AWGN, influence of restricted bandwidth can easily be evaluated since a Gaussian noise process always remains Gaussian. However, as soon as phase noise such as laser phase noise is added, evaluation becomes complicated and very comprehensive as the filter input signal is now nongaussian process. Moreover, this nongaussian process will additionally be changed by filtering. This problem of coherent optical communication systems is discussed in more detail in the following Chapter.

3 LASER NOISE

The gain of coherent optical communication systems can seriously be deteriorated by the noise of transmitter and local laser, especially by the *laser phase noise*. In addition to shot noise of photodiodes, thermal noise of resistors and electrical amplifiers called the circuit noise and polarization fluctuations caused by the fiber, laser phase noise is also a dominant source of noise in coherent communication systems. In particular, coherent optical receivers with optical or electrical synchronous detection are strongly influenced by laser phase noise; for example, PSK homodyne and heterodyne receivers which require a phase-locked loop (PLL).

Unlike coherent detection, conventional optical communication systems with intensity modulation of light and direct detection (IM/DD) are absolutely insensitive to laser phase noise. This, of course, is a significant advantage of using direct detection system instead of coherent detection systems. Modulation in direct detection system is easily performed by turning the laser light on and off as per the digital binary information signal to be trans- mitted. Hence, this kind of modulation is usually called on-off keying (OOK) or in a more devaluated way "smoke-sign modulation". A special knowledge of laser physics is not required and a simple black-box consideration of laser source is quite adequate for such applications.

In order to analyze and optimize coherent optical communication systems, this simplified black-box consideration is not adequate. The analysis of coherent optical homodyne and heterodyne receivers requires a much more detailed knowledge of the laser physics, in particular a fundamental knowledge of the statistical properties of laser noise. For this purpose, a brief review of the basic principle of laser operation is given in Section 3.1. Physical reasons for laser noise have been given and discussed in Section 3.2. With growing interest in realizing high-quality coherent communication links in the early eighties, physical reasons of laser noise had been examined very intensively. The results were primarily focused on physical than on communication aspects and published in many technical papers, reviews and books e.g., [39, 96-98, 211, 212, 258, 277, 278, 280, 290]. In this book, communication aspects of laser noise have been primarily considered.

Section 3.3 derives and discusses most important statistical properties of laser noise by means of a simple model that especially allows to highlight the communication aspects of laser noise. Laser phase noise, laser frequency noise and so-called phase-change noise are considered in detail. In addition, harmonic oscillations disturbed by phase noise are also discussed.

Section 3.4 briefly examines the effects of laser relaxation oscillations and Section 3.5 presents some interesting and generally valid aspects of filtering signals perturbed by phase noise. The concluding Section 3.6 presents some important technical solutions to reduce the effects of laser phase noise, especially if semiconductor lasers are used in coherent commu- nication systems.

3.1   PRINCIPLE OF LASER

Considering lasers, several different types have to be distinguished such as gas lasers, solid state lasers and semiconductor lasers. In particular, semiconductor laser diodes are indispensable key components of advanced optical fiber communication systems. Despite the difference in their size, material and pumping scheme, principle of operation is same for these lasers.

3.1.1   ABSORPTION, SPONTANEOUS AND STIMULATED EMISSION

Each atom is characterized by a perfectly specified number of electrons which are distributed on a fixed number of orbits. As an example, Fig. 3.1 shows two orbits with energy W_1 and W_2. The orbit with lower energy W_1 is called the *ground state* or *equilibrium state* and with higher energy W_2 the *excited state*. If the temperature is absolutely zero (0 K), then all atoms of a molecule are in the ground state. Normally, atomic system is also in the ground state for an indoor temperature of 293 K or 20° C.

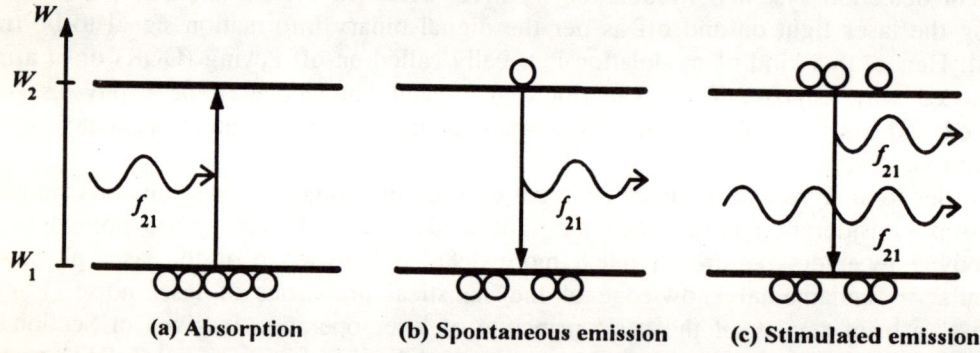

(a) Absorption   (b) Spontaneous emission   (c) Stimulated emission

Fig. 3.1: Absorption and emission in two-level system

The difference W_{21} in energy follows the well-known physical relation

$$W_{21} = W_2 - W_1 = hf_{21} \tag{3.1}$$

Here, $h = 6.6625 \cdot 10^{-34}$ Ws$^2 = 4.135 \cdot 10^{-15}$ eVs is the Planck's constant and f_{21} the frequency of light.

(i) ABSORPTION

An atom in the ground state is only able to pick up or absorb energy. If this energy equals the energy difference W_{21}, then this atom is transferred from the ground state to the excited state as shown in Fig. 3.1a. Usually, absorption is accomplished within a radiation field. In that case, only photons with frequency f_{21} can be absorbed by the atom. If, for example, a large number of atoms are located within the radiation field, then the intensity of radiation decreases exponentially with the absorption [268].

In an atom, absorption process strongly depends on the energy density ρ of external radiation field and number of electrons N_1 in the ground state W_1. Absorption decreases the number of ground state electrons with rate

$$R_a = -\left(\frac{dN_1}{dt}\right)_a = B_{12}\rho N_1 \tag{3.2}$$

Here, B_{12} represents the Einstein's absorption coefficient.

(II) SPONTANEOUS EMISSION

As the excited state is an unstable state, an atom which has been excited by absorption only remains in this state for a certain time called the lifetime τ_{21}. Afterwards, this atom returns to the ground state by itself spontaneously as shown in Fig. 3.1b. Thereby, a photon of energy W_{21} and frequency f_{21} is emitted. As this process occurs without any external stimulation, it is called spontaneous emission.

Normally, an atom contains several different energy levels. Therefore, spontaneous emission yield light radiation of many different frequencies. Considering a light wave at single frequency, a random and absolutely uncorrelated phase can be observed. This imply that spontaneous emission generate noncoherent light.

In a light emitting diode (LED), spontaneous emission is the result of recombination of excited electrons in the conduction band (excited energy level) and holes in the valence band (ground level). Since the light of an LED is noncoherent, LEDs are exclusively used in direct detection system. In contrast, they are absolutely unusable in coherent communication systems.

Spontaneous emission is not influenced by an external radiation field. As an statistical process, spontaneous emission only depend on the population N_2 of excited electrons in state W_2. By using the Einstein's coefficient A, emission rate R_{sp} of spontaneous emission is given by

$$R_{sp} = -\left(\frac{dN_2}{dt}\right)_{sp} = AN_2 = \frac{N_2}{\tau_{21}} \tag{3.3}$$

The spontaneous emission rate R_{sp} describes the temporal removal of electrons in the excited state W_2 and, hence, the temporal increase of spontaneously generated photons. This rate as seen from Eq. (3.3) is directly proportional to the population N_2 and inversely proportional to the average lifetime $\tau_{21} = 1/A$ of excited electrons in state W_2 [268]. If lifetime is high, excited electrons only return to ground state very rarely and the spontaneous emission rate R_{sp} is low. In contrast to this, emission rate R_{sp} becomes high when the lifetime is short.

(iii) STIMULATED EMISSION

In contrast to absorption, there are two different mechanisms for emission: one is the spontaneous emission as described above and the other is stimulated or induced emission as shown in Fig. 3.1c. Unlike spontaneous emission which is a random process, stimulated emission is a deterministic process providing the supply of external energy. Stimulated emission only occur if a stimulating photon of energy W_{21} and frequency f_{21} impinges upon the system. Thereby, a new photon of same frequency, phase, polarization and direction of propagation as the stimulating photon is generated. Thus, the incoming photon triggers the generation of an additional photon. Stimulating photon itself is not influenced by this process which means that the optical waves of both photons are coherent. Thus, stimulated emission of radiation offer the possibility to amplify a light wave true in phase and frequency.

The physical process of amplification of a light wave by stimulated emission is, therefore, the reason for the well-known abbreviation *LASER* which stands for *Light Amplification by Stimulated Emission of Radiation*. In LED, light generation is primarily performed by spontaneous emission, whereas in a laser light is predominantly generated by stimulated emission. Unlike spontaneous emission, stimulated emission is directly proportional to the energy density ρ of external radiation field. In addition, emission rate R_{st} of stimulated emission increases, of course, with the increase in number N_2 of excited electrons. Hence,

$$R_{st} = -\left(\frac{dN_2}{dt}\right)_{st} = B_{21}\rho N_2 \tag{3.4}$$

Here, B_{21} is the Einstein's stimulated coefficient.

(iv) BUDGET OF EMISSION AND ABSORPTION

Indeed, absorption and stimulated emission are mutual processes. Moreover, they are absolutely homogeneous that is,

$$B_{12} = B_{21} = B \tag{3.5}$$

The relationship between the coefficients A and B is given by

$$A = 8\pi h \lambda_{21}^{-3} n^3 B = 8\pi h \left(\frac{f_{21}}{c_0}\right)^3 n^3 B \tag{3.6}$$

which was given by Einstein [268]. Here, λ_{21} is the wavelength of emitted light wave, n the refraction index and $c_0 = 3 \cdot 10^8$ m/s the velocity of light in vacuum (i.e., free-space).

Considering a system to be composed of a large number of atoms, some atoms are in the ground state and the others in the excited state. In this atomic system, numerous absorption and emission processes occur (Fig. 3.2). Each absorption process attenuates the incident light wave, whereas each stimulated emission process amplify it. In addition, a large number of spontaneous emission processes are superimposed randomly. Unlike light waves generated by stimulated emission, light waves caused by spontaneous emission exhibit a random phase and are likewise transmitted in all directions. The small number of spontaneously emitted photons which accidentally have the same direction as the photons generated by stimulated emission, is the fundamental physical reason of laser phase and laser amplitude or laser intensity noise.

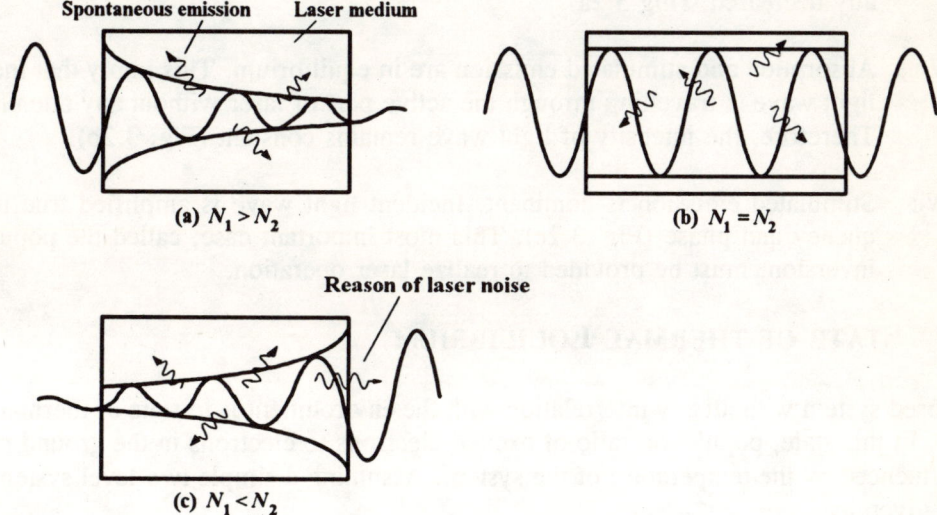

Fig 3.2: Influence of number of electrons in ground state and excited state on incident light wave

In a closed system, absorption and emission are in state of thermal equilibrium. This means that

$$R_a = R_{st} + R_{sp} \tag{3.7}$$

Using Eqs. (3.2) to (3.4), it becomes

$$B_{12}\rho N_1 = B_{21}\rho N_2 + AN_2 \tag{3.8}$$

Laser operation i.e., amplification of incident light wave provides that stimulated emission is dominant compared with absorption. To determine which of these two processes is actually dominant, the ratio

$$\frac{R_{st}}{R_a} = \frac{B_{21}\rho N_2}{B_{12}\rho N_1} = \frac{N_2}{N_1} \tag{3.9}$$

of stimulated emission rate to absorption rate is a well-suited parameter. It becomes clear that this ratio depends only on the population ratio of electrons in the excited state and ground state. On the basis of above ratio, following classification can be made:

$N_1 > N_2$:   Absorption is dominant. Hence, the intensity of incident light wave is exponentially decreased   (Fig 3.2a).

$N_1 = N_2$:   Absorption and stimulated emission are in equilibrium. This imply that incident light wave is travelling through the active part of laser without any attenuation. Therefore, the intensity of light wave remains constant (Fig. 3.2b).

$N_1 < N_2$:   Stimulated emission is dominant. Incident light wave is amplified true in frequency and phase (Fig. 3.2c). This most important case, called the population inversion, must be provided to realize laser operation.

3.1.2   STATE OF THERMAL EQUILIBRIUM

A closed system without any interrelation with the environment is in state of thermal equilibrium. In this state, population ratio of excited electrons to electrons in the ground state is only influenced by the temperature of the system. Assuming a simple two-level system, this ratio is given by

$$\frac{N_2}{N_1} = \exp\left(-\frac{W_{21}}{k_B T}\right) = \exp\left(-\frac{hf_{21}}{k_B T}\right) \tag{3.10}$$

Here, $k_B = 1.38 \cdot 10^{-23}$ Ws/K $= 8.62 \cdot 10^{-5}$ eV/K is the Boltzmann's constant and T the temperature measured in Kelvin.

Example 3.1

With a difference in energy $W_{21} = 0.827$ eV ($f_{21} = 200$ THz, $\lambda_{21} = 1.5$ μm) at an indoor temperature of $T = 293$ K (i.e., 20 °C), electron-population ratio given in Eq. (3.10) yields $N_2/N_1 = \exp(-32.74) = 6.02 \cdot 10^{-15}$. If the temperature is very high, for example $T = 1273$ K (i.e., 1000 °C), then we obtain $N_2/N_1 = \exp(-7.54) = 5.33 \cdot 10^{-4}$.

It becomes clear from the above example that even for a very high and unrealistic temperature of 1000 °C, electrons are almost in the ground state. Thus, a population inversion which is required for amplification of a light wave can never be achieved within a closed system in state of thermal equilibrium. For this, state of equilibrium must be disturbed first.

This, for example, can easily be accomplished by supplying an external energy which is usually called "pumping" (Fig. 3.3). In a semiconductor laser diode, inversion is simply achieved by the injection current.

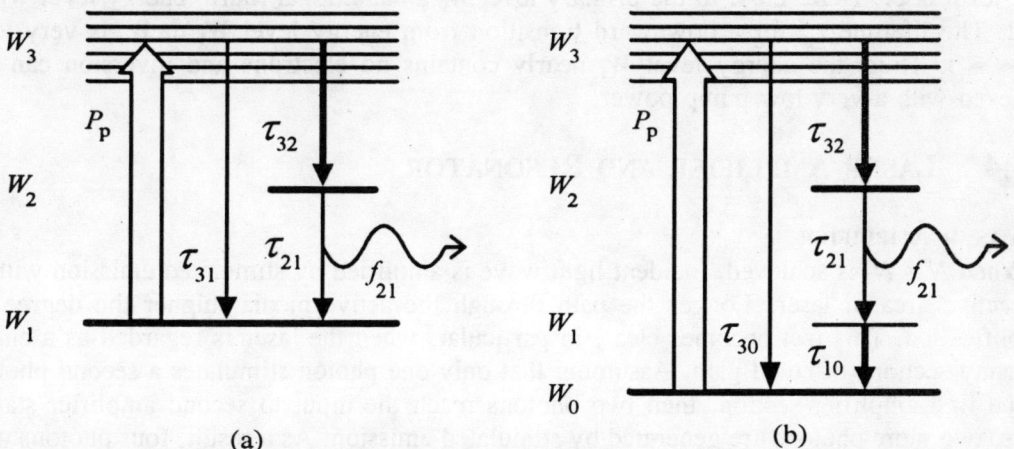

Fig. 3.3: Realization of population inversion by means of (a) three- and (b) four-level laser

3.1.3 POPULATION INVERSION

As mentioned above, population inversion means to disturb the state of thermal equilibrium of closed system by an external energy supply to achieve electron population ratio $N_2/N_1 > 1$. For this, a three- or four-level instead of a simple two-level structure and an appropriate pump energy have to be considered.

Besides the ground state W_1, two excited states W_2 and W_3 are used in a three-level structure. Regarding the lifetime of electrons in the excited states, special conditions must be fulfilled: Firstly, the lifetime τ_{31} of electrons crossing from the excited energy level W_3 to the ground level W_1 should be very large, whereas the lifetime τ_{32} for downward transition from level W_3 to W_2 should be very small ($\tau_{32} \ll \tau_{31}$). Thus, a transition from state W_3 to W_1 is very rare, while a transition from W_3 to W_2 is very probable. In addition, a pass over from level W_3 to W_2 should be without any radiation. Secondly, to collect a sufficiently large number of electrons at the excited energy level W_2 and, consequently, to achieve the desired population inversion, level W_2 must be a so-called meta-stable energy level with a high lifetime τ_{21}.

With an external energy source, electrons at level W_1 are pumped to the higher energy state W_3 as shown in Fig. 3.3a. For this purpose, a pump power P_P is required. To reach a high efficiency level, W_3 should be a rather broad energy level. Due to very low lifetime $\tau_{32} \ll \tau_{31}$, most of the pumped electrons pass over to the meta-stable level W_2 within a very short time. If the pump power P_P is appropriately high enough, then the required population inversion $N_1 < N_2$ is achieved.

In a three-level laser, inversion can be obtained only when the pump power exceeds a threshold power P_{th}. It is a serious drawback of three-level system. Consequently, four- or multilevel lasers are used. As an example, Fig. 3.3b shows the simplified structure of a four-level laser. Here, close to the primary level W_0 an additional fourth energy level W_1 is used. The lifetime τ_{10} for a downward transition from energy level W_1 to W_0 is very low ($\tau_{10} \ll \tau_{31}$). Thus, the energy level W_1 nearly contains no electrons and inversion can be achieved with a very low pump power.

3.1.4 LASER AMPLIFIER AND RESONATOR

(i) LASER AMPLIFIER

When $N_1 < N_2$ is achieved, incident light wave is amplified by stimulated emission within the active area of laser. Longer the path through the active media, higher the degree of amplification. This fact becomes clear, in particular, when the laser is regarded as a chain of many sections of equal gain. Assuming that only one photon stimulates a second photon in the first amplifier section, then two photons reach the input to second amplifier stage. Here, two more photons are generated by stimulated emission. As a result, four photons will be there at the output of the second, eight photons at the output of the third and sixteen photons at the output of the fourth amplifier stage (avalanche effect). This process is called one way amplification or traveling wave amplification. Population N_2 of the excited electrons at state W_2 decreases considerably and, hence, the efficiency of amplification. Finally, amplification reaches to a saturation value.

Instead of increasing the active path length of light in the laser, a higher pump power level P_P also yields an improvement in amplification. Thereby, the number N_2 of electrons at level W_2 increases and saturation occurs at a longer path length.

(ii) LASER CAVITY

In order to improve light amplification by increasing the length of optical path in the active area, this area is normally embedded inside an optical resonator called the laser cavity. By means of high reflecting plates at both sides of the resonator (e.g., two mirrors), length of the actual optical path becomes a multiple of cavity length. Due to a very large number of reflections (optical feedback), light amplification becomes very efficient.

In the practical realization of a laser cavity, plates with a reflectivity of somewhat less than 100% are used to extract and transmit a certain part of the laser light. The laser process usually starts with spontaneously emitted photons and reaches real laser operation within a very short time provided the pump energy is sufficiently large.

(iii) LASER EFFICIENCY

The optical output power P_o of a laser, which is extracted from the laser resonator, is used to transmit information through an optical fiber or open space. To generate this optical power, electrical pump power P_P is required as mentioned above. The efficiency of a laser is defined as

$$\eta = \frac{P_o}{P_P} \qquad\qquad (3.11)$$

where η varies from 0.01% to 0.1% for He-Ne-lasers, $\approx 20\%$ for CO_2-lasers and $\approx 30\%$ for semiconductor lasers.

3.1.5 SPECTRAL FEATURES OF LASERS

(i) EMISSION SPECTRUM OF RADIATING ELECTRON TRANSITION

Up-to-now, we have assumed that an atom is characterized by discrete energy levels. Because of this, each downward transition from higher to lower energy level always yields a photon of well-defined frequency (Figs. 3.1 and 3.3). However, this is actually not true since atomic energy states are continuous. As a consequence, each radiating transition from the exited state W_2 to the ground state W_1 exhibits a characteristic emission power spectral density $G_{21}(f)$. As shown in Fig. 3.4, this spectrum is primarily determined by its centre frequency f_{21} and 3 dB bandwidth Δf_{21}.

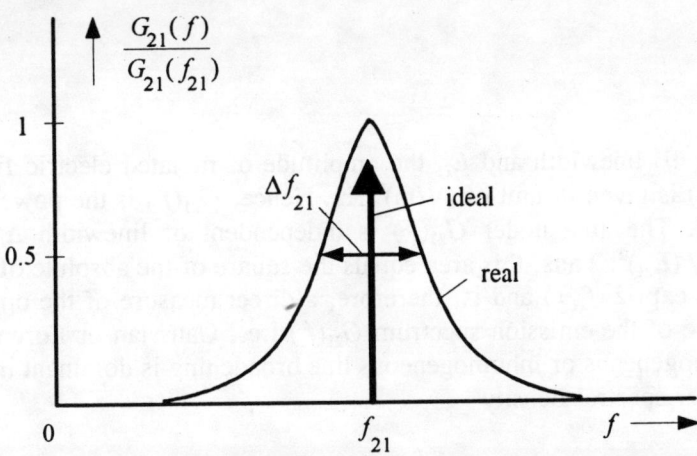

Fig. 3.4: Emission spectrum $G_{21}(f)$ of radiating energy transition under ideal and real conditions

At the centre frequency f_{21}, radiated power reaches maximum since most of the electrons are located in the centre of continuous energy level W_2 with very high probability. In contrast, the probability to find an electron in some distance from this centre decreases rapidly with the increase in the distance (Fermi distribution). Therefore, optical power spectral density decreases for frequencies $f > f_{21}$ and $f < f_{21}$.

The physical reason for the variations in the energy level is given by various interrelations between the different atoms of an atomic system, caused by pressure, pushes, temperature fluctuations and other internal mechanisms. In a gas laser, for example, fast moving radiating gas atoms give rise to optical Doppler effect which also broaden the energy levels.

Line broadening caused by the above effects is called the *inhomogeneous line broadening*. In this case, shape of the emission spectrum $G_{21}(f)$ is Gaussian.

Electrons which are located at the excited energy level W_2 return to the ground level W_1 after an average lifetime τ_{21} as already explained in the previous Section. Due to variations in lifetime, an uncertainty $\Delta W_{21} \approx h/\tau_{21}$ in the transition energy in accordance with the well-known uncertainty relation can be observed. This uncertainty gives rise to a broaden emission spectrum $G_{21}(f)$ as shown in Fig. 3.4. This type of broadening due to inherent uncertainty is frequently called the *natural* or *homogeneous line broadening* [268, 275]. In this case, shape of the emission spectrum follows the equation

$$G_{21}(f) = \hat{E}_{21}^2 \frac{2}{\pi \Delta f_{21}} \frac{1}{1 + \left(\dfrac{f - f_{21}}{\Delta f_{21}/2}\right)^2} \tag{3.12}$$

which is called the *Lorentzian line shape* [268, 275]. In the above equation

$$\Delta f_{21} = \frac{1}{\tau_{21}} \tag{3.13}$$

represents the 3 dB linewidth and $\hat{E}_{21}$ the amplitude of radiated electric field. It should be noted that $G_{21}(f)$ is given in unit of $(V/m)^2/Hz$. Hence, $G_{21}(f)$ is the power spectral density of electric field. The area under $G_{21}(f)$ is independent of linewidth Δf_{21} and yields the constant quantity $(\hat{E}_{21})^2$. Thus, this area equals the square of the absolute of complex electric field $\underline{E}(t) = \hat{E}_{21} \exp(j2\pi f_{21} t)$ and is, therefore, a direct measure of the optical power.

The real shape of the emission spectrum $G_{21}(f)$ i.e., Gaussian or Lorentzian depends on whether the homogeneous or inhomogeneous line broadening is dominant in the system. The normalized power spectral density

$$g_f(f) = \frac{G_{21}(f)}{\hat{E}_{21}^2} = \frac{2}{\pi \Delta f_{21}} \frac{1}{1 + \left(\dfrac{f - f_{21}}{\Delta f_{21}/2}\right)^2} \tag{3.14}$$

corresponds to the probability density function (pdf) of the random frequency f. Hence, the function $g_f(f)$ given in unit of per Hz represents a direct measure of determining the occurrence probabilities of all possible frequencies within the broaden energy transition W_{21}. The probability of a transmitted photon with frequency f in the range of df will be $g_f(f)df$. Thus, photons with frequency close to the centre frequency f_{21} are transmitted very frequently, whereas photons with higher or lower frequencies are transmitted very rarely. The normalized power spectral density $g_f(f)$ is called the *normalized line shape*. As a pdf, this function is characterized by

$$\int_0^{+\infty} g_f(f)\,df = 1 \tag{3.15}$$

provided that $\Delta f_{21}/f_{21} \ll 1$.

(ii) SPECTRAL FEATURES OF AMPLIFIED LIGHT WAVE

In principle, amplification of an incident light wave by means of stimulated emission is only possible within the emission spectrum $G_{21}(f)$ of a radiating energy transition. Light waves with frequencies outside this spectrum are not amplified and, consequently, they are insignificant as compare to the transmitted laser light wave. According to occurrence probability $g_f(f)\,df$, light waves with frequency close to the centre frequency f_{21} are amplified much more efficiently than those light waves of frequency not close to f_{21}. This fact becomes clear particularly when the optical amplifier is considered in multiple stages as in Section 3.1.4. Assuming that only one photon of frequency f_{21} and only one photon of double-frequency $2f_{21}$ enter the input to the first amplifier stage. We also assume that each amplifier section amplifies a light wave of frequency f_{21} by a factor of four and a light wave of frequency $2f_{21}$ by a factor of two only. Hence, we observe four photons of frequency f_{21} and two photons of frequency $2f_{21}$ at the input to the second amplifier stage. At the inputs to the third and fourth amplifier stages, we obtain 16 photons and 64 photons of frequency f_{21} and only 4 photons and 8 photons of frequency $2f_{21}$ respectively.

It is clear from this that the optical amplification strongly depends on the frequency of incident light wave. Further, bandwidth of emitted light wave at the output of last amplifier stage decreases rapidly with the increase in number of stages or reflections inside the laser cavity. Hence, spectral bandwidth Δf_g of amplified light wave is usually some order of magnitudes smaller than the bandwidth Δf_{21} of normal light wave caused by a simple energy transition ($\Delta f_g \ll \Delta f_{21}$).

(iii) MODES OF LASER CAVITY

Owing to the discrete size of laser cavity (i.e., optical resonator) or the active area of laser, only a limited number of discrete and specific light waves called *laser modes* are able to propagate. These modes are perfectly specified by standing waves inside the cavity. All other waves do not exist. Due to three dimensions of a cavity, three orthogonal directions for transmitting light will be there in principle.

As an example, the principle physical structure of *gain-guided semiconductor laser* realization is shown in Fig. 3.5. The width d of the optical cavity is usually in the order of some one tenth of μm and cavity length L is in the order of some 100 μm. The average length is about 300 μm. In x-direction, active area of this laser diode is confined by an electrode. This is used for the injection current to provide required pump energy and, hence, achieve required population inversion. The width w of this electric stripe guide is approximately in the order of some μm. In the y-direction, active area of this laser in confined by n-dotted emitter layer and p-dotted buried substrate, both characterized by a refraction index less than the refraction index of the active layer.

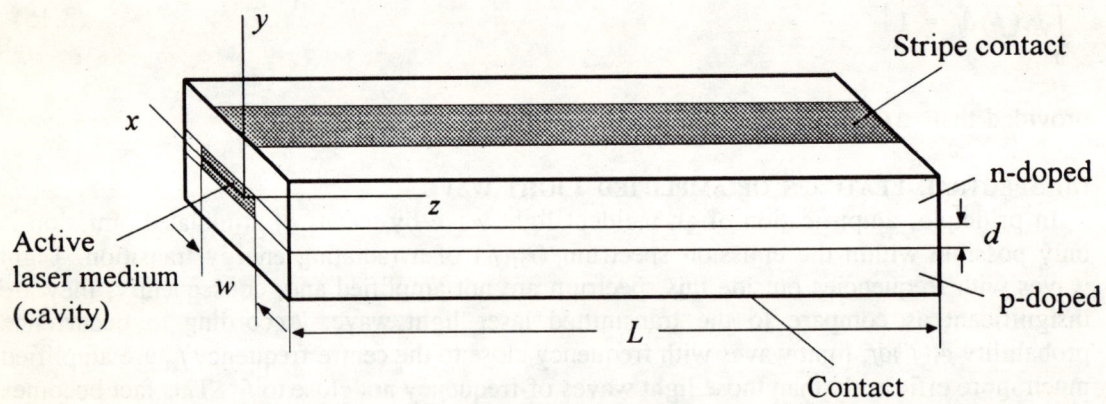

Fig. 3.5: Principle physical structure of gain-guided semiconductor laser

An alternate solution to confine the active area of laser is to realize the cavity as passive optical waveguide offering the advantage of much stronger optical guiding. Here, active area is confined by means of a change in the refraction index also in x-direction. This type of laser is called *index-guided laser*.

In both types of lasers mentioned above, optical power is emitted from the facets only which can be regarded as semi-transparent mirrors. Since the principle structure of the above lasers is similar to a Fabry-Perot resonator, these special kinds of lasers are frequently called Fabry-Perot lasers.

The modes which are able to travel in x- and y-directions are called the lateral and the transverse modes respectively. In optical communications, only the longitudinal or the axial modes in z direction are of practical importance. For an active media of refraction index n, axial mode a and wavelength λ_a for axial modes, following simple relationship can be derived:

$$2L = |a|\frac{\lambda_a}{n} = |a|\frac{c_0}{nf_a} \tag{3.16}$$

It should be noted that the wavelength λ_a is related to vacuum. In the above equation, mode numbers $a \in \{\cdots -2, -1, 0, 1, 2, \cdots\}$ are in accordance with the number of possible half waves inside the confinement of laser in the z-direction. It becomes clear that frequency f_a or wavelength λ_a of a laser strongly depend on the mode number a. Considering the frequency separation δf_a between two neighbouring axial modes, we obtain

$$\delta f_a = f_{a+1} - f_a = \frac{c_0}{2nL} \qquad (3.17)$$

Thus, the frequency separation δf_a is constant within the emission spectrum of a laser. As seen from Eq. (3.17), this separation is small if the length of laser cavity is large and vice versa. In contrast to constant frequency separation, spectral wavelength separation

$$\delta \lambda_a = \lambda_a - \lambda_{a+1} = \lambda_a \lambda_{a+1} \frac{1}{2nL} \approx \frac{\lambda_a^2}{2nL} \qquad (3.18)$$

depends on the wavelength itself. In brief, spectral separation $\delta \lambda_a$ is small for modes which are close to the central wavelength λ_a and large for modes which are not as close.

Example 3.2

Consider a cavity of length $L=200~\mu$m and refraction index $n=3.6$. At the central wavelength $\lambda_a=1.5~\mu$m, spectral separation from Eq. (3.18) is $\delta \lambda_a=1.56$ nm. It corresponds to a frequency separation of $\delta f_a=208$ GHz. In comparison to the bandwidth of typical information signal (for example, 5 MHz for analog TV), this is really a very high frequency separation.

(iv) Spectral Features of Transmitted Laser Light Wave

In Fig. 3.6, power spectral density $G_{21}(f)$ of light wave caused by a simple transition from the excited energy level W_2 to the ground energy level W_1 and emission spectrum $G_g(f)$ of an amplified (gained) light wave due to stimulated emission are shown.

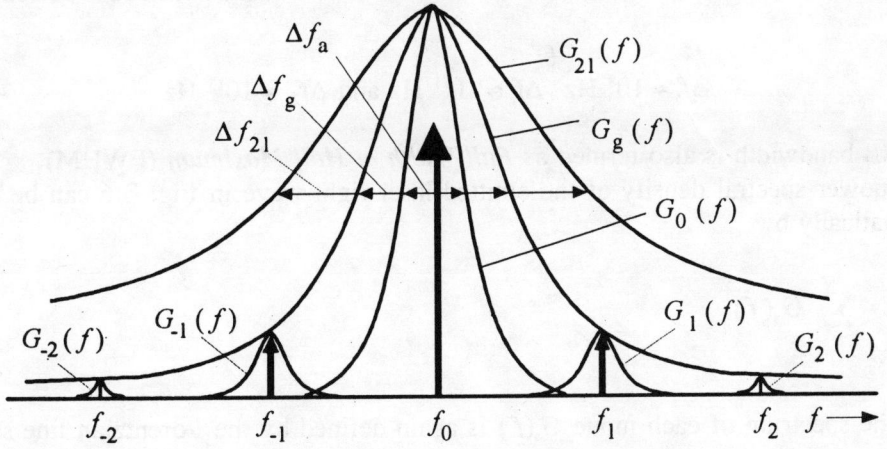

Fig. 3.6: Power spectral density of light wave

In addition, this figure also shows the modes which are determined by the size of laser cavity as mentioned above. The emission spectrum $G_{21}(f)$ of the energy transition exhibits a broad Lorentzian line shape with a linewidth Δf_{21} as given in Eq. (3.12). Due to selective amplification in the active area, 3 dB bandwidth Δf_g of amplified emission spectrum $G_g(f)$ is much smaller than the 3 dB bandwidth of emission spectrum $G_{21}(f)$ i.e., $\Delta f_g \ll \Delta f_{21}$. Within the emission spectrum $G_g(f)$, only a certain and limited number of waves called the laser modes are able to exist. The laser modes are characterized by standing waves within the discrete size of the laser cavity. In absence of special steps of realization, a laser generally oscillates in many different axial and transverse modes. In Fig. 3.6, axial modes are represented by Dirac delta functions. It becomes clear that this type of laser is, of course, not a monochromatic laser. As it oscillates in many different modes, we call this type of laser a *multimode laser*.

The stimulated and amplified light waves (the modes) are accompanied by spontaneously emitted light waves. Thus, the stimulated waves are disturbed by random light waves called the *laser noise*. This most fundamental noise in coherent optical communication systems arises from spontaneous emission and results in a serious broadening of linewidth of each mode. In Fig. 3.6, this linewidth is represented by Δf_a. Laser noise i.e., laser phase and amplitude or intensity noise will be discussed in Sections 3.2 and 3.3.

Similar to the line shape of spectrum $G_{21}(f)$ of an energy transition, shape of power spectral density $G_a(f)$ of the modes are also Lorentzian (see also Section 3.3.4). From Fig. 3.6, following relationship can be given:

$$\Delta \lambda_a < \Delta \lambda_g < \Delta \lambda_{21} \tag{3.19}$$

Example 3.3

For commercially available semiconductor diode lasers without any special technological steps to reduce the emission bandwidth, 3 dB bandwidths mentioned above are in the order of

$$\Delta f_a \approx 10^8 \text{ Hz}, \ \Delta f_g \approx 10^{11} \text{ Hz and } \Delta f_{21} \approx 10^{13} \text{ Hz}.$$

The 3 dB bandwidth is also termed as *Full Width at Half Maximum* (FWHM).

The power spectral density of the emitted laser light wave in Fig. 3.6 can be expressed mathematically by

$$G(f) = \sum_a G_a(f) \tag{3.20}$$

where the spectrum of each mode $G_a(f)$ is again defined by the Lorentzian line shape i.e.,

$$G_{\mathbf{a}}(f) = \hat{E}_{\mathbf{a}}^2 \; \frac{2}{\pi \Delta f_{\mathbf{a}}} \; \frac{1}{1 + \left(\dfrac{f - f_{\mathbf{a}}}{\Delta f_{\mathbf{a}}/2}\right)^2} \qquad (3.21)$$

Here, $\hat{E}_{\mathbf{a}}$ represents the amplitude of electric field.

(v) SINGLE-MODE LASER

The transmission capacity of optical communication systems, especially of coherent optical communication systems, depends on the spectral emission bandwidth of the lasers (transmitter and local lasers). Smaller the bandwidth, higher the possible bit rate and larger the possible repeaterless transmission span. In the ideal case, laser should be a real mono-chromatic source with one single frequency and zero linewidth i.e., $G(f) = \delta(f - f_0)$ is a Dirac delta function at frequency f_0.

In order to get such a laser, first step is to realize a single-mode laser and second step is to reduce its residual linewidth. In principle, single-mode operation can be accomplished either by decreasing the length L of the laser cavity or decreasing the pump power level P_{P} as shown in Fig. 3.7.

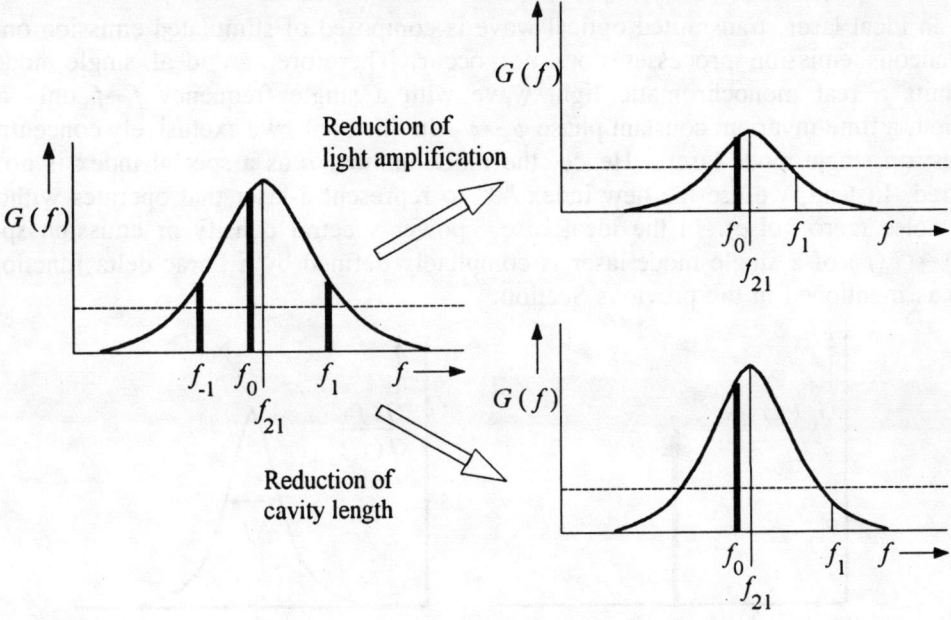

Fig. 3.7: Influence of laser cavity length and amplification on laser modes

It is observed from this figure that if pump power is low enough, only one single mode is able to exist. In case of all other waves, pump power is below the threshold power. If pump power is increased, laser will start operating again in multimode. With an appropriate short laser cavity, frequency separation of the modes can be made sufficiently large (Eq. 3.17). This again results in single-mode operation. In addition, Fig. 3.7 illustrates that the centre frequency f_0 of a single-mode laser may be different from the frequency f_{21} of energy transition.

In order to achieve highly reliable single-mode operation, more efficient solution is generally used. DFB (Distributed Feedback) lasers and DBR (Distributed Bragg Reflection) lasers, principally based on change in laser structure (Section 3.6), offer a very stable and highly efficient single-mode operation. In high bit rate direct detection system, DFB lasers are particularly most popular. Even if stable single-mode operation is achieved, a single-mode laser is still not a real monochromatic laser since the linewidth $\Delta\lambda_a$ of the mode is still not zero. To enhance system performance, further reduction of the remaining laser linewidth is required particularly when a coherent optical PSK homodyne system is needed (Sections 5.1.2 and 7.2.2). By using appropriate technologies, laser linewidths in the kHz range can, for example, be achieved with semiconductor laser diodes (Section 3.6).

3.2   REASONS AND FORMATION OF LASER NOISE

In an ideal laser, transmitted optical wave is composed of stimulated emission only i.e., spontaneous emission processes does not occur. Therefore, an ideal single-mode laser transmits a real monochromatic light wave with a single frequency $f_a \rightarrow f_0$ only and, in addition, a time-invariant constant phase $\phi_a \rightarrow \phi_0$. In this book, we exclusively concentrate our interest on single-mode lasers. Hence, the mode number a as a special index is no longer required. Instead, we use the new index "0" to represent a laser that operates without any laser noise (zero noise). In the ideal case, power spectral density or emission spectrum $G_a(f) \rightarrow G_0(f)$ of a single-mode laser is completely defined by a Dirac delta function (Fig. 3.8a) as mentioned in the previous Section.

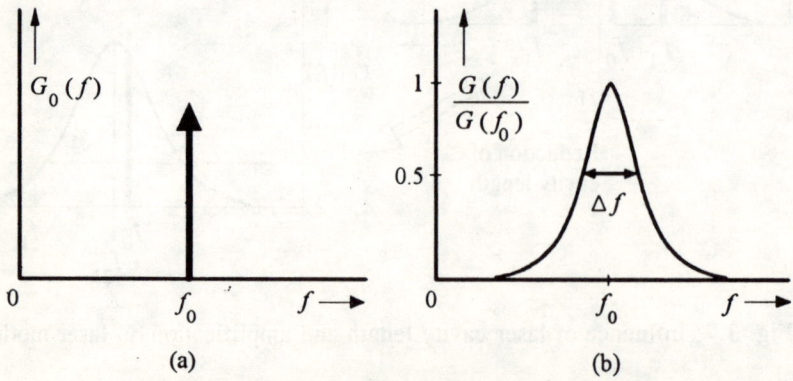

(a)   (b)

Fig. 3.8: Emission spectrum of (a) ideal and (b) real single-mode laser

Using the complex representation, time dependence of the electric field at the output of ideal single-mode laser can be described by the equation

$$E_0(t) = \hat{E}_0 \, e^{j2\pi f_0 t} = \hat{E}_0 \, e^{j\phi_0} \, e^{j2\pi f_0 t} \qquad (3.22)$$

It simply represents a harmonic oscillation. The electric field in addition of time as given by Eq. (3.22) is also a function of position. In the study of laser noise, parameter of location is insignificant. However, it has to be considered in the next Chapter when polarization effects are discussed.

(i) REASONS OF LASER NOISE

As mentioned in the previous Section, laser noise arises from spontaneous emission processes which are unavoidable in a laser. In gas lasers, physical reason of spontaneous emission is primarily the local fluctuations of laser mirrors. These fluctuations are in turn caused by the changes in temperature and external mechanical disturbances.

Spontaneous emission yield a time-varying amplitude $|E(t)| = \hat{E}(t)$ and, in addition, a time-varying phase $\phi(t)$. As a result, laser emission spectrum is substantially broaden as shown in Fig. 3.8b. The random phase $\phi(t)$ is called as *laser phase noise* and the random amplitude $|E(t)|$ as *laser amplitude* or *laser intensity noise*. The electric field of the transmitted laser wave, which is disturbed by laser noise, can be expressed as

$$E(t) = \hat{E}(t) \, e^{j(\phi_0 + \phi(t))} \, e^{j2\pi f_0 t} \qquad (3.23)$$

The fundamental measure to assess the strength of laser noise is given by the spectral laser linewidth $\Delta f_a \rightarrow \Delta f$, which is termed as 3 dB linewidth or linewidth at FWHM. This characteristic linewidth is defined by

$$G\left(f_0 \pm \frac{\Delta f}{2}\right) = 0.5 G(f_0) \qquad (3.24)$$

Formation of laser noise that is the conversion of single spontaneous emission process into time-varying random amplitude $|E(t)|$ and phase $\phi(t)$, is the topic of discussion in the following Subsection.

(ii) FORMATION OF LASER NOISE

By making use of a simple model, formation of laser noise will now be discussed in more detail [96-98]. In a single-mode laser with discrete energy states, spontaneous emission as well as stimulated emission only generate optical waves at frequency f_0. Phases of the stimulated waves are all synchronous, whereas they are random and absolutely uncorrelated in case of spontaneously emitted light waves. Each spontaneous emission event yields a spontaneous wave $\underline{E}_{spi}(t) = \hat{E}_{spi}\sin(2\pi f_0 t + \phi_{spi})$ which is superimposed onto the stimulated wave given in Eq. (3.22).

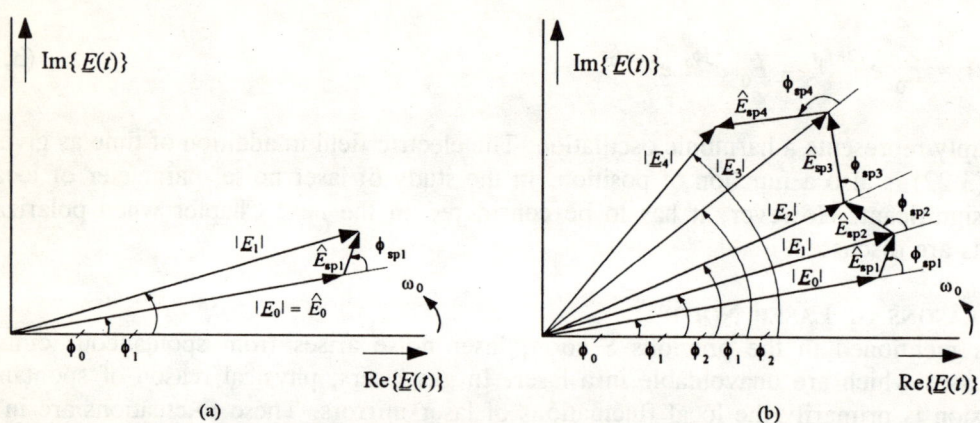

Fig. 3.9: Model of distortion of stimulated optical wave by (a) one and (b) four spontaneous emission events

As shown in Fig. 3.9a, stimulated optical wave is disturbed by a single spontaneous emission event only. This figure presents the complex electric field vectors of both stimulated optical wave $\underline{E}_0(t)$ and spontaneously emitted light wave $\underline{E}_{sp1}(t)$. The spontaneous phase ϕ_{sp1} is absolutely random and each value lies in the range $-\pi$ to π and can occur with same probability. Fig. 3.9b shows in successive stages the formation of optical laser wave $\underline{E}(t)$ now disturbed by four spontaneous emission events. It becomes evident from Figs. 3.9a and 3.9b that the resulting optical wave at the output is disturbed in phase (laser phase noise) as well as in amplitude (laser amplitude or intensity noise). Since spontaneously generated waves are not amplified to the same extent as stimulated waves, random amplitudes of spontaneous waves are always some order of magnitudes lower than the amplitude of the stimulated wave. Hence, the following relationship is valid:

$$\hat{E}_{spi} \ll \hat{E}_0 \qquad i \in N \tag{3.25}$$

As a consequence, laser amplitude or intensity noise is usually neglected i.e., $|\underline{E}(t)| \approx \hat{E}_0$. On the other hand, laser phase noise remains a predominant source of noise. The resulting laser phase ϕ_1 after one spontaneous emission event from Fig. 3.9 is

$$\phi_1 = \phi_0 + \arctan\left(\frac{\hat{E}_{sp1}\sin(\phi_{sp1})}{\hat{E}_0 + \hat{E}_{sp1}\cos(\phi_{sp1})}\right) \approx \phi_0 + \frac{E_{sp1}}{\hat{E}_0}\sin(\phi_{sp1}) \tag{3.26}$$

This equation has taken into account the relationship derived in Eq. (3.25). After I spontaneous emission events (Fig. 3.9b), laser phase is given by

$$\phi_I \approx \phi_0 + \sum_{i=1}^{I} \frac{\hat{E}_{spi}}{\hat{E}_0} \sin(\phi_{spi}) \qquad (3.27)$$

Next, we have to derive the relationship between discrete number I of spontaneous emission processes and continuous time t since laser phase noise $\phi(t)$ is a function of time. This relationship can be easily obtained by taking into account the spontaneous emission rate R_{sp} as given in Eq. (3.3). It gives

$$I = I(t) = N_2(0) - N_2(t) = R_{sp}t \qquad t \geq 0 \qquad (3.28)$$

It means, on an average I spontaneous emission processes occur or I spontaneous photons are generated in time t. Further, disturbance to the stimulated laser light wave due to spontaneous emission starts at $t = 0$. The electric field of optical wave at the output of laser can now be described by

$$E(t) = \hat{E}_0 \, e^{j\phi_0} \, e^{j\phi(t)} \, e^{j2\pi f_0 t} \qquad (3.29)$$

with

$$\phi(t) \approx \sum_{i=1}^{I(t)} \frac{\hat{E}_{spi}}{\hat{E}_0} \sin(\phi_{spi}) \qquad I(t) = R_{sp}t \qquad (3.30)$$

In order to make it more clear, results of computer simulation experiment based on $I = 10^5$ spontaneous emission events are shown in Fig. 3.10. It shows the electric field of a laser light wave disturbed by laser noise. In the simulation, the ratio $\hat{E}_{spi}/\hat{E}_0$ has been chosen to be constant (10^{-3}). In practice, this ratio is actually lower. For instant, numerical values in the order of 10^{-5} are usually obtained for semiconductor lasers. With an amplitude ratio of 10^{-3}, much lower number of simulated spontaneous emission events are required. Thus, much lower simulation run time is needed to obtain recognizable effects and achieve representative results. Moreover, the results obtained would be very similar and in principle be the same as with 10^{-5}.

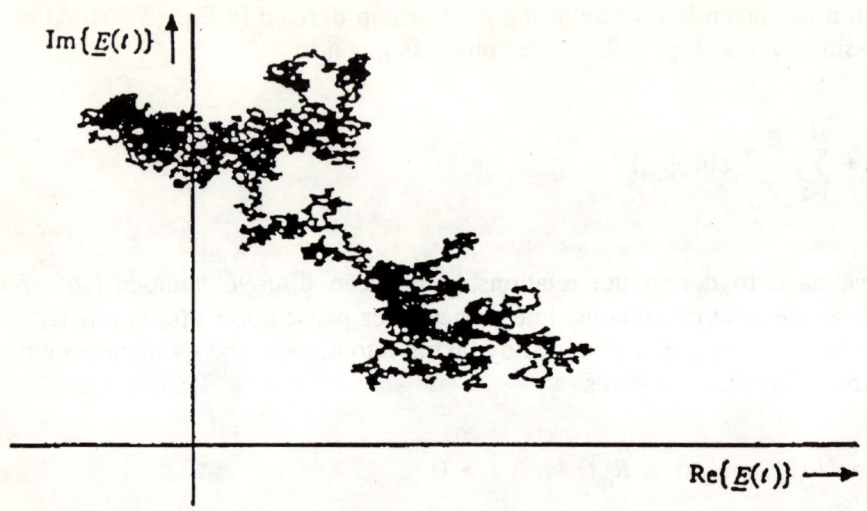

Fig. 3.10: Simulation results on the disturbance of stimulated light wave by spontaneous emission

3.3   STATISTICAL PROPERTIES OF LASER

In the first part of this Section, fundamental statistical features of laser phase noise $\phi(t)$ (Section 3.3.1) and other important laser random processes are derived and discussed in detail. Statistical properties of the *phase noise change* $\Delta\phi(t, \Delta T)$ and temporal derivation $d\phi(t)/dt$ which is called the *laser frequency noise* are discussed in Sections 3.3.2 and 3.3.3 respectively.

The most important statistical quantities which have to be derived are the expected value (or mean value), variance (i.e., square of standard deviation), probability density function (pdf), autocorrelation function (acf) and power spectral density (psd). In coherent optical communication systems, "harmonic" random processes $\sin(\phi(t))$, $\cos(\phi(t))$ and $\exp(j(\phi(t)))$ which include phase noise $\phi(t)$ as argument are of fundamental importance in particular. They are considered in detail in Section 3.3.4.

In order to determine the statistical properties of laser, communication-based considerations have been kept in view. Indeed, the more profound physical consideration of phase noise is useful to study laser as an optoelectronic component. However, this consideration is not well-suited in the frame of this book. The statistical results obtained in the following Sections are summarized at the end of each Section. It gives a small collection of important formulas, which will turn out to be a powerful tool to analyze and optimize coherent optical communication systems in Chapter five.

3.3.1   STATISTICS OF LASER PHASE NOISE

(i) EXPECTED VALUE

The expected value (mean value) of random laser phase $\phi(t)$ from Eq. (3.30) is given by

$$E\{\phi(t)\} = \frac{1}{\hat{E}_0} \sum_{i=1}^{I(t)} E\{\hat{E}_{spi} \sin(\phi_{spi})\} = \frac{1}{\hat{E}_0} \sum_{i=1}^{I(t)} E\{\hat{E}_{spi}\} \, E\{\sin(\phi_{spi})\} = 0 \qquad (3.31)$$

As spontaneous amplitudes $\hat{E}_{spi}$ and phases ϕ_{spi} are statistically independent random variables, expected value of the product in the first part of Eq. (3.31) can be separated in two as given in the second part. The phases ϕ_{spi} of a spontaneous emission event are uniformly distributed in the range $-\pi$ to $+\pi$ and the random variables $\sin(\phi_{spi})$ are symmetrical about zero. The expected value of the random variables $\sin(\phi_{spi})$ and, therefore, $\phi(t)$ will be zero.

(ii) VARIANCE

By using the fundamental statistical relation

$$\sigma_\phi^2(t) = E\{\phi^2(t)\} - E^2\{\phi(t)\} \qquad (3.32)$$

variance of $\phi(t)$ can be determined as follows:

$$\sigma_\phi^2(t) = \frac{1}{\hat{E}_0^2} E\left\{\left[\sum_{i=1}^{I(t)} \hat{E}_{spi}\sin(\phi_{spi})\right]^2\right\} = \frac{1}{\hat{E}_0^2}\sum_{i=1}^{I(t)}\sum_{j=1}^{I(t)} E\{\hat{E}_{spi}\hat{E}_{spj}\} E\{\sin(\phi_{spi})\sin(\phi_{spj})\}$$

$$(3.33)$$

$$= \frac{1}{\hat{E}_0^2}\sum_{i=1}^{I(t)} E\{\hat{E}_{spi}^2\} \, E\{\sin^2(\phi_{spi})\} = \frac{1}{2}\frac{E\{\hat{E}_{spi}^2\}}{\hat{E}_0^2} R_{sp}t = \frac{1}{2}\frac{P_{sp}}{P_0} R_{sp}t = K_\phi t$$

In the above equation, transformation from a single summation to double is based on the consideration that the expected values $E\{\sin(\phi_{spi}) \sin(\phi_{spj})\}$ are zero for $i \neq j$. The next change in Eq. (3.33), which finally yields the result, takes into account Eq. (3.28) and the fact that the expected values of $\sin^2(\phi_{spi})$ is 0.5. In the last step, P_{sp} and P_0 represent the average light power of the spontaneous and stimulated emission processes respectively ($P_{sp} \ll P_0$). The constant of proportionality K_ϕ is related to the laser linewidth Δf (Fig. 3.8 and Eq. 3.24) as follows:

$$K_\phi = 2\pi \Delta f \qquad\qquad\qquad\qquad\qquad\qquad\qquad\qquad (3.34)$$

This relationship will be derived in Section 3.3.4. It becomes clear from Eq. (3.33) that the variance of laser phase noise is time-varying. Hence, it is called a *nonstationary random process*. This typical statistical feature of phase noise is illustrated in Fig. 3.11 by means of three different sample processes simulated by using computer. Nonstationarity of laser phase noise becomes clear immediately when these processes are considered at different point of times t_1 and $t_2 > t_1$.

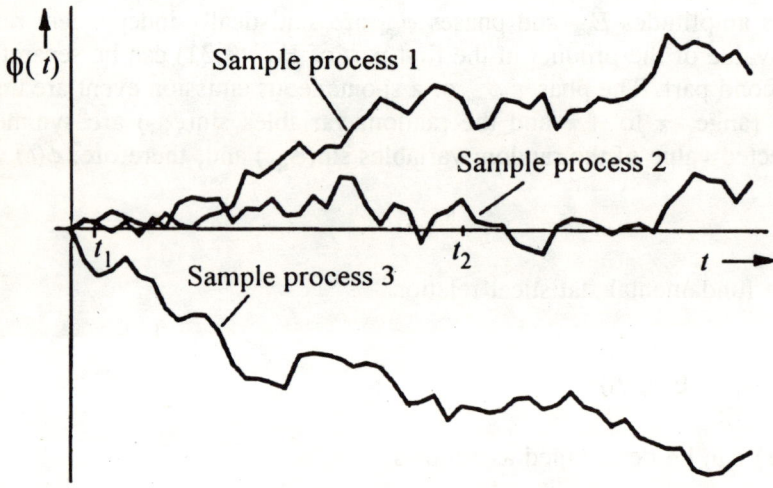

Fig. 3.11: Typical sample processes of laser phase noise $\phi(t)$

(iii) PROBABILITY DENSITY FUNCTION (pdf)

As seen from Eq. (3.30), laser phase noise is formally generated by a sum of sine functions with spontaneous emission phases ϕ_{spi} as arguments which are statistically independent and uniformly distributed in the range of $-\pi$ to $+\pi$. Due to inherent nonlinearity of the sine function, pdf of the random variables $\sin(\phi_{spi})$ are not uniformly distributed (Section 3.3.4). The sum given in Eq. (3.30) is characterized by two special properties: First, this sum contains a very large number of terms. Considering a typical spontaneous emission rate $R_{sp} = 10^{12}$ s^{-1} for instant, then this sum includes 1000 terms after a very short time interval of $t = 1$ ns only. It should be remembered that this time interval corresponds to the point of time when the laser source starts operation. The second characteristic feature is the statistical independence of the individual sum terms due to independence of phases ϕ_{spi} as mentioned earlier.

Both features permit the use of well-known central limit theorem of statistics [206]. This theorem states that if the random variables are independent (here $\sin(\phi_{spi})$), then the probability density function of their sum tends to follow a normal curve as the number of terms is sufficiently large. With this theorem, we obtain that laser phase noise is a *Gaussian random process* or normal process.

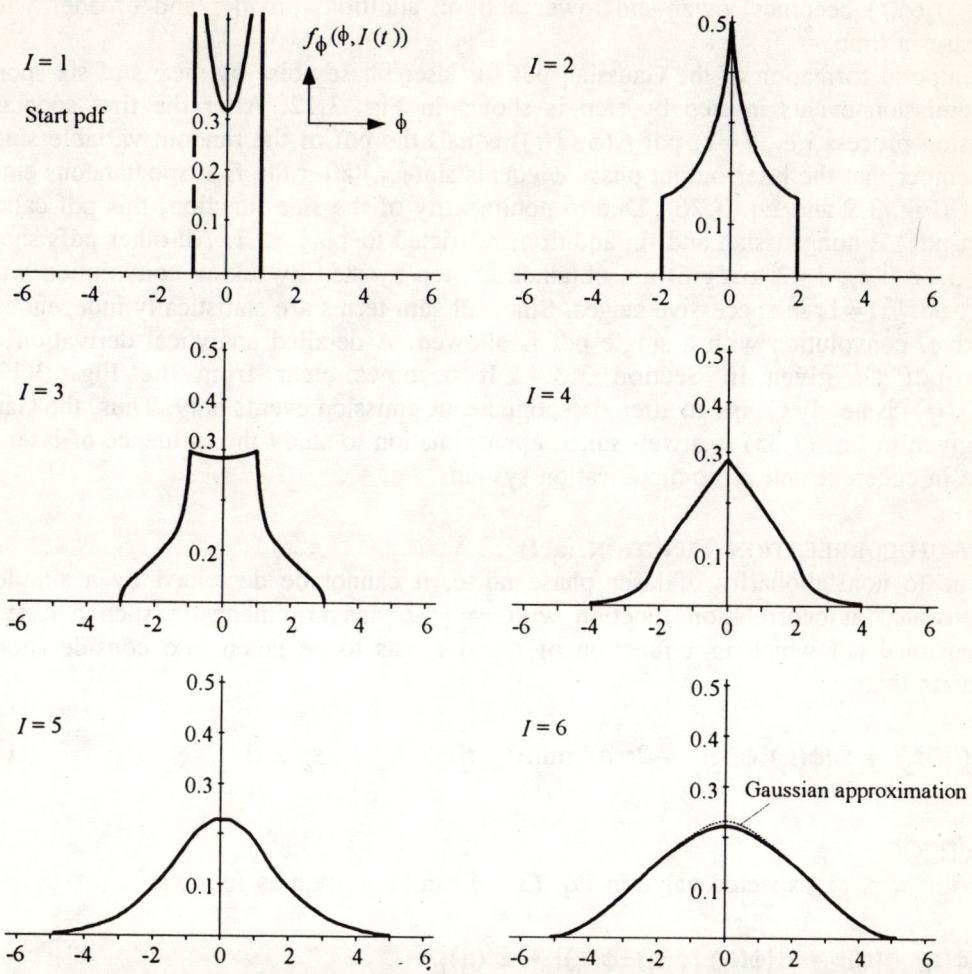

Fig. 3.12: Probability density function $f_\phi(\phi, I(t))$ of laser phase noise ϕ after $I = R_{sp}t$ spontaneous emission events. The Gaussian approximation in case of $I = 6$ follows Eq. (3.35) with $\sigma_\phi^2 = 2\pi\Delta ft = 0.5\ R_{sp}t = I/2 = 3$. In this figure, pdf is normalized to a constant field ratio $\hat{E}_{spi}/\hat{E}_0$.

Using Eqs. (3.31), (3.33) and (3.34), pdf of laser phase noise is given by

$$f_\phi(\phi, t) = \frac{1}{\sqrt{2\pi}\sigma_\phi} \exp\left(-\frac{\phi^2}{2\sigma_\phi^2}\right) = \frac{1}{2\pi\sqrt{\Delta ft}} \exp\left(-\frac{\phi^2}{4\pi\Delta ft}\right) \qquad t \geq 0 \qquad (3.35)$$

Due to time dependence of phase noise variance $\sigma_\phi^2(t)$, pdf $f_\phi(\phi,t)$ is also a function of time. As given in Eq. (3.33), variance $\sigma_\phi^2(t)$ of laser phase noise increases linearly with time. Thus, $f_\phi(\phi,t)$ becomes lower and lower and, in addition, broader and broader with the increase in time.

Temporal formation of the Gaussian pdf for laser phase noise by means of six spontaneous emission events in step by step is shown in Fig. 3.12. After the first spontaneous emission process i.e., $I=1$, pdf $f_\phi(\phi, I(t))$ equals the pdf of the random variable $\sin(\phi_{sp1})$. Remember that the laser output phase ϕ equals $\sin(\phi_{sp1})$ after the first spontaneous emission event (Fig. 3.9 and Eq. 3.26). Due to nonlinearity of the sine function, this pdf called the "start-pdf" is nongaussian and, in addition, restricted to $|\phi_1| \leq 1$. All other pdfs shown in Fig. 3.12 (i.e., $I=2$ to $I=6$) are obtained in step by step by taking convolution with the "start-pdf" ($I=1$) in successive stages. Since all sum terms are statistically independent and additive, convolution with a single pdf is allowed. A detailed analytical derivation of the "start-pdf" is given in Section 3.3.4. It becomes clear from the Fig. 3.12 that $f_\phi(\phi, I(t))$ is nearly Gaussian after six spontaneous emission events only. Thus, the Gaussian pdf given in Eq. (3.35) is a well-suited approximation to study the influence of laser phase noise in coherent optical communication systems.

(iv) AUTOCORRELATION FUNCTION (acf)

Due to nonstationarity of laser phase noise, it cannot be described by a simple one-dimensional autocorrelation function with $\tau=|t_2-t_1|$ as argument. In such a case, two-dimensional acf which is a function of t_1 and t_2 has to be taken into consideration. We maintain that

$$R_\phi(t_1, t_2) = E\{\phi(t_1)\, \phi(t_2)\} = 2\pi\Delta f \min(t_1, t_2) \qquad t_1, t_2 \geq 0 \qquad (3.36)$$

Proof:

With $t_1 \leq t_2$, expected value in Eq. (3.36) can be written as follows:

$$E\{\phi(t_1)\, \phi(t_2)\} = E\{\phi(t_1)\, [\phi(t_2)-\phi(t_1)] + \phi^2(t_1)\}$$
$$= E\{\phi(t_1)\, [\phi(t_2)-\phi(t_1)]\} + E\{\phi^2(t_1)\} \qquad (3.37a)$$

The phase noise difference $\phi(t_2) - \phi(t_1)$ is independent of the previous temporal phase $\phi(t_1)$ since spontaneous emission events are statistically independent. For this reason, first expected value in the second line of Eq. (3.37a) is zero. Thus,

$$E\{\phi(t_1)\, \phi(t_2)\} = E\{\phi^2(t_1)\} = \sigma_\phi^2(t_1) = 2\pi\Delta f t_1 \qquad t_1 \leq t_2 \qquad (3.37b)$$

By changing t_1 and t_2, that is $t_1 \rightarrow t_2$ and $t_2 \rightarrow t_1$, reader may prove Eq. (3.36) for $t_2 \leq t_1$.

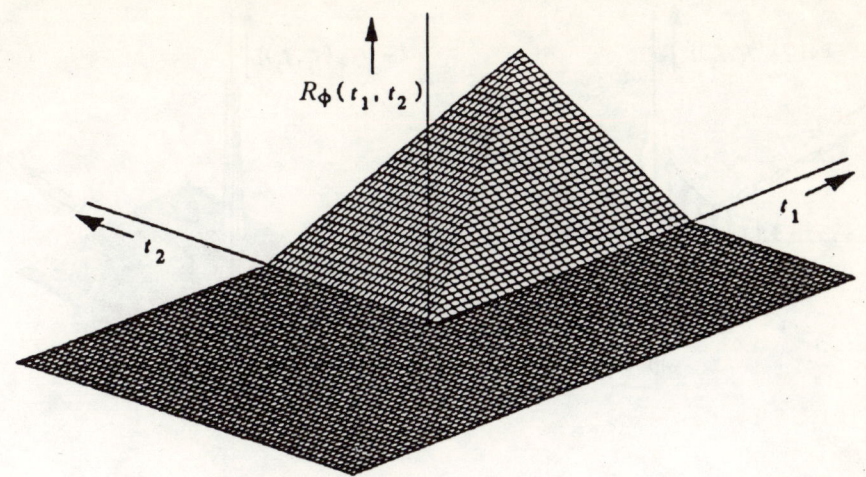

Fig. 3.13: Autocorrelation function $R_\phi(t_1, t_2)$ of nonstationary laser phase noise $\phi(t)$

(v) POWER SPECTRAL DENSITY (psd)

The power spectral density $\underline{G}_\phi(f_1, f_2)$ of the laser phase noise $\phi(t)$ is a complex quantity. It is obtained from the real acf $R_\phi(t_1, t_2)$ by taking two-dimensional Fourier transform. It is given by

$$
\underline{G}_\phi(f_1, f_2) = \int\limits_{-\infty}^{+\infty} \int\limits_{-\infty}^{+\infty} R_\phi(t_1, t_2)\, e^{-j2\pi f_1 t_1}\, e^{-j2\pi f_2 t_2}\, dt_1\, dt_2
$$

$$
= \frac{\Delta f}{4\pi}\left[\frac{2}{f_1^2 + f_2^2}\delta(f_1 + f_2) - \frac{1}{f_2^2}\delta(f_1) - \frac{1}{f_1^2}\delta(f_2)\right] + j\frac{\Delta f}{4\pi}\frac{1}{f_1 f_2 (f_1 + f_2)}
$$

(3.38)

The real and imaginary parts of $\underline{G}_\phi(f_1, f_2)$ are shown in Fig. 3.14. It can be seen from this figure that the psd of laser phase noise decreases rapidly at high frequencies. Thus, the influence of laser phase noise is much more in the low frequency range. Usually, a psd is primarily characterized by its 3 dB bandwidth or FWHM bandwidth. However, in case of nonstationary phase noise, it is impossible to define a noise equivalent bandwidth neither by defining a 3 dB cut-off frequency nor by determining an equivalent cube or cylinder with a volume in accordance with the volume of two-dimensional function $\underline{G}_\phi(f_1, f_2)$. The reasons for this problem are $1/f^2$ shape of the psd and poles at the origin $f=0$.

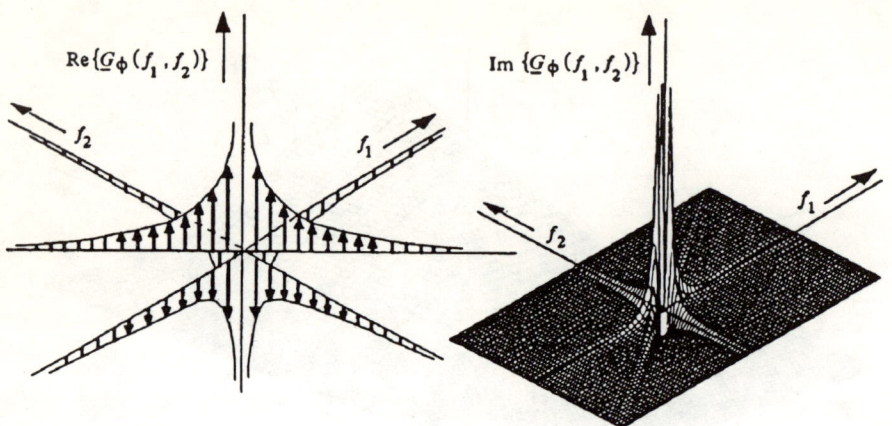

Fig. 3.14: Complex power spectral density $\underline{G}_\phi(f_1, f_2)$ of nonstationary laser phase noise $\phi(t)$

Statistical properties of laser phase noise $\phi(t)$ are summarized as follows:

(i)   nonstationary

(ii)   expected value: $E\{\phi(t)\} = 0$

(iii)   variance: $\sigma_\phi^2(t) = 2\pi\Delta f\, t \qquad t \geq 0$

(iv)   pdf: $f_\phi(\phi, t) = \dfrac{1}{2\pi\sqrt{\Delta ft}}\, \exp\left(-\dfrac{\phi^2}{4\pi\Delta ft}\right) \qquad t \geq 0$

(v)   acf: $R_\phi(t_1, t_2) = 2\pi\Delta f\, \min(t_1, t_2) \qquad t_1, t_2 \geq 0$

(vi)   psd: $\underline{G}_\phi(f_1, f_2) \;\bullet\!\!-\!\!\circ\; R_\phi(t_1, t_2)$

Finally, it may be mentioned that laser phase noise exhibits the same statistical features as the *brownian motion* of molecular particles.

3.3.2   STATISTICS OF PHASE NOISE DIFFERENCE

The random process

$$\Delta\phi(t, \Delta T) = \phi(t+\Delta T) - \phi(t) \tag{3.39}$$

which is defined by a temporal change in laser phase during the time interval ΔT will be called a phase noise difference or a phase noise change in this book. This noise is not a physical noise such as shot noise of photodiode or phase noise of laser, but to be considered in coherent optical communication systems. The time interval ΔT is in general arbitrary, but usually equal the bit duration. A profound knowledge of the statistical features of phase noise change $\Delta\phi(t, \Delta T)$ is especially required if the statistical features of laser frequency noise $d\phi(t)/dt$ have to be determined (Section 3.3.3) or coherent optical DPSK heterodyne system has to be analyzed (Section 5.3.3).

(i) EXPECTED VALUE

Like laser phase noise $\phi(t)$, phase noise difference $\Delta\phi(t, \Delta T)$ is also a random process of zero mean. Thus, its expected value is also zero i.e.,

$$E\{\Delta\phi(t, \Delta T)\} = E\{\phi(t+\Delta T)\} - E\{\phi(t)\} = 0 \qquad (3.40)$$

(ii) VARIANCE

Using Eqs. (3.33) and (3.36), its variance can be determined as

$$\sigma_{\Delta\phi}^2 = E\{[\phi(t+\Delta T) - \phi(t)]^2\}$$

$$= E\{\phi^2(t+\Delta T)\} - 2E\{\phi(t+\Delta T)\ \phi(t)\} + E\{\phi^2(t)\} \qquad (3.41)$$

$$= 2\pi\Delta f\left[(t+\Delta T) - 2\ \min(t+\Delta T,\ t) + t\right] = 2\pi\Delta f|\Delta T|$$

Unlike the variance of $\phi(t)$ given in Eq. (3.33), variance of $\Delta\phi(t, \Delta T)$ is not a function of time. In fact, it is a function of constant time difference ΔT only. It may be mentioned that each phase is actually a phase change. Thus, laser phase itself is also a phase change from a fixed reference phase $\phi(0)$ at $t=0$ when the phase noise process has started. Hence,

$$\phi(t) = \Delta\phi(0, t) \qquad (3.42)$$

This means that $\Delta T \rightarrow t$ has to be considered as a variable of time in phase noise, whereas ΔT is a constant in case of phase noise difference.

As mentioned at the beginning of this Section, phase noise difference is important in the analysis of coherent optical DPSK transmission system. In Section 5.3.3, it has been derived that the standard deviation $\sigma_{\Delta\phi}$ of $\Delta\phi(t, \Delta T)$ must be less than 0.24 to achieve a bit error rate of less than 10^{-10}. For example, at a bit rate of 155 Mbit/s (i.e., $\Delta T=T=1/155\ \mu s$), Eq. (3.41) gives a maximum allowed laser linewidth of about 1.42 MHz.

(iii) PROBABILITY DENSITY FUNCTION (pdf)

In Section 3.3.1, a Gaussian pdf was derived for laser phase $\phi(t)$. The phase noise change $\Delta\phi(t, \Delta T)$ is also a *Gaussian random process* since $\Delta\phi(t, \Delta T)$ is a linear combination of two random laser phases as defined in Eq. (3.39). Therefore, we have

$$f_{\Delta\phi}(\Delta\phi) = \frac{1}{2\pi\sqrt{\Delta f|\Delta T|}} \exp\left(-\frac{\Delta\phi^2}{4\pi\Delta f|\Delta T|}\right) \tag{3.43}$$

By comparing this result with Eq. (3.35), a close relationship between $\Delta\phi(t, \Delta T)$ and $\phi(t)$ becomes evident again. As laser linewidth Δf increases, $f_{\Delta\phi}(\Delta\phi)$ becomes broader. It approaches to Dirac delta function i.e., $f_{\Delta\phi}(\Delta\phi) \rightarrow \delta(\Delta\phi)$ when laser linewidth Δf equals zero. In this unrealistic ideal case, laser phase remains constant and no difference between $\phi(t)$ and $\phi(t + \Delta T)$ can be observed. Thus, $\Delta\phi(t, \Delta T)=0$.

(iv) AUTOCORRELATION FUNCTION (acf)

The acf of $\Delta\phi(t, \Delta T)$ can be calculated by

$$R_{\Delta\phi}(t_1, t_2) = E\{\Delta\phi(t_1, \Delta T)\, \Delta\phi(t_2, \Delta T)\}$$

$$= 2\pi\Delta f \begin{cases} |\Delta T|-|\tau| & \text{if } |\tau| = |t_2-t_1| \leq |\Delta T| \\ 0 & \text{otherwise} \end{cases} \tag{3.44}$$

$$= R_{\Delta\phi}(\tau) = R_{\Delta\phi}(-\tau)$$

Proof:

Using Eqs. (3.36), (3.39) and, in addition, the linear property of expected values, above equation can be expressed as

$$R_{\Delta\phi}(t_1, t_2) = E\{ [\phi(t_1+\Delta T) - \phi(t_1)] [\phi(t_2+\Delta T) - \phi(t_2)] \}$$

$$= 2\pi\Delta f [\min(t_1+\Delta T, t_2+\Delta T) - \min(t_1+\Delta T, t_2) \tag{3.45}$$

$$- \min(t_2+\Delta T, t_1) + \min(t_1, t_2)]$$

Now we have to consider all possible interrelations between the points of time t_1, $t_1+\Delta T$, t_2 and $t_2+\Delta T$. Evaluation of all these cases and their substitution in Eq. (3.45) yield Eq. (3.44).

In contrast to the autocorrelation function $R_\phi(t_1, t_2)$ of nonstationary phase noise $\phi(t)$, auto-correlation function $R_{\Delta\phi}(t_1, t_2) = R_{\Delta\phi}(t_2-t_1) = R_{\Delta\phi}(\tau)$ of $\Delta\phi(t, \Delta T)$ only depends on the time difference $\tau = |t_2 - t_1|$. For this reason, $\Delta\phi(t, \Delta T)$ is called a *stationary random process*. Thus, its autocorrelation function is simply a one-dimensional function in variable τ. The typical triangular shape of $R_{\Delta\phi}(\tau)$ is shown in Fig. 3.15a.

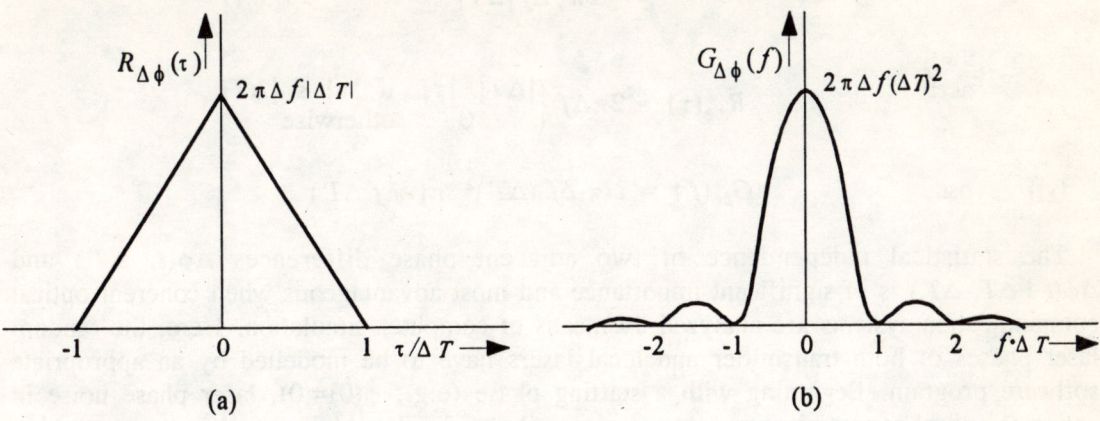

Fig. 3.15: (a) Autocorrelation function $R_{\Delta\phi}(\tau)$ and (b) power spectral density $G_{\Delta\phi}(f)$ of the phase noise difference $\Delta\phi(t, \Delta T)$

(v) POWER SPECTRAL DENSITY (psd)

Due to stationarity of $\Delta\phi(t, \Delta T)$, psd $G_{\Delta\phi}(f)$ can easily be obtained by taking one-dimensional Fourier transform of $R_{\Delta\phi}(\tau)$. We obtain

$$G_{\Delta\phi}(f) \quad \circ\!\!\!-\!\!\!\circ \quad R_{\Delta\phi}(\tau)$$

$$G_{\Delta\phi}(f) = 2\pi\Delta f(\Delta T)^2 \mathrm{si}^2(\pi f \Delta T) \tag{3.46a}$$

where

$$\mathrm{si}(x) = \frac{\sin(x)}{x} \tag{3.46b}$$

The power spectral density $G_{\Delta\phi}(f)$ is shown in Fig. 3.15b. Like phase noise, phase noise difference is also dominant at low frequencies.

Statistics of phase noise difference $\Delta\phi(t, \Delta T) = \phi(t+\Delta T) - \phi(t)$ are summarized as follows:

(i)   stationary

(ii)   expected value: $E\{\Delta\phi\} = 0$

(iii)   variance: $\sigma_{\Delta\phi}^2 = 2\pi\Delta f \, |\Delta T|$

(iv)   pdf: $f_{\Delta\phi}(\Delta\phi) = \dfrac{1}{2\pi\sqrt{\Delta f |\Delta T|}} \, \exp\left(-\dfrac{\Delta\phi^2}{4\pi\Delta f |\Delta T|}\right)$

(v)   acf: $R_{\Delta\phi}(\tau) = 2\pi\Delta f \begin{cases} |\Delta T| - |\tau| & \text{if } |\tau| \leq |\Delta T| \\ 0 & \text{otherwise} \end{cases}$

(vi)   psd: $G_{\Delta\phi}(f) = 2\pi\Delta f \,(\Delta T)^2 \, \text{si}^2(\pi f \Delta T)$

The statistical independence of two adjacent phase differences $\Delta\phi(t, \Delta T)$ and $\Delta\phi(t+\Delta T, \Delta T)$ is of significant importance and most advantageous when coherent optical communication systems are analyzed by means of computer simulation. Here, the random laser phases of both transmitter and local lasers have to be modelled by an appropriate software program. Beginning with a starting phase (e.g., $\phi(0)=0$), laser phase noise in coherent optical communication systems can easily be simulated by a step-by-step generation of independent Gaussian distributed phase noise changes to be added in successive stages i.e., $\phi([n+1]\Delta T)=\phi(n\cdot\Delta T)+\Delta\phi(n\cdot\Delta T, \Delta T)$. Thereby, a sample of phase noise after every ΔT is obtained. The parameter ΔT can be chosen arbitrarily for the desired simulation accuracy and run time. In order to generate the Gaussian distributed phase noise changes $\Delta\phi(n\cdot\Delta T, \Delta T)$, a simple computer-based random generator can be used. Finally, it may be mentioned that computer simulation represents a very powerful and least expensive tool to analyze coherent optical communication systems. In this book, several simulation results are given.

3.3.3   STATISTICS OF LASER FREQUENCY NOISE

In this Section, statistics of laser frequency noise $\omega(t) = d\phi(t)/dt$ will be discussed. Any temporal derivation is generally defined by its differential quotient i.e., $\Delta\phi(t, \Delta T)/\Delta T$ with $\Delta T \to 0$. As $\Delta\phi(t, \Delta T)$ is a stationary random process, frequency noise also becomes a *stationary random process*. Spontaneous emission events which are responsible for $\phi(t)$ are, consequently, responsible also for $\omega(t)$. Fluctuations of laser centre frequency f_0 due to slow changes in temperature or in injection current are not included in $\omega(t)$. These fluctuations have to be considered separately. In this book, we always presume that the laser centre frequency f_0 has been sufficiently stabilized by means of an appropriate frequency control circuit; for example, by a standard automatic frequency control (AFC).

(i) EXPECTED VALUE

As temporal derivation is a linear mathematical operation, expected value of $\omega(t)$ equals the expected value of $\Delta\phi(t, \Delta T)$. It leads to

$E\{\omega(t)\} = 0$

(ii) VARIANCE

Direct analytical evaluation of variance of $\omega(t)$ similar to Eqs. (3.33) and (3.41) is not possible. Thus, an alternative approach given by calculating the area under psd of $\omega(t)$ must be taken. Since this area is infinite (see acf and psd below), variance is also infinite i.e.;

$$\sigma_\omega^2 \to \infty \tag{3.48}$$

(iii) PROBABILITY DENSITY FUNCTION (pdf)

As $\Delta\phi(t, \Delta T)$ and $\omega(t)$ are interrelated by a linear mathematical operation, $\omega(t)$ is also Gaussian distributed. To describe the pdf by an equation is, however, not useful since the variance is infinity.

(iv) AUTOCORRELATION FUNCTION (acf)

In principle, acf $R_\omega(\tau)$ as well as psd $G_\omega(f)$ of $\omega(t)$ could be derived directly from the statistical characteristics of $\phi(t)$ as given in Section 3.3.1. However, it becomes very comprehensive and difficult since the nonstationarity of $\phi(t)$ must be considered. In comparison to this, acf can be calculated much more easily when the calculation is based on the stationary process $\Delta\phi(t, \Delta T)$. Moreover, it is advantageous to derive the psd first (see Subsection below). Then, acf can simply be obtained by taking its inverse Fourier transform.

$$R_\omega(\tau) = \text{E}\{\omega(t)\ \omega(t+\tau)\} = 2\pi\Delta f\ \delta(\tau) \circ\!\!-\!\!\bullet\ G_\omega(f) \tag{3.49}$$

It is shown in Fig. 3.16a.

(v) POWER SPECTRAL DENSITY (psd)

The psd of $\omega(t)$ is given by

$$G_\omega(f) = G_\omega = 2\pi\Delta f \tag{3.50}$$

which is a constant (Fig. 3.16b). Thus, laser frequency noise is a *white noise process*.

Proof:

Starting with the equation

$$\Delta\omega(t, \Delta T)\ \Delta\omega(t+\tau, \Delta T) = [\omega(t+\Delta T)-\omega(t)]\ [\omega(t+\tau+\Delta T)-\omega(t+\tau)] \tag{3.51}$$

Its autocorrelation function can be written as

$$R_{\Delta\omega}(\tau) = 2R_{\omega}(\tau) - R_{\omega}(\tau + \Delta T) - R_{\omega}(\tau - \Delta T) \qquad (3.52a)$$

The psd is obtained by taking Fourier transform on both sides. Thus,

$$|j2\pi f|^2 G_{\Delta\phi}(f) = 2 G_{\omega}(f) - G_{\omega}(f) e^{j2\pi f\Delta T} - G_{\omega}(f) e^{-j2\pi f\Delta T} \qquad (3.52b)$$

The factor $j2\pi f$ inside the absolute lines on the left side of above equation corresponds to the frequency response of a differentiator which converts the random process $\Delta\phi(t, \Delta T)$ at the input to a differential process $\Delta\omega(t)$ at the output. Substitution of psd $G_{\Delta\phi}(f)$ from Eq. (3.46a) and further simplification give Eq. (3.50).

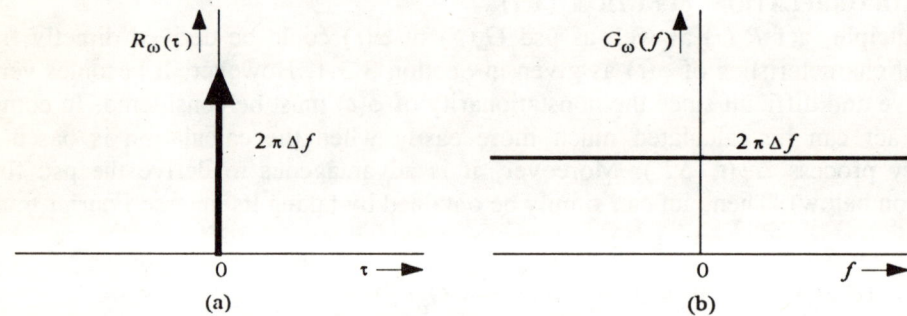

Fig. 3.16: (a) Autocorrelation function and (b) power spectral density of laser frequency noise $\omega(t)$

Comparing the psd of $\omega(t)$ with the psd of $\phi(t)$, a typical feature of laser noise can be recognized: laser phase noise is predominant in the low frequency range though frequency noise is a white noise process. The reason for this feature can be explained by using the integral representation

$$\phi(t) = \int_0^t \omega(\tau) \, d\tau \qquad (3.53)$$

which corresponds to low-pass filtering. Thus, laser phase noise is formally obtained at the output of a low-pass filter provided that white frequency noise is given at the input. Thereby, low frequencies will pass through the filter while high frequencies will be suppressed. As a result, the laser phase noise is a low-frequency process while the laser frequency noise is a white noise process. Finally, it will be mentioned that the integral representation given in Eq. (3.53) is in accordance with the fact that laser phase noise can simply be generated by a step-by-step addition of statistically independent and Gaussian distributed phase noise differences (Section 3.3.2).

Statistical properties of laser frequency noise $\omega(t)$ are summarized as follows:

(i)   stationary and white

(ii) expected value: $E\{\omega\} = 0$

(iii) variance: $\sigma_\omega^2 \to \infty$

(iv) pdf: Gaussian

(v)   acf: $R_\omega(\tau) = 2\pi\Delta f\,\delta(\tau)$

(vi) psd: $G_\omega(f) = 2\pi\Delta f$

3.3.4 STATISTICS OF HARMONIC OSCILLATIONS WITH PHASE NOISE

The purpose of this Section is to derive the statistical properties of "harmonic" random process

$$w(t) = \cos\left(2\pi f_0 t + \phi_0 + \phi(t)\right) = \cos\left(\eta_\psi(t) + \phi(t)\right) = \cos\left(\psi(t)\right) \tag{3.54}$$

Table 3.1: Phase noise in coherent optical communication systems

Signal	$\phi(t)$	$\eta_\psi(t)$	$\sigma_\psi^2(t) = \sigma_\phi^2(t)$	Remark
Carrier wave of transmitter laser	$\phi_t(t)$	$2\pi f_t\, t + \phi_{t0}$	$2\pi\Delta f_t\, t$	$\phi_t(t)$: phase noise of transmitter laser
Light wave of local laser	$\phi_l(t)$	$2\pi f_l\, t + \phi_{l0}$	$2\pi\Delta f_l\, t$	$\phi_l(t)$: phase noise of local laser
IF signal in heterodyne receiver	$\phi_l(t) - \phi_t(t)$	$2\pi[f_l - f_t]\, t + \phi_{l0} - \phi_{t0}$	$2\pi[\Delta f_l + \Delta f_t]\, t$	$\phi_t(t) = \phi_r(t)$ (see Section 2.4.2)
Baseband signal in homodyne receiver	$[\phi_l(t) - \phi_t(t)] - \phi_{\mathrm{PLL}}(t)$	0	see Section 5.1.3	$\phi_{\mathrm{PLL}}(t)$: controlled local laser phase

As in the previous Sections, expected value, variance, pdf, acf and psd of random process $w(t)$ will be obtained and discussed.

Harmonic oscillations corrupted by phase noise exists at various points in coherent optical communication systems. Hence, the statistical properties of this random process are of prime importance to study such systems (Chapter five). Depending on the position, centre frequency f_0, constant phase ϕ_0 and phase noise $\phi(t)$ of the random process $w(t)$ have to be replaced by the related quantities of local meaning (Table 3.1).

In the random process $\psi(t)$, phase $2\pi f_0 t + \phi_0 = \eta_\psi(t)$ is a deterministic signal instead of a random process. Hence, $\eta_\psi(t)$ can be regarded as a time-varying expected value of random phase $\psi(t) = \eta_\psi(t) + \phi(t)$. The phase noise $\phi(t)$ itself is a Gaussian process with zero mean as described in Section 3.3.1.

In order to obtain clear expressions and simplify calculations, time dependence of above signals and random processes will not be declared in our further considerations i.e., $\phi(t) \to \phi$, $\psi(t) \to \psi$, $\eta_\psi(t) \to \eta_\psi$, $\sigma_\psi(t) \to \sigma_\psi$, $\sigma_\phi(t) \to \sigma_\phi$ and so on.

(i) EXPECTED VALUE

As random variable ϕ and, hence, ψ are Gaussian distributed, expected value of new random variable $w = \cos(\psi)$ is given by

$$\eta_w = E\{w\} = E\{\cos(\psi)\} = \int_{-\infty}^{+\infty} \cos(\psi)\, f_\psi(\psi)\, d\psi \tag{3.55}$$

$$= \frac{1}{\sqrt{2\pi}\sigma_\phi} \int_{-\infty}^{+\infty} \cos(\psi)\, e^{-(\psi-\eta_\psi)^2/2\sigma_\phi^2}\, d\psi = \cos(\eta_\psi)\, e^{-\sigma_\phi^2/2}$$

Thus, the expected value η_w increases when σ_ϕ decreases. In absence of laser phase phase noise (i.e., $\sigma_\phi = 0$), expected value η_w reaches its maximum value $\eta_w = \cos(\eta_\psi)$.

The AWGN may either increase or decrease a given signal (two-sided distortion), whereas phase noise always decrease the signal (one-sided distortion). This important feature of phase noise becomes clear when an eye patten disturbed by phase noise is considered in Chapter seven. Using Eq. (3.55), following interesting special cases can be derived:

$$\eta_\psi = 0: \quad w = \cos(\phi) \to \eta_w = e^{-\sigma_\phi^2/2} \tag{3.56}$$

$$\eta_\psi = -\frac{\pi}{2}: \quad w = \sin(\phi) \to \eta_w = 0 \tag{3.57}$$

$$w = e^{j\phi} \quad \to \eta_w = e^{-\sigma_\phi^2/2} \tag{3.58}$$

(ii) VARIANCE

Variance of random process w from Eq. (3.55) is given by

$$\sigma_w^2 = E\{w^2\} - \eta_w^2 = \frac{1}{\sqrt{2\pi}\,\sigma_\phi} \int_{-\infty}^{+\infty} \cos^2(\psi)\, e^{-(\psi-\eta_\psi)^2/2\sigma_\phi^2}\, d\psi \; - \; \eta_w^2$$

$$= \frac{1}{2}\left[1 - e^{-\sigma_\phi^2}\right]\left[1 - \cos(2\eta_\psi)e^{-\sigma_\phi^2}\right]$$

(3.59)

The following special cases are again of practical importance:

$$\eta_\psi = 0: \quad w = \cos(\phi) \;\rightarrow\; \sigma_w^2 = \frac{1}{2}\left[1 - e^{-\sigma_\phi^2}\right]^2$$

(3.60)

$$\eta_\psi = -\frac{\pi}{2}: \quad w = \sin(\phi) \;\rightarrow\; \sigma_w^2 = \frac{1}{2}\left[1 - e^{-\sigma_\phi^2}\right]\left[1 + e^{-\sigma_\phi^2}\right]$$

(3.61)

$$w = e^{j\phi} \;\rightarrow\; \sigma_w^2 = \left[1 - e^{-\sigma_\phi^2}\right]$$

(3.62)

The expected value of phase noise ϕ is zero and maximum gradient of sine function occurs at point $\phi = 0$. As a result, variance σ_w^2 of the random process $w = \sin(\phi)$ is always higher than variance of $w = \cos(\phi)$. If σ_ϕ approaches to infinity, then both variances are equal and approach to $\sigma_w^2 = 0.5$. This corresponds to the normalized average power of a sinusoidal or cosinusoidal harmonic oscillation. It may be mentioned that same result can also be obtained when uniformly distributed phase ϕ in the range $-\pi$ to $+\pi$ (or in the range 0 to 2π) instead of Gaussian distributed with $\sigma_\phi \rightarrow \infty$. To keep the disturbance by laser phase noise low, σ_w^2 should be as small as possible. In the ideal case which practically can never occur, $\sigma_\phi = 0$.

(iii) PROBABILITY DENSITY FUNCTION (pdf)

In order to derive the pdf $f_w(w)$ of random variable $w = \cos(\psi) = \cos(\eta_\psi + \phi)$, it is advantageous to employ the statistical transformation

$$f_w(w) = f_\phi(\phi)\, \frac{1}{\left|\dfrac{d[\cos(\eta_\psi + \phi)]}{d\phi}\right|} \quad \text{with} \quad \phi = \arccos(w) - \eta_\psi$$

(3.63)

It is valid for functions of one random variable only [206]. Due to the periodicity of cosine function $w = \cos(\psi)$ and the ambiguity of the inverse cosine function $\psi = \arccos(w)$, above transformation cannot be used without modifications. To avoid the ambiguity, we first have to split the cosine function and, consequently, the Gaussian pdf $f_\phi(\phi)$ as shown in Fig. 3.17. Second, we derive the pdf of random variable $w = \cos(\phi)$ presuming $\eta_\psi = 0$. This result will be frequently required in the analysis of coherent optical communication systems in Chapter five. Every section of cosine function is weighted by the Gaussian pdf of random phase ϕ. Thus, the pdf $f_w(w)$ is determined by a sum of an infinite number of pdf components. The cosine function and Gaussian pdf are even functions. So, it is possible to restrain $\phi \geq 0$ and multiply the resulting pdf with a factor of 2. The different phases in the sections outlined in Fig. 3.17 are defined by

$$\phi_k = \arccos(w) + k2\pi \tag{3.64}$$

$$\phi_l = -\arccos(w) + (l+1)2\pi \tag{3.65}$$

where k and l are integers.

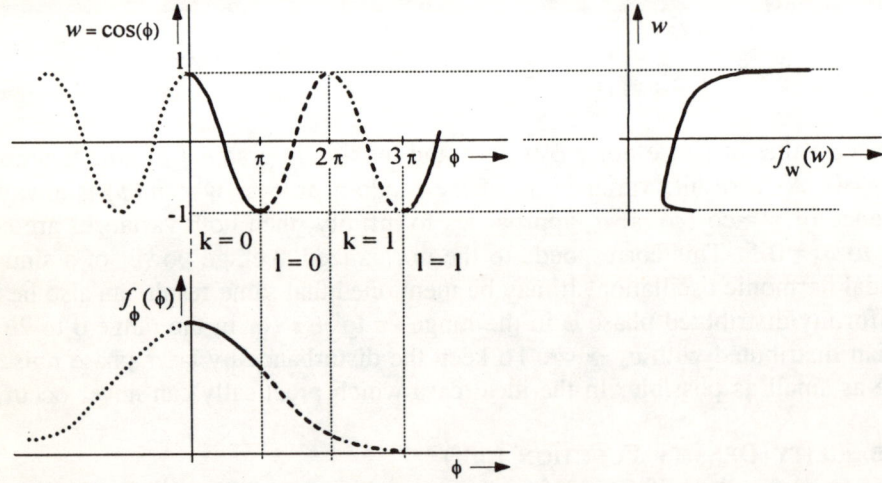

Fig. 3.17: Principle to evaluate the pdf $f_w(w)$ of random variable $w = \cos(\phi)$

With the above considerations and using the relationship

$$\left| \frac{d[\cos(\phi)]}{d\phi} \right|_{\phi = \phi_k} = \left| \frac{d[\cos(\phi)]}{d\phi} \right|_{\phi = \phi_l} = \sqrt{1 - w^2} \tag{3.66}$$

pdf $f_w(w)$ can be written as

$$f_w(w) = \frac{1}{\sqrt{1-w^2}} \left[\sum_{k=-\infty}^{+\infty} f_\phi(k2\pi + \arccos(w)) + \sum_{l=-\infty}^{+\infty} f_\phi((l+1)2\pi - \arccos(w)) \right] \qquad (3.67)$$

After going through some straightforward mathematical operations, we finally obtain

$$f_w(w) = \frac{2(1+r(w))}{\sqrt{2\pi(1-w^2)}\,\sigma_\phi} \exp\left[-\frac{\arccos^2(w)}{2\sigma_\phi^2} \right]. \qquad (3.68)$$

where

$$r(w) = 2\sum_{m=1}^{+\infty} \exp\left[-2\left(\frac{m\pi}{\sigma_\phi}\right)^2 \right] \cosh\left[\frac{m2\pi \arccos(w)}{\sigma_\phi^2} \right] \qquad (3.69)$$

represents a rest function or residual function [61]. In Eq. (3.69), we have not differentiated between k and l and taken a common letter m as sum index.

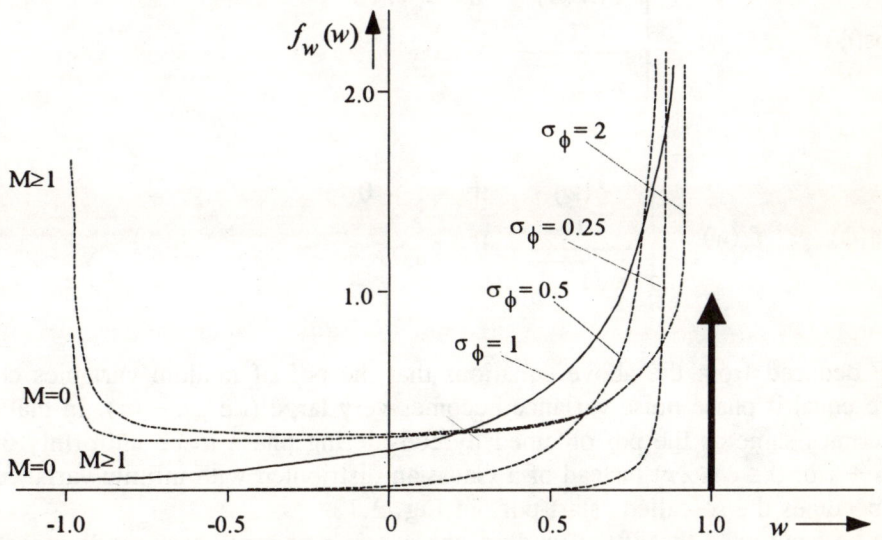

Fig. 3.18: Probability density function $f_w(w)$ of random variable $w = \cos(\phi)$

The pdf $f_w(w)$ for different values of σ_ϕ is shown in Fig. 3.18. Index M represents the number of sum terms used to evaluate the rest function $r(w)$. It is seen that for variances $\sigma_\phi < 1$, rest function $r(w)$ is negligible i.e., $r(w) \ll 1$. In that case, $f_w(w)$ from Eq. (3.68) can be approximated by taking $r(w) = 0$ or $M = 0$. In coherent optical communication systems, for example in homodyne system, $\sigma_\phi < 0.24$ is required to achieve a BER of less than 10^{-10}. Thus, neglecting of rest function $r(w)$ is practically allowed. It becomes clear again from Fig. 3.18 that phase noise always attenuates the signal since $\cos(\phi) \leq 1$ always appears as a random factor in the signal.

By replacing $\arccos(w) \rightarrow \arccos(w) - \eta_\psi$ in Eq. (3.67), η_ψ which has been neglected by now can be considered. After some mathematical manipulations, we obtain

$$f_w(w) = \frac{2}{\sqrt{2\pi(1-w^2)}\,\sigma_\phi} \exp\left[-\frac{\arccos^2(w)}{2\,\sigma_\phi^2}\right] \cdot$$
$$\cdot \sum_{m=-\infty}^{+\infty} \exp\left[-\frac{(m2\pi-\eta_\psi)^2}{2\sigma_\phi^2}\right] \cosh\left[\frac{(m2\pi-\eta_\psi)\,\arccos(w)}{\sigma_\phi^2}\right] \qquad (3.70)$$

Following boundary cases can be derived from this generally valid result:

$$w = \cos(\phi) \quad \rightarrow \quad f_w(w) = \begin{cases} \delta(w-1) & \text{if } \sigma_\phi = 0 \\[2mm] \dfrac{1}{\pi\sqrt{1-w^2}} & \text{if } \sigma_\phi \rightarrow \infty \end{cases} \qquad (3.71)$$

$$w = \sin(\phi) \quad \rightarrow \quad f_w(w) = \begin{cases} \delta(w) & \text{if } \sigma_\phi = 0 \\[2mm] \dfrac{1}{\pi\sqrt{1-w^2}} & \text{if } \sigma_\phi \rightarrow \infty \end{cases} \qquad (3.72)$$

It can be deduced from the above equations that the pdf of random variables $\cos(\phi)$ and $\sin(\phi)$ are equal if phase noise variance becomes very large (i.e., $\sigma_\phi \rightarrow \infty$). In that case, pdf $f_w(w)$ becomes same as the pdf obtained by considering phase to be uniformly distributed $(-\pi \leq \phi \leq +\pi$ or $0 \leq \phi \leq 2\pi)$ instead of a Gaussian distributed with infinite variance. Hence, this pdf becomes the so-called "start-pdf" in Fig. 3.12.

It may be mentioned that if the random phase ϕ is a nonstationary random process such as the laser phase noise $\phi(t)$, then the variance $\sigma_\phi^2 = \sigma_\phi^2(t) = 2\pi\Delta ft$ as well as $f_w(w)$ of the random variable $w = \cos(\psi) = \cos(\eta_\psi + \phi)$ are functions of time. Starting with $t = 0$, shape of $f_w(w)$ changes very rapidly first implying that random process $w(t)$ exhibits a strong non-stationary behaviour. Later on, when the phase noise variance σ_ϕ^2 becomes sufficiently large, changes in the shape of $f_w(w)$ becomes negligibly small. As soon as $\sigma_\phi^2 > 2$, $f_w(w)$ can be considered as time-invariant and $w(t)$ a stationary random process. For example, with a laser of linewidth $\Delta f = 1$ MHz, $\sigma_\phi^2 = 2$ is achieved after $t = 4/(2\pi\Delta f) \approx 640$ ns only.

(iv) AUTOCORRELATION FUNCTION (acf)

Irrespective of whether the random process $w(t)$ is stationary or nonstationary, acf of $w(t)$ can be determined by:

$$R_w(t_1, t_2) = E\left\{\cos(\eta_\psi(t_1) + \phi(t_1)) \cos(\eta_\psi(t_2) + \phi(t_2))\right\}$$

$$= \frac{1}{2} E\left\{\cos\left(\underbrace{\eta_\psi(t_1) + \eta_\psi(t_2) + \phi(t_1) + \phi(t_2)}_{u}\right)\right\} + \frac{1}{2} E\left\{\cos\left(\underbrace{\eta_\psi(t_1)) - \eta_\psi(t_2) + \phi(t_1) - \phi(t_2)}_{v}\right)\right\} \quad (3.73)$$

$$= \frac{1}{2}\left[\cos\left(\eta_\psi(t_1) + \eta_\psi(t_2)\right)\exp\left(-\frac{1}{2}\sigma_u^2\right) + \cos\left(\eta_\psi(t_1) - \eta_\psi(t_2)\right)\exp\left(-\frac{1}{2}\sigma_v^2\right)\right]$$

The new random variables u and v which are Gaussian distributed with zero mean are characterized by their variances

$$\sigma_u^2 = \sigma_\phi^2(t_1) + \sigma_\phi^2(t_2) + 2E\{\phi(t_1)\phi(t_2)\}$$

$$= R_\phi(t_1, t_1) + R_\phi(t_2, t_2) + 2R_\phi(t_1, t_2) \quad (3.74)$$

and

$$\sigma_v^2 = \sigma_\phi^2(t_1) + \sigma_\phi^2(t_2) - 2E\{\phi(t_1)\phi(t_2)\}$$

$$= R_\phi(t_1, t_1) + R_\phi(t_2, t_2) - 2R_\phi(t_1, t_2) \quad (3.75)$$

If we substitute σ_u and σ_v in Eq. (3.73) from the Eqs. (3.74) and (3.75) respectively, we get

$$R_w(t_1, t_2) = \frac{1}{2}\exp\left[-\frac{1}{2}\left(R_\phi(t_1, t_1) + R_\phi(t_2, t_2)\right)\right]\cdot\left[\cos\left(\eta_\psi(t_1) + \eta_\psi(t_2)\right)\exp(-R_\phi(t_1, t_2))\right.$$

$$\left. + \cos\left(\eta_\psi(t_1) - \eta_\psi(t_2)\right)\exp(+R_\phi(t_1, t_2))\right] \quad (3.76)$$

This is a general result valid for both nonstationary as well as stationary phase noise $\phi(t)$. Let us now consider special random processes $\cos(\phi(t))$, i.e., $\eta_\psi=0$, $\sin(\phi(t))$, i.e., $\eta_\psi=\pi/2$ and $\exp(j\phi(t))$. The phase noise $\phi(t)$ itself may be either nonstationary or stationary. For nonstationary phase noise, statistical properties of $\phi(t)$ have been derived in Section 3.3.1. Using Eqs. (3.34) and (3.36), we have

$$R_\phi(t_1, t_2) = K_\phi \min(t_1, t_2) = 2\pi\Delta f \min(t_1, t_2) \tag{3.77}$$

Depending on the physical meaning of random variable w, linewidth Δf has to be replaced by Δf_t, Δf_l or $\Delta f_{IF} = \Delta f_t + \Delta f_l$. In case of stationary phase noise such as rest-phase noise in a homodyne receiver (Chapter five) with an optical phase-locked loop (OPLL), above equation simplifies to

$$R_w(t_1, t_2) = R_w(t_2-t_1) = R_w(\tau) = \frac{1}{2}\exp(-R_\phi(0))\left[\exp(+R_\phi(\tau))+\exp(-R_\phi(\tau))\right]$$

$$= \exp(-\sigma_\phi^2)\cosh(R_\phi(\tau)) \tag{3.78}$$

Taking Eqs. (3.76) and (3.77) for nonstationary processes and Eq. (3.78) for stationary processes, following results are obtained:

Table 3.2: Autocorrelation function $R_w(t_1, t_2)$ of harmonic oscillations disturbed by phase noise

$\phi(t)$	$w(t)$	$R_w(t_1, t_2)$	$R_w(\tau),\ \tau=t_2-t_1$	
Nonstationary	$\cos(\phi(t))$	$\exp[\frac{1}{2}K_\phi(t_1+t_2)\cosh(K_\phi\min(t_1,t_2))]$	$\frac{1}{2}\exp(-\frac{1}{2}K_\phi\|\tau\|)$ $t_1\to\infty,\ t_2\to\infty$	(3.79)
$E\{\phi(t)\}=0$	$\sin(\phi(t))$	$\exp[-\frac{1}{2}K_\phi(t_1+t_2)\sinh(K_\phi\min(t_1,t_2))]$	$\frac{1}{2}\exp(-\frac{1}{2}K_\phi\|\tau\|)$ $t_1\to\infty,\ t_2\to\infty$	(3.80)
$\sigma_\phi^2=K_\phi t$	$\exp(j\phi(t))$	---	$\exp(-\frac{1}{2}K_\phi\|\tau\|)$	(3.81)
Stationary	$\cos(\phi(t))$	---	$\exp(-\sigma_\phi^2)\cdot$ $\cosh[R_\phi(\tau)]$	(3.82)
$E\{\phi(t)\}=0$	$\sin(\phi(t))$	---	$\exp(-\sigma_\phi^2)\cdot$ $\sinh[R_\phi(\tau)]$	(3.83)
$\sigma_\phi^2=\text{const.}$	$\exp(j\phi(t))$	---	$\exp(-\sigma_\phi^2)\cdot$ $\exp[R_\phi(\tau)]$	(3.84)

The variables of time t_1 and t_2 are related to time when the laser has been switched on and spontaneous emission and phase noise processes have started. Hence, they are always positive in Eqs. (3.79) to (3.81). As laser phase noise is a nonstationary process, "harmonic" processes $\cos(\phi(t))$ and $\sin(\phi(t))$ are also nonstationary. Therefore, their acfs are time-varying function. In the boundary case $t_1 \to \infty$ and $t_2 \to \infty$, that is a "long time" after the laser has been switched on, nonstationary processes $\cos(\phi(t))$ and $\sin(\phi(t))$ become stationary (see Subsection on pdf above). Thus, stationarity is sufficiently fulfilled when t_1, $t_2 \gg 1/(2\pi\Delta f)$. In that case, acf of the random processes $\cos(\phi(t))$ and $\sin(\phi(t))$ are function of the time difference $\tau = |t_2-t_1|$ only.

Irrespective of whether the phase noise is stationary such as the phase noise in PLL or nonstationary such as the laser phase noise itself, random process $\exp(j\phi(t))$ is always stationary (Eqs. 3.81 and 3.84).

Proof:
Consider the random process $\exp(j\phi(t))$. Its acf can simply be determined by the following equation:

$$w(t) = e^{j\phi(t)} \quad \to \quad R_w(t_1, t_2) = E\{w(t_1)\, w^*(t_2)\} = E\{e^{j\phi(t_1)} e^{-j\phi(t_2)}\}$$

$$= E\{e^{j[\phi(t_1)-\phi(t_2)]}\} = E\{e^{j\Delta\phi}\} = e^{-\frac{1}{2}\sigma_{\Delta\phi}^2} \tag{3.85}$$

$$= e^{-\frac{1}{2}K_\phi|t_2-t_1|} = e^{-\frac{1}{2}K_\phi|\tau|}$$

Hence, the information of nonstationarity in the processes $\cos(\phi(t))$ and $\sin(\phi(t))$ is lost when the complex process $\exp(j\phi(t))$ is considered. The acf $R_w(\tau)$ of the random process $\exp(j\phi(t))$ is shown in Fig. 3.19a.

(v) POWER SPECTRAL DENSITY (psd)
The psd $G_w(f_1, f_2)$ of nonstationary random process can be obtained by two-dimensional Fourier transform of acf $R_w(t_1, t_2)$, whereas the psd $G_w(f)$ of stationary random process by a simple one-dimensional Fourier transform of acf $R_w(\tau)$.

In this Subsection, we focus our interest mainly on the random process $w = \exp(j\phi(t))$ which is an important harmonic random process in coherent optical communication systems. The physical representation of this process is the emission spectrum of laser light wave $\underline{E}(t)$ disturbed by phase noise $\phi(t)$ provided $f_0 = 0$ and $\phi_0 = 0$ (Eq. 3.29). This spectrum can simply be obtained by shifting the psd $G_w(f)$ to the laser centre frequency f_0 i.e., $G_w(f) \to G_w(f-f_0)$. Using Eq. (3.81), we get

$$G_w(f) = \int_{-\infty}^{+\infty} e^{-\frac{1}{2}K_\phi|\tau|} e^{-j2\pi f\tau} \, d\tau = \frac{4}{K_\phi} \frac{1}{1 + \left(\dfrac{4\pi f}{K_\phi}\right)^2} \tag{3.86}$$

The spectral bandwidth at 50% intensity points i.e., 3 dB bandwidth

$$\Delta f = \frac{K_\phi}{2\pi} = \frac{1}{4\pi} \frac{P_{sp}}{P_0} R_{sp} \quad \text{with } G_w\!\left(\frac{\Delta f}{2}\right) = \frac{1}{2} G_w(0) \tag{3.87}$$

is called the *laser linewidth*. The interrelation between Δf and K_ϕ given in the above expression has been used in the previous Sections. The relationship between linewidth Δf and power ratio P_{sp}/P_0 i.e.; the ratio of average spontaneous emission power to the power level of the stimulated wave has already been given in Eq. (3.33). Combining Eqs. (3.86) and (3.87), we finally obtain the well-known result:

$$G_w(f) = \frac{2}{\pi\Delta f} \frac{1}{1 + \left(\dfrac{f}{\Delta f/2}\right)^2} \tag{3.88}$$

This important formula, frequently called the *Lorentzian line shape*, describes the normalized emission spectrum of laser corrupted by phase noise. Here, the spectrum is shifted by laser centre frequency f_0 to the origin $f=0$. Emission spectrum $G_w(f)$ is shown in Fig. 3.19b.

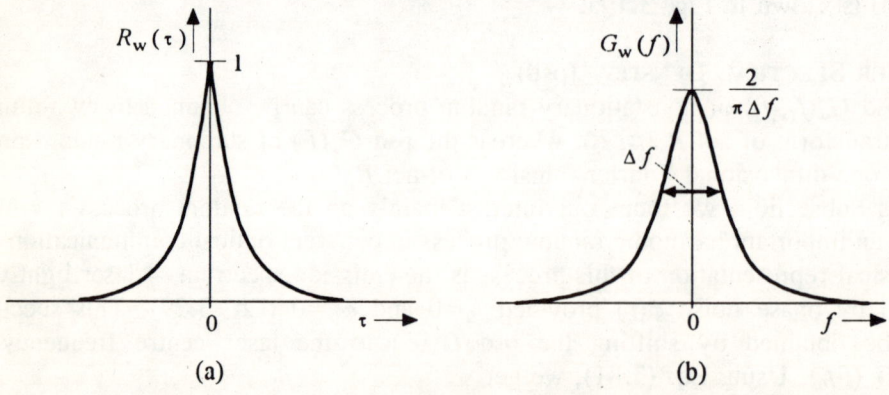

(a)   (b)

Fig. 3.19 (a) Autocorrelation function $R_w(\tau)$ and (b) power spectral density $G_w(f)$ of random process $w = \exp(j\phi(t))$

First measurement on the Lorentzian line shape of an AlGaAs injection semiconductor laser was carried out by Fleming and Mooradian in 1981 [56]. As shown by Eq. (3.87), linewidth Δf that varies inversely with the laser output power P_0 was observed. It implies that laser linewidth is smaller when laser power is higher. This is generally valid irrespective of whether the laser is semiconductor, gas or solid-state type. Obviously, laser linewidth Δf should be as small as possible to achieve a high quality, high bit rate data transmission. The linewidth measured by Fleming and Mooradian was surprisingly about 50 times more than as predicted from Eq. (3.87). The reason for it will be explained briefly in the next Section.

3.4 RELAXATION OSCILLATIONS

In the previous Section, formation of laser phase and laser amplitude noise has been explained and discussed by means of a simple model shown in Fig. 3.9. Based on this model, we have derived the statistical properties of phase noise, phase noise difference, laser frequency noise and harmonic oscillations disturbed by phase noise.

Except a minor modification which will be performed in this Section, justification of this model is given on the basis of good agreement of results obtained from this model and either by measurements [190, 229] or exact physical laser theory [98]. Here, the term exact laser theory means the mathematical solution of coupled differential equations with "laser phase" and "laser intensity" as the variables.

Like other physical models, for example, the atomic model by Niels Bohr, model of laser phase and amplitude noise is an approximate model, but very useful and advantageous to obtain a fast and good insight into the principle interrelations. Further, this model allows to obtain a first description by formulas. A detailed description of the actual microscopic processes in the system is not possible by using this simple model. However, this detailed knowledge is usually not required to analyze coherent optical communication systems for their behaviour in data transmission. For this, simple model given in Section 3.2 (Fig. 3.9) is normally sufficient and, in addition, often more powerful and practical based than a system analysis based on exact laser or quantum theory.

Concurrently to the development of coherent optical communication systems, intense practical and theoretical investigations had been performed in the area of semiconductor lasers, in particular to examine the spectral properties of laser diodes. A broader laser linewidth than predicted by the standard laser theory was measured. Further, a significant deviation in the Lorentzian line shape was observed. This deviation results in *satellite peaks* in the emission spectrum of laser.

In the following, reasons for line broadening and origin of satellite oscillations will be explained. For this, we shall take into consideration our simple noise model again. However, we have to perform now some appropriate modifications. A detailed mathematical analysis is not required and will not be covered in this book [39, 97, 98, 212, 258, 278, 289].

Up-to-now we have considered that the laser phase noise and amplitude or laser intensity noise are absolutely statistically independent. This is actually not correct since a close coupling of phase and amplitude noise can always be observed in semiconductor laser diodes (Fig. 3.20).

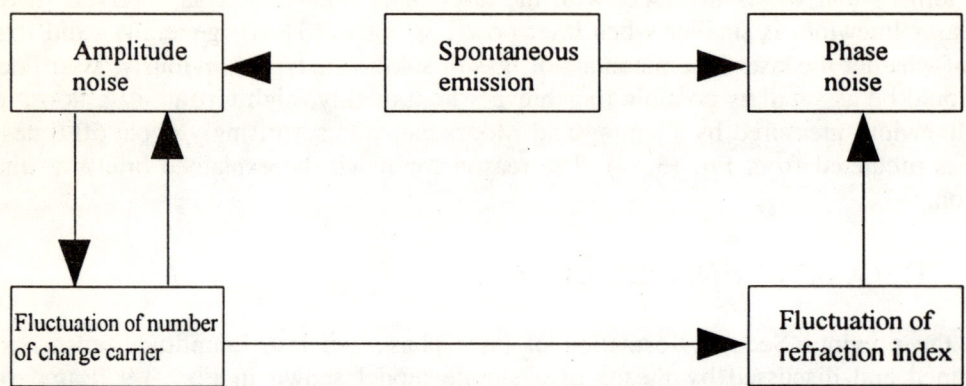

Fig. 3.20: Model of coupled phase and amplitude noise in laser

As shown in this figure, spontaneous emission is responsible for both laser phase and amplitude noise. Each spontaneously generated photon yields a *direct phase error* in the stimulated laser light wave since the phase of each photon is uniformly distributed in the range $-\pi$ to $+\pi$. Despite of this direct disturbance to the laser phase, an *additional delayed phase error* occurs as a response to the inherent direct amplitude disturbance. In order to restore the original steady state amplitude or intensity, laser performs relaxation oscillations for about 1 ns duration. During this characteristic time, real part Re$\{\underline{n}\}$ as well as imaginary part Im$\{\underline{n}\}$ of the refraction index $\underline{n}$ are changed due to fluctuation in the number of charge carriers, which again are caused by the amplitude fluctuations. Finally, the loop is closed by the spontaneous emission processes which are responsible for both phase as well as amplitude noise (Fig. 3.20).

The imaginary part Im$\{\underline{n}\}$ of the complex refraction index $\underline{n}$ is a measure of gain or loss in the active area of laser. Consequently, it is also a measure of the resulting amplification in this area. A change in the imaginary part Im$\{\underline{n}\}$ causes a change in the amplification, which finally yields the regeneration of the amplitude and, therefore, the intensity. After the relaxation oscillations are over, primary intensity is restored.

Due to relaxation oscillations of laser, emission spectrum exhibits satellite peaks after every f_{re} around the laser centre frequency f_0. The characteristic frequency f_{re} is called the *frequency of relaxation* and is in the order of 1 GHz to 2 GHz [97].

The real part Re$\{\underline{n}\}$ of complex refraction index $\underline{n}$ represents a measure of dispersion in the active area of laser. Hence, the real part Re$\{\underline{n}\}$ determines the phase and frequency of extendable laser waves in the laser cavity. It may be mentioned that real part Re$\{\underline{n}\}$ of refraction index $\underline{n}$ has been used in Section 3.1 (represented by n) where the imaginary part was not required.

A change in the real part Re{$\underline{n}$} causes delayed indirect phase changes which occur, in addition, to the direct phase changes due to the spontaneous emission events. Hence, the overall phase noise of a laser is increased two fold and linewidth is additionally broaden as shown in Fig. 3.21.

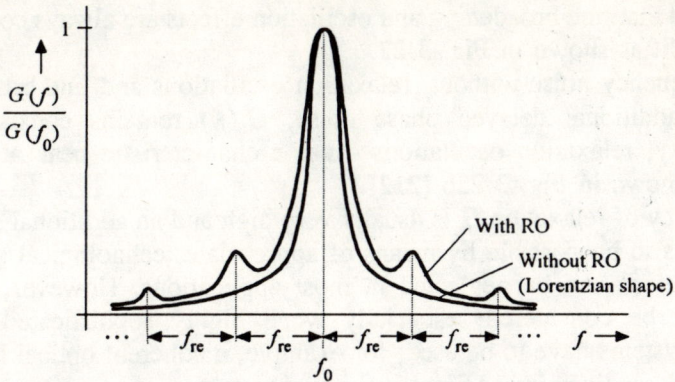

Fig. 3.21: Emission spectrum of laser including the interactions of laser phase and laser amplitude noise with and without relaxation oscillations (RO)

Relaxation oscillations influence the laser phase noise and the laser frequency noise. To make it more clear, Fig. 3.22a shows the variance $\sigma_\phi^2(t)$ of the laser phase noise and Fig. 3.22b the power spectral density $G_\omega(f)$ of the laser frequency noise.

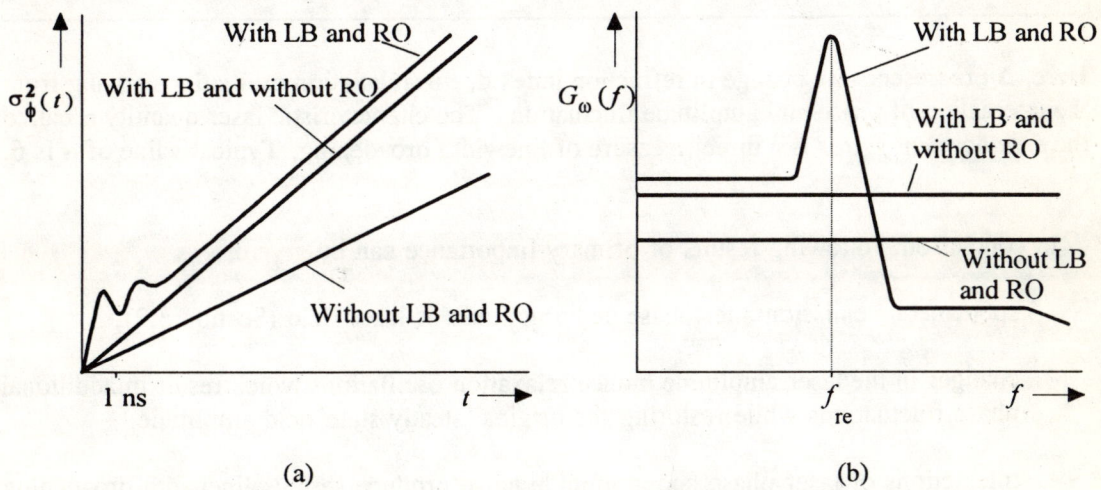

(a)   (b)

Fig. 3.22: (a) Variance $\sigma_\phi^2(t)$ of laser phase noise and (b) power spectral density $G_\omega(f)$ of laser frequency noise including laser phase-to-amplitude conversion (RO: Relaxation Oscillations, LB: Line Broadening)

In the absence of relaxation oscillations and line broadening, $\sigma_\phi^2(t)$ increases linearly with time (Section 3.3.1). In addition, delayed phase changes caused by changes in the real part Re$\{n\}$ increases phase noise and $\sigma_\phi^2(t)$. Thus, $\sigma_\phi^2(t)$ now increases more rapidly with time. Finally, relaxation oscillations due to the interactions of laser phase and amplitude noise introduce variations in variance $\sigma_\phi^2(t)$ during the relaxation time of about 1 ns (Fig. 3.22a). It should be noted that line broadening and oscillation effects are always combined physically and cannot be split as shown in Fig. 3.22.

In case of frequency noise without relaxation oscillations and line broadening, $G_\omega(f)$ is constant. With additional delayed phase noise, $G_\omega(f)$ remains constant, but increases somewhat. Finally, relaxation oscillations cause a characteristic peak at the frequency of relaxation f_{re} as shown in Fig. 3.22b [212].

As the frequency of relaxation f_{re} is usually very high and an additional shift to still higher frequencies seems to be possible by means of appropriate technological steps, influence of relaxation oscillations can be neglected in most applications. However, the inherent line broadening must be considered, especially when highly sophisticated coherent optical communication systems have to be used; for example, a coherent optical homodyne system. Here, the actual laser linewidth Δf is of prime importance.

In order to determine the laser linewidth mathematically, modification in Eq. (3.87) based on our simple model is required. It can be shown that this modification is given by the following simple substitution [97]:

$$\Delta f \rightarrow \Delta f(1 + \alpha^2) \quad \text{with} \quad \alpha = \frac{\text{Re}\{\Delta n\}}{\text{Im}\{\Delta n\}} \tag{3.89}$$

Here, Δn represents the change in refraction index due to relaxation oscillations arising from the interaction of phase and amplitude fluctuations. The characteristic laser quantity α called the *enhancement factor* is a direct measure of linewidth broadening. Typical value of α is 6.

In conclusion, following results of primary importance can be given:

-   spontaneous emission alter phase and amplitude of laser field (Section 3.2),

-   changes in the laser amplitude induce relaxation oscillations which result in additional phase fluctuations while restoring the original steady state field amplitude,

-   interactions of laser phase and amplitude noise produce serious linewidth broadening and side peaks (satellite peaks) in the laser emission spectrum.

3.5   INFLUENCE OF FILTERING

Each communication system irrespective of whether it is analog or digital is disturbed by various sources of noise. In order to reduce the influence of noise, filters are normally used in both types of systems. To determine the residual influence of the filter output noise, statistical properties of this noise have to be derived by means of statistical theory. In a conventional system without phase noise, it can be carried out easily since the signal at the filter input exhibits the following simple structure:

> Structure of signal in *conventional communication systems*:
>
> Filter input signal = K · information signal + Gaussian noise

Here K is a constant. The systems with the above signal structure are frequently termed as Additive White Gaussian Noise (AWGN) systems. In an optical *direct detection system* with $K = R_0 \cdot M \cdot P_r$, filter input signal is the noisy signal at the input of the low-pass filter shown in Fig. 2.1a. Assuming a linear filter, filter response to the signal (i.e., information) and noise can be determined separately. As the filter input noise is usually Gaussian distributed, the noise at the filter output also has the same distribution. Hence, determining expected value (which is frequently zero) and standard deviation of the filter output noise is sufficient. The relevant formulas are given in Fig. 3.23a.

With a nongaussian noise at the filter input, calculation of statistical properties of the filter output noise becomes complicated (Fig. 3.23b). This problem arises, for example, if *coherent optical communication systems* with $K \sim \sqrt{P_r P_l}$ have to be analyzed. Here, the information signal is additionally corrupted by laser phase noise. Thus, the characteristic signal structure is:

> Structure of signal in *coherent optical communication systems*:
>
> Filter input signal = K · information · phase noise term + Gaussian noise

In contrast to the characteristic signal structure of conventional communication systems, an additional *phase noise term* has to be considered now. This term is *multiplicative* and *nongaussian* in nature.

In coherent optical communication systems with synchronous or coherent detection (for example, an optical homodyne system), noisy filter input signal is given by the baseband signal at the input of baseband filter (low-pass filter) shown in Fig. 2.1b. Here, the relevant phase noise term is mathematically represented by the cosine of phase noise i.e., $\cos(\phi(t))$. It will be explained in more detail in Section 5.1. Since the cosine function is a nonlinear function, corresponding probability density function of the random process $\cos(\phi(t))$ is non-gaussian (Section 3.3.4).

In an optical homodyne system, random process $\cos(\phi(t))$ is always a strongly correlated random process due to the required phase-locked loop circuit. Here, the so-called filter problem that is the determination of the statistical properties of the filter output noise with a nongaussian input noise can be avoided since strongly correlated phase noise changes very slowly. Hence, the phase noise term can be taken outside the convolution integral required to evaluate the filter output signal and noise (Section 5.1).

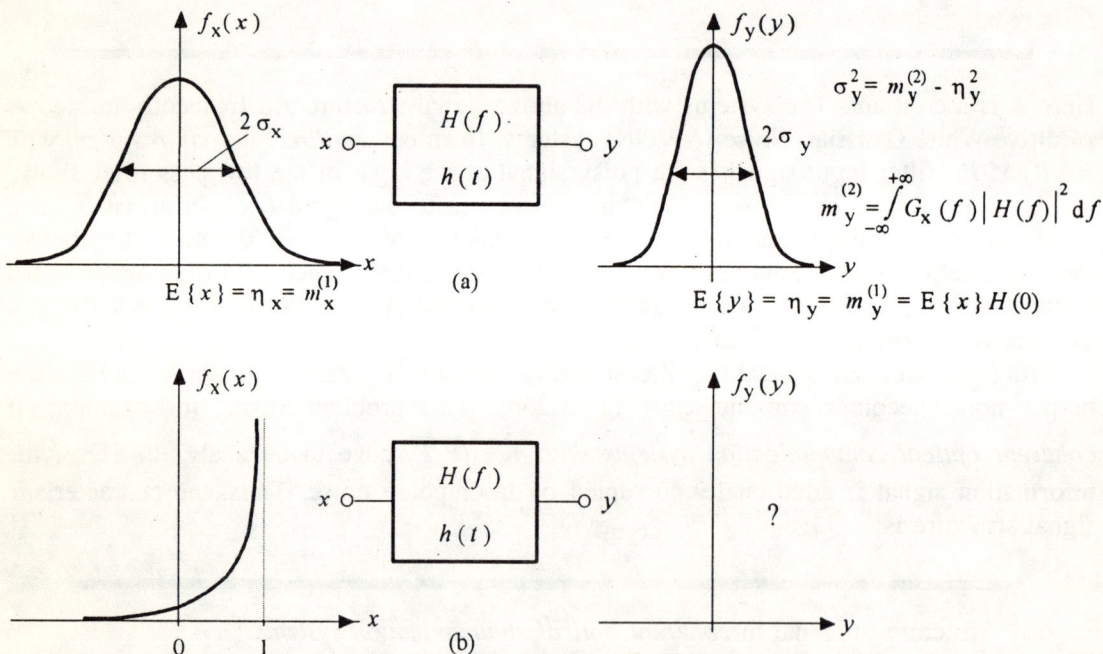

Fig. 3.23: Probability density function at the output of filter in case of
(a) Gaussian and (b) nongaussian noise at the input

In coherent transmission systems with noncoherent detection, for example, ASK hetero-dyne system with envelope detection, relevant filter input signal corresponds to noisy intermediate frequency signal at the input of IF filter. Here, the phase noise term can be expressed in terms of complex signal $\exp(j\phi(t))=\cos(\phi(t))+j\sin(\phi(t))$. As noncoherent detection system does not require a phase control circuit, phase noise term $\exp(j\phi(t))$ is a relatively uncorrelated random process. It is not allowed to take out this term outside the convolution integral. Therefore, it is unavoidable to solve the filter problem (Section 5.3).

A filter only influences expected value and standard deviation of the pdf if the filter input noise is Gaussian. However, the shape of pdf may get fundamentally changed if the noise at the input is nongaussian. To obtain a complete description of this unknown nongaussian pdf, statistical moments of higher order are also required in addition to expected value and standard deviation.

In the following, various methods to solve the filter problem are presented and discussed in detail. As the numerical calculation is rather comprehensive, this book will focus mainly on the principle relations. A more detailed analysis is given e.g., in [37, 38, 95, 158]. Readers, who are not particularly interested in this theoretical Section may switch over to next Chapter without loss of continuity.

The mathematical basis of most of the methods presented in this Section is the sampling theorem or digitalization of corresponding filter such as shown in Fig. 3.24 [165].

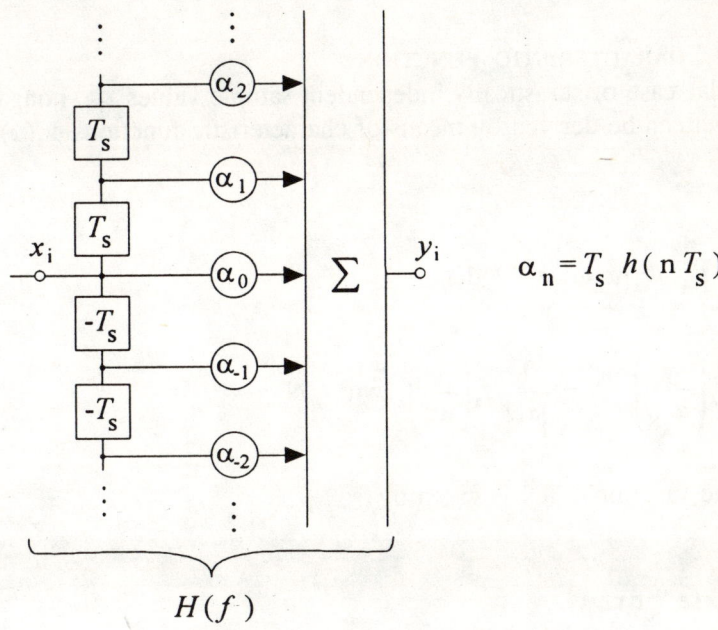

Fig. 3.24: Digitalization of filter $H(f)$

The sampled signal at the output of a digitalized filter after neglecting transit times can be described by

$$y(iT_s) = y_i = \sum_{n=-\infty}^{+\infty} x_{i-n} \alpha_n \qquad (3.90)$$

In this equation, T_s represents the sampling period which is required to fulfil the sampling theorem, $\alpha_n = T_s h(nT_s)$ the sampled impulse response of the filter $H(f)$ weighted by T_s and $x_{i\text{-}n} = x([i\text{-}n]T_s)$ the sample values of the nongaussian distributed signal at the input of filter. To fulfil the sampling theorem, filter input signal $x(t)$ as well as filter frequency response (system function) $H(f)$ must be band-limited. Only in that case, sampled filter output signal $y(iT_s)$ exactly equals the actual output signal $y(t)$ at the sampling points iT_s. Since this condition is normally not fulfilled in practice, an arbitrary cut-off frequency must be defined for the desired accuracy of calculation.

All the mathematical methods presented in the following Subsections are valid in principle for stationary as well as nonstationary random processes at the filter input. However, in order to avoid generality, a stationary random process will be considered. In that case, statistical properties of the random variables $x_{i\text{-}n}$ and y_i are independent of sampling time $(i\text{-}n)T_s$ and iT_s respectively. For a nonstationary random process $x(t)$, calculation can be performed in the same way, but we have to incorporate some modifications which can be easily accomplished.

(i) METHOD OF CHARACTERISTIC FUNCTION

In the particular case of statistically independent sample values $x_{i\text{-}n}$, nongaussian pdf $f_y(y)$ at the filter output can be derived by means of characteristic functions $\Phi_x(\omega)$ of input signal $x(t)$. We obtain

$$
f_y(y) = \frac{1}{2\pi} \int_{-\infty}^{+\infty} \prod_{n=-N}^{+N} \Phi_x(\alpha_n \omega) \, e^{-j\omega y} \, d\omega
$$

$$
= \frac{1}{|\alpha_{-N}|} f_x\left(\frac{y}{\alpha_{-N}}\right) \cdot \ \cdots \ \cdot \frac{1}{|\alpha_N|} f_x\left(\frac{y}{\alpha_N}\right) \quad \text{with} \quad N \to \infty
$$

(3.91)

The characteristic function $\Phi_x(\omega)$ is given by

$$
\Phi_x(\omega) = \int_{-\infty}^{+\infty} f_x(x) e^{j\omega x} dx
$$

(3.92)

It corresponds to a Fourier transform of $f_x(x)$ with respect to the variable x. As seen from Eq. (3.91), $f_y(y)$ at the filter output can be determined in three steps: First, characteristic functions $\Phi_x(\omega)$ of the random variables $x_{i\text{-}n}$ weighted by α_n have to be evaluated, second characteristic functions have to be multiplied with one another and third Fourier inversion must be performed to obtain the desired pdf $f_y(y)$. It should be noted that multiplication of characteristic functions is related to convolution of the corresponding probability density functions.

The method of characteristic function is a relatively simple solution of the filter problem, but requires the statistical independence of random variables x_{i-n} as mentioned above. The practical problem of this method is especially the high processing time required to evaluate Eq. (3.91) by means of computer. However, computer run time can at least partly be reduced when only a finite number N of sum terms instead of an infinite number are used in the calculation. It will impair the accuracy of calculation to some extent. Moreover, accuracy is also influenced by the bandwidth of filter $H(f)$. Higher the filter bandwidth, faster the impulse response $h(t)$ decreases with time and higher is the accuracy of calculation.

(ii) METHOD OF NARROW-BAND APPROXIMATION

When the bandwidth of filter $H(f)$ is small in comparison to the bandwidth of filter input signal $x(t)$, impulse response $h(t)$ of the filter decreases very slowly while the acf $R_x(x)$ of the fast random input process $x(t)$ decreases very rapidly. In that case, sum in Eq. (3.90) includes an excessive number of terms during the impulse response of filter. Moreover, these terms are only weakly correlated. Hence, the necessary conditions for applying the central-limit theorem of statistics are fulfilled and the pdf of filter output process is approximately Gaussian [206]. It is sufficient to evaluate both the characteristic quantities which are required to describe a Gaussian pdf i.e., expected value η_y and standard deviation σ_y. Both quantities can easily be determined from the power spectral density $G_x(f)$ of the filter input process $x(t)$ and the filter transfer function $H(f)$ as given in Fig. 3.23. The method of narrow-band approximation requires filters of narrow bandwidth; for example, an intermediate frequency filter of bandwidth $B_{IF} \ll \Delta f$ if heterodyning is applied. Here, Δf represents the bandwidth of the IF signal at the input to filter due to laser phase noise alone. If phase noise is zero, then $\Delta f = 0$. However, the required relationship $B_{IF} \ll \Delta f$ is frequently not fulfilled in practical realizations.

(iii) METHOD OF STATISTICAL MOMENTS

In this method, all statistical moments $m_y^{(k)} = E\{y^k\}$, $k \in \mathbb{N}$ of the random filter output signal $y(t)$ are derived to calculate the pdf $f_y(y)$ by developing an appropriate series of sum terms. Here, three different cases have to be distinguished:

First, the filter input process $x(t)$ is a *Gaussian process* with *statistically independent* samples. In that case, calculation of pdf $f_y(y)$ is very easy since the moments of output process only depend on the moments of input process of same order (i.e., same k). Moreover, pdf $f_y(y)$ is also a function of the filter coefficients α_n. Since a Gaussian input process $x(t)$ always yields a Gaussian output process $y(t)$, only the moments of first and second order are actually required to describe the pdf $f_y(y)$. The filter output process $y(t)$ can also be completely described by the acf $R_y(\tau)$ or psd $G_y(f)$ as both functions already contains moments required to describe a Gaussian pdf i.e., expected value $\eta_y = m_y^{(1)}$ and variance $\sigma_y^2 = m_y^{(2)} - \eta_y^2$.

Second, the process $x(t)$ is a *nongaussian process* and samples are *statistically independent*. In that case, statistical moments of order k of the output process $y(t)$ are a function of filter coefficients α_n and all input moments of same *and* lower order. Method of statistical moments can be applied here also [37]. However, acf $R_y(\tau)$ and psd $G_y(f)$ are merely no longer sufficient to describe the pdf $f_y(y)$ completely.

The third case of importance is when *neither x(t) is Gaussian process nor the sample values are statistically independent.* Here, this method cannot be applied in general since further statistical information on the filter output process $y(t)$ are now required in addition to the moments $m_y^{(k)}$. For example, the expected value $E\{y(t_1)^k y(t_2)^l\}$ which represents an autocorrelation function of higher order. Again, acf $R_y(\tau)$ and psd $G_y(f)$ are not sufficient to describe the random process $y(t)$ completely. For this, all linear and nonlinear statistical interrelations in $y(t)$ must additionally be known, in particular all the moments and autocorrelation functions of higher order. Even if all these statistical quantities are known, calculation of the pdf $f_y(y)$ remains restricted to some special cases only (compare [37] and [38]). One such case of prime practical interest is, for example, given by the random process $x(t)=\cos(\phi(t))$. As already mentioned, this process is nongaussian due to nonlinear cosine function, whereas the random phase $\phi(t)$ itself is Gaussian. By means of this process, which is of interest in coherent optical communication systems, method of statistical moments will be explained in more detail.

First, the moments $m_x^{(k)}$ of the random process $x(t)=\cos(\phi(t))$ at the input to the filter have to be calculated. These are given by

$$
m_x^{(k)} = E\{x^k\} = E\{\cos^k(\phi)\} = \frac{1}{\sqrt{2\pi}\,\sigma_\phi} \int_{-\infty}^{+\infty} \cos^k(\phi)\exp\left(-\frac{\phi^2}{2\sigma_\phi^2}\right) d\phi
$$

$$
= \left(\frac{1}{2}\right)^k \sum_{j=0}^{k} \binom{k}{j}\exp\left(-\frac{(k-2j)^2\sigma_\phi^2}{2}\right)
$$

(3.93)

Second, the acf

$$
R_x(\tau) = \frac{1}{2}e^{-\sigma_\phi^2}\left[e^{R_\phi(\tau)} + e^{-R_\phi(\tau)}\right] = e^{-\sigma_\phi^2}\cosh(R_\phi(\tau))
$$

(3.94)

of the filter input process $x(t)$ is required. It may be remembered that $R_x(\tau)$ has already been derived in Section 3.3.4 (Eq. 3.77), where $x(t)=\cos(\phi(t))$ was represented by $w(t)$. To simplify calculations, following abbreviation will be used:

$$
R_\phi([i-n]T_s) := R_\phi(i-n)
$$

(3.95)

The *first order moment* of random process $y(t)$ at the filter output is given by

$$m_y^{(1)} = E\left\{\sum_{n=-\infty}^{+\infty} x_{i-n}\alpha_n\right\} = \sum_{n=-\infty}^{+\infty} E\{x_{i-n}\}\alpha_n = m_x^{(1)} \sum_{n=-\infty}^{+\infty} \alpha_n = m_x^{(1)} H(0) \qquad (3.96)$$

As the random process $x(t)$ is assumed to be stationary, moments $m_x^{(1)} = \eta_x$ and $m_y^{(1)} = \eta_y$ are independent of sampling time iT_s. In the above equation, $H(0)$ represents the DC frequency response of the filter $H(f)$ at $f=0$ (compare Fig. 3.23). Using Eqs. (3.93) and (3.96), we obtain

$$m_y^{(1)} = e^{-\frac{1}{2}\sigma_\phi^2} H(0) \qquad (3.97)$$

For the *second order moment*, following expression can be derived:

$$m_y^{(2)} = \int_{-\infty}^{+\infty} G_x(f) |H(f)|^2 \, df \quad \text{with} \quad G_x(f) \circ\!\!-\!\!\bullet R_x(\tau) \qquad (3.98)$$

Here, the calculation is based only on the psd $G_x(f)$ or the acf $R_x(\tau)$ of random process $x(t) = \cos(\phi(t))$. As an alternative, calculation can also be based on acf $R_\phi(\tau)$ of Gaussian distributed random phase $\phi(t)$. In that case, we will get

$$m_y^{(2)} = \frac{1}{2}e^{-\sigma_\phi^2} \sum_{n=-\infty}^{+\infty} \sum_{m=-\infty}^{+\infty} \left[e^{R_\phi(n-m)} + e^{-R_\phi(n-m)}\right] \alpha_n \alpha_m \qquad (3.99)$$

Similarly, *third order moment* can be determined. With some simple trigonometrical relations, following comprehensive formula is obtained:

$$m_y^{(3)} = \frac{1}{4}e^{-\frac{3}{2}\sigma_\phi^2} \sum_{n=-\infty}^{+\infty} \sum_{m=-\infty}^{+\infty} \sum_{l=-\infty}^{+\infty} p(n,m,l) \, \alpha_n \alpha_m \alpha_l \qquad (3.100a)$$

where

$$p(n,m,l) = e^{+R_\phi(n-m) + R_\phi(n-l) + R_\phi(m-l)} + e^{+R_\phi(n-m) - R_\phi(n-l) - R_\phi(m-l)}$$

$$+ e^{-R_\phi(n-m) - R_\phi(n-l) + R_\phi(m-l)} + e^{-R_\phi(n-m) + R_\phi(n-l) - R_\phi(m-l)}$$

(3.100b)

Moments of higher order can, in principle, be determined in the same way. The computer execution time increases rapidly if the order of moment is higher than three. It becomes clear from the above equation that terms $R_\phi(n-m)$, $R_\phi(n-l)$ and $R_\phi(m-l)$ frequently exhibit the same numerical value during the calculation of the sum with respect to n, m and l. Hence, the next step would be to reduce this redundancy existing manifold in the higher order moments.

When the required number of moments with desired accuracy are available, pdf $f_y(y)$ can be determined by developing an appropriate series of sum terms. For this, various series are available. As an example, characteristic functions and pdf calculated by means of Taylor series are given by the following equation [206]:

$$\Phi_y(\omega) = 1 + \sum_{k=1}^{+\infty} \frac{m_y^{(k)}}{k!}(j\omega)^k \quad \text{and} \quad f_y(y) = \frac{1}{2\pi}\int_{-\infty}^{+\infty}\Phi_y(\omega)\, e^{-j\omega y}\, d\omega$$

(3.101)

Unfortunately, convergence of this series is very weak and a large number of moments are required. In contrast, the Gram-Charlier [36], the Edgeworth [46] and the Cornish-Fisher series are more convenient. A more detailed analysis of above series is not given in this book [53].

(iv) METHOD OF SHAPE FILTER

The idea of shape filter method is to trace back the correlation of input random process $x(t)$ characterized by dependent sample values x_i to a random process $u(t)$ first characterized by a psd $G_u(f)$ due to a band-limited white noise process and second by statistically independent samples u_i. The filter model including real filter and virtual shape filter are illustrated in Fig. 3.25. The task of shape filter $H_s(f)$ is to generate actual random process $x(t)$ at the input of real filter $H(f)$ by means of a virtual random process $u(t)$, defined by the statistical properties mentioned above. Hence, the *virtual shape filter* $H_s(f)$ is only a mathematical tool to determine the pdf $f_y(y)$, but not a filter which really exists in coherent optical communication systems.

Let us consider that $H_s(f)$ is mathematically existing. The pdf $f_y(y)$ at the output of real filter $H(f)$ can be calculated using the method of moments discussed above. It may be remembered that statistical independence of samples at the filter input have been the only presumption for applying this method. Here, statistical independence is fulfilled by the virtual samples u_i at the input of shape filter, whereas the samples x_i at the input of real filter are statistically dependent. To clarify the calculation, characteristic features of $u(t)$, $x(t)$, and $y(t)$ are summarized in Table 3.2.

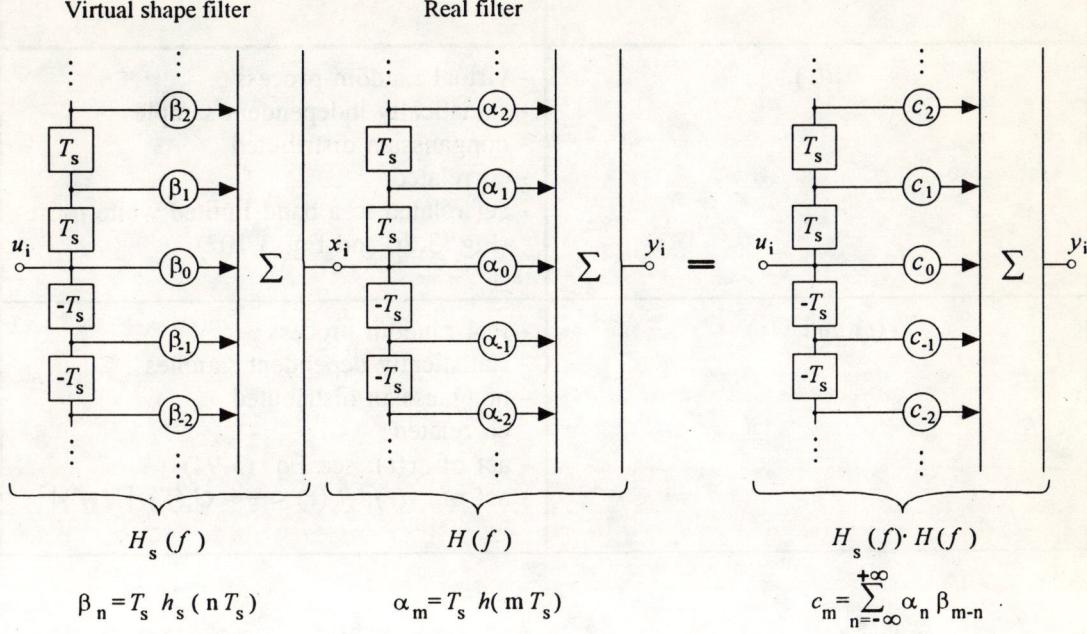

$$\beta_n = T_s \, h_s(nT_s) \qquad \alpha_m = T_s \, h(mT_s) \qquad c_m = \sum_{n=-\infty}^{+\infty} \alpha_n \beta_{m-n}$$

Fig. 3.25: Model of filter to calculate probability density function $f_y(y)$

Since newly defined random variables u_i are independent and psd of $u(t)$ is related to band-limited white noise, acf $R_u(\tau)$ and psd $G_u(f)$ are given by (Fig. 3.26):

$$R_u(\tau) = \sigma_u^2 \, \text{si}\!\left(\frac{\pi \tau}{T_s}\right) \; \circ\!\!-\!\!\bullet \; G_u(f) = \begin{cases} \sigma_u^2 T_s & \text{if } |f| \leq \dfrac{1}{2T_s} \\[3mm] 0 & \text{otherwise} \end{cases} \qquad (3.102)$$

From this equation, it can be easily deduced that the random process $u(t)$ is band-limited by $|f| \leq 1/(2T_s)$. Thus, $u(t)$ is completely described by its samples $u_i = u(iT_s)$ using the sampling theorem. Moreover, samples u_i are statistically independent as acf $R_u(\tau)$ is zero at every T_s (Fig. 3.26a).

Table 3.2: Statistical properties of random processes $u(t)$, $x(t)$ and $y(t)$ given in Fig. 3.25

Random process	Statistical properties
$u(t)$	- virtual random process - statistically independent samples - nongaussian distributed - correlated - acf related to a band-limited white noise (Fig. 3.26 and Eq. 3.102)
$x(t)$ and $y(t)$	- real random process - statistically dependent samples - nongaussian distributed - correlated - acf of $x(t)$: see Eq. (3.94) - acf of $y(t)$: $R_y(\tau) \circ\!\!-\!\!\bullet\ G_x(f) \cdot |H(f)|^2$

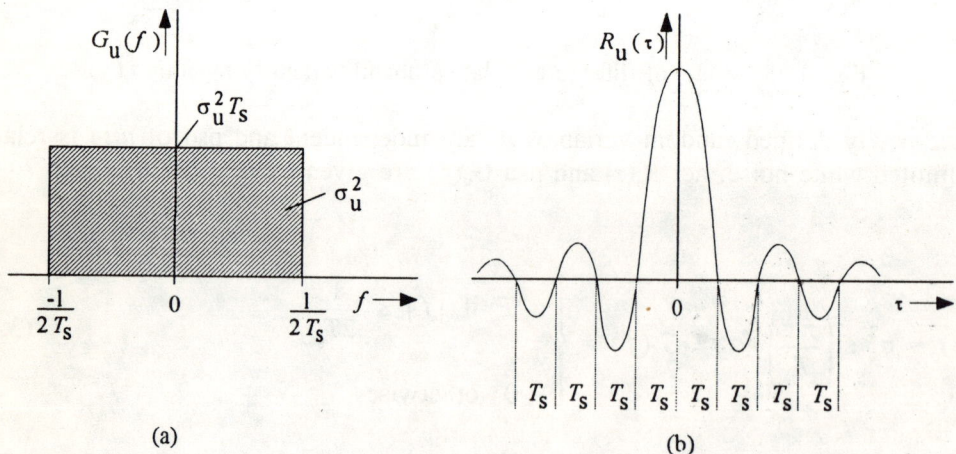

Fig. 3.26: (a) Power spectral density and (b) autocorrelation function of virtual random process $u(t)$ at the input of shape filter shown in Fig. 3.25

The random process $u(t)$ is now defined, whereas the filter coefficients β_m of shape filter $H_s(f)$ and $f_u(u)$ or moments $m_u^{(k)}$ are still undefined. To determine the filter coefficients β_m, amplitude response

$$|H_s(f)| = \sqrt{\frac{G_x(f)}{G_u(f)}} \;\rightarrow\; h_s'(t) \;\circ\!\!-\!\!\bullet\; |H_s(f)| \;\rightarrow\; \beta_m' = T_s h_s'(mT_s) \qquad (3.103)$$

of shape filter defined in terms of power spectral densities $G_x(f)$ and $G_u(f)$ has to be considered. The phase response of shape filter still remains undetermined. As a result, impulse response $h_s(t)$ and coefficients β_m of shape filter cannot be unambiguously defined. Therefore, these quantities are marked with dash in Eq. (3.103). They only define the amplitude response of filter clearly, but not the phase response. Assuming a Gaussian input process $u(t)$, phase response of shape filter can, however, be chosen arbitrarily because a linear filter never changes the shape of Gaussian pdf.

Next, the moments $m_u^{(k)}$ of virtual input process $u(t)$ have to be determined. This can be done by taking into consideration the moments $m_x^{(k)}$ of real random process $x(t)$ and coefficients β_m of shape filter. Since the samples u_i are considered to be statistically independent, method of moments can be applied. The output moments $m_x^{(k)}$ will depend only on the input moments $m_u^{(k)}$ of same order (i.e., same k) and vice versa.

Finally, the moments $m_y^{(k)}$ and pdf $f_y(y)$ of the real filter output process

$$y_i = \sum_{m=-\infty}^{+\infty} u_m\, c_{i-m} \quad \text{with} \quad c_{i-m} = \sum_{n=-\infty}^{+\infty} \alpha_n\, \beta_{i-n-m} \qquad (3.104)$$

can be determined by taking into consideration the moments $m_u^{(k)}$ of the random process $u(t)$, common filter coefficients c_m (Fig. 3.25) and at last an appropriate series; for example, the Gram-Charlier series.

In the implementation of this method on computer, a relationship between the pdf $f_y(y)$ at the output of real filter and filter coefficients β_m of shape filter has been observed [37, 38]. If the input process $u(t)$ is Gaussian, then this interrelation does not exist. This means that all statistical moments of order higher than k=2 are functions of filter coefficients β_m, whereas the moments of first and second order which are sufficient to describe a Gaussian pdf completely are independent of β_m. Hence, the filter response to a nongaussian input process cannot be completely described statistically when input pdf and, in addition, input acf or psd are known. These functions only determine the amplitude response of shape filter and not the phase response.

As mentioned earlier, acf and psd only contain information about first and second moments. Since these moments are dominant with respect to location and shape of each pdf, error caused by incorrect moments of higher order remains relatively low [37].

(v) METHOD OF INTEGRAL

This method is based on the integration of multi-dimensional joint density function, briefly called the joint pdf. For determining the pdf $f_y(y)$, joint pdf has to be derived first. For this, a set of $(2N+1)$ equations is required where $(2N+1)$ equals the number of random variables on the right side of the following equation:

$$y_i = \sum_{n=-N}^{+N} x_{i-n} \alpha_n \qquad (3.105)$$

It can be seen from Eq. (3.90) that N is infinite. However, if the impulse response $h(t)$ of the filter $H(f)$ is practically restricted to a certain time interval, finite N can be employed. Of course, the accuracy of calculation will somewhat be reduced. For a straight forward calculation, unambiguous sets of equations are particularly suited since these are characterized by only one single and perfectly specified solution with respect to all variables on the right hand side of set. Hence, special care has to be taken in case of random process $x(t) = \cos(\phi(t))$ which inherently exhibits an ambiguous inverse function (see Section 3.3.4).

With new auxiliary random variables h_{-N+1} to h_N, following set of $(2N+1)$ equations is obtained:

$$
\begin{aligned}
y_i &= \sum_{n=-N}^{+N} x_{i-n}\,\alpha_n \\
h_{-N+1} &= x_{-N+1} \\
&\vdots \\
h_0 &= x_0 \\
&\vdots \\
h_N &= x_N
\end{aligned}
\qquad (3.106)
$$

It may be mentioned again that a large number of sets of equations are usually available, but only sets with a single perfectly specified solution yield a proper result. In the above equation, both sides exactly contain $(2N+1)$ random variables. Using statistical transformation approach of multi-dimensional random variables [206], Eq. (3.106) yields the following multi-dimensional joint pdf:

$$
f_{y,\,h_{-N+1}\cdots h_N}(y,\, h_{-N+1} \cdots h_N) =
$$

$$
\frac{f_{x_{-N}\cdots x_N}(x_{-N}^{(1)} \cdots x_N^{(1)})}{\left| \det J\left(x_{-N}^{(1)} \cdots x_N^{(1)}\right) \right|} + \cdots + \frac{f_{x_{-N}\cdots x_N}(x_{-N}^{(k)} \cdots x_N^{(k)})}{\left| \det J\left(x_{-N}^{(k)} \cdots x_N^{(k)}\right) \right|}
\qquad (3.107)
$$

Here, $x_n^{(1)}$ to $x_n^{(k)}$ with $-N \leq n \leq N$ represent the set of solutions for $(2N+1)$ random variables on the right hand side of Eq. (3.106). With an unambiguous set of equations, exactly one single perfectly specified solution instead of k solutions exists for each random variable (i.e., k=1). In that case, joint pdf given in Eq. (3.107) is restricted to one single sum term only. The denominators in Eq. (3.107) are given by the determinant of Jacobian matrix [206]

$$
J = \begin{bmatrix}
\dfrac{\delta y}{\delta x_{-N}} & \cdots & \dfrac{\delta y}{\delta x_N} \\
\cdot & & \cdot \\
\cdot & & \cdot \\
\cdot & & \cdot \\
\dfrac{\delta h_N}{\delta x_{-N}} & \cdots & \dfrac{\delta h_N}{\delta x_N}
\end{bmatrix}
\tag{3.108}
$$

Due to proper selection of set of equations (3.106), inverse set exhibits a perfectly specified solution for each of $(2N+1)$ variables. In this particular case, absolute of det J of Jacobian matrix exactly yields 1. Thus, the joint pdf simplifies to

$$
f_{y, h_{-N+1} \cdots h_N}(y, h_{-N+1} \cdots h_N) = f_{x_{-N} \cdots x_N}\left(\frac{1}{\alpha_{-N}}\left[y - \sum_{n=N+1}^{N} x_n \alpha_n \right], h_{-N+1} \cdots h_N \right)
\tag{3.109}
$$

Finally, we have to integrate (hence, the name method of integral) with respect to the random variables h_{-N+1} to h_N to obtain the desired pdf of random variable y. It is given by

$$
f_y(y) = \int_{-\infty}^{+\infty} \cdots \int_{-\infty}^{+\infty} f_{x_{-N} \cdots x_N}\left(\frac{1}{\alpha_{-N}}\left[y - \sum_{n=N+1}^{N} x_n \alpha_n \right], h_{-N+1} \cdots h_N \right) dh_{-N+1} \cdots dh_N
\tag{3.110}
$$

This equation gives an exact solution of pdf $f_y(y)$ i.e., no approximation is included. The problem in evaluating Eq. (3.110) primarily is due to multi-dimensional joint-pdf in the integral which is to be determined. However, in case of random process $x(t) = \cos(\phi(t))$, this problem can be easily solved.

Beginning with an arbitrary "start-phase", for example, $\phi([i-N]T_a) = \phi_{i-N}$, all other phases can be determined by adding a phase difference step-by-step: i.e., $\phi_{i-N+1} = \phi_{i-N} + \Delta\phi_{i-N}$. Hence, the joint pdf given by Eq. (3.110) can be replaced by a joint pdf of new random variables $\Delta\phi_{i\text{-}N}$. As the phase differences are statistically independent (see Section 3.3.2), joint pdf can be determined by a product of individual pdfs. By this approach, a compact analytical solution for the pdf $f_y(y)$ at the filter output is obtained. It will be taken up again when ASK heterodyne system is discussed in Section 5.3.1.

The interrelation between the method of integral and the method of statistical moments is given by the equation

$$m_y^{(k)} = \int_{-\infty}^{+\infty} \cdots \int_{-\infty}^{+\infty} \left(\sum_{n=-N}^{N} x_n \alpha_n \right)^k f_{x_{-N} \cdots x_N}(x_{-N} \cdots x_N) \, dx_{-N} \cdots dx_N \qquad (3.111)$$

where $(2N+1)$ instead of $2N$ integrals have to be considered (compare Eq. 3.110).

(vi) METHOD OF QUASI-CONSTANT FREQUENCY

This approximation method, proposed first by Garrett and Jacobsen, presumes that the frequency noise $\dot{\phi}(t) = \omega(t) = \omega_c$ is nearly constant during the time interval

$$\Delta t_H = \frac{1}{h(0)} \int_{-\infty}^{+\infty} h(t) \, dt = \frac{H(0)}{h(0)} = \frac{1}{\Delta f_H} \approx T \qquad (3.112)$$

which is usually in the order of bit period T in practical systems [75, 110]. Physically, Δt_H represents the rectangular equivalent pulse width of filter time response $h(t)$ and Δf_H the double-sided rectangular equivalent bandwidth of filter frequency response $H(f)$. Thus, $\Delta f_H \cdot \Delta t_H = 1$. For a Gaussian filter $H(f) = \exp[-\pi(f/\Delta f_H)^2]$ with a bandwidth $\Delta f_H = 1/T$, rectangular equivalent pulse width exactly yields $\Delta t_H = T$.

As shown in Fig. 3.27, constant frequency ω_c during Δt_H is related to phase which increases or decreases (ω_c can also be negative) linearly with time i.e., $\phi(t) = \omega_c t + \phi_c$. It may be mentioned that ω_c is a zero mean, Gaussian distributed random variable that statistically changes its value from one time interval (or one bit) to the following interval (the next bit). The variance of Gaussian pdf is $\sigma_{\omega c}^2 = E\{[\Delta\phi(t, \Delta t_H)/\Delta t_H]^2\} = 2\pi\Delta f \Delta t_H$, where Δf represents the resulting laser linewidth of both transmitter and local lasers.

In this method, filter output process $y(t)$ is first determined by convolution. For the special random process $x(t) = \exp(j\phi(t))$, we get

$$y(t) = x(t) \star h(t) = \int_{-\infty}^{+\infty} e^{j\phi(t-\tau)} h(\tau) \, d\tau = H(\omega_c) \, e^{j\phi_c} \, e^{j\omega_c t} \qquad (3.13)$$

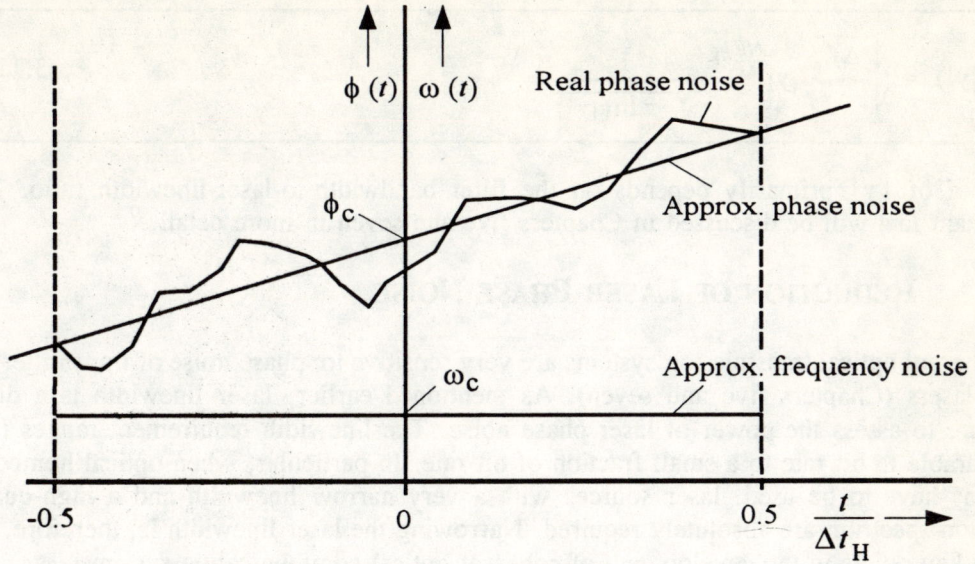

Fig. 3.27: Approximation of laser phase and frequency noise

The process $y(t)$ is a "harmonic" process changing frequency from one time interval to the other. In coherent optical communication systems with noncoherent detection (for example envelope detection), absolute random process $|y(t)|$ is of prime importance. Since $|y(t)|$ is a function of random variable ω_c, its pdf can easily be determined by applying the statistical transformation of random variables. We obtain

$$f_{|y|}(|y|) = f_{\omega_c}(\omega_c) \frac{1}{\left| \dfrac{dH(\omega_c)}{d\omega_c} \right|_{\omega_c = H^{-1}(|y|)}} \tag{3.114}$$

The above equation represents a simple, but powerful method to evaluate the pdf of $|y|$ provided random frequency $\phi(t)$ is nearly constant within the filter time response $h(t)$. Hence, this equation is valid for $\Delta f_H \gg \Delta f$ (i.e., filter bandwidth $\gg$ laser linewidth).

Example 3.5

Consider a filter with Gaussian frequency response $H(f) = \exp[-\pi(f/\Delta f_H)^2]$ and a complex nongaussian random process $x(t) = \exp(j\phi(t))$ at the filter input with $\sigma_{\omega c}^2 = 2\pi\Delta f\Delta t_H$, pdf of the nongaussian random process $|y|$ at the filter output is

$$f_{|y|}(|y|) = \sqrt{\frac{\Delta f_H}{\Delta f}} \, |y|^{\frac{\Delta f_H}{\Delta f} - 1} \, \frac{1}{\sqrt{-\pi \ln(|y|)}} \qquad\qquad (3.115)$$

The pdf of $|y|$ primarily depends on the filter bandwidth-to-laser-linewidth ratio. This important fact will be discussed in Chapters five and seven in more detail.

3.6   REDUCTION OF LASER PHASE NOISE

Coherent optical transmission systems are very sensitive to phase noise of transmitter and local lasers (Chapters five and seven). As mentioned earlier, laser linewidth is a direct measure to assess the power of laser phase noise. The linewidth requirement ranges from comparable to bit rate to a small fraction of bit rate. In particular, when optical homodyne systems have to be used, laser sources with a very narrow linewidth and a high-quality emission spectrum are absolutely required. Narrowing the laser linewidth is, therefore, one of the key points in the development of coherent optical communication systems.

Despite of linewidth narrowing, high output power and uniform FM response are additional requirements of a transmitter laser. The latter one is important particularly when direct modulation instead of external modulation is used. Local laser should be tuneable in wavelength, especially when high-density multichannel system have to be employed. Throughout the tuning range, fixed output power and constant linewidth should be maintained.

This Section is focused on the techniques to reduce the influence of laser noise. Some typical physical approaches for narrowing the laser linewidth are briefly discussed. A more detailed discussion is given, for example, in [50, 56, 83, 149]. In addition to these physical approaches concerning the laser source, various other system design techniques can also be employed to reduce the influence of phase noise; for example, error-correction coding.

It may be mentioned that physical approaches described below are exclusively based on laser diodes since they are the most promising sources for practical systems due to their small size, mechanical stability and potential for integration. Furthermore, laser diodes can be ASK, FSK and DPSK modulated by direct modulation of the injection current.

(i) DFB AND DBR LASER

The abbreviations DFB and DBR stand for *distributed feedback* and *distributed-bragg reflection*. Both types of lasers use a periodical disturbed waveguide as reflector instead of a plane reflector (Fabry-Perot laser). A plane reflector (for example, a mirror) is broadband with respect to the optical spectrum, whereas a periodical disturbed waveguide acts as a wavelength selective grating. The grating period is defined by the ratio of emission wavelength to refraction index of active area. Stable oscillation is only possible when all reflections interfere constructively. One specific mode is selected from the possible modes, while all others are suppressed.

In a DFB laser, corrugated feedback grating is located within the active area, whereas in a DBR laser this grating is located outside i.e., directly in front of the active area.These lasers are in general single mode and characterized by highly stable wavelength. Hence, these are commonly used as high quality and powerful single-mode lasers. However, their drawback is that a recognizable reduction of the laser linewidth in the order of some MHz can only be achieved by increasing the light power.

(ii) EXTERNAL CAVITY LASER

A substantial reduction of linewidth can be achieved by means of an external passive cavity, in addition, to the real active resonator inside the laser. Higher the length ratio of passive to active cavity, more is the improvement in linewidth narrowing. To obtain single-mode operation, grating is mostly used as external reflector. For example, a linewidth of about 10 kHz was achieved by applying an external cavity with a length of 20 cm [149]. Acoustooptic disturbances inside the external resonator represent a significant drawback. Further, external cavity lasers are unsuitable for optical integration.

(iii) LASER WITH OPTICAL REFLECTION

It is well-known that the reflected light usually disturbs the emission spectrum of laser and decreases the overall system performance. This is not valid in general since reflected light may also decrease the linewidth of a laser and, hence, improve the system performance. The phase relation between emitted and reflected light waves decides whether the first or second process is dominant. To reflect the light, grating, mirror or fiber itself can be used. In the last case, reduction of laser linewidth is accidental and mostly not reproducible. Although linewidths in the order of 100 kHz can be achieved, this method is not suitable for commercial applications.

(iv) COUPLED-CAVITY LASER

Usually, a laser contains single optical cavity to generate the light wave. In principle, this resonator (cavity) can be divided in two subcavities which have to be coupled. It is called a *cleaved-coupled-cavity laser* or briefly a C^3-laser. Both cavities are separated by a small slit (gap) only. Length of both resonators approximately equals the length of a single cavity in a conventional laser. Therefore, size of a C^3-laser is essentially smaller than a laser with external cavity. The laser linewidth is, in principle, decreased by constructive interference between the fields of both cavities. Linewidths of less than 1 MHz are possible. In addition, wavelength stabilization is improved. The gain in linewidth reduction and stability primarily depends on the size of the slit between the resonators. However, realization and reproduction of slits of a well-defined size represent a technological problem.

(v) CHANNEL CODING

Besides the physical approaches mentioned above, channel coding or error correction coding is another well-suited possibility to reduce the influence of laser phase noise. Of course, coding can also be used, in addition, to the physical approaches. The physical approaches are related to laser sources, whereas coding is related to the system design.

It is well-known from coding theory that channel coding, such as block coding or convolution coding, can be used to correct bit errors, improve signal-to-noise ratio or both. Channel coding makes use of redundancy. If information bit rate is fixed, then redundancy increases the overall bit rate on the channel (e.g., the fiber), which is frequently called the channel bit rate. The ratio of information bit rate to channel bit rate is termed as code rate. As the coding increases channel bit rate, system bandwidth must be increased appropriately. However, a higher bandwidth results in a stronger influence of the additive noise (i.e., shot noise and thermal noise) and system performance degrades (first effect). On the other hand, coding allows to correct bit errors and system performance improves (second effect). If the second effect dominates the first effect, coding improves the overall system performance. For a fixed bit error rate, this improvement results in a lower required signal-to-noise ratio or a lower required optical power at the receiver input. This gain in receiver sensitivity or signal-to-noise ratio is termed as *coding gain*. Irrespective of whether the system is coherent or direct detection, a coding gain can always be achieved.

In a coherent optical communication system, third effect will also be observed in addition to earlier effects mentioned above. As will be discussed in Chapter five, influence of laser phase noise decreases rapidly with the decrease in the linewidth-bit duration product ΔfT. Hence, higher the bit rate $1/T$, lower is the influence of laser phase noise. For a fixed information bit rate, channel coding always requires an increased channel bit rate and reduces the influence of phase noise. The coding gain obtained by error correcting coding offers an improvement in system performance with respect to signal-to-noise ratio as well as in laser linewidth requirements. By applying coding, maximum permissible linewidth may be approximately ten times broader than without coding [69].

(vi) PHASE NOISE CANCELLATION CIRCUITS

The phase noise in a coherent receiver can theoretically be suppressed completely when a reference signal having exactly the same phase noise as the carrier signal is available. Since phase noise is already existing in the transmitter, reference signal must be transmitted simultaneously to the information signal. This, for example, can be realized by means of an optical subcarrier generated by a simple frequency shift of the unmodulated laser wave using an acoustooptic modulator. In the receiver, phase noise is cancelled by generating the square of the sum of both reference and modulated carrier signal. As it is rather difficult to produce and transmit a reference signal which exactly contains and maintains a copy of the phase noise, phase noise cancellation circuits have a limited scope.

4   POLARIZATION FLUCTUATIONS

Ordinary single-mode fibers do not normally preserve the polarization state of the propagating light. As a consequence, polarization fluctuations can be observed at the output of fiber. Conventional optical communication systems with intensity modulation of light and direct detection (IM/DD) are insensitive to polarization fluctuations, which is one of their significant advantage. In contrast, polarization fluctuations are of great concern in coherent optical communication systems. As already mentioned in Chapter two, polarization fluctuations in the received light wave will produce fluctuations in the photodiode current and, therefore, in the detected signal at the input of sample and hold circuit. In the worst-case, polarization fluctuations may even extinguish the detected signal completely. Thus, either a polarization stabilization or an active control of *state of polarization* (SOP) is indispensable. It will not be true in polarization-diversity receiver.

In order to find appropriate techniques to solve the polarization problem, a profound knowledge about the reasons of fluctuations and their effects on the polarization of optical wave is required. This important topic has been discussed in detail in Sections 4.1 and 4.2. Readers who are not especially interested in the theoretical analysis of polarization fluctuations may skip these Sections.

Finally, Section 4.3 presents various technical approaches to stabilize or to control the polarization of a light wave. In this Section, polarization-diversity receiver which represents an attractive alternative to control and stabilization circuits is discussed.

It may be mentioned that the optical wave is frequently referred as light wave or simply "light". However, most light waves used in optical communication systems have their wave-lengths in the range 800 nm to 1.5 μm. These light waves are nonvisible and, hence, not actually a light in true sense.

4.1   POLARIZATION PROPAGATION IN SINGLE-MODE FIBER

4.1.1   EIGENMODES

In an ideal circular single-mode fiber, other types of fibers such as the graded-index fiber are unsuitable for coherent optical communication systems and will not be considered here, two orthogonal independent *degenerate modes* (i.e., the orthogonal polarizations) are always existing. These modes are degenerated since both are defined by the same propagation constant or same velocity of propagation. In general, the electric field in an optical fiber is always a linear superposition of these *two eigenpolarizations* or *eigenmodes*. As eigenmodes are independent, they propagate without any mutual disturbance.

In a real single-mode fiber, various asymmetries can be observed; for example, non-circular fiber core or asymmetrical lateral pressures. Thereby, the degeneration of both eigenmodes is removed and they propagate with different velocities instead of same velocity. Thus, these modes actually become two different modes.

Besides non-circular fiber core and asymmetrical lateral stress, various other internal and external imperfections can be observed along the fiber. Typical additional imperfections are bendings, torsions and asymmetrical distributions of the refraction index. Moreover, variations in temperature may also influence the quality of light propagation through the optical fiber since different thermal coefficients of expansion with respect to fiber core and jacket yield again asymmetrical internal stresses. Independent of type, all these perturbations break the circular waveguide geometry and influence the velocity of propagation of both the eigenmodes.

In addition to change in the velocity of propagation, fiber distortion such as mentioned above also yield an undesired *mode coupling*. Because of this, a reciprocal exchange of energy takes place. As a result, both modes influence each other and they are no longer independent. They are not eigenmodes now since only independent modes are called eigenmodes. However, it can be shown that coupled modes can always be expressed in terms of two independent eigenmodes again, which now also take into account the perturbations of the fiber (Section 4.1.2). The orthogonal polarizations of these new defined eigenmodes differ from the orthogonal polarizations of the eigenmodes which were existing in the absence of fiber perturbations. The orthogonal directions of polarization are called the *principal axes*. If, for example, fiber exhibits an elliptical core, principal axes are the small and the large half-axes of the elliptical cross-section of the fiber core (Section 4.3).

This Section first provides that mode coupling does not exist. This means that one eigenmode propagates more slowly as compared to other in the single-mode fiber. To focus our discussion on the polarization effects only, we presume that the principal axes of the eigenmodes are same as the x- and y-directions of a rectangular cartesian coordinate system. In this coordinate system, propagation constants are represented by β_x and β_y respectively. The axis of fiber that is the direction of propagation is the z-direction. We also presume that the fiber is perturbed only by an axial asymmetry which does not change with the position variable z.

An important quantity to assess how polarization is changed or maintained in a single-mode fiber is given by the difference

$$\Delta \beta = \beta_x - \beta_y \tag{4.1}$$

of both propagation constants β_x and β_y [195]. Normalizing this difference by $2\pi/\lambda$, where λ denotes the wavelength of the light, we obtain the dimensionless quantity

$$D = \frac{\Delta \beta \; \lambda}{2\pi} \tag{4.2}$$

called the *double refraction* or *birefringence*. Thus, single-mode fibers with different velocities of propagation of both their eigenmodes are normally called *birefringent fibers*.

In the following, we examine the polarization propagation as a function of double refraction $\Delta\beta$ and location z. To avoid generality, we consider a plane optical wave with a linear polarization at the input to the fiber i.e., $z=0$. This wave can be described by the electric field

$$\vec{E}(t) = \begin{pmatrix} E_x(t) \\ E_y(t) \end{pmatrix} = \hat{E}\, e^{j2\pi ft} \begin{pmatrix} \cos(\theta) \\ \sin(\theta) \end{pmatrix} = \hat{E}\, e^{j2\pi ft}\, \vec{e}$$

$$= \underline{\hat{E}\cos(\theta)\, e^{j2\pi ft}\, \vec{e}_x} + \underline{\hat{E}\cos(\theta)\, e^{j2\pi ft}\, \vec{e}_y} \qquad (4.3)$$

eigenmode x $\qquad\qquad$ eigenmode y

Here, $f = c/\lambda$ represents the frequency of light and $\hat{E}$ the field amplitude which is assumed to be constant. Thus, laser phase and amplitude noise are neglected in this Chapter. The linear polarization of the fiber input field is defined by the *unit polarization vector $\vec{e}$* with $|\vec{e}| = 1$. The direction (i.e., the orientation) of this vector is determined by the angle θ called the *polarization* angle. Finally, $\vec{e}_x$ and $\vec{e}_y$ represent the unit vector in x- and y-directions respectively.

The locus curve (i.e., the polarization) of the real part of electric field vector $\vec{E}(t)$ in the xy-plane is shown in Fig. 4.1. At the input to the fiber ($z=0$), this vector describes a straight line in accordance with the linear input polarization discussed above.

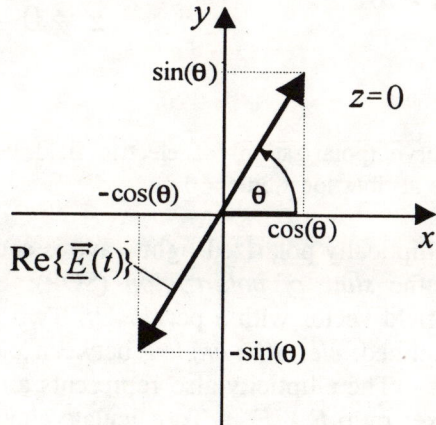

Fig. 4.1: Locus curve (i.e., polarization) of electric field vector of linearly polarized light wave at the fiber input ($z=0$)

When we consider the shape of the locus curve at an arbitrary other fiber location $z \neq 0$, a polarization that is usually not linear can be observed. Instead, it is generally elliptical with an electric field given by

$$\vec{E}(z,t) = \begin{pmatrix} \underline{E}_x(z,t) \\ \underline{E}_y(z,t) \end{pmatrix} = \hat{E} \ e^{j2\pi ft} \begin{pmatrix} \cos(\theta) \ e^{-j\beta_x z} \\ \sin(\theta) \ e^{-j\beta_y z} \end{pmatrix} = \hat{E} \ e^{j2\pi ft} \ \underline{\vec{e}}(z)$$

(4.4)

$$= \underline{\hat{E} \ \cos(\theta) \ e^{j2\pi ft} \ e^{-j\beta_x z} \ \vec{e}_x} + \underline{\hat{E} \ \sin(\theta) \ e^{j2\pi ft} \ e^{-j\beta_y z} \ \vec{e}_y}$$

eigenmode x eigenmode y

In contrast to Eq. (4.3), *unit polarization vector* $\underline{\vec{e}}(z)$ is complex and a function of location in addition. Optical waves characterized by Eq. (4.4) are termed as *elliptically polarized*.

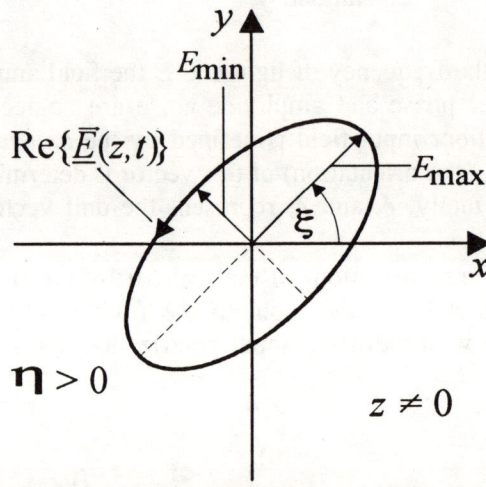

Fig. 4.2: Locus curve (polarization) of electric field vector of elliptically polarized light wave at fiber location $z \neq 0$

Typical locus curve of an elliptically polarized light wave is shown in Fig. 4.2. It is called the *polarization ellipse* or the *state of polarization* (SOP). Each polarization ellipse is periodically orbited by the field vector with a period $1/f$. Two characteristic parameters of polarization ellipse can be defined: *elevation angle* ξ between x-direction and large half-axis of the ellipse and *ellipticity* η. The ellipticity also represents an angle since η is defined by the arctangent of the half-axes ratio E_{min}/E_{max}. To calculate elevation angle Θ and ellipticity η, phase difference

$$\Delta\phi(z) = \Delta\beta z$$

(4.5)

between both orthogonal field components $\underline{E}_x(z, t)$ and $\underline{E}_y(z, t)$ of the electric field given in Eq. (4.4) is required. This phase difference changes with location z. Using some simple trigonometrical relations, elevation angle ξ and ellipticity η can finally be expressed by [60]

$$\eta = \pm \arctan\left[\frac{E_{min}}{E_{max}}\right] = \pm \arctan\left[\frac{\sin(2\theta)\,\sin(\Delta\phi(z))}{1 + \sqrt{1 - \sin^2(2\theta)\,\sin^2(\Delta\phi(z))}}\right] \qquad (4.6)$$

and

$$\xi = \frac{1}{2}\arctan\left[\frac{\sin(2\theta)\,\cos(\Delta\phi(z))}{\cos(2\theta)}\right] \qquad (4.7)$$

The range of η and ξ are given by $-\pi/4 \leq \eta \leq +\pi/4$ and $-\pi/2 \leq \xi \leq +\pi/2$ respectively. The ellipticity η is positive if the electric field vector orbits the polarization ellipse anticlockwise and negative if the rotation is clockwise. The view of the observer is directed opposite to the direction of propagating light wave. This means that the observer is looking into the fiber core from the receiver side. By taking into consideration ellipticity η, another important parameter of the polarization ellipse called the *degree of polarization* can be defined. It is given by

$$P = \frac{1 - \tan^2(\eta)}{1 + \tan^2(\eta)} = \sqrt{1 - \sin^2(2\theta)\,\sin^2(\Delta\phi(z))} \qquad (4.8)$$

The degree of polarization is extended over the range 0 to 1. Here, $P = 0$ represents circular polarization, $0 < P < 1$ elliptical polarization and $P = 1$ linear polarization.

When the characteristic quantities η, ξ and P are analyzed in more detail, a periodicity in the variable z of location can be observed. Its period

$$L_b = \frac{2\pi}{\Delta\beta} \qquad (4.9)$$

represents a measure of length and is called the *beat length*. It becomes clear that after a distance of every L_b, same polarization as given at the fiber input ($z = 0$) can always be observed provided, of course, that no additional fiber imperfections occur along the fiber length.

In order to illustrate the variations of polarization during propagation through a single-mode fiber of length L_b, Fig. 4.3 shows the locus curve of the real part of the electric field vector given by Eq. (4.4). At the fiber input ($z=0$), a linear polarization is presumed. Further, an equal energy division between the orthogonal linear eigenmodes is also presumed. Thus, the angle of linear polarization at the fiber input is $\theta = \pi/4$ as shown in Fig. 4.3. It becomes evident from this figure that the polarization of light exhibits large variations along the fiber. These variations are deterministic if the fiber deformations are fixed and random for deformations fluctuating with time and temperature, as in practical systems.

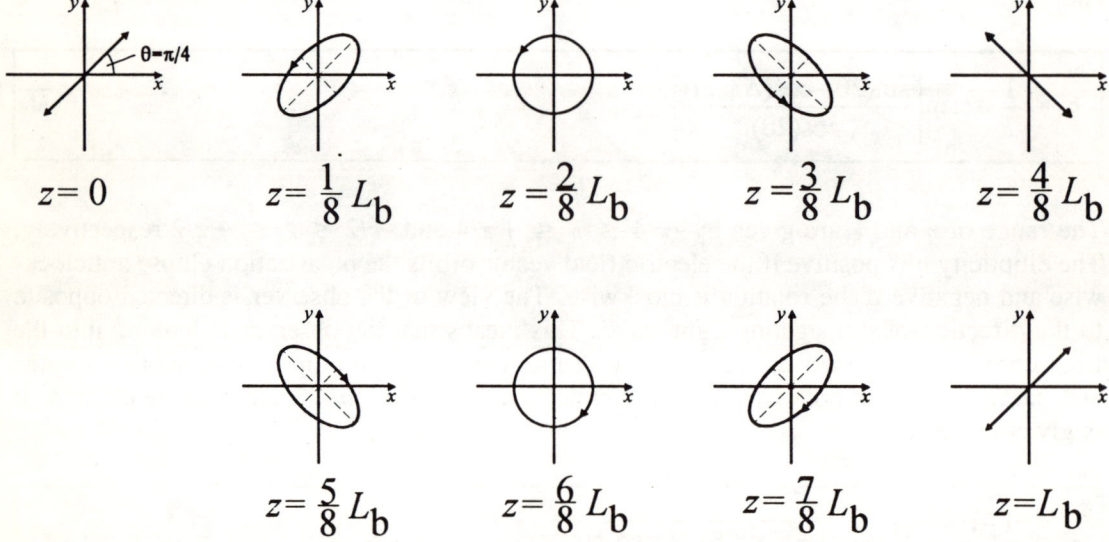

Fig. 4.3: Polarization in fiber at different locations z provided linear polarization with $\theta = \pi/4$ exists at fiber input ($z=0$)

Let us now consider that one of the eigenmodes is excited more than the other at the fiber input. In such case, polarization during propagation is changed. As an example, Fig. 4.4 provides that eigenmode of x-direction is more excited which means $\theta < \pi/4$. It becomes clear from this figure that the variations in state of polarization are less than that of given in Fig. 4.3. A circular polarization does not occur at all.

When only one of linear orthogonal eigenmodes is excited at the fiber input, a special case of significant importance is obtained. Here, the angle of linear input polarization is determined by $\theta = k\pi/2$ where $k \in \{\cdots -1, 0, 1 \cdots\}$. In that case, all characteristic quantities of polarization ellipse, namely η, ξ and P are now independent of variable z of location. Thus, a linearly polarized light at the fiber input always remains linearly polarized along the entire fiber length. Eigenmodes always define two characteristic orthogonal directions (i.e., the principal axes) which are suitable to maintain polarization along the fiber in principle.

However, it first requires that only one eigenmode is excited at the fiber input and second no additional unknown perturbations occur during propagation.

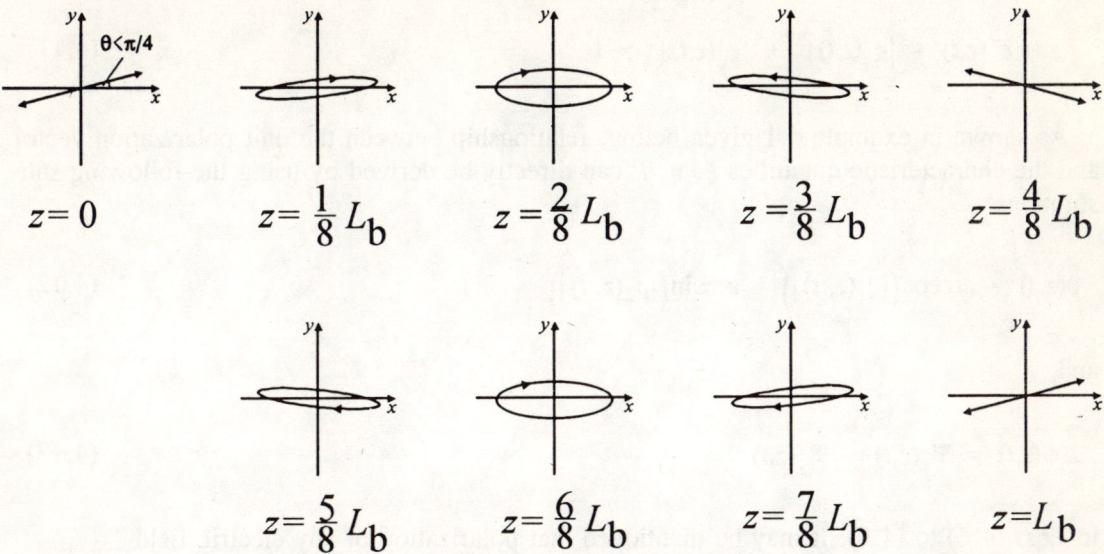

Fig. 4.4: Polarization in fiber at different locations z provided linear polarization with $\theta < \pi/4$ exists at the fiber input ($z = 0$)

In practice, various perturbations at different locations along the fiber can be observed, which all yield a certain fiber deformation. Each deformation changes the direction of the principal axes. Hence, even if one single eigenmode is excited at the fiber input, polarization usually remains constant in a short piece of fiber only. As soon as an additional fiber deformation occurs, new principal axes which are normally different from the principal axes at the fiber input determine the further polarization transmission. Therefore, stable transmission of polarization is usually not possible when standard single-mode fibers are used in the system. For this reason, fiber has to be modified by special technological processes (Section 4.3).

In order to summarize above discussions, Table 4.1 illustrates several types of polarizations and their relationship to the characteristic quantities viz., elevation angle ξ, ellipticity η and degree of polarization P. Shape of polarization ellipse and, consequently, the numerical values of their characteristic parameters are primarily defined by the unit polarization vector. In general, this vector is described by

$$\vec{e}(z,t) = \begin{pmatrix} e_x(z,t) \\ e_y(z,t) \end{pmatrix} = \begin{pmatrix} |e_x(z,t)| \; e^{j\Psi_x(z,t)} \\ |e_y(z,t)| \; e^{j\Psi_y(z,t)} \end{pmatrix} \qquad (4.10)$$

This means that it is now a function of time as well as location. As this vector is a unit vector, following feature is always satisfied:

$$\vec{e}(z,t)\ \vec{e}^*(z,t)\ =\ |e_x(z,t)|^2\ +\ |e_y(z,t)|^2\ =\ 1 \tag{4.11}$$

As shown in example 4.1 given below, relationship between the unit polarization vector and the characteristic quantities ξ, η, P can directly be derived by using the following substitutions

$$\theta(z,t)\ =\ \arccos\left(|e_x(z,t)|\right)\ =\ \arcsin\left(|e_y(z,t)|\right) \tag{4.12}$$

and

$$\Delta\phi(z,t)\ =\ \Psi_x(z,t)\ -\ \Psi_y(z,t) \tag{4.13}$$

in Eqs. (4.6) to (4.8). It may be mentioned that polarization of any electric field

$$\vec{E}(z,t)\ =\ \begin{pmatrix} E_x(z,t) \\ E_y(z,t) \end{pmatrix}\ =\ \hat{E}\ e^{j2\pi ft}\ \vec{e}(z,t) \tag{4.14}$$

can be described completely by the unit polarization vector, irrespective of type and number of fiber imperfections.

Example 4.1

Consider a linearly polarized light wave at the input to a single-mode fiber and different propagation constants as the only imperfection of the fiber. In that case, unit polarization vector is given by

$$\vec{e}(z,t)\ =\ \begin{pmatrix} e_x(z,t) \\ e_y(z,t) \end{pmatrix}\ =\ \begin{pmatrix} \cos(\theta)\ e^{-j\beta_x z} \\ \sin(\theta)\ e^{-j\beta_y z} \end{pmatrix}\ =\ \vec{e}(z)$$

Hence, the following relations can be obtained: $|e_x(z,t)|=\cos(\theta)$, $|e_y(z,t)|=\sin(\theta)$, $\Psi_x(z,t)=\beta_x z$, and $\Psi_y(z,t)=\beta_y z$.

Table 4.1: Typical polarizations and their relationship to fundamental quantities ξ, η, P (Ψ_0: arbitrary, but constant angle)

	Polarization ellipse	Unit vector of polarization $\vec{e}$	Ellipticity η	Angle of elevation ξ	Polarization degree P	Coordinates of Poincaré (S_1, S_2, S_3)
(1)		$\begin{pmatrix} \frac{1}{\sqrt{2}}e^{j\Psi_0} \\ \frac{1}{\sqrt{2}}e^{j(\Psi_0 - 90°)} \end{pmatrix}$	45° anti-clockwise	undefined	0	$(0,\ 0,\ 1)$
(2)		$\begin{pmatrix} \frac{1}{\sqrt{2}}e^{j\Psi_0} \\ \frac{1}{\sqrt{2}}e^{j(\Psi_0 + 90°)} \end{pmatrix}$	-45° clockwise	undefined	0	$(0,\ 0,\ -1)$
(3)		$\begin{pmatrix} \frac{\sqrt{3}}{2}e^{j\Psi_0} \\ \frac{1}{2}e^{j(\Psi_0 - 90°)} \end{pmatrix}$	30°	0°	0.5	$\left(0.5,\ 0,\ \frac{\sqrt{3}}{2}\right)$
(4)		$\begin{pmatrix} \frac{1}{\sqrt{2}}e^{j\Psi_0} \\ \frac{1}{\sqrt{2}}e^{j(\Psi_0 - 45°)} \end{pmatrix}$	22.5°	45°	$1/\sqrt{2}$	$\left(0,\ \frac{1}{\sqrt{2}},\ \frac{1}{\sqrt{2}}\right)$
(5)		$\begin{pmatrix} \frac{1}{2}e^{j\Psi_0} \\ \frac{\sqrt{3}}{2}e^{j(\Psi_0 - 90°)} \end{pmatrix}$	30°	90°	0.5	$\left(-0.5,\ 0,\ \frac{\sqrt{3}}{2}\right)$
(6)		$\begin{pmatrix} 1\,e^{j\Psi_0} \\ 0 \end{pmatrix}$	0°	0°	1	$(1,\ 0,\ 0)$
(7)		$\begin{pmatrix} -\frac{1}{\sqrt{2}}e^{j\Psi_0} \\ \frac{1}{\sqrt{2}}e^{j\Psi_0} \end{pmatrix}$	0°	-45°	1	$(0,\ 1,\ 0)$

As mentioned earlier, in coherent optical communication systems polarization of the received light wave at the output end of fiber is of significant importance. In that case, variable z of location is fixed and equals the length L of the fiber ($z=L$). Therefore, it is no longer required to represent the fiber location explicitly. The unit polarization vector $\vec{e}(z, t)=\vec{e}(L, t)=\vec{e}(t)$ is only a function of time now due to the time dependence of different fiber perturbations; for example, temporal change in internal stress or temperature. In addition, various time-invariant deformations can be observed; for example, elliptical fiber core. Thus, polarization state is unstable at the fiber output and, therefore, unpredictable.

When all geometrical fiber deformations are assumed to be time-invariant and polarization ellipse to be unchanged, electric field vector at the input to the receiver ($z=L$) still changes direction. Since this vector orbits the polarization ellipse periodically, orientation and length of this vector are consequently changed also. Therefore, polarization matching of both received and local laser light waves in coherent optical receiver is unavoidable even in this ideal case. For this reason, appropriate polarization matching techniques are required (Section 4.3).

Another very useful tool to describe polarization fluctuations also exits. Every type of polarization is completely defined by three parameters: angle of elevation ξ, ellipticity η and rotating direction (clockwise or anticlockwise). These parameters can be expressed in terms of three orthogonal coordinates S_1, S_2, and S_3 which are called the *Stokes parameters*. It may be remembered that the degree of polarization P is an additional characteristic quantity, but not really a required parameter since P is a function of ξ or η as given in Eq. (4.8). Taking into consideration the normalized Stokes parameters

$$S_1 = |e_x(z,t)|^2 - |e_y(z,t)|^2 = \cos(2\eta)\cos(2\xi) \tag{4.15}$$

$$S_2 = e_x(z,t)e_y^*(z,t) + e_x^*(z,t)e_y(z,t) = \cos(2\eta)\sin(2\xi) \tag{4.16}$$

$$S_3 = -j\left(e_x(z,t)e_y^*(z,t) - e_x^*(z,t)e_y(z,t)\right) = \sin(2\eta) \tag{4.17}$$

characteristic quantities of each state of polarization can now be projected on the surface of a sphere called the *Poincaré sphere* [220]. Thereby, each possible state of polarization corresponds to exactly one characteristic point $\mathcal{P}:=\mathcal{P}(S_1, S_2, S_3)=\mathcal{P}(2\xi, 2\eta)$ on the surface of this sphere as shown in Fig. 4.5.

For linearly polarized ($\eta=0$) light wave, characteristic point $\mathcal{P}$ is always located on the "Equator" of the Poincaré sphere. The points H and V represents the horizontal and vertical polarizations respectively, whereas the points P and M represent linear polarization at an angle of elevation $\xi=\pm45°$. If an optical wave is circularly polarized, then point $\mathcal{P}$ is either located at the "North Pole" (anticlockwise rotation) or "South Pole" (clockwise rotation) of the sphere. Elliptically polarized optical waves of same angle of elevation ξ are located on meridian, while same ellipticity η are always located on parallel. If both the states of polarization $\vec{\varepsilon}_1(z,t)$ and $\vec{\varepsilon}_2(z,t)$ are orthogonal, i.e., $\vec{\varepsilon}_1(z,t)\cdot\vec{\varepsilon}_2^*(z,t)=0$, then they are diametrically opposite.

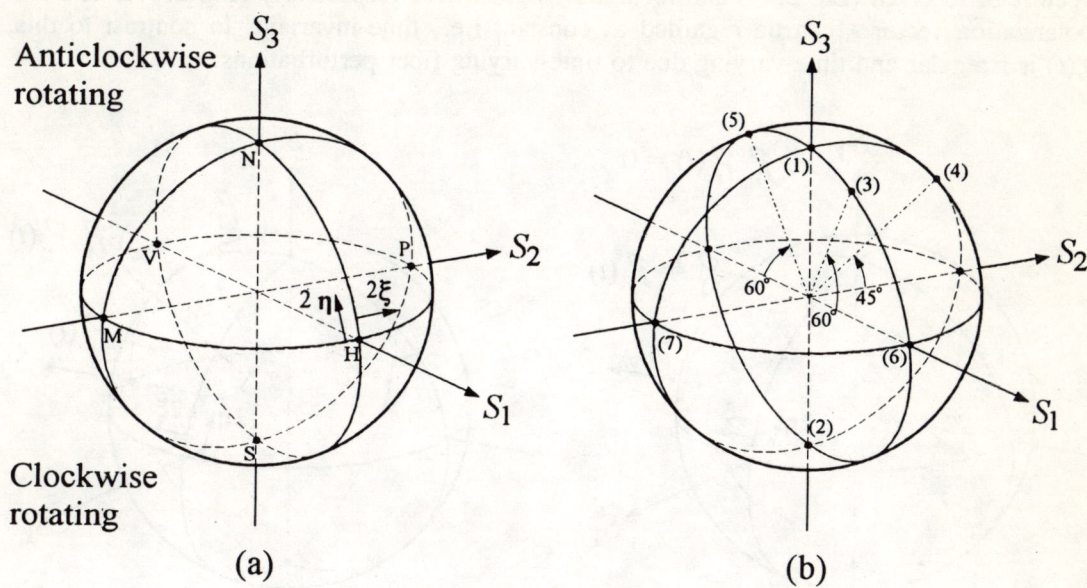

Fig. 4.5: Representation of state of polarization on Poincaré sphere (a) typical states of polarization and (b) states of polarization corresponding to Table 4.1

As mentioned earlier, performance of coherent optical communication system is considerably influenced by the states of polarization of both received light wave $\vec{E}_r(t)$ and local laser wave $\vec{E}_l(t)$. In the ideal case, both states of polarization are same and time-invariant. As discussed in Chapter two, each deviation from this ideal case results in amplitude and phase fluctuations $a_p(t)$ and $\phi_p(t)$ respectively in the photodiode current $i_{PD}(t)$ given by Eq. (2.61). Only in the ideal case, both fluctuations are always zero i.e., $\phi_p(t)=0$ and $a_p(t)=a_{p,max}=1$. It must be remembered that $a_p(t)$ is a multiplicative factor in the photodiode current $i_{PD}(t)$. For this reason, $a_p(t)$ should be as large as possible in order to increase the signal power. Hence, any deviation from the maximum value $a_{p,max}=1$ can be regarded as an attenuation. In practice, however, this deviation is time-varying since $a_p(t)$ is changing with time. In the worst-case i.e., $a_p(t)=0$, both states of polarization are orthogonal. Here, the effective part of photodiode current $i_{PD}(t)$ becomes zero (Eq. 2.61).

Phase fluctuations $\phi_p(t)$ caused by polarization fluctuations always occur in addition (i.e., additive) to the phase noise $\phi(t)$ of both transmitter and local lasers. However, phase fluctuations $\phi_p(t)$ are a rather slow random process as compared to laser phase noise $\phi(t)$. In practice, polarization matching is always required. Here, phase fluctuations $\phi_p(t)$ can usually be ignored in comparison with laser phase noise.

As discussed in Chapter two (Eq. 2.58), amplitude fluctuations $a_p(t)$ are determined by the absolute of the scaler product $\vec{\underline{e}}_l \cdot \vec{\underline{e}}_r(t)$, where $\vec{\underline{e}}_r(t)$ and $\vec{\underline{e}}_l$ represent the unit polarization vectors of received (Eq. 2.44) and local laser light waves respectively (Eq. 2.47). The unit polarization vector $\vec{\underline{e}}_l$ can be regarded as constant i.e., time-invariant. In contrast to this, $\vec{\underline{e}}_r(t)$ is irregular and time-varying due to time-varying fiber perturbations.

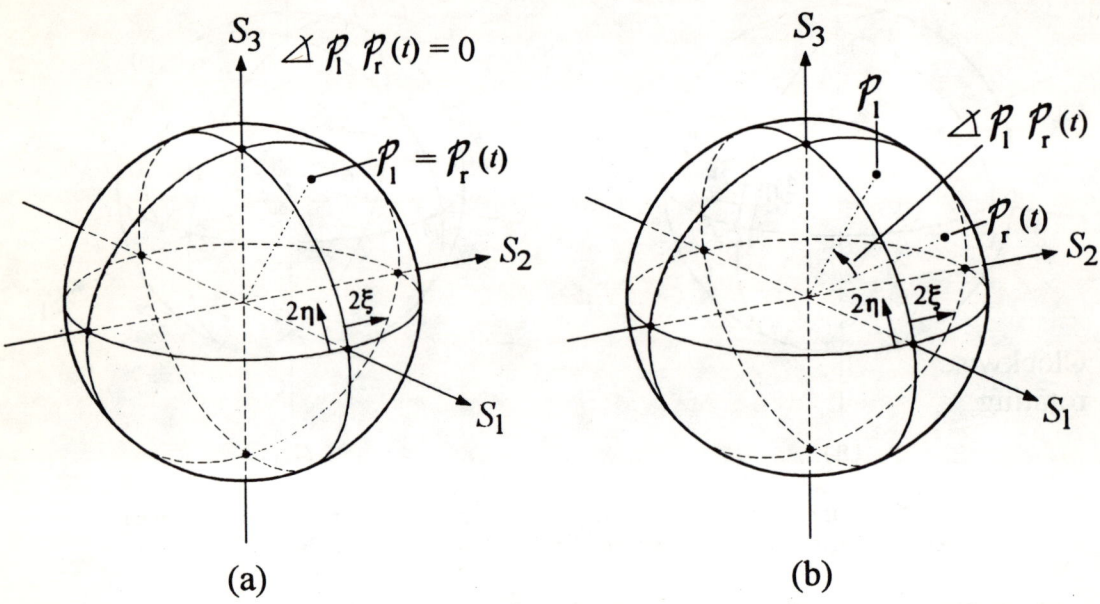

(a)   (b)

Fig. 4.6: States of polarization of received and local laser light waves in (a) ideal and (b) real case

Considering the Poincaré sphere again, amplitude fluctuations $a_p(t)$ can be quickly analyzed. For this, unit polarization vectors $\vec{\underline{e}}_l$ and $\vec{\underline{e}}_r(t)$ must be described by two points on the Poincaré sphere (Fig. 4.6) using Eqs. (4.15) to (4.17). These points are

$$\mathcal{P}_l = \mathcal{P}_l(S_{1l}, S_{2l}, S_{3l}) \quad \text{and} \quad \mathcal{P}_r(t) = \mathcal{P}_r(S_{1r}(t), S_{2r}(t), S_{3r}(t)) \tag{4.18}$$

The point $\mathcal{P}_r(t)$ of unit polarization vector of $\vec{\underline{e}}_r(t)$ is irregularly moving on the surface of the sphere according to its time dependence, whereas $\mathcal{P}_l$ remains fixed.

The spatial distance between the points $\mathcal{P}_r(t)$ and $\mathcal{P}_l$ expressed using the angle $\triangle \mathcal{P}_r(t)\mathcal{P}_l$ represents a direct measure to assess the strength of the amplitude fluctuations $a_p(t)$. After some mathematical simplifications, we obtain

$$a_p(t) = |\vec{\underline{e}}_r(t) \, \vec{\underline{e}}_l^*| = \cos\left(\frac{1}{2}\triangle \mathcal{P}_r(t) \, \mathcal{P}_l\right) \tag{4.19}$$

where

$$\Delta \mathcal{P}_r(t) \, \mathcal{P}_l = S_{11} S_{1r}(t) + S_{21} S_{2r}(t) + S_{31} S_{3r}(t) \qquad (4.20)$$

Performance of coherent optical communication systems rapidly deteriorates with the increase of angle $\Delta \mathcal{P}_r(t)\mathcal{P}_l$. Hence, as the spatial separation between both states of polarization increases, system performance deteriorates. When both states of polarization are same i.e., $\underline{\varepsilon}_r(t) = \underline{\varepsilon}_l$, corresponding points $\mathcal{P}_r(t)$ and $\mathcal{P}_l$ lie on top of each other as shown in Fig. 4.6a. Thereby, $a_p(t)$ reaches its maximum value $a_{p,max} = 1$ irrespective of the location of common points on the sphere.

Table 4.2: Influence of deviation in states of polarization of received and local laser waves on $a_p(t)$

$\Delta \mathcal{P}_r \mathcal{P}_l$	a_p	Remarks
0°	1	ideal case, same state of polarization
10°	0.996	very little influence
15°.	0.991	very little influence
20°	0.985	little influence
50°	0.906	little influence
90°	0.707	strong influence
180°	0	extinguishing of signal, orthogonal polarizations

To make it more clear, Table 4.2 shows the influence of the spatial angle $\Delta \mathcal{P}_r(t)\mathcal{P}_l$ on $a_p(t)$. Here, $\Delta \mathcal{P}_r(t)\mathcal{P}_l := \Delta \mathcal{P}_r \mathcal{P}_l$ is assumed to be constant. According to this table, influence of deviation in state of polarization is practically negligible, when the spatial angle is less than about 15°. In that case, deviation in $a_p(t)$ from its maximum value $a_{p,max} = 1$ is less than 1%. Even if this angle is relatively large, for example 50°, the deviation remains less than 10%. On the other hand, a considerable influence of deviation is observed if the angle $\Delta \mathcal{P}_R(t)\mathcal{P}_L$ exceeds 50°. Due to various disturbances in layed fiber cables, angle deviations of more than 50° are quite realistic. Therefore, either an appropriate polarization matching technique or a polarization-diversity receiver is absolutely required when a coherent optical communication system has to be employed.

As explained above, the Poincaré sphere is a powerful tool offering a clear description of polarization fluctuations. Moreover, the Poincaré sphere simplifies the analysis of influence of deviation in the state of polarization of received and local laser light waves on the performance of coherent optical transmission system. In combination with an appropriate software, the Poincaré sphere also represents a well-suited tool in polarization measurement equipment to determine the statistics of polarization fluctuations or study the polarization behaviour of a single-mode fiber cable.

4.1.2   THEORY OF MODE COUPLING

In the previous Section, we have discussed the effects of deviation in velocity of propagation of both orthogonal polarizations. The main reasons for a velocity deviation are deformations of the fiber geometry, for example, an elliptical fiber core. Both polarizations have propagated independently without any mode coupling. However, this is not actually true since mode coupling is normally existent in standard fiber.

The main purpose of this Section is to describe the reasons and effects of mode coupling which are caused, for example, by additional fiber perturbations. Due to mode coupling, a reciprocal energy exchange can be observed. As a result, both the modes i.e., orthogonal polarizations are no longer independent. Hence, coupled modes are not eigenmodes as eigenmodes always propagate independently as discussed in the previous Section.

In this Section, efforts are made to give an answer to the question: Is it possible to find out again two independent eigenmodes propagating without any energy exchange irrespective of type of fiber perturbation? Let us suppose that these new eigenmodes are existent. Then we have to focus our interest on two further questions: Which are the new orthogonal orientations of the new eigenmodes and what are their velocities of propagation? It is clear that the new orthogonal eigenmodes as well as their velocities of propagation will primarily depend on the fiber characteristics especially on its geometrical deformations [171].

In order to avoid generality and simplify calculation, we first take a certain well-specified deformation (for example, an elliptical fiber core) that corresponds to two independent eigenmodes propagating with different velocities. Next, we assume a second (or even more) geometrical fiber perturbation which deforms the unperturbed fiber core in addition. As a result, both the modes (mode x and mode y) are coupled now and they are no longer independent eigenmodes.

We consider that the first fiber deformation (for example, the elliptical fiber core as assumed above) as well as all further perturbations are independent of z-direction i.e., the fiber axis. The following results are valid only for geometrical deformations which are independent of location. Since this is normally valid over a very short distance, the results are applicable for short piece of fiber only.

Linear superposition of mode x and mode y to generate the optical fiber wave is again clarified by the following equations:

$$\vec{E}(z,\ t) = \begin{pmatrix} E_x(z,\ t) \\ E_y(z,\ t) \end{pmatrix} = E_x(z,\ t)\ \vec{e}_x + E_y(z,\ t)\ \vec{e}_y \tag{4.21}$$

$$= \hat{E}_x\ \underline{a}(z)\ e^{j2\pi ft}\ \vec{e}_x + \hat{E}_y\ \underline{b}(z)\ e^{j2\pi ft}\ \vec{e}_y$$

$$\uparrow \qquad\qquad\qquad \uparrow$$
$$\text{mode } x \leftarrow \text{dependent} \rightarrow \text{mode } y$$
$$\downarrow \qquad\qquad\qquad \downarrow$$

$$\vec{H}(z,\ t) = \begin{pmatrix} H_x(z,\ t) \\ H_y(z,\ t) \end{pmatrix} = \hat{H}_y\ \underline{a}(z)\ e^{j2\pi ft}\ \vec{e}_y - \hat{H}_x\ \underline{b}(z)\ e^{j2\pi ft}\ \vec{e}_x \tag{4.22}$$

$$= H_y(z,\ t)\ \vec{e}_y - H_x(z,\ t)\ \vec{e}_x$$

Eq. (4.21) describes the electric field and Eq. (4.22) the magnetic field of the light wave in the fiber. The modes are termed as mode x and mode y since the directions of the electric field components $\underline{E}_x(z,\ t)$ and $\underline{E}_y(z,\ t)$ are $\vec{e}_x$ and $\vec{e}_y$ respectively.

In previous Section 4.1.1, a plane wave is considered for both the modes which propagate in z-direction i.e., in the direction of fiber axis. The electric and magnetic field vectors of mode x and mode y are orthogonal and are located in the xy-plane (Fig. 4.7b). In addition, they are functions of time t and location z, but independent of location x and y. The light wave in a fiber is actually not a plane wave due to restricted size of the fiber core and the fact that the core is surrounded by a jacket with a refraction index lower than the index of the core (this is the fundamental condition for guided waves). However, in the discussion on principle of mode coupling, error introduced by taking a plane wave is less important. Instead, we take the advantages of plane waves to reach a clear description of mode coupling.

In order to describe the mutual energy exchange due to mode coupling, Eqs. (4.21) and (4.22) include two terms $\underline{a}(z)$ and $\underline{b}(z)$ which still have to be determined. These terms are complex since they additionally describe the periodical dependence of the fiber wave on the location z. It should be noted that $\underline{a}(z)$ and $\underline{b}(z)$ are dependent and cannot be chosen arbitrarily. In a fiber without loss, optical power flow $\vec{S}$ always remains constant i.e., independent of location z. We get

$$\vec{S} = \frac{1}{2} Re\{\vec{E}(z, t) \times \vec{H}^*(z, t)\} = \vec{S}_x(z) + \vec{S}_y(z) = (S_x(z) + S_y(z)) \vec{e}_z$$

$$= \left(\underbrace{|\underline{a}(z)|^2 \frac{1}{2} \hat{E}_x \hat{H}_y}_{S_{x,max}} + \underbrace{|\underline{b}(z)|^2 \frac{1}{2} \hat{E}_y \hat{H}_x}_{S_{y,max}} \right) \vec{e}_z = \underbrace{(|\underline{a}(z)|^2 + |\underline{b}(z)|^2)}_{= 1} S \vec{e}_z \qquad (4.23)$$

$$= S \vec{e}_z \neq \vec{S}(z)$$

It is seen from the above equation that the terms $\underline{a}(z)$ and $\underline{b}(z)$ which describe the energy exchange as a function of location z are related by $|\underline{a}(z)|^2 + |\underline{b}(z)|^2 = 1$. Hence, no power is lost in the fiber as considered above. It should be noted that the sign "x" in the first line of Eq. (4.23) represents a vector product [28]. The optical power flow $\vec{S}$ given in Eq. (4.23) is measured in AV/m². Hence, this quantity physically describes the effective optical power floating through unit area perpendicular to the direction of propagation (z-direction). As shown in Eq. (4.23), power flow can be divided into two parts: one is the power flow $\vec{S}_x(z)$ of mode x and the other power flow $\vec{S}_y(z)$ of mode y. The maximum power flows $S_{x,max}$ and $S_{y,max}$ of both modes are equal. As $S_{x,max}$ and $S_{y,max}$ can never be more than the total power flow S of the light wave $\vec{E}(0, t)$ at the fiber input (i.e., $z=0$), $S_{x,max}$ and $S_{y,max}$ are identical i.e., $S_{x,max}=S_{y,max}=S$.

In the following, $\underline{a}(z)$ and $\underline{b}(z)$ will be determined as a function of geometrical deformations which disturb the fiber. Afterwards, we shall analyze the field Eqs. (4.21) and (4.22). Finally, we shall study whether the field $\vec{E}(z, t)$ given in Eq. (4.21) can be divided into two independent modes again or not. For this purpose, we first take the equivalent approach

$$\vec{E}(z, t) = \quad E_u(z, t) \vec{e}_u \quad + \quad E_v(z, t) \vec{e}_v$$

$$= \underbrace{\hat{E}_u e^{-j\beta_u z} e^{j2\pi ft} \vec{e}_u}_{\uparrow \text{ eigenmode } x} \quad + \quad \underbrace{\hat{E}_v e^{-j\beta_v z} e^{j2\pi ft} \vec{e}_v}_{\uparrow \text{ eigenmode } y} \qquad (4.24)$$

$$\text{eigenmode } x \leftarrow \text{independent} \rightarrow \text{eigenmode } y$$

$$\vec{H}(z, t) = \quad H_v(z, t) \vec{e}_v \quad + \quad H_u(z, t) \vec{e}_u$$

$$= \underbrace{\hat{H}_v e^{-j\beta_u z} e^{j2\pi ft} \vec{e}_v}_{\uparrow} \quad + \quad \underbrace{\hat{H}_u e^{-j\beta_v z} e^{j2\pi ft} \vec{e}_u}_{\uparrow} \qquad (4.25)$$

$$\text{eigenmode } x \leftarrow \text{independent} \rightarrow \text{eigenmode } y$$

which replaces the field Eqs. (4.21) and (4.22). The unknown parameters in this new approach are the orthogonal directions $\vec{e}_u$ and $\vec{e}_v$ which define the principal axes of the new eigenmodes and their propagation constants β_u and β_v respectively. These quantities can be determined by a simple comparison of the electric field given in Eqs. (4.21) and (4.24), provided $\underline{a}(z)$ and $\underline{b}(z)$ have already been determined and substituted.

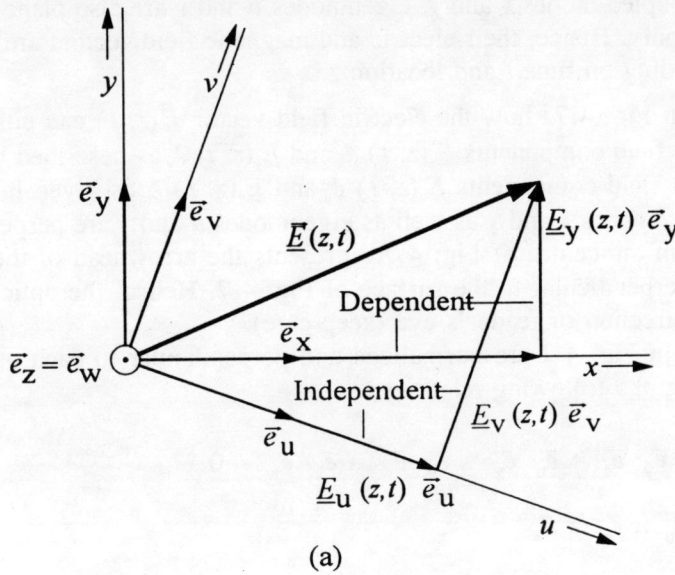

(a)

Mode x: $\underline{E}_x(z,t)\,\vec{e}_x$, $\underline{H}_y(z,t)\,\vec{e}_y$ Eigenmode u: $\underline{E}_u(z,t)\,\vec{e}_u$, $\underline{H}_v(z,t)\,\vec{e}_v$

Mode y: $\underline{E}_y(z,t)\,\vec{e}_y$, $-\underline{H}_x(z,t)\,\vec{e}_x$ Eigenmode v: $\underline{E}_v(z,t)\,\vec{e}_v$, $-\underline{H}_u(z,t)\,\vec{e}_u$

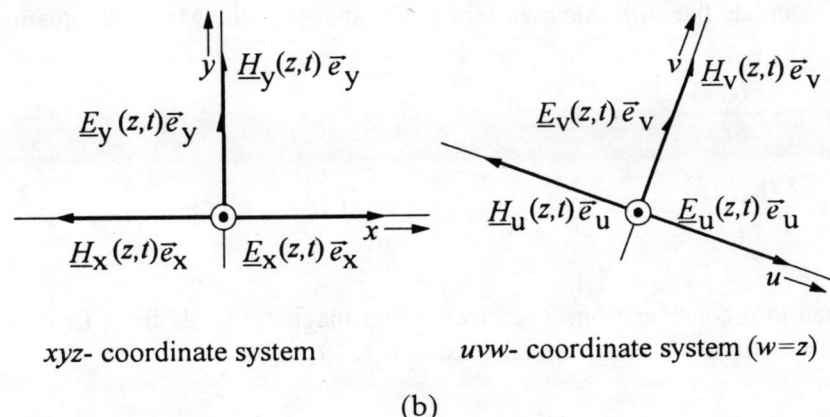

xyz- coordinate system   *uvw*- coordinate system ($w=z$)

(b)

Fig. 4.7: (a) Splitting of electric field vector $\vec{\underline{E}}(z, t)$ in modes x and y and eigenmodes u and v (b) orthogonality of modes

The amplitudes $\hat{E}_u$ and $\hat{E}_v$ of electric field depend on the input field of fiber being considered. These are completely determined by the boundary conditions (Section 4.2). The amplitudes $\hat{H}_u$ and $\hat{H}_v$ of magnetic field can directly be calculated by means of electric field since electric and magnetic fields are strongly correlated. They often differ only by a constant depending on fiber characteristics.

Similar to the coupled modes x and y, eigenmodes u and v are also plane waves propagating in z-direction only. Hence, their electric and magnetic field vectors are again located in the xy-plane depending on time t and location z.

It is illustrated in Fig. 4.7a how the electric field vector $\vec{E}(z, t)$ can either be separated into two dependent field components $\underline{E}_x(z, t)\,\vec{e}_x$ and $\underline{E}_y(z, t)\,\vec{e}_y$ as described by Eq. (4.21) or in two independent field components $\underline{E}_u(z, t)\,\vec{e}_u$ and $\underline{E}_v(z, t)\,\vec{e}_v$ as given in Eq. (4.24). As shown in Fig. 4.7b, modes x and y as well as eigenmodes u and v are perpendicular to each other. The dot $\odot$ in the centre of Fig. 4.7 represents the arrowhead of the unit vectors $\vec{e}_z$ and $\vec{e}_w$ which are perpendicular to the surface of Fig. 4.7. Hence, the optical wave directly propagates in the direction of reader's eye (keep care).

All unit vectors in Fig. 4.7 are normalized and perpendicular to each other. Thus, they are characterized by the following relations:

$$\vec{e}_x \cdot \vec{e}_y = \vec{e}_x \cdot \vec{e}_z = \vec{e}_y \cdot \vec{e}_z = \vec{e}_u \cdot \vec{e}_v = \vec{e}_u \cdot \vec{e}_w = \vec{e}_v \cdot \vec{e}_w = 0 \qquad (4.26)$$

$$\vec{e}_x \times \vec{e}_y = \vec{e}_z = \vec{e}_u \times \vec{e}_v = \vec{e}_w \qquad (4.27)$$

$$|\vec{e}_x| = |\vec{e}_y| = |\vec{e}_z| = |\vec{e}_u| = |\vec{e}_v| = |\vec{e}_w| = 1 \qquad (4.28)$$

In the above equations, symbols "$\times$" and "$\cdot$" represent the vector product and the scaler product respectively [28].

In order to calculate the still unknown terms $\underline{a}(z)$ and $\underline{b}(z)$, the Maxwell equations

$$\text{rot}\big(\vec{E}(z, t)\big) = -\frac{\delta \vec{\underline{B}}(z, t)}{\delta t} \qquad (4.29)$$

$$\text{rot}\big(\vec{H}(z, t)\big) = \frac{\delta \vec{\underline{D}}(z, t)}{\delta t} \qquad (4.30)$$

have to be taken into consideration. The electric and magnetic fields from Eqs. (4.21) and (4.22) must be substituted in the material equations [219]

$$\underline{\vec{D}}(z, t) = \epsilon \; \vec{E}(z, t) = \epsilon_r \epsilon_{r_0} \; \vec{E}(z, t) \tag{4.31}$$

$$\vec{B}(z, t) = \mu \; \vec{H}(z, t) = \mu_0 \; \vec{H}(z, t) \tag{4.32}$$

where $\underline{\vec{D}}(z, t)$ and $\vec{B}(z, t)$ represent the electric and magnetic power flux respectively. Both the Maxwell and material equations are already adapted to dielectric waveguide i.e., optical fiber. Thus, Eq. (4.32) is valid only for a non-magnetic material i.e., $\mu_r = 1$ and Eq. (4.29) does not contain an electric current component [219].

In a completely isotropic material, the relative dielectric constant ϵ_r is either a real (i.e., material without any loss) or a complex quantity (i.e., material with loss). In that case, electric field $\vec{E}(z, t)$ and electric power flux $\vec{D}(z, t)$ vectors point in the same direction and are parallel. However, fiber is usually non-isotropic due to various inherent asymmetries as already mentioned above. In such cases, relative dielectric constant ϵ_r is not a simple scaler quantity. Instead it is a tensor [28, 219], describing how the field is changing in different directions in the fiber. Mathematically, a tensor describes the linear dependence of the components of two vectors. In the particular case of fiber, this tensor describes the linear correlation of electric field $\vec{E}(z, t)$ and electric power flux $\vec{D}(z, t)$ vectors. Due to non-isotropic fiber characteristics, both vectors are normally not parallel.

For studies based on system aspects, we are more interested on the effects of mode coupling than in the atomic reasons of mode coupling. Thus, it is not required for the reader to study tensor calculation in detail. Instead, it is sufficient to know that a tensor of dielectric constants contains nine (complex) scaler quantities similar to a 3×3 matrix, two tensors are added in the same way as two matrixes and a tensor is in principle multiplied by a vector in the same way as a matrix is multiplied by a vector [219]. The result of such a product is again a vector.

All nine components of a tensor are only defined unambiguously in a specified coordinate system. Here, this system is the rectangular xyz-coordinate system. If this system is changed, for example rotated, then the numerical values of the tensor components are also changed. Hence, a tensor of dielectric constants varies with the change in the coordinate system. Optical fiber wave physically remains unchanged irrespective of kind of mathematical coordinate system chosen. Therefore, each change in coordinate system requires an appropriate tensor transformation which, however, will not be considered in the frame of this book. A more detailed description of tensor mathematics can be found in the literature; for example, in [28, 219].

In order to simplify the further calculations, it is useful to divide the tensor of dielectric constants

$$[\epsilon]_r = [\epsilon]_0 + [\epsilon]_m \tag{4.33}$$

into two parts. Thereby

$$[\epsilon]_0 = \begin{bmatrix} \epsilon_{xx} & 0 & 0 \\ 0 & \epsilon_{yy} & 0 \\ 0 & 0 & \epsilon_{zz} \end{bmatrix} \tag{4.34}$$

describes a fiber without any mode coupling and

$$[\epsilon]_m = \begin{bmatrix} \epsilon'_{xx} & \epsilon_{xy} & \epsilon_{xz} \\ \epsilon_{yx} & \epsilon'_{yy} & \epsilon_{yz} \\ \epsilon_{zx} & \epsilon_{zy} & \epsilon'_{zz} \end{bmatrix} \tag{4.35}$$

takes into account all the perturbations which are responsible for mode coupling. As provided at the beginning of this Section, $[\epsilon]_0$ takes into account the first fiber deformation (for example, the elliptical fiber core) and $[\epsilon]_m$ all additional perturbations of the fiber geometry. To distinguish the three diagonal tensor components of the first deformation and additional perturbations, these components are identified by dash in Eq. (4.35). It may be mentioned that a tensor of dielectric constants is normally symmetrical i.e., $\epsilon_{xy} = \epsilon_{yx}$, $\epsilon_{xz} = \epsilon_{zx}$ and $\epsilon_{yz} = \epsilon_{zy}$.

After this brief description of the tensor of dielectric constants, we are now able to replace the field components in the Maxwell Eqs. (4.29) and (4.30) by the field equations (4.21) and (4.22). With some mathematical operations, following coupled system of differential equations is obtained:

$$\frac{d}{dz}\begin{bmatrix} a(z) \\ b(z) \end{bmatrix} = -j \begin{bmatrix} N_{11} & N_{12} \\ N_{21} & N_{22} \end{bmatrix} \cdot \begin{bmatrix} a(z) \\ b(z) \end{bmatrix} \tag{4.36}$$

This system of differential equations (DEQ) is called a normal, linear and homogeneous DEQ system in the variable of location z [28]. The four coefficients N_{11}, N_{12}, N_{21} and N_{22} which can be either real or complex (dependent on the type of fiber distortion) are a direct result of the substitution mentioned above. We obtain

$$N_{11} = \pi f \left(\epsilon_0 \frac{\hat{E}_x}{\hat{H}_y} \vec{e}_x [\epsilon]_r \vec{e}_x + \mu_0 \frac{\hat{H}_y}{\hat{E}_x} \right) = \beta_x \sqrt{1 + \frac{\vec{e}_x [\epsilon]_m \vec{e}_x}{\vec{e}_x [\epsilon]_0 \vec{e}_x}}$$

(4.37)

$$= \beta_x \sqrt{1 + \frac{\epsilon'_{xx}}{\epsilon_{xx}}} = \beta'_x$$

$$N_{12} = \pi f \epsilon_0 \frac{\hat{E}_y}{\hat{H}_y} \vec{e}_x [\epsilon]_r \vec{e}_y = 0.5 \sqrt{\beta'_x \beta'_y} \frac{\vec{e}_x [\epsilon]_m \vec{e}_y}{\sqrt{(\vec{e}_x [\epsilon]_r \vec{e}_x)(\vec{e}_y [\epsilon]_r \vec{e}_y)}}$$

(4.38)

$$= 0.5 \sqrt{\beta'_x \beta'_y} \frac{\epsilon_{xy}}{\sqrt{\left(\epsilon_{xx} + \epsilon'_{xx}\right)\left(\epsilon_{yy} + \epsilon'_{yy}\right)}}$$

$$N_{21} = \pi f \epsilon_0 \frac{\hat{E}_x}{\hat{H}_x} \vec{e}_y [\epsilon]_r \vec{e}_x = 0.5 \sqrt{\beta'_x \beta'_y} \frac{\vec{e}_y [\epsilon]_m \vec{e}_x}{\sqrt{(\vec{e}_x [\epsilon]_r \vec{e}_x)(\vec{e}_y [\epsilon]_r \vec{e}_y)}}$$

(4.39)

$$= 0.5 \sqrt{\beta'_x \beta'_y} \frac{\epsilon_{yx}}{\sqrt{\left(\epsilon_{xx} + \epsilon'_{xx}\right)\left(\epsilon_{yy} + \epsilon'_{yy}\right)}} = N_{12}^*$$

$$N_{22} = \pi f \left(\epsilon_0 \frac{\hat{E}_y}{\hat{H}_x} \vec{e}_y [\epsilon]_r \vec{e}_y + \mu_0 \frac{\hat{H}_x}{\hat{E}_y} \right) = \beta_y \sqrt{1 + \frac{\vec{e}_y [\epsilon]_m \vec{e}_y}{\vec{e}_y [\epsilon]_0 \vec{e}_y}}$$

(4.40)

$$= \beta_y \sqrt{1 + \frac{\epsilon'_{yy}}{\epsilon_{yy}}} = \beta'_y$$

In the above equations, we have already taken into account that both the vector operations $\vec{e}_x [\epsilon]_0 \vec{e}_y$ and $\vec{e}_y [\epsilon]_0 \vec{e}_x$ yield zero.

Next we consider the coefficients N_{11}, N_{12}, N_{21} and N_{22} in more detail. It becomes clear from Eqs. (4.37) to (4.40) that these coefficients are primarily determined by the tensors $[\epsilon]_0$ and $[\epsilon]_m$. Thus, they are determined by fiber material and type of geometrical perturbations. In addition, they are influenced by the frequency f of light. As will be shown below, both propagation constants β_x and β_y also are determined by the fiber characteristics only. Hence, it is useful to evaluate and tabulate N_{11}, N_{12}, N_{21}, and N_{22} for some typical fiber deformations which are usually present in practice; for example, torsion, bending, transverse and axial pressure [189, 234]. The coefficients N_{12} and N_{21} are responsible for mode coupling. If the tensor of dielectric constants $[\epsilon]_r$ is real, then these coefficients are equal and real also. However, if $[\epsilon]_r$ is complex, then both coefficients are also complex and related by $N_{21} = N_{12}^*$.

When both the tensor components ϵ_{xy} and ϵ_{yx} and coefficients N_{12} and N_{21} are zero, mode coupling does not exist between mode x and mode y. Hence, these modes remain the characteristic eigenmodes of the single-mode fiber. Certainly, their propagation constants β_x and β_y and consequently velocities of propagation are changed when the diagonal components ϵ'_{xx}, ϵ'_{yy} and ϵ'_{zz} are changed. Propagation constants β_x and β_y are only related to the first primary fiber deformation, whereas β'_x and β'_y also consider the additional perturbations. From Eqs. (4.37) and (4.40), β'_x and β'_y are equal to the coefficients N_{11} and N_{22} respectively. If N_{12} and N_{21} are zero as considered above, then the coupled DEQ systems can be expressed in terms of two independent equations. In that case, first and second equations are only functions of $\underline{a}(z)$ and $\underline{b}(z)$ respectively. Moreover, the coefficients of coupled DEQ system in Eq. (4.36) and, therefore, the propagation of optical wave is not influenced by the six tensor elements ϵ_{zz}, ϵ'_{zz}, ϵ_{xz}, ϵ_{zx}, ϵ_{yz} and ϵ_{zy}. Therefore, mode x and mode y still propagate in the z-direction and their electric field vectors still contain no z-component. For the propagation constants β_x, β_y, β'_x, and β'_y as well as the ratios of field amplitudes $\hat{E}_x/\hat{H}_y$ and $\hat{E}_y/\hat{H}_x$, following relations can be derived:

$$\beta'_x = 2\pi f \mu_0 \frac{\hat{H}_y}{\hat{E}_x} = 2\pi f \epsilon_0 \vec{e}_x \, [\epsilon]_r \, \vec{e}_x \, \frac{\hat{E}_x}{\hat{H}_y} = 2\pi f \epsilon_0 \left(\epsilon_{xx} + \epsilon'_{xx} \right) \frac{\hat{E}_x}{\hat{H}_y}$$

$$= 2\pi f \sqrt{\mu_0 \, \epsilon_0 \left(\epsilon_{xx} + \epsilon'_{xx} \right)} = \underline{2\pi f \sqrt{\mu_0 \, \epsilon_0 \, \epsilon_{xx}}} \, \sqrt{1 + \epsilon'_{xx}/\epsilon_{xx}}$$
$$\uparrow$$
$$\beta_x$$

(4.41)

$$\beta'_y = 2\pi f \mu_0 \frac{\hat{H}_x}{\hat{E}_y} = 2\pi f \epsilon_0 \, \vec{e}_y \, [\epsilon]_r \, \vec{e}_y \, \frac{\hat{E}_y}{\hat{H}_x} = 2\pi f \epsilon_0 \left(\epsilon_{yy} + \epsilon'_{yy} \right) \frac{\hat{E}_y}{\hat{H}_x}$$

$$= 2\pi f \sqrt{\mu_0 \, \epsilon_0 \left(\epsilon_{yy} + \epsilon'_{yy} \right)} = \underline{2\pi f \sqrt{\mu_0 \, \epsilon_0 \, \epsilon_{yy}}} \, \sqrt{1 + \epsilon'_{yy}/\epsilon_{yy}}$$
$$\uparrow$$
$$\beta_y$$

(4.42)

It becomes clear again from the above equations that the propagation constants β_x and β_y are related to the first fiber deformation only (which is described by $[\epsilon]_0$), whereas the new propagation constants β'_x and β'_y now also take into account all additional fiber perturbations described by the tensor $[\epsilon]_m$. If $[\epsilon]_m = 0$, then $\beta'_x = \beta_x$ and $\beta'_y = \beta_y$.

In order to solve the coupled DEQ system given in Eq. (4.36), exponential formulation

$$
\begin{bmatrix} \underline{a}(z) \\ \underline{b}(z) \end{bmatrix} = C_i e^{-j\beta_i z} \begin{bmatrix} e_{ix} \\ e_{iy} \end{bmatrix} = C_i e^{-j\beta_i z} \, \vec{e}_i \tag{4.43}
$$

can be applied, where C_i is first taken as an arbitrary constant. Later on, this constant will be determined by means of appropriate boundary conditions. Taking now the coupled DEQ system and substituting $\underline{a}(z)$ and $\underline{b}(z)$ from the above expression, two independent solutions of equal significance are obtained. Finally, the general solution is simply obtained by a linear superposition of both the solutions. We get

$$
\begin{bmatrix} \underline{a}(z) \\ \underline{b}(z) \end{bmatrix} = C_1 e^{-j\beta_1 z} \begin{bmatrix} e_{1x} \\ e_{1y} \end{bmatrix} + C_2 e^{-j\beta_2 z} \begin{bmatrix} e_{2x} \\ e_{2y} \end{bmatrix}
$$
$$
= C_1 e^{-j\beta_1 z} \, \vec{e}_1 + C_2 e^{-j\beta_2 z} \, \vec{e}_2 \tag{4.44}
$$

We call constants β_1 and β_2 the *eigenvalues* and unit vectors $\vec{e}_1$ and $\vec{e}_2$ the *eigenvectors* of coupled DEQ system. For the polarization propagation in single-mode fiber, these quantities are most important as they define the four characteristic parameters β_u, β_v, $\vec{e}_u$, and $\vec{e}_v$ of new eigenmodes $\underline{E}_u(z, t) \, \vec{e}_u$ and $\underline{E}_v(z, t) \, \vec{e}_v$ given in Eq. (4.24). By using exponential formulation given in Eq. (4.43), these parameters are given by

$$
\beta_1 = \beta_u = \frac{1}{2} \left[(N_{11} + N_{22}) + \sqrt{(N_{11} - N_{22})^2 + 4 |N_{12}|^2} \right] \tag{4.45}
$$

$$
\beta_2 = \beta_v = \frac{1}{2} \left[(N_{11} + N_{22}) - \sqrt{(N_{11} - N_{22})^2 + 4 |N_{12}|^2} \right] \tag{4.46}
$$

$$
\vec{e}_1 = \vec{e}_u = \frac{1}{\sqrt{1 + \left| \dfrac{\beta_u - N_{11}}{N_{12}} \right|^2}} \begin{bmatrix} 1 \\ \dfrac{\beta_u - N_{11}}{N_{12}} \end{bmatrix} \tag{4.47}
$$

and

$$\vec{e}_2 = \vec{e}_v = \cfrac{1}{\sqrt{1 + \left|\cfrac{\beta_v - N_{11}}{N_{12}}\right|^2}} \begin{bmatrix} 1 \\ \cfrac{\beta_v - N_{11}}{N_{12}} \end{bmatrix} \tag{4.48}$$

Both eigenvectors $\vec{e}_1 = \vec{e}_u$ and $\vec{e}_2 = \vec{e}_v$ are perpendicular to each other and, in addition, normalized i.e., their length equals 1. Mathematically, this relationship can be expressed as

$$\vec{e}_i \cdot \vec{e}_j = \delta_{ij} = \begin{cases} 1 \text{ if } i = j \\ 0 \text{ if } i \neq j \end{cases} \tag{4.49}$$

where $i \in \{1, 2\}$ and $j \in \{1, 2\}$. The symbol δ_{ij} is the Kronecker symbol [28]. Considering both propagation constant β_u and β_v and, in addition, the coefficients N_{11}, N_{12}, N_{21}, and N_{22} of the coupled DEQ system given in (4.36), following ratios can be deduced:

$$\frac{\beta_u - N_{11}}{N_{12}} = \frac{N_{21}}{\beta_u - N_{22}} \quad \text{and} \quad \frac{\beta_v - N_{11}}{N_{12}} = \frac{N_{21}}{\beta_v - N_{22}} \tag{4.50}$$

The most important parameters to assess the stability of polarization propagation in a single-mode fiber are given by the difference

$$\Delta\beta_{uv} = \beta_u - \beta_v = \sqrt{(N_{11} - N_{22})^2 + 4\,|N_{12}|^2} = \sqrt{\Delta\beta'^2 + 4\,|N_{12}|^2} \tag{4.51}$$

between the propagation constants β_u and β_v and beat length

$$L_b = \frac{2\pi}{\Delta\beta_{uv}} \tag{4.52}$$

Difference $\Delta\beta_{uv}$ given in Eq. (4.51) now also includes all fiber perturbations which are responsible for mode coupling. This is in contrast to the difference $\Delta\beta = \Delta\beta_{xy} = \beta_x - \beta_y$ given in Eq. (4.1).

The efficiency of undesired mode coupling in a fiber is primarily determined by the coefficient N_{12}. Again, it should be noted that mode coupling does not occur when N_{12} approaches to zero. It becomes clear from Eq. (4.51) that the influence of mode coupling is less when difference $\Delta\beta'$ in propagation constants is large. Since $\Delta\beta'$ strongly depends on

$\Delta\beta$ (Eqs. 4.37 and 4.40) $\Delta\beta'$ can be increased by increasing $\Delta\beta$. If $\Delta\beta'$ is large in comparison with $2N_{12}$, then the following simple approximation can be used:

$$\Delta\beta_{uv} \approx \Delta\beta' \tag{4.53}$$

In order to prevent or even to reduce undesired mode coupling, the ratio $|\Delta\beta'/N_{12}|$ should be as large as possible. If polarization maintaining single-mode fibers are to be realized, then this fact has to be taken into account in particular. In practice, a high ratio $|\Delta\beta'/N_{12}|$ can be achieved by realizing a very strong desired principal fiber deformation; for example, a highly elliptical fiber core of very strong, but equal fiber torsion. Thereby, this perfectly specified principal deformation will always be dominant in comparison with all other undesired geometrical perturbations along the fiber (Section 4.3).

The principal fiber deformation is a well-defined and desired fiber distortion, whereas all other perturbations are of random in nature and undesired. In this particular case of prime importance, polarization propagation is completely defined and characterized by the principal fiber deformation only. Thus, state of polarization is maintained along the entire fiber length provided only one of the two eigenmodes has been stimulated at the fiber input. When the ratio $|\Delta\beta'/N_{12}|$ is low, second mode is also stimulated within a short distance from the fiber input since mode coupling is not negligible now. As already shown in Figs. 4.3 and 4.4, this results in a continuous change of polarization along the fiber.

In order to make it more clear, Figs. 4.8a to 4.8f illustrate how the optical power flow S in a fiber is divided into the power flows of both the coupled modes: mode x and mode y. For this, power flow is taken as a function of location z and, in addition, as a function of both the coefficients $N_{12}=N_{21}^*$ and N_{22}. Assuming a fiber without any loss, resulting optical power flow $S=S_x(z)+S_y(z)$ always remains constant independent of location z. Here, $S_x(z)=|\underline{a}(z)|^2 \cdot S_{x,max}$ and $S_y(z)=|\underline{b}(z)|^2 \cdot S_{y,max}$ represents the power flow of mode x and mode y respectively.

The Fig. 4.8 provides that coefficient N_{22} is constant. Hence, this coefficient has been used as a normalization factor. This figure also provides that mode x and mode y are equally stimulated at the fiber input. Thus, $S_x(0)=S_y(0)=S/2$. As already mentioned, coefficient N_{12} is responsible for mode coupling since no mode coupling occurs when N_{12} equals zero. On the other hand, as N_{12} increases, influence of mode coupling also increases.

Considering Figs. 4.8b, 4.8d and 4.8f (left side), coefficient N_{12} is three times larger than that of in Figs. 4.8a, 4.8c and 4.8e (right side). Thus, the influence of undesired mode coupling is much more serious in the figures on left side.

We first focus our discussion on Figs. 4.8a and 4.8b where $N_{22}=N_{11}$. The ratio $|\Delta\beta'/N_{12}|$ which should be as large as possible to realize polarization maintaining fibers is assumed to be zero. Therefore, mode coupling is most efficient in Figs. 4.8a and 4.8b. It becomes evident that the energy of mode x and mode y is completely interchanged twice within the beat length L_b. Thereby, in distances of $z=nL_b+L_b/4$ (where n=0, 1, 2, 3, $\cdots$) from the fiber input (i.e., $z=0$), power flow of mode y reaches its maximum value $S_y(z)=S$, while the power flow $S_x(z)=0$ of mode x approaches to zero.

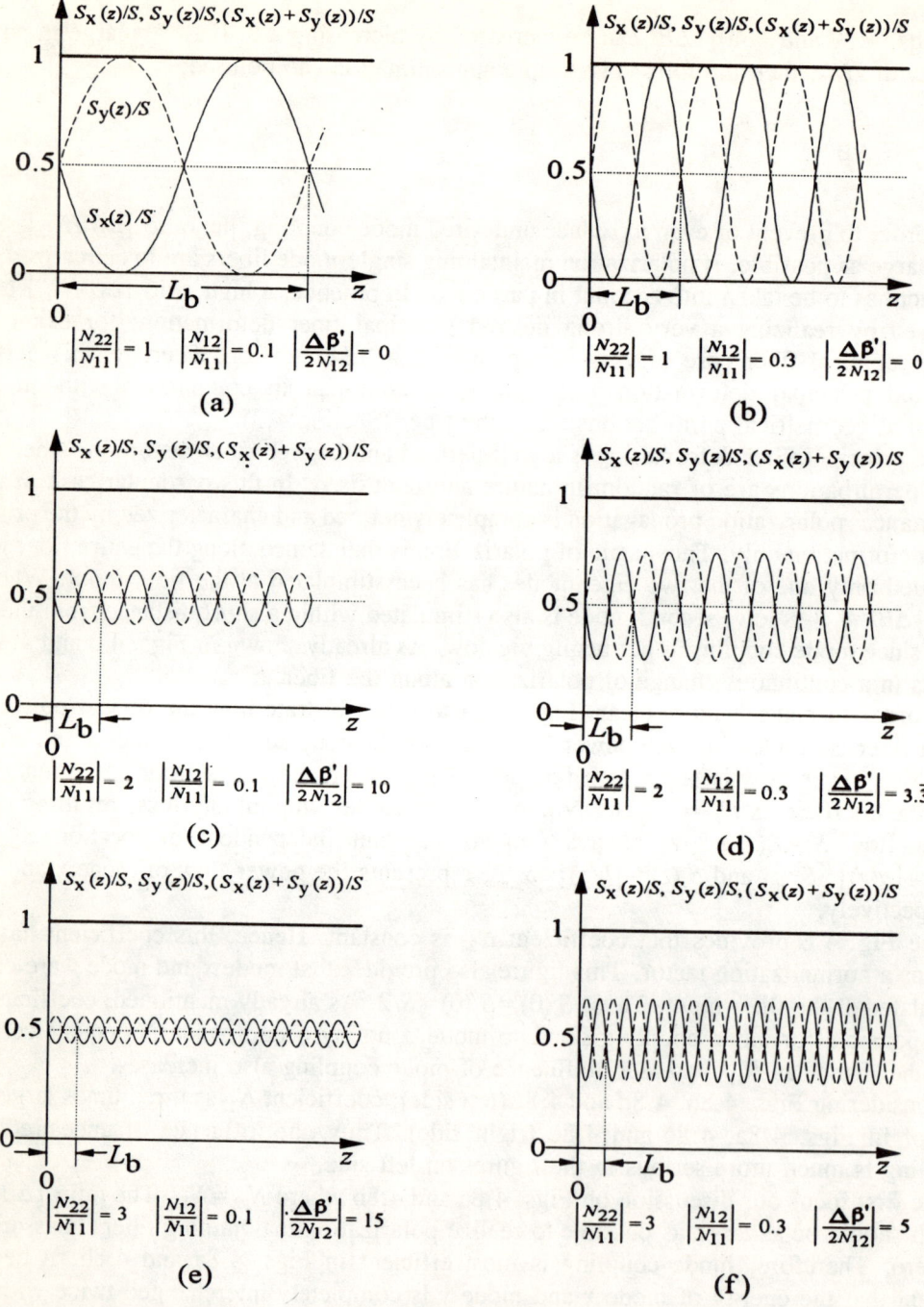

Fig. 4.8: Division of constant power flow $S = S_x(z) + S_y(z)$ in both coupled and orthogonal modes (polarizations) of single-mode fiber

In contrast to this, power flow of mode x reaches maximum and power flow of mode y is zero at locations $z = nL_b + 3L_b/4$. At distances of $z = nL_b/2$, power distribution always equals the distribution at the fiber input. It becomes clear from Figs. 4.8a and 4.8b that the distance of complete energy interchange decreases if the mode coupling coefficient N_{12} increases. It should be noted that the period of a complete energy interchange is determined by the beat length $L_b = 2\pi/\Delta\beta_{uv}$ given in Eq. (4.52).

Next, N_{11} and N_{22} are considered to be unequal in Figs. 4.8c to 4.8f. In that case, $\Delta\beta'$ and, consequently, the ratio $|\Delta\beta'/N_{12}|$ are no longer zero. As a result, influence of mode coupling is less efficient as compared to Figs. 4.8a and 4.8b, where $\Delta\beta'$ and $|\Delta\beta'/N_{12}|$ are taken to be zero. In contrast to Figs. 4.8a and 4.8b, energy does not interchange completely. Instead, energy of mode x and mode y is interchanged only partially. Therefore, power flow of mode x and mode y neither reaches maximum S nor minimum zero provided that both the modes have been stimulated equally at the fiber input as considered above (compare with [223]).

The mutual energy interchange between the coupled modes x and y decreases with the increase in the ratio $|\Delta\beta'/N_{12}|$. Thus, lesser the mode coupling coefficient N_{12}, lower the amplitudes of both the periodical power flows $S_x(z)$ and $S_y(z)$. Comparing Figs. 4.8d and 4.8f with Figs. 4.8c and 4.8e, this important fact becomes quite clear. On the other hand, mutual energy interchange decreases as the difference $\Delta\beta'$ (or $\Delta\beta$) increases. This becomes clear by comparing Figs. 4.8e and 4.8f with Figs. 4.8a and 4.8b.

The lowest energy interchange can be observed in Fig. 4.8e where the ratio $|\Delta\beta'/N_{12}|$ reaches its maximum value. Here, both the modes are only coupled very weakly and their power flows which are close to the average $S/2$ are only changed insignificantly. This means that mode coupling is practically negligible in Fig. 4.8e.

In the ideal case $|\Delta\beta'/N_{12}| \to \infty$ or $|\Delta\beta/N_{12}| \to \infty$ which can never be achieved in practice, no mode coupling exists and both the power flows remain constant. When both the modes are equally stimulated at the fiber input, this constant is given by the average $S/2$. If only one single mode is exactly stimulated, for example, mode x, then this mode retains the total and maximum power flow $S_x(z) = S_{x,max} = S$ along the entire fiber length. Thereby, mode y remains unstimulated i.e., $S_y(z) = 0$.

Now, both the terms $\underline{a}(z)$ and $\underline{b}(z)$ are completely determined and we are able to describe any energy interchange of coupled modes in a single-mode fiber. In the following, we have to come back again to the calculation of characteristic eigenmodes of single-mode fiber. For this purpose, we first consider Eq. (4.21) again where the substitution of the terms $\underline{a}(z)$ and $\underline{b}(z)$ must be made. In the second step, we shall compare this field equation with the eigenmode formulation given in Eq. (4.24). This will give us a generally valid solution for determining the optical field as well as characteristic eigenmodes of single-mode fiber. After discussing the generally valid solutions (see below), some special cases of practical importance will be used to explain the general results of this Section in more detail.

(i) GENERAL SOLUTION

In order to obtain a general solution, we have to combine solution given in Eq. (4.43) for coupled DEQ system (Eq. 4.36) and field Eq. (4.21) which describes the electric field vector of an optical fiber wave. By applying some straightforward mathematical operations, we obtain

$$\underline{\vec{E}}(z,\ t) = C_1 \hat{E}_x\ e^{-j\beta_1 z}\ e^{j2\pi ft}\ \vec{e}_1\ +\ C_2 \hat{E}_y\ e^{-j\beta_2 z}\ e^{j2\pi ft}\ \vec{e}_2$$

$$= \hat{E}_u\ e^{-j\beta_u z}\ e^{j2\pi ft}\ \vec{e}_u\ \ \ +\ \hat{E}_v\ e^{-j\beta_v z}\ e^{j2\pi ft}\ \vec{e}_v \qquad (4.54)$$

$$= \underline{E_u(z,\ t)\ \vec{e}_u}\ \ \ \ \ \ \ \ \ \ \ +\ \underline{E_v(z,\ t)\ \vec{e}_v}$$

$$\uparrow \uparrow$$

eigenmode u $\leftarrow$ **independent** $\rightarrow$ **eigenmode** v

which is exactly related to the eigenmode formulation given in Eq. (4.24). From the above equation, it becomes immediately clear that both the eigenvalues β_1 and β_2 and eigenvectors $\vec{e}_1$ and $\vec{e}_2$ of the coupled DEQ system are absolutely identical to the related characteristic quantities β_u, β_v, $\vec{e}_u$ and $\vec{e}_v$ of both the new eigenmodes (compare Eq. 4.24). As seen in Eq. (4.54), constants C_1 and C_2 which are still undefined have been combined with the electric field amplitudes $\hat{E}_x$ and $\hat{E}_y$. As a result, the amplitudes $\hat{E}_u$ and $\hat{E}_v$ have to be determined now by appropriate boundary conditions instead of constants C_1 and C_2. A typical and frequently used boundary condition is given by a fixed and well-determined light wave at the input to the fiber at $z=0$.

The amplitudes of the electric (Eq. 4.54) and magnetic fields (Eq. 4.25) are related by the following equation:

$$\frac{\hat{E}_u}{\hat{H}_v} = \frac{2\pi f\mu_0}{\beta_u} \qquad \qquad \frac{\hat{E}_v}{\hat{H}_u} = \frac{2\pi f\mu_0}{\beta_v} \qquad (4.55)$$

As seen from Eq. (4.54), it is always possible, in principle, to find two eigenmodes which propagate independently without any mode coupling either along the entire fiber or even along a small piece of fiber. The characteristic quantities β_u, β_v, $\vec{e}_u$ and $\vec{e}_v$ of both the eigenmodes are completely determined by the fiber material, type of fiber deformation and frequency of light as described in Eqs. (4.45) to (4.48). Remembering again our question in the beginning of this Section whether characteristic eigenmodes are existing in a geometrically perturbed single-mode fiber or not, this question can now be answered by yes.

It has been assumed so far that all four coefficients N_{11}, N_{12}, N_{21} and N_{22} of the coupled DEQ system (Eq. 4.36) are real. In such a case, eigenvectors $\vec{e}_u$ and $\vec{e}_v$ of the new eigenmodes (i.e., eigenmode u and eigenmode v) are real also. It can be simply proved from Eqs. (4.47) and (4.49). Thus, no phase difference exists between the vector components of each eigenvector. For this reason, eigenmodes of the optical fiber wave are called *linearly polarized eigenmodes* since the electric field vector $\underline{E}_u(z,\,t)\vec{e}_u$ of eigenmode u is always moving on a *linear locus curve* which is same as the u-axis of rectangular uv-coordinate system at a certain location z. Similarly, the electric field vector $\underline{E}_v(z,\,t)\vec{e}_v$ of eigenmode v is also moving on a linear locus curve which is, however, perpendicular to the locus curve of $\underline{E}_u(z,\,t)\vec{e}_u$. Thus, the locus curve of $\underline{E}_v(z,\,t)\vec{e}_v$ is same as the v-axis of the coordinate system as shown in Fig. 4.9 below.

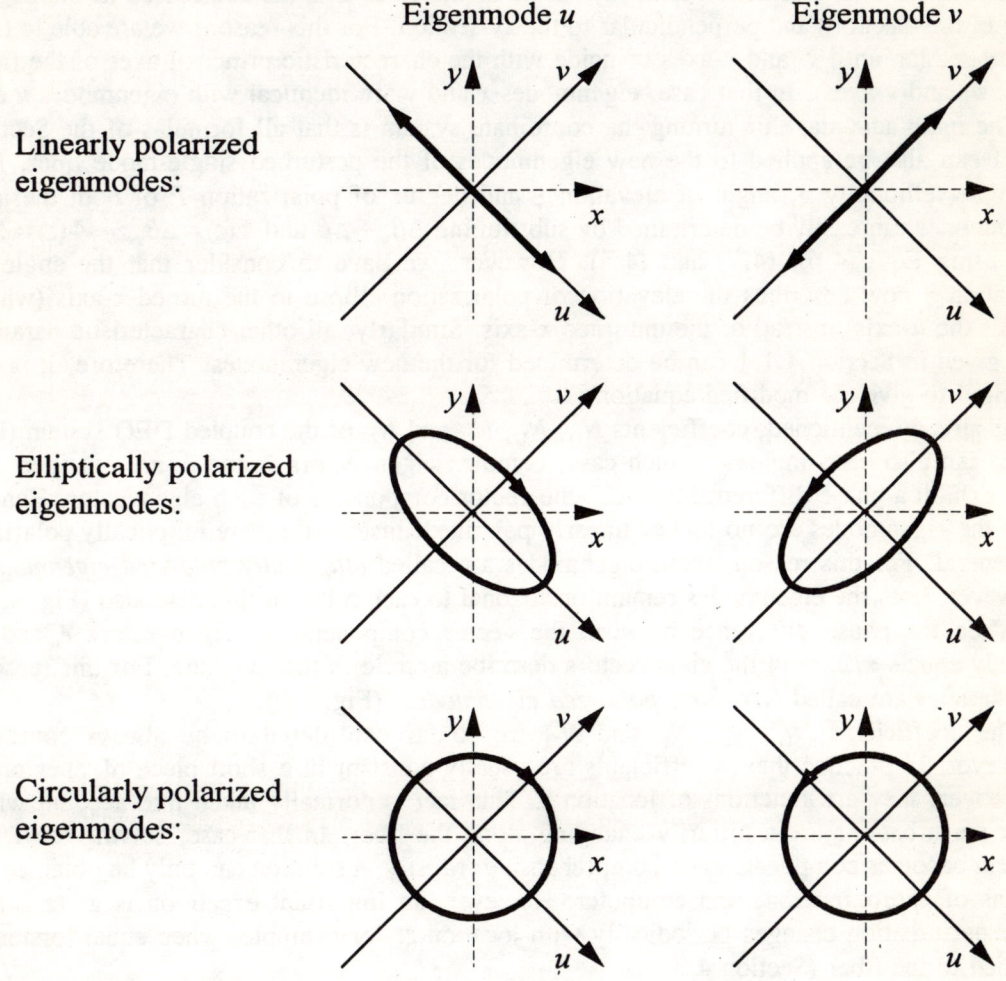

Fig. 4.9: Possible polarizations of both the orthogonal eigenmodes of single-mode fiber

It is, of course, not really necessary that a superposition of two linearly polarized eigenmodes yields again a linearly polarized wave. Because a difference in phase $\Phi(z) = \Delta\beta_{uv}z$ normally exists between the complex field vectors $\underline{E}_u(z,\,t)\vec{e}_u$ and $\underline{E}_v(z,\,t)\vec{e}_v$, polarization of superimposed optical wave usually varies with location z. This typical behaviour of single-mode fibers has already been discussed in Section 4.1.1 (Figs. 4.3 and 4.4). There, the linearly polarized eigenmodes are same as the eigenmode x and eigenmode y and the related locus curves of the perpendicular field vectors are identical to the x- and y-axes of the xy-coordinate system.

It should be noted that the orientation of the rectangular xy-coordinate system which is always located in the cross-section of the fiber core (xy-section), can be chosen arbitrarily. In contrast to this, z-direction is always same as the fiber axis (as considered in the beginning of this Section) and perpendicular to the xy-section. For this reason, we are able to turn the xy-section until x- and y-axes coincide with the characteristic principal axes of the fiber (i.e., u- and v-axes). In that case, eigenmodes x and y are identical with eigenmodes u and v. The main advantage of turning the coordinate system is that all formulas of the Section 4.1.1 can also be applied to the new eigenmodes of the perturbed single-mode fiber. For example, ellipticity η, angle of elevation ξ and degree of polarization P of both the new eigenmodes can easily be determined by substituting $\Delta\beta_{uv} \rightarrow \Delta\beta$ and $\Phi(z) = \Delta\beta_{uv}z \rightarrow \Phi(z) = \Delta\beta z$ and using Eqs. (4.6), (4.7) and (4.8). However, we have to consider that the angle of elevation ξ now describes the elevation of polarization ellipse to the turned x-axis (which equals the u-axis) instead of the unturned x-axis. Similarly, all other characteristic parameters given in Section 4.1.1 can be determined for the new eigenmodes. Therefore, it is not required to give the modified equations.

As already mentioned, coefficients N_{11}, N_{12}, N_{21} and N_{22} of the coupled DEQ system (Eq. 4.36) can also be complex. In such case, complex eigenvectors $\vec{e}_u$ and $\vec{e}_v$ are obtained and they exhibit a phase difference between the vector components of each eigenvector. Hence, both the eigenmodes are no longer linearly polarized, instead they are elliptically polarized in general. For this reason, these eigenmodes are called *elliptically polarized eigenmodes*. However, both the eigenmodes remain orthogonal to each other in this case also (Fig. 4.9).

When the phase difference between the vector components of eigenvectors $\vec{e}_u$ and $\vec{e}_v$ exactly equals $\pi/2$, both the eigenvectors describe a circle in the uv-plane. For this reason, eigenmodes are called *circularly polarized eigenmodes*. (Fig. 4.9).

The coefficients N_{11}, N_{12}, N_{21} and N_{22} are so far considered to be always constant. However, in practice these coefficients are usually constant in a short piece of fiber only. Moreover, they are functions of location z. This fact is normally taken into account when fiber perturbations are arbitrarily changing along the fiber. In that case, solving the DEQ system becomes complicated and comprehensive. Mostly, a solution can only be obtained by means of approximations and computer. However, an important exception is given when fiber perturbation changes periodically with location z; for example, when equal torsion is applied to the fiber (Section 4.3).

(ii) FIBER WITHOUT MODE COUPLING ($N_{12} = 0$)

Consider an ideal fiber without any mode coupling. In that case, all the coefficients of tensor $[\epsilon]_m$ are zero except the diagonal components ϵ'_{xx}, ϵ'_{yy}, and ϵ'_{zz}. As a result, mode coupling coefficients N_{12} and N_{21} of the coupled DEQ system (Eq. 3.36) are also zero. Thus, the DEQ system can be split into two independent uncoupled differential equations which can easily be solved by applying a simple exponential formulation. The solution obtained is the following:

$$\underline{a}(z) = C_1 \, e^{-j\beta_1 z} \, \vec{e}_1 \tag{4.56}$$

$$\underline{b}(z) = C_2 \, e^{-j\beta_2 z} \, \vec{e}_2 \tag{4.57}$$

This solution is a special case of the general solution given in Eq. (4.44). Hence, eigenvalues and eigenvectors of the DEQ system can again be determined by Eqs. (4.45) to (4.48) above. With $N_{12} = 0$, these equations get simplified to

$$\beta_1 = \beta_u = N_{11} = \beta_x' \tag{4.58}$$

$$\beta_2 = \beta_v = N_{22} = \beta_y' \tag{4.59}$$

$$\vec{e}_1 = \vec{e}_u = \vec{e}_x \tag{4.60}$$

$$\vec{e}_2 = \vec{e}_v = \vec{e}_y \tag{4.61}$$

Next we have to take into account the field Eq. (4.21). Making substitutions for $\underline{a}(z)$, $\underline{b}(z)$ and both the eigenvalues and eigenvectors as given by Eqs. (4.56) to (4.61), we get

$$\vec{E}(z, \, t) = \underline{\hat{E}_x \, e^{-j\beta_x' z} \, e^{j2\pi ft} \, \vec{e}_x} \quad + \quad \underline{\hat{E}_y \, e^{-j\beta_y' z} \, e^{j2\pi ft} \, \vec{e}_y} \tag{4.62}$$

$$\uparrow \qquad\qquad\qquad\qquad\qquad \uparrow$$

eigenmode u ← independent → eigenmode v
= eigenmode x = eigenmode y

With the exception of new propagation constants β'_x and β'_y, above solution exactly yields the field Eq. (4.21). Thus, both the characteristic eigenmodes remain same. As the mode coupling does not disturb the optical fiber wave, change of the principal axes does not occur (Eqs. 4.60 and 4.61). However, the propagation constants and, so, the velocities of propagation are changed ($\beta_x \rightarrow \beta'_x$ and $\beta_y \rightarrow \beta'_y$) due to various geometrical fiber perturbations which are analytically defined by the diagonal tensor components ϵ'_{xx}, ϵ'_{yy}, and ϵ'_{zz}. It should be remembered that the above solution is a special case of general solution. This particular case is, however, not observed in practice since geometrical perturbations normally do not only influence the diagonal components of $[\epsilon]_m$. Instead, the other tensor components are also influenced which result in mode coupling.

(iii) FIBER WITHOUT PERTURBATIONS

In this Section, fiber without perturbations is assumed to be a fiber which is deformed by one well-defined and desired deformation only; for example, a highly elliptical fiber core. No additional undesired geometrical perturbations exist in the fiber. Thus, the term "without perturbations" mathematically means that all the nine components of tensor $[\epsilon]_m$ are zero, while $[\epsilon]_0$ describes the elliptical fiber core (Eq. 4.33). As a result, no mode coupling and also no change in velocity of propagation of eigenmodes are there. Hence, both the modes given in Eq. (4.21) always remain the eigenmodes along the entire fiber length. The propagation constants β_x and β_y are determined by the diagonal components of tensor $[\epsilon]_0$ and frequency of light from Eqs. (4.41) and (4.42). Therefore, the electric field from Eq. (4.21) is given by

$$\vec{E}(z,\ t) = \underline{\hat{E}_x\ e^{-j\beta_x z}\ e^{j2\pi ft}\ \vec{e}_x} \quad + \quad \underline{\hat{E}_y\ e^{-j\beta_y z}\ e^{j2\pi ft}\ \vec{e}_y} \tag{4.63}$$

$$\uparrow \qquad\qquad\qquad \uparrow$$

$$\text{eigenmode } x \ \leftarrow \ \textbf{independent} \ \rightarrow \ \text{eigenmode } y$$

(iv) STIMULATION OF SINGLE EIGENMODE

Let us now assume that additional geometrical imperfections impair the polarization propagation of single-mode fiber. In that case of practical importance, it is again possible to determine two new eigenmodes propagating independently without any undesired mode coupling. As explained earlier, these new eigenmodes (i.e., eigenmodes u and v) are completely and unambiguously determined by the eigenvalues and eigenvectors of the coupled DEQ system as given in Eq. (4.36).

Assuming further that only one eigenmode is stimulated at the fiber input (for example, eigenmode u), then only this single mode propagates along the entire fiber. In contrast, the second eigenmode always remains unstimulated. Thus, the electric field is completely determined by the field of eigenmode u. If eigenmode v is stimulated, then the electric field of the optical fiber wave becomes same as the field of this mode. In conclusion, we obtain

$$\vec{E}(z,\ t) = \begin{cases} \hat{E}_u\ e^{-j\beta_u z}\ e^{j2\pi ft}\ \vec{e}_u & \text{if eigenmode } u \text{ is stimulated} \\ \hat{E}_v\ e^{-j\beta_v z}\ e^{j2\pi ft}\ \vec{e}_v & \text{if eigenmode } v \text{ is stimulated} \end{cases} \tag{4.64}$$

As seen from above equation, polarization of an optical fiber wave is always same as the polarization of the eigenmode which has been stimulated at the fiber input. Since this polarization does not alter with location z, state of polarization remains unchanged along the entire fiber length. However, this special type of polarization maintaining fiber first requires that all fiber deformations are known, second they are completely described by both the tensors $[\epsilon]_0$ and $[\epsilon]_m$ and third additional fiber perturbations do not occur.

When the fiber perturbations change randomly with time, principal axes $\vec{e}_u$ and $\vec{e}_v$ of both the eigenmodes and, consequently, the fiber polarization also changes randomly with time. Since this is generally true in practice, commercially available single-mode fibers are normally not suitable to maintain polarization.

In order to realize *polarization maintaining fibers*, this particular problem must be solved. If a fiber is extremely disturbed by a fixed and perfectly specified deformation, for example, a highly-elliptical fiber core or a very strong, but equal torsion (Section 4.3), then all the other random perturbations are practically negligible in comparison with this dominant fiber deformation. In such a case, polarization propagation is completely characterized and determined by the dominant fiber deformation only. Hence, state of polarization always remains maintained provided only one eigenmode is stimulated at the fiber input.

4.2 MATRIX OF POLARIZATION PROPAGATION

In this Section, boundary conditions which are usually determined by the optical wave at the input to the fiber (i.e., $z=0$), will be considered in more detail. For this purpose, stimulating optical wave

$$\vec{E}(0,\ t) = \begin{pmatrix} E_x(0,\ t) \\ E_y(0,\ t) \end{pmatrix} = \begin{pmatrix} E_{x0} \\ E_{y0} \end{pmatrix} e^{j2\pi ft} \tag{4.65}$$

will be assumed at the fiber input. Substituting this equation in the general solution given in Eq. (4.54) for the electric field of an optical fiber wave, following result is obtained after some mathematical operations [60]:

$$\vec{E}(z,\ t) = \begin{pmatrix} E_x(z\ t) \\ E_y(z,\ t) \end{pmatrix} = \begin{pmatrix} m_{11} & m_{12} \\ m_{21} & m_{22} \end{pmatrix} \begin{pmatrix} E_{x0} \\ E_{y0} \end{pmatrix} e^{-j0.5(N_{11}+N_{22})z}\ e^{j2\pi ft} \tag{4.66}$$
$$\uparrow$$
$$(m_{ij})$$

The matrix (m_{ij}) is usually termed as the *polarization propagation matrix*. This matrix is defined by four coefficients which can be calculated by substituting Eq. (4.65) in Eq. (4.54). We obtain

$$m_{11} = \cos(0.5\Delta\beta_{uv}z) - j\frac{\Delta\beta_{uv}}{\Delta\beta'}\sin(0.5\Delta\beta_{uv}z) = m_{22}^* \tag{4.67}$$

$$m_{12} = j\frac{2N_{12}}{\Delta\beta_{uv}} \sin(0.5\Delta\beta_{uv}z) = -m_{21}^* \qquad (4.68)$$

By taking into account matrix (m_{ij}), we can determine the electric field of an optical fiber wave at any location z as a function of any stimulating optical wave at the fiber input ($z=0$). Subsequently, all its characteristic parameters can be determined by applying the appropriate formulas given in Section 4.1.1; for example, polarization ellipse, ellipticity, angle of elevation and degree of polarization.

With the help of matrix (m_{ij}), each single-mode fiber can be represented as a simple two-port network characterized by the four matrix coefficients m_{11}, m_{12}, m_{21}, and m_{22} and complex factor $\exp[-j0.5(N_{11}+N_{22})z]$. All kinds of geometrical imperfections are included in this network representation .(Fig. 4.10).

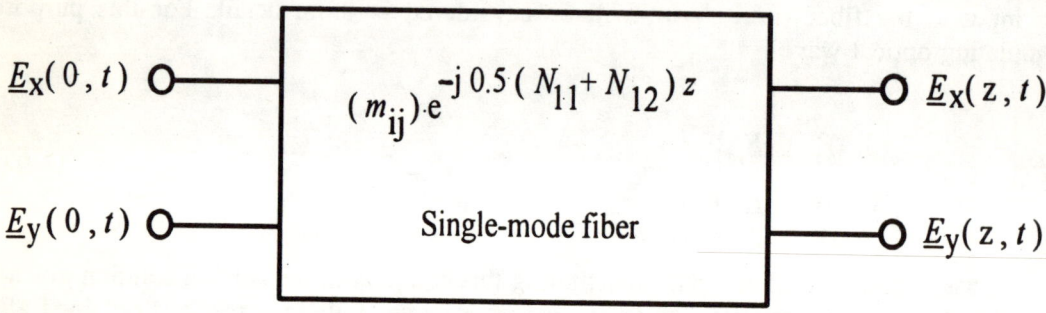

Fig. 4.10: Two-port network representation of single-mode fiber

4.3   REDUCTION OF POLARIZATION FLUCTUATIONS

4.3.1   POLARIZATION MAINTAINING SINGLE-MODE FIBERS

As explained in the previous Sections, two independent fundamental reasons for polarization fluctuations in a single-mode fiber are

- different velocities of propagation of both the orthogonal eigenmodes (Section 4.1.1) and

- mode coupling (Section 4.1.2).

When only one of the two eigenmodes is excited at the fiber input, a stable transmission of polarization is possible in principle. In presence of mode coupling during transmission, second eigenmode is also excited. As a result, state of polarization of the optical fiber wave changes as a function of location z. Typical variations of polarization have already been shown in Figs. 4.3 and 4.4.

Further, we have discussed in Section 4.1.2 that always two independent and orthogonal eigenmodes are existing, even if transmission is disturbed by mode coupling. If only one of these "new" eigenmodes is stimulated at the fiber input, then a polarization maintaining transmission is possible again. However, all geometrical fiber perturbations and deformations must be absolutely time-invariant. Otherwise, the principal axes of the eigenmodes are temporally changed and polarization fluctuations occur again since both the eigenmodes are now stimulated in the fiber. Because this is normally true in practice, ordinary single-mode fibers are not suitable to preserve state of polarization.

For this reason, the goal of realizing polarization maintaining fibers, frequently called single-mode single-polarization fibers or briefly SMSP fibers, must be twofold: First, all undesired perturbations along the fiber must be minimized. This means that the mode coupling coefficient N_{12} has to be decreased as far as possible. Second, the difference $\Delta\beta$ in propagation constant of both the eigenmodes given in Eq. (4.1) or (4.51) has to be maximized [208]. Maximization of $\Delta\beta$ or minimization of the beat length L_b (i.e, maximization of birefringence) can practically be achieved by applying a perfectly specified dominant fiber deformation, for example a highly elliptical fiber core or a strong and equal fiber torsion. This desired geometrical fiber deformation is mathematically well-defined and described by the tensor $[\epsilon]_0$ of dielectric constants given in Eq. (4.34). In comparison to all other undesired fiber perturbations which normally change with location and time randomly, desired deformation should be absolutely dominant and, in addition, independent of location and time. Only in that particular case, all undesired and random fiber perturbations are negligible. As a result, eigenmodes and polarization propagation are completely characterized by the dominant fiber deformation only. Hence, state of polarization is absolutely maintained provided only one eigenmode is excited at the fiber input. Since maximization of birefringence is the basis of most polarization maintaining fibers, these fibers are frequently called *birefringent fibers*.

Depending on the type of realized desired deformation, different eigenvalues and eigenvectors can be obtained (Eqs. 4.45 to 4.48). As a consequence, eigenmodes may exhibit different kinds of polarization. For this reason, polarization maintaining fibers can be classified as follows:

- single-mode fibers with linearly polarized eigenmodes due to
  - an axial asymmetrical fiber core (e.g., elliptical-core fibers) or
  - an axial asymmetrical pressure on the fiber (e.g., stress-induced fibers),

- single-mode fibers with circularly polarized eigenmodes due to equal torsion of fiber and

- absolutely polarization maintaining single-mode fibers.

In the following, above fibers will be discussed in more detail.

(i) SINGLE-MODE FIBERS WITH LINEARLY POLARIZED EIGENMODES

These fibers are also called linear highly birefringent fibers since the difference $\Delta\beta$ in the propagation constants and, therefore, the birefringence is extremely large. Linear birefringent fibers are characterized by a well-defined deformation of the fiber core; for example, a highly elliptical fiber core as already mentioned. Fibers with a non-circular core have been first studied in 1978 [221]. Further examinations were primarily focused on elliptical fiber cores [2, 47, 157, 265]. An important and common theoretical result of all these studies was that the beat length L_b which has to be minimized is inversely proportional to $(\Delta n)^2$ where

$$\Delta n = \frac{\left(n_1^2 - n_2^2\right)}{2n_1^2} \tag{4.69}$$

represents the relative refraction index change between fiber core (n_1) and cladding (n_2). Thus, a short beat length L_b requires a large index change Δn. On the other hand, a large Δn is always combined with two significant drawbacks: first, fiber attenuation increases and second core diameter to provide single-mode operation decreases. Practically, beat lengths in the order of 1 mm can be achieved by applying elliptical fiber cores. In comparison, commercially available single-mode fibers without polarization maintaining mechanism exhibit a beat length in the range of some 10 cm to 2 m. This range is also valid for all kinds of undesired random fiber perturbations which are always existing in addition to the desired elliptical fiber core. Taking into account a first order approximation, a characteristic ratio $|\Delta\beta/(2\ N_{12})|$ in the range of 100 to 1000 can be achieved in the polarization maintaining fibers with elliptical fiber core (Fig. 4.8).

An improvement, especially in the fiber attenuation can be achieved when the fiber is deformed by an axial asymmetrical pressure [107, 121, 124, 222, 242, 249, 261]. Practically, this transverse stress can easily be realized by different temperature coefficients of expansion of fiber core and cladding. Thereby, beat lengths similar to elliptical core fibers can be obtained. However, fiber attenuation is much less as compared to the earlier case.

All single-mode linearly birefringent fibers show a common disadvantage of prime importance. If two fibers are to be connected, then the principal axes of both eigenmodes in the fibers must be exactly on top of each other. Otherwise, both the eigenmodes will be excited in the connected fiber and polarization fluctuations will occur again.

(ii) SINGLE-MODE FIBERS WITH CIRCULARLY POLARIZED EIGENMODES

An essential improvement in the polarization matching of coupled fibers can be achieved by using a circularly birefringent fiber which will be examined now in more detail [17, 118, 159, 173, 273]. Circularly birefringent fibers are realized by an equal torsion of the fiber about its axis as shown in Fig. 4.11. To ensure that the fiber torsion is the dominant deformation along the fiber length, degree of torsion should be sufficiently large. Only in that case, all other undesired and random perturbations are negligibly small and polarization transmission is completely characterized by the fiber torsion only. For an ideal circular single-mode fiber with uniform torsion, coefficients N_{11}, N_{12}, N_{21} and N_{22} of the coupled DEQ system given in Eq. (4.36) are determined by [234]

$$N_{11} = N_{22} = \beta_0 = 2\pi f \sqrt{\mu_0 \epsilon_0} \tag{4.70a}$$

$$N_{12} = -jc\gamma = N_{21}^* \tag{4.70b}$$

Here, β_0 represents the propagation constant for an ideal circular single-mode fiber without any torsion. Twist rate γ given in unit of rad/m and constant c (e.g., $c \approx 0.07$ [225, 273]) describe the degree of torsion for a given fiber length.

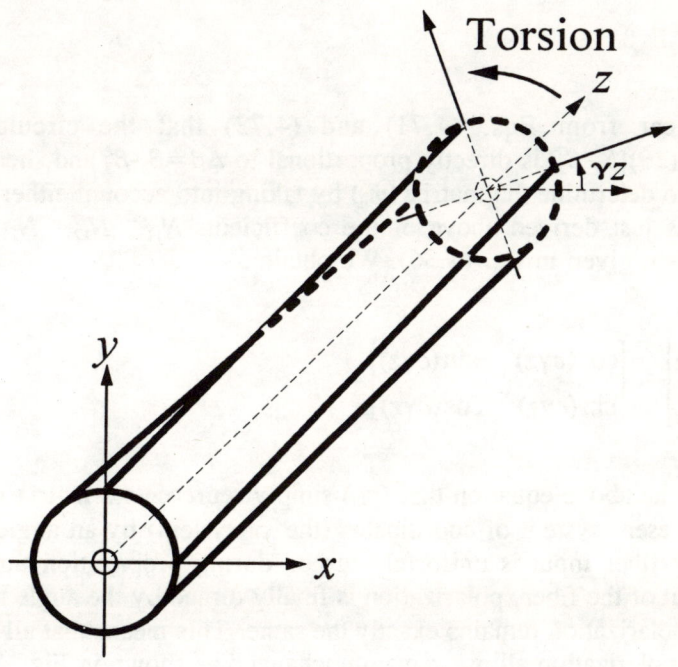

Fig. 4.11: Torsion of ideal circular single-mode fiber

The coefficients N_{12} and N_{21} which are responsible for mode coupling are now not real, rather they are imaginary. This important fact results in a phase difference of exactly $\pi/2$ between the vector components (x- and y-components) of both eigenvectors $\vec{e}_u$ and $\vec{e}_v$ (see below). Therefore, both eigenmodes become circularly polarized as demanded above.

In order to determine eigenvalues and eigenvectors using Eqs. (4.45) to (4.48), four coefficients given in Eq. (4.70) are substituted. We get

$$\beta_1 = \beta_u = \beta_0 + c\gamma \tag{4.71}$$

$$\beta_2 = \beta_v = \beta_0 - c\gamma \tag{4.72}$$

$$\vec{e}_1 = \vec{e}_u = \frac{1}{\sqrt{2}} \begin{bmatrix} 1 \\ +j \end{bmatrix} \tag{4.73}$$

and

$$\vec{e}_2 = \vec{e}_v = \frac{1}{\sqrt{2}} \begin{bmatrix} 1 \\ -j \end{bmatrix} \tag{4.74}$$

It becomes clear from Eqs. (4.71) and (4.72) that the circular birefringence $D = \lambda/(2\pi)\Delta\beta = \lambda/(2\pi)[\beta_1 - \beta_2]$ is directly proportional to $\Delta\beta = \beta_1 - \beta_2$ and, hence, twist rate γ. Finally, we have to determine the matrix (m_{ij}) by taking into account either the eigenvectors and eigenvalues as just derived above or the coefficients N_{11}, N_{12}, N_{21} and N_{22} of the coupled DEQ-system given in Eq. (4.36). We obtain

$$(m_{ij}) = \begin{bmatrix} m_{11} & m_{12} \\ m_{21} & m_{22} \end{bmatrix} = \begin{bmatrix} \cos(c\gamma z) & -\sin(c\gamma z) \\ \sin(c\gamma z) & \cos(c\gamma z) \end{bmatrix} \tag{4.75}$$

It is evident from the above equation that (m_{ij}) simply represents a matrix of transformation which turns the present system of coordinates (the xy-system) by an angle $c\gamma z$. Therefore, polarization at the fiber input is uniformly turned during propagation through the optical fiber. At the output of the fiber, polarization is finally turned by the angle mentioned above. Thereby, type of polarization remains exactly the same. This means that all the characteristic parameters of the polarization ellipse remain unchanged as shown in Fig. 4.12 i.e., η and P are not a function of z.

The main advantage of a polarization maintaining fiber with circularly polarized eigenmodes is given by its simple technique of connection. In contrast to a fiber with linearly polarized eigenmodes, no adjustment of the principal axes is required [170]. High stability of polarization transmission is second advantage since the influence of undesired geometrical fiber perturbations is much lower than that of linearly polarized eigenmodes. Moreover, realization of fibers with circularly polarized eigenmodes can be achieved more easily.

It is obvious that fiber torsion is practically limited to a certain degree. Hence, all additional fiber perturbations cannot usually be neglected. If, for example, a fiber has a highly elliptical core in addition, then eigenmodes are to be determined afresh. However, each additional geometrical fiber distortion makes calculation more and more comprehensive [234 - 238].

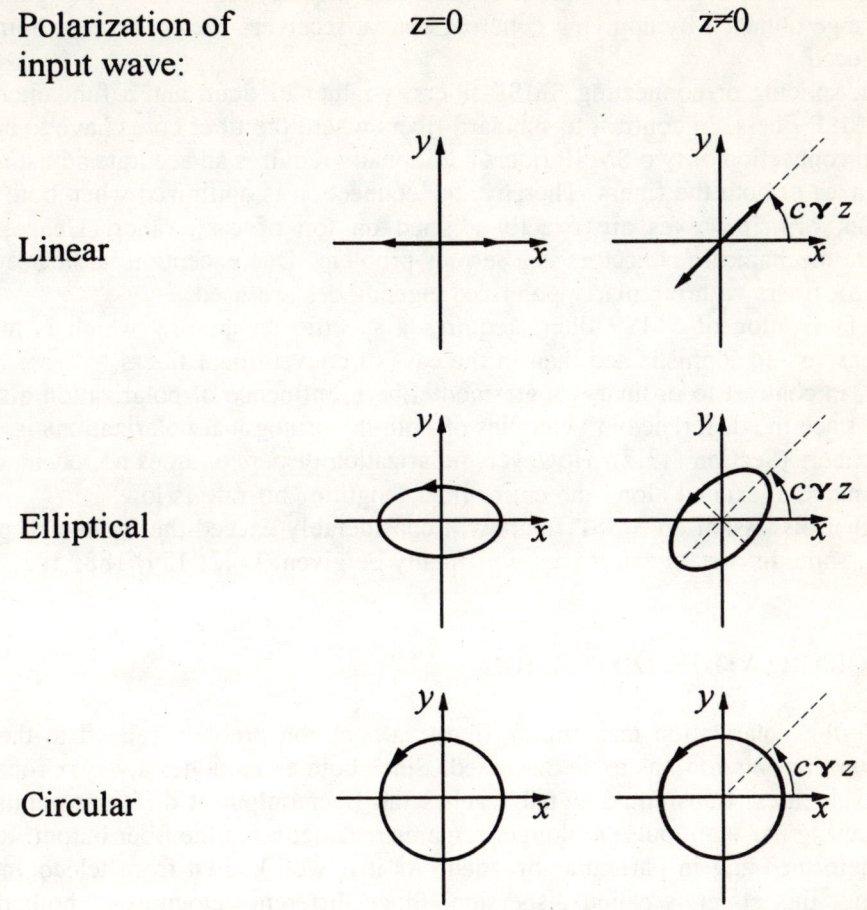

Fig. 4.12: Transmission of polarization along single-mode fiber with circularly polarized eigenmodes

(iii) ABSOLUTELY POLARIZATION MAINTAINING SINGLE-MODE FIBERS

In contrast to the polarization maintaining single-mode fibers discussed above, which actually have two modes, this type of fiber really exhibits one single mode only. Here, the second mode is completely suppressed by means of a frequency shift. Thereby, the undesired mode is shifted beyond the cut-off frequency where a propagation is not possible. However, cut-off frequencies of both the eigenmodes are very close to each other [106, 191]. Hence, even a small undesired fiber deformation may either activate the second mode again or suppress both the modes. For this reason, realization of absolutely SMSP fibers is, of course, a very difficult task.

In order to summarize the results obtained in this Section, following important features of polarization maintaining single-mode fibers are given:

First, the attenuation of SMSP fiber is always more than the attenuation of ordinary single-mode fiber i.e., a fiber without polarization maintaining. Hence, the gain in transmission range obtained by applying coherent optical receivers (heterodyne or homodyne) is partly reduced.

Second, splicing or connecting SMSP fibers is rather difficult and a fundamental drawback of SMSP fibers. In contrast to standard fibers where the fiber cores have to be adjusted only, each connection of two SMSP fibers additionally requires an accurate adjustment of the principal axes of both the fibers. Thereby, the connection is optimized when both core and, in addition, principal axes are exactly aligned on top of each other. Hence, realizing appropriate mechanical connectors is a serious problem. One exception is, how-ever, given when SMSP fibers with circularly polarized eigenmodes are used.

Third, fabrication of SMSP fibers requires a specific technology which is much more comprehensive and sophisticated than in the case of conventional fibers.

Fourth, in contrast to ordinary single-mode fibers, influence of polarization dispersion is increased since the difference in velocities of both the orthogonal polarizations is very large in SMSP fibers (Section 4.3.2). However, polarization dispersion does not occur when only one eigenmode is excited along the entire fiber length or bit rate is low.

As further discussion on SMSP fibers will considerably exceed the frame and purpose of this book, some important references will finally be given: [122, 159, 188, 192, 196, 208, 224].

4.3.2 POLARIZATION DISPERSION

Considering polarization maintaining fibers, a common problem related to the speed of information transmission has to be discussed. Since both eigenmodes always propagate with different velocities, transmitted signal reaches the fiber output at different points of time. Thus, a rectangular input pulse no longer remains rectangular at the fiber output. Rather, the pulse is deformed and in particular broaden. As it is well-known from telecommunication engineering, this effect is called dispersion. Since different velocities of both the polarizations are the reason for dispersion in SMSP fibers, this type of dispersion is referred as *polarization dispersion*.

Irrespective of type, dispersion always degrades the performance of digital communication systems since a broaden pulse normally yields undesired intersymbol interference (Section 2.5.1). It is clear that the influence of intersymbol interference strongly increases with the increase in the speed of information transmission i.e., the bit rate. Normally, polarization dispersion is negligible in comparison with waveguide dispersion and chromatic dispersion. When bit rate increases, polarization dispersion becomes more and more evident. As confirmed by various measurements, polarization dispersion must be taken into account when modulation frequencies exceed about 4 GHz [29, 108, 168, 172, 177, 178, 214 - 216]. Since modern optical communication systems are going to operate in the multigigabit range (i.e., 10 Gbit/s to 100 Gbit/s), polarization dispersion will become more and more serious problem. Further, influence of dispersion also increases with the increase in the transmission distance. Hence, polarization dispersion is, in particular, a very serious problem in long-range multigigabit systems; for example, in transoceanic transmission links.

The polarization dispersion is primarily determined by the difference in velocities of both the eigenmodes. Therefore, higher this difference more is the polarization dispersion. As the difference in velocities is proportional to the difference $\Delta\beta$ in the propagation constants, it is high, in particular, when SMSP fibers are used. It should be remembered that a large $\Delta\beta$ or a high birefringence D is a fundamental feature of most SMSP fibers. When only one eigenmode is stimulated at the fiber input and no undesired fiber perturbations occur in addition, polarization dispersion does not exist. However, this is generally not true in practice.

4.3.3 POLARIZATION CONTROL

When polarization is not stabilized by means of polarization maintaining fiber, alternate techniques have to be employed to match the states of polarization of received and local laser light waves. One important approach is to adjust the polarization by appropriate components which are able to influence the polarization of light wave. These optical components are called *retarders*. A retarder enables to change the polarization of light wave by a specified amount. It is, in principle, independent of whether the polarization of local laser or received light wave is controlled. Due to practical reasons, it is, how-ever, more useful to place the retarder in the light path of the local laser since each retarder also exhibits a certain amount of attenuation. In system performance, an additional attenuation of the received light wave will be much more evident than a somewhat decreased effective local laser power.

Retarders are based either on electrooptical, magnetooptical or mechanical effects. In the latter, fiber polarization is changed by applying external pressure to the fiber surface. Thereby, fiber is specifically deformed and polarization is changed as described in Section 4.2. An equal pressure of specified amount and, consequently, an equal fiber deformation can be obtained either by employing electromagnetic- or piezo-based fiber squeezers [272].

In order to match the polarizations of local laser and received light waves, it is required that the local laser polarization which is normally linear is converted to desired state of polarization. Hence, changing polarization by retarder must always be a well-defined process. In case of electro- and magnetooptical retarders, this task is performed by controlling an electric driving current which is required to generate the electric or magnetic field [136, 145]. Retarders based on mechanical pressure can also be controlled by an electric current. Depending on the magnitude of electric current, mechanical force is generated and a piece of fiber of certain length is deformed accordingly.

Deformation of the fiber is also the basis of another type of retarder called *Lefevre polarizer* [151]. Here, the fiber is carried around three cylindrical bodies (discs) which are sequentially located one after another. Number of coils (usually two, four and two on the first, second and third disc respectively) and diameter of the discs are chosen in such a way that the first and also the third disc represent a $\lambda/4$-plate and the second disc a

$\lambda/2$-plate at a fixed wavelength (for example, 1.5 μm). State of polarization can be changed by turning the discs. Thereby, the fiber and, hence, the orientation of the principal axes are changed. Lefevre polarizers allow to transfer each type of polarization to every other desired type at the output. However, such polarizers can only be controlled by hand and not automatically. Therefore, these polarizers are frequently used in laboratory experiments, but they are unsuitable for commercial use.

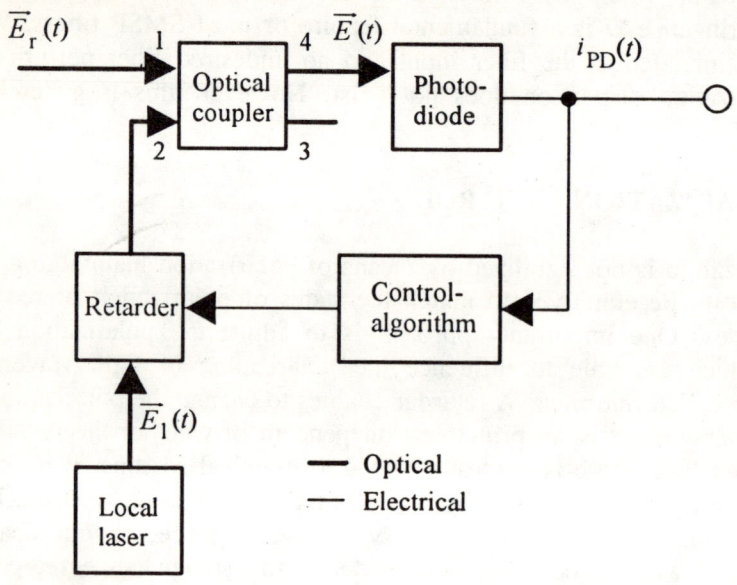

Fig. 4.13: Simplified block diagram of polarization control circuit in single-diode coherent receiver

Each retarder acts as a control unit in polarization control circuit. In addition, each polarization control circuit requires a measure to determine the deviation of local laser polarization and polarization of received light wave. For this purpose, either the photodiode current $i_{PD}(t)$ or the intermediate frequency signal $i_{IF}(t)$ can be used. As explained in Section 4.1.1, photodiode current $i_{PD}(t)$ reaches its maximum value when the polarizations of local laser and received light waves are exactly same. This ideal case is characterized by two identical points $\mathcal{P}_r(t) = (S_{1r}(t), S_{2r}(t), S_{3r}(t))$ and $\mathcal{P}_l(t) = (S_{1l}(t), S_2(t), S_3(t))$ which are on top of each other on the Poincaré sphere. Any deviation from this ideal case decreases the effective part of the photodiode current $i_{PD}(t)$ given in Eq. (2.60). In the worst-case, where both the polarizations are perpendicular to each other, effective part of the photodiode current will become zero.

Maximization and stabilization of the amplitude of photodiode current (Eq. 2.60) must, therefore, be the goal of each polarization control circuit. If, for example, the photodiode current decreases, then the retarder must change the polarization of local laser wave so long as both the polarizations are matched again. Finally, both the Poincaré coordinates $\mathcal{P}_r(t)$ and $\mathcal{P}_l(t)$ are congruent and the photodiode current reaches its maximum value again.

Polarization matching always requires a certain time. Since polarization fluctuations are normally a very slow random process, no special strong demands are placed on the speed of polarization control circuit.

Each polarization control circuit requires at least two retarders since the dynamic range of a single retarder is limited and, therefore, only a limited range of states of polarization can be matched. With stable local laser polarization i.e., only polarization of received light wave is randomly changing, two retarders are sufficient. If the deviation in the state of polarization becomes very large, then polarization matching by two retarders cannot be achieved continuously [182]. For this reason, three retarders are required which enable to realize an *endless-polarization control circuit* without any unsteadiness [160, 161, 182, 183]. This circuit offers the possibility to convert every type of polarization at the input to any other type at the output. Thus, each deviation in Poincaré coordinates $\mathcal{P}_r(t)$ and $\mathcal{P}_l(t)$ can be controlled continuously, even if $\mathcal{P}_r(t)$ is moving around the whole Poincaré sphere many times. This case is usually true in practice. When the state of polarization of local laser wave also fluctuates, a fourth retarder is required in addition to match both the random polarizations.

Control of all the retarders is accomplished by managing the related electric driving currents. This operation requires a complex and comprehensive control algorithm. The main task of this algorithm is to generate the appropriate driving currents for the retarders depending on the random changes in the photodiode current $i_{PD}(t)$. It can be performed, for example, by using a microprocessor. In that case, required algorithm is realized by a software program.

A polarization control circuit reduces the influence of polarization fluctuations in the optical frequency range. This requires optical components such as retarders. However, realization of optical components is usually more difficult than a realization of electrical components. In addition, polarization control by means of permanent fiber deformation may finally yield a fiber break.

4.3.4 POLARIZATION-DIVERSITY RECEIVER

As mentioned above, a polarization control circuit requires some critical optical components (for example, the retarder) which is a drawback. With a polarization-diversity receiver, this problem is shifted to the electrical frequency domain. Here, the influence of polarization fluctuations is reduced by a separate detection of both horizontal and vertical electric field components $E_x(t)$ and $E_y(t)$ of the superimposed light wave $\vec{E}(t)$. For this purpose, two heterodyne receivers have to be realized; one for vertical field component and the other for horizontal component (Fig. 4.14). The output signals of both the receivers are finally combined. In the ideal case, this receiver configuration called the *polarization-diversity receiver* is able to extinguish the influence of polarization fluctuations completely.

In comparison to polarization control circuit, polarization-diversity receiver requires lesser number of optical components, but a higher number of electrical components. As an example, Fig. 4.14 shows the simplified block diagram of an ASK heterodyne polarization-diversity receiver. This receiver includes one local laser, one optical coupler, a polarization beam splitter, two photodiodes, two identical electric signal paths (x-path and y-path), one sum up device and finally a low-pass filter. Similar to standard coherent receivers, detected signal $d(t)$ at the output of the receiver is applied to a sample and hold circuit and finally to a decision circuit.

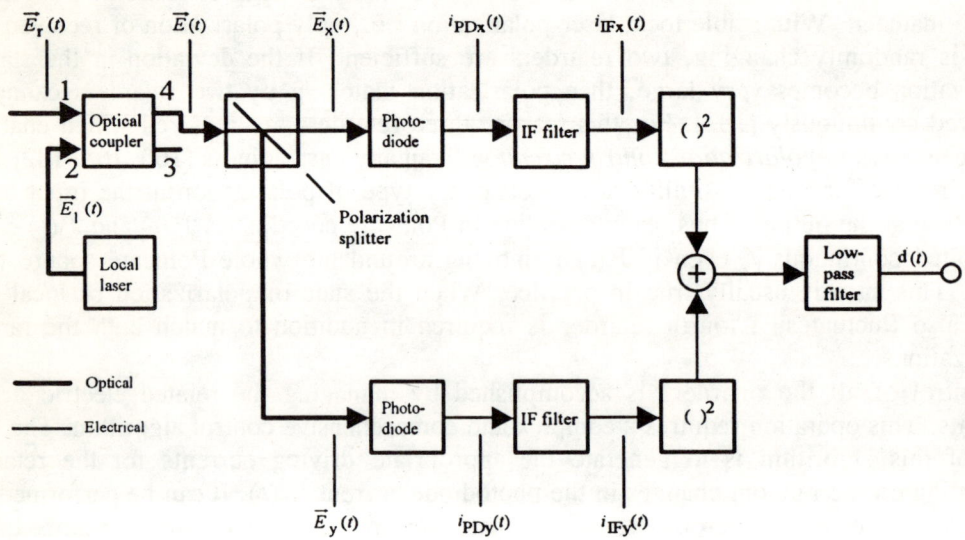

Fig. 4.14: Principle block diagram of polarization-diversity receiver

The diversity receiver shown in Fig. 4.14 requires a stable and linear local laser polarization which shows an inclination of 45° to the horizontal x-axis. Thus, both the field components (x- and y-components) of the local laser light wave $\vec{E}(t)$ are equal. Using Eqs. (2.62) and (2.63), intermediate frequency signals $i_{\text{IFx}}(t)$ and $i_{\text{IFy}}(t)$ are given by

$$i_{\text{IFx}} = a_{\text{Px}}(t) \, R_0 \, \sqrt{P_r P_1} \, s(t) \, \cos(2\pi f_{\text{IF}} t + \phi_{\text{Px}}(t)) \tag{4.76}$$

$$i_{\text{IFy}} = a_{\text{Py}}(t) \, R_0 \, \sqrt{P_r P_1} \, s(t) \, \cos(2\pi f_{\text{IF}} t + \phi_{\text{Py}}(t)) \tag{4.77}$$

For simplicity, laser phase noise and additive Gaussian noise (i.e., shot noise and thermal noise) has not been taken into consideration. Both the above equations provide that the IF filter does not influence the signal shape of input signal. Hence, this filter is only used to suppress the baseband parts of the photodiode current (compare Eq. 2.60). The coupling ratio k of the optical coupler is assumed to be 0.5 (3 dB coupler).

As shown in Fig. 4.14, both the IF signals $i_{\text{IFx}}(t)$ and $i_{\text{IFy}}(t)$ are squared, added and finally applied to a low-pass filter. Thereby, the double-frequency components (two times the IF) are canceled out. The detected signal $d(t)$ at the output of the low-pass filter is given by

$$d(t) = \frac{1}{2} K_{sq} R_0^2 P_r P_l s^2(t) \underbrace{\left[a_{Px}^2(t) + a_{Py}^2(t) \right]}_{= 0.5} \tag{4.78}$$

$$= \frac{1}{4} K_{sq} R_0^2 P_r P_l s^2(t)$$

The new parameter K_{sq} measured in unit of 1/Ampere is a constant of proportionality for both the squaring devices. As seen from Eq. (4.78), detected signal $d(t)$ is proportional to the square of normalized information signal $s(t)$. Since this signal is unipolar binary ASK signal which is either 0 or 1 (Eq. 2.30), the squaring process does not affect the result. Further, superposition of random factors $a_{Px}^2(t)$ and $a_{Py}^2(t)$ yields the constant 0.5. This will be proved in the following:

Proof:

As described in Section 2.4.3, polarization unit vectors $\vec{e}_r(t)$ and $\vec{e}_l(t)$ of the received and the local laser light waves are given by

$$\vec{e}_r(t) = \begin{bmatrix} e_{rx}(t) \\ e_{ry}(t) \end{bmatrix} \quad \text{where } \vec{e}_r(t) \, \vec{e}_r^*(t) = |e_{rx}(t)|^2 + |e_{ry}(t)|^2 = 1 \tag{4.79}$$

$$\vec{e}_l = \begin{bmatrix} e_{lx} \\ e_{ly} \end{bmatrix} = \begin{bmatrix} \dfrac{1}{\sqrt{2}} \\ \dfrac{1}{\sqrt{2}} \end{bmatrix} \quad \text{where } \vec{e}_l(t) \, \vec{e}_l^*(t) = 1 \tag{4.80}$$

As shown in Fig. 4.14, both the x-components are responsible for signal in the upper path, while the y-components in the lower path. The random amplitudes $a_{Px}(t)$ and $a_{Py}(t)$ when calculated from Eq. (2.59) give rise to

$$a_{Px}^2(t) + a_{Py}^2(t) = \frac{1}{2} |e_{rx}(t)|^2 + \frac{1}{2} |e_{ry}(t)|^2 = \frac{1}{2} \tag{4.81}$$

Thus, polarization fluctuations in the received light wave $\vec{E}_r(t)$ can be completely suppressed by a polarization-diversity receiver such as shown in Fig. 4.14. Similar results are obtained when FSK, DPSK or PSK modulation schemes instead of ASK are used [144, 180, 184].

Taking into account the additive Gaussian noise also, it can be shown that the receiver sensitivity of diversity receiver is only negligibly lower than that of conventional coherent receiver without polarization diversity.

4.3.5   POLARIZATION SWITCHING

Polarization switching represents a simple, but powerful solution to reduce the influence of polarization fluctuations. The most common technique is given by *data-induced polarization switching* which will be briefly discussed. Polarization switching can either be applied to the transmitter or receiver. Data-induced polarization switching is applicable only for FSK systems.

The principle block diagram is shown in Fig. 4.15 wherein the switching process is performed in the transmitter by a passive birefringent component. This component is placed right after the transmitter laser which is binary modulated in frequency by modulating the injection current or by an external frequency modulator. To avoid high insertion loss, it is advantageous to choose a piece of polarization maintaining fiber as a birefringent component. Polarization switching requires a linearly polarized local laser light which has to be launched at 45° with respect to the principal axes of the polarization maintaining fiber (Fig. 4.3). This means that the optical power is equally divided between the two eigenmodes.

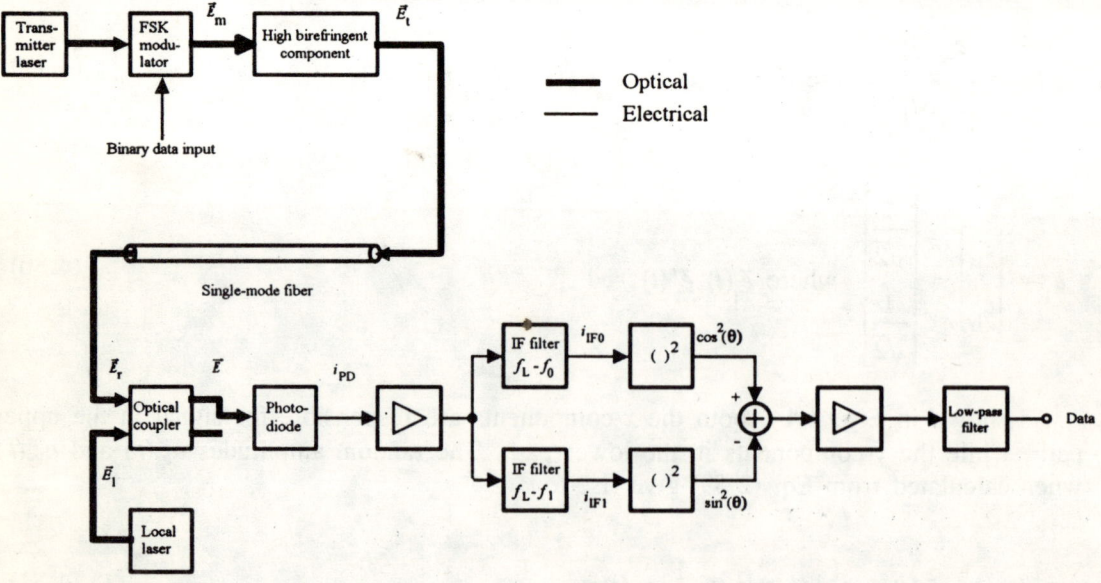

Fig. 4.15: Data-induced polarization switching in transmitter of coherent communication system

As explained in Section 4.1, both eigenmodes of a polarization maintaining fiber propagate with different velocities. Hence, state of polarization at the output of the birefringent component will generally differ from the linear polarization at the input. Depending on the phase difference between the eigenmodes (mode x and mode y), state of polarization at the output can be either linear, circular or elliptical. Since this difference in phase is also a function of frequency of light, it will change with the binary FSK modulation. If the product of time delay of both the eigenmodes and FSK frequency deviation is chosen to be 0.5, then the phase difference will changed by 180^0 as the frequency is switched. Because of this, polarization modulation is simultaneously added to the frequency modulation.

The optical signal which is now modulated in frequency *and* polarization is transmitted through a standard single-mode fiber to a conventional heterodyne receiver with a FSK demodulator; for example, a single- or dual-filter configuration with envelope or synchronous detection. However, due to polarization switching there is a permanent mismatch in the state of polarizations between the received light wave and local laser light wave that results in an intrinsic loss of 3 dB.

The advantage of data-induced polarization switching is that the passive birefringent component is the only optical component required to overcome the polarization problem. No polarization control or diversity technique is necessary in addition. The possibility of applying polarization switching in the transmitter reduces the cost of receivers which, for example, is very advantageous if coherent optical multichannel distribution systems have to be realized.

Due to a strong polarization dispersion in the birefringent component in the transmitter intersymbol interference can be observed which may limit the system performance and permissible bit rate.

4.3.6   COMPARISON

In the previous Sections, four different methods to solve the polarization problem have been considered in detail. The simplest, but most expensive solution is given by the polarization maintaining fiber (e.g., SMSP fiber). Due to higher cost, SMSP fibers are normally not used in long-haul optical fiber communication systems where the fiber length is usually in the range of few km to some hundreds of km. In addition, SMSP fibers exhibit higher attenuation than the standard single-mode fibers. Consequently, possible applications of SMSP fibers are restricted to a length of few cm only. The most common application is, in particular, given by SMSP fiber pigtails for polarization dependent optical components such as optical waveguide phase modulator or optical coupler.

The most competitive polarization handling methods for coherent optical communication systems are polarization diversity, polarization control and data-induced polarization switching. The first two methods require a number of additional optical and electrical components, while the polarization switching only requires a birefringent component in the transmitter or receiver, which can easily be realized by a rather short polarization maintaining fiber of length about 100 m [185]. If the birefringent component is placed in the transmitter, then a standard heterodyne receiver can be employed. For this reason, polarization switching represents the least expensive solution to overcome the polarization problem. However, data-induced polarization switching is only applicable for FSK systems. The data-induced polarization switching is well-suited for multisubscriber systems such as coherent optical multichannel video distribution networks. However, in the multiple-access network, each optical transceiver requires a birefringent component in the transmitter.

As far as losses are concerned, polarization control is the most efficient since the insertion loss of only the retarder has to be considered. In the case of fiber retarder, this loss is typically less than 0.5 dB. Therefore, polarization control is well-suited for coherent optical long-range links where the sensitivity is a critical parameter. In comparison, polarization diversity and polarization switching exhibit an inherent power loss of 3 dB. In the former, this loss is additionally increased by the loss of optical components; for example, by the loss of polarization beam splitters. In the latter, power loss of the birefringent component which is about 0.5 dB [185] has to be considered in addition. In the polarization-diversity receiver, inherent loss of 3 dB can, however, be avoided if two local lasers (one for each of the diversity receiver branches) are used.

Finally, the response time should be compared. The response time is rather slow if a polarization-diversity receiver or a polarization switching component is used. In contrast, response time is in the order of some magnitudes higher if a polarization-control circuit is used in the system where state of polarization is tracked by means of a feedback loop and control algorithm. Hence, if a fast acquisition speed is required, polarization-diversity and polarization switching are the most promising solutions.

Summarizing the discussion above, it becomes clear that the most efficient technique to solve the polarization problem primarily depends on the applications of the coherent optical communication systems.

5   SYSTEM ANALYSIS AND OPTIMIZATION

Based on the fundamentals of Chapter two, this Chapter now discusses the analysis and the optimization of various coherent optical communication systems in detail. For this purpose, modulation and demodulation schemes are used as characteristic features to distinguish and classify the different systems. The following optical communication systems will be studied in this Chapter:

- coherent ASK and PSK homodyne systems (Section 5.1),

- coherent ASK, FSK and PSK heterodyne systems (Section 5.2),

- noncoherent ASK, FSK and DPSK heterodyne systems (Section 5.3).

Finally, the conventional optical direct detection system with intensity modulation will be reviewed briefly as a well-suited reference for system comparison (Section 5.4).

It should be noted that in this Chapter, the terms "coherent" and "noncoherent" are exclusively used in conjunction with the demodulation technique applied in the receiver. As it is well-known, each communication system is disturbed by various sources of noise. This Chapter takes into account the shot noise of the photodiode(s), the thermal (electronic) noise of the resistors and amplifiers (mainly of the front-end amplifier) as well as the phase noise of both transmitter and local lasers. To discuss system degradation by noise and other system imperfections, this Chapter also considers the influence of filters such as IF and baseband filters on noise and signal (intersymbol interference).

In the *system analysis*, main task is to calculate the probability of error as it represents the most powerful measure to assess system performance and transmission quality and also as a well-suited parameter for system comparison (Chapter seven). To evaluate the probability of error or the bit error rate (BER), signals at different points in the receiver are to be determined which will depend on modulation scheme and noise (see Chapter two). The detected signal $d(t)$ at the input to the sample and hold circuit is of prime importance since the quality of this signal directly influences the BER and, hence, the performance of digital communication system.

System analysis requires the examination of probability density function (pdf) of sample values $d(\nu T + t_0)$ of the detected signal to calculate the BER. In addition, pdf and its statistical characteristics viz., expected value and standard deviation give a clear insight of complex interconnections of various subsystems in a coherent optical communication system. Probability density function, expected value and standard deviation allow us to identify the most important parameters in a system. Furthermore, the different effects of the sources of noise become clear in particular. Results of a change in system parameters can also be assessed very fast.

Another very powerful measure in system performance analysis is the eye pattern. As explained in Section 2.5, eye pattern can be measured by an oscilloscope. In theoretical system analysis, eye pattern can be easily calculated mathematically. It can also be simulated by an appropriate computer simulation program. The eye pattern highlights typical features of the system and especially allows to assess the effects of intersymbol interference in the detected signal $d(t)$.

In this book, *system optimization* is always performed with a goal to minimize the probability of bit error. An exact analytical optimization is, however, often not possible. In such a case, optimization is performed by employing appropriate numerical, iterative methods in conjunction with a computer. To minimize computer processing time, simplifying approximations are taken.

5.1   HOMODYNE SYSTEMS

In homodyne systems, modulated optical signal is converted directly into an electrical baseband signal by means of an optical coupler, a local laser and one or two photodiodes. This conversion is true in frequency and phase. Thus, the electrical baseband signal is simply a frequency translated replica of the original optical incoming signal.

Optical homodyning requires that frequency *and* phase of both received and local laser waves are exactly same. This means that both waves have to be *synchronous* or *coherent*. To realize this very strong demand, a highly accurate *optical phase-locked loop* (OPLL) circuit is required. For this reason, principle of an OPLL is briefly explained in Section 5.1.3.

The process of converting a modulated high frequency signal (HF signal) into baseband by means of a high frequency synchronous local carrier wave is usually called synchronous or coherent detection. Since both modulated HF signal and synchronous local carrier are optical waves in optical homodyne receiver, we call it *optical synchronous detection* or *optical coherent detection*. Hence, optical communication systems with homodyne receivers are strictly to be called as coherent optical communication systems with optical coherent detection. Here, the first "coherent" is related to a coherent laser source, whereas the second "coherent" corresponds to the synchronous demodulation process. It should be noted that this terminology is not an accepted standard. Indeed, heterodyne and homodyne systems are also frequently called as coherent systems in general, irrespective of whether the demodulation process is coherent or noncoherent. Thus, this simplified terminology is related to the laser source only which should be coherent as far as possible.

Principle block diagram of a homodyne system suitable for ASK and PSK modulation schemes is shown in Fig. 5.1. It must be noted that a high frequency carrier signal modulated in phase cannot be detected by synchronous demodulation in general since a simple frequency shift of the high frequency spectrum to baseband normally does not yield the original spectrum of the transmitted information. As it is known from analog telecommunication techniques, this is normally valid only for amplitude modulated signals. For example, a sinusoidal carrier modulated with a sinusoidal signal in phase yields a so-called Bessel spectrum at carrier frequency f_C. Synchronous detection shifts this spectrum true in phase and frequency to the origin $f = 0$. However, this shifted spectrum (Bessel spectrum at $f = 0$) is clearly not the spectrum of sinusoidal information signal.

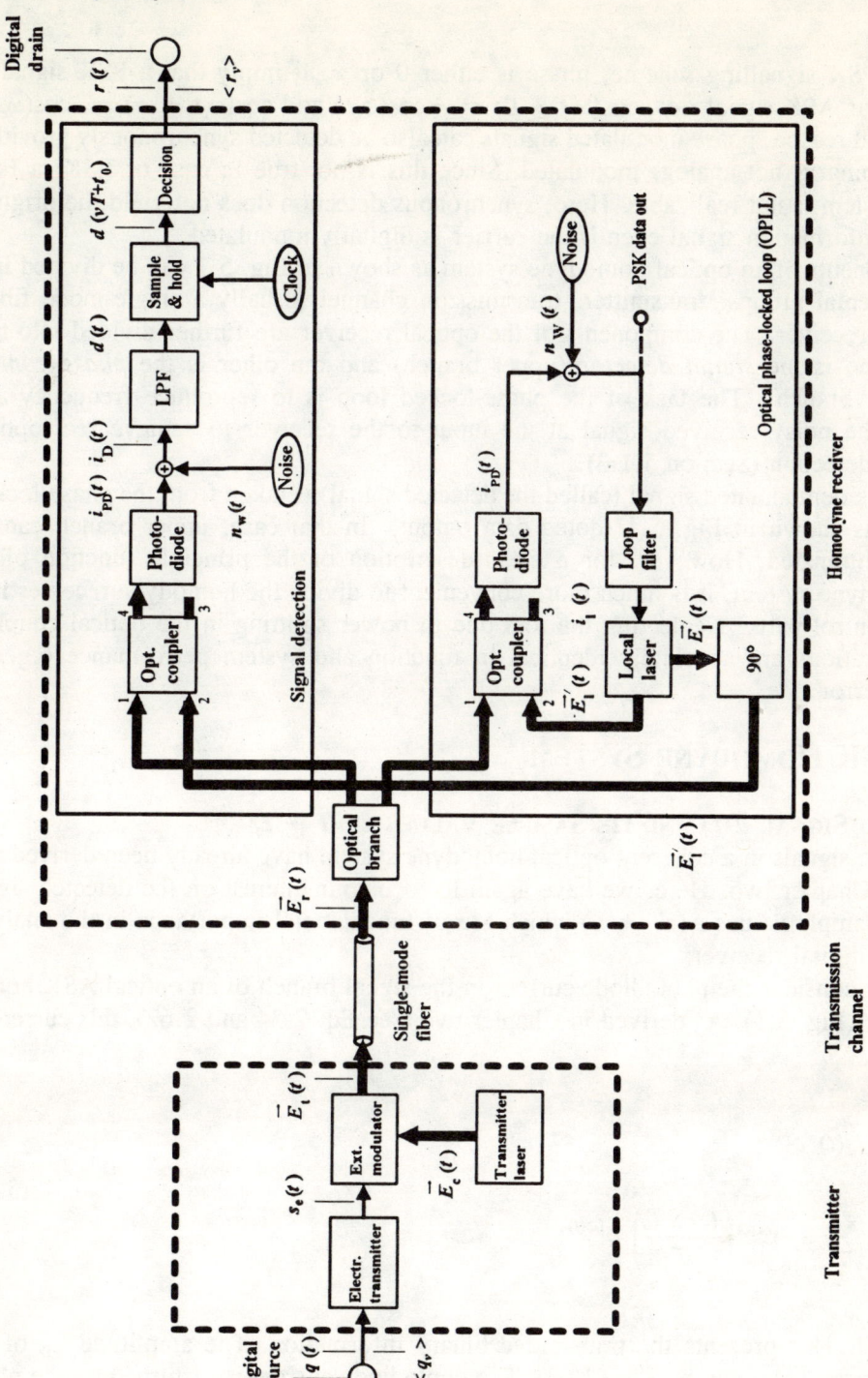

Fig. 5.1: Block diagram of coherent optical communication system with homodyne detection

In binary PSK signalling scheme, phase is either 0 or π. It imply that a PSK signal is same as bipolar ASK signal i.e., $\cos(\omega_c t + 0) = +\cos(\omega_c t)$ and $\cos(\omega_c t + \pi) = -\cos(\omega_c t)$. For this special reason, phase modulated signals can also be detected synchronously provided the phase is binary (not analog) modulated. Since this is not true in case of FSK, a FSK homodyne system is not realizable. Here, synchronous detection does not yield the original spectrum of information signal even if the carrier is digitally modulated.

The components of an optical homodyne system as shown in Fig. 5.1 can be divided into three fundamental groups: transmitter, transmission channel (usually a single-mode fiber) and coherent receiver. The components of the optical receiver are further divided into two subgroups: one is the *signal detector* (upper branch) and the other is the *phase control circuit* (lower branch). The task of the phase-locked loop is to reproduce frequency and phase from the noisy received signal at the input to the receiver to achieve real optical synchronous detection (Section 5.1.3).

Usually, the demodulated signal (called the detected signal) is taken from the phase-locked loop circuit as shown in Fig. 5.1 (doted data output). In that case, upper branch can be completely eliminated. However, for a clear description of the principle function of an optical homodyne system, it is much more convenient to divide the homodyne receiver into signal and control parts. Neglecting the loss due to power splitting in the optical coupler, both configurations are absolutely identical in function and system performance e.g., in terms of bit error rate.

5.1.1 ASK HOMODYNE SYSTEM

(i) DETECTED SIGNAL $d(t)$ AND ITS SAMPLE VALUES $d(\nu T + t_0)$

Most of the signals in a coherent optical homodyne system have already been derived and discussed in Chapter two. Here, we have again to focus our interest on the detected signal $d(t)$ and its sample values $d(\nu T + t_0)$ which are of fundamental importance in the analysis of an optical digital receiver.

Let us first consider the photodiode current in the signal branch of an optical ASK homodyne receiver (Fig. 5.1). As derived in Chapter two (see Eq. 2.34 and 2.62), this current is given by

$$i_{PD}(t) = \hat{i}_{PD}\, s(t)\, e^{j\phi(t)}$$

$$= \hat{i}_{PD} \sum_{\nu=-\infty}^{\infty} s_\nu\, \text{rect}\!\left(\frac{t-\nu T}{T}\right) e^{j\phi(t)} \tag{5.1}$$

where $s_\nu \in \{0, 1\}$ represents the transmitted binary information. The amplitude $\hat{i}_{PD}$ of the photodiode current is given by Eq. (2.63). The photodiode current is disturbed by the phase noise

$$\phi(t) = [\phi_t(t) - \phi_l(t)] - \phi_{PLL}(t) \qquad (5.2)$$

where $\phi_t(t)$ and $\phi_l(t)$ are the phase noise of transmitter and local lasers respectively. In addition, phase $\phi(t)$ also includes a noiseless part $\phi_{PLL}(t)$ of the local laser phase which is controlled by the OPLL circuit. In the ideal case which cannot be achieved in practice, control phase $\phi_{PLL}(t)$ equals the resulting phase noise $\phi_t(t) - \phi_l(t)$. As a consequence, the residual phase noise $\phi(t)$ also called the rest-phase noise becomes zero. However, in practical systems a residual phase noise $\phi(t) \neq 0$ is unavoidable.

In order to analyze homodyne systems and especially to calculate the bit error rate, statistical properties of the residual phase noise $\phi(t)$ have to be determined first. As will be seen in Section 5.1.3, residual phase noise $\phi(t)$ is a *Gaussian stationary* random process with *zero mean*. Its standard deviation σ_ϕ is a function of both the laser linewidths Δf_t and Δf_l of transmitter and local lasers and constant noise power spectral density G_c of the shot noise and the thermal noise. Further, σ_ϕ is also a function of the characteristic parameters of the OPLL circuit (see Eq. 5.57).

The detected signal at the output of the low-pass filter (Fig. 5.1) is

$$d(t) = i_{PD}(t) \star h_B(t) + n(t)$$

$$= \hat{i}_{PD} \int_{-\infty}^{+\infty} s(\tau) \cos(\phi(\tau)) \, h_B(t-\tau) \, d\tau + n(t) \qquad (5.3)$$

Here, $h_B(t)$ is the impulse response of the low-pass filter (baseband filter), $s(t)$ the information signal (Eq. 2.34), and $n(t)$ the additive, band-limited and Gaussian distributed receiver noise due to shot noise of photodiodes and thermal noise. This noise is deter-mined by its standard deviation $\sigma_n = \sigma_{hom}$ given in Eq. (2.78).

As $n(t)$ and $\cos(\phi(t))$ are random processes, sample values $d(\nu T + t_0)$ of the detected signal are random variables. It should be noted that both random processes can be regarded as stationary (Section 3.3.4). Hence, all their statistical properties are independent of sampling time $\nu T + t_0$. For this reason, it is sufficient to consider a certain, fixed sampling time; for example, $t = t_0$. The results obtained are valid in general i.e., for each sampling time $\nu T + t_0$.

Comparing the rectangular equivalent pulse duration Δt_B of the low-pass filter impulse response $h_B(t)$ and the rectangular equivalent correlation width Δt_w of the autocorrelation function $R_w(\tau)$ of the random process $w(t) = \cos(\phi(t))$, following important relationship can be observed in most practical systems [54]:

$$\Delta t_\text{B} = \frac{1}{h_\text{B}(0)} \int\limits_{-\infty}^{+\infty} h_\text{B}(t)\,\mathrm{d}t \;\; < \;\; \Delta t_\text{w} = \frac{1}{R_\text{w}(0)} \int\limits_{-\infty}^{+\infty} R_\text{w}(\tau)\,\mathrm{d}\tau \tag{5.4}$$

Because of this inequality, random process $w(t)=\cos(\phi(t))$ is strongly correlated during the pulse duration Δt_B of the low-pass filter. Therefore, this process can be considered as a time-invariant during the time interval Δt_B (i.e., $\phi(t){\to}\phi(t_0)$). For this reason, term $\cos(\phi(t_0))$ can be taken outside the integral in Eq. (5.3). However, this constant is changing from one time interval to the other and $\cos(\phi(t_0))$ is actually a random variable. Considering the sampled detected signal $d(t_0)$, we now obtain the following simple expression:

$$d(t_0) = \hat{i}_\text{PD}\, a(t_0)\, \cos(\phi(t_0)) + n(t_0) \tag{5.5}$$

where

$$a(t_0) = \int\limits_{-\infty}^{+\infty} s(\tau)\, h_\text{B}(t_0-\tau)\,\mathrm{d}\tau \quad \text{with} \quad 0 \le a(t_0) \le 1 \tag{5.6}$$

Thus, $a(t_0)$ is same as sampled value $d(t_0)$ normalized with respect to $\hat{i}_\text{PD}$ and in absence of any noise i.e., $n(t)=0$ and $\phi(t)=0$. Since $a(t_0)$ is a function of impulse response $h_\text{B}(t)$, transmitted information $s(t)$ and, consequently, transmitted symbol sequence $<q_\nu>$, the noiseless sample value $a(t_0)$ is impaired by intersymbol interference. Thus, the sampled value $a(t_0)$ at the current symbol $q_\nu=q_0$ is essentially influenced by the neighbouring symbols (compare Section 2.5.1). The limited range of $a(t_0)$ given in the above equation is related to a Gaussian baseband filter $H_\text{B}(f)$ as given in Eq. (2.77).

In order to simplify the further calculation, it is useful to perform the following substitutions and normalizations:

$$d := \frac{d(t_0)}{\hat{i}_\text{PD}} \qquad a := a(t_0) \qquad \phi := \phi(t_0)$$

$$n := \frac{n(t_0)}{\hat{i}_\text{PD}} \;\; \to \;\; \sigma := \frac{\sigma_n}{\hat{i}_\text{PD}} = \frac{\sigma_\text{hom}}{\hat{i}_\text{PD}}$$

Now Eq. (5.5) can be written in the following simple form:

$$d = a \cos(\phi) + n = aw + n \quad \text{with} \quad |w| \le 1 \quad \text{and} \quad 0 \le a \le 1 \tag{5.7}$$

As this fundamental equation will be used frequently in our further considerations, the physical meaning of its different quantities is given below again:

d: normalized sample value of detected signal,

a: noiseless part of d disturbed by intersymbol interference (a includes the transmitted information),

$\cos(\phi)$:   phase noise factor,

n: additive Gaussian receiver noise.

It becomes clear from the above equation that the sample values of the detected signal are influenced by two random variables: one is the phase noise term $w = \cos(\phi)$ and the other is additive Gaussian noise n. In addition, the detected sampled signal d is also a function of transmitted information which is included in the term a.

(ii) PROBABILITY DENSITY FUNCTION $f_d(d)$

Since both the random variables w (phase noise term) and n (additive Gaussian noise) are statistically independent, probability density function (pdf) of the detected sample value d can easily be obtained by convolution:

$$
\begin{aligned}
f_d(d) &= \frac{1}{|a|} f_w\!\left(\frac{d}{a}\right) \star f_n(d) \\[2mm]
&= \int_{-a}^{+a} \frac{1}{|a|} f_w\!\left(\frac{w}{a}\right) f_n(d-w)\, dw \;=\; \int_{-1}^{+1} f_w(w)\, f_n(d-aw)\, dw
\end{aligned}
\tag{5.8}
$$

Here, the last integral takes into account that the pdf $f_w(w)$ is limited on $|w| \le 1$ as shown in Fig. 3.18. If $a=0$, then the normalized pdf $(1/|a|) \cdot f(w/a)$ which is included in the first integral changes to the Dirac delta function $\delta(w)$. In this particular case, pdf $f_d(d)$ is same as the Gaussian pdf $f_n(n)$ of the sampled additive noise n. Substituting the pdfs $f_w(w)$ and $f_n(n)$ from Eqs. (3.68) and (2.72) respectively and after some further simplifications, we obtain [61]:

$$f_d(d) = \frac{1}{\pi \sigma_\phi \sigma} \int\limits_{0}^{+\infty} \exp\left(-\frac{\Psi^2}{2\sigma_\phi^2}\right) \exp\left(-\frac{a^2}{2\sigma^2}\left[\frac{d}{a} - \cos(\Psi)\right]^2\right) d\Psi \qquad (5.9)$$

In this equation, σ_ϕ represents the standard deviation of the stationary rest-phase noise $\phi(t)$, and σ the normalized standard deviation of the additive, stationary and Gaussian circuit noise $n(t)$. The constant of normalization is given by the amplitude $\hat{\imath}_{PD}$ of the photodiode current $i_{PD}(t)$ as mentioned earlier.

The pdf $f_d(d)$ for permanent "0" (i.e., $s_\nu=0$ for all ν and, thus, $a=0$) and permanent "1" (i.e. $s_\nu=1$ for all ν and, thus, $a=1$) are shown in Figs. 5.2a and 5.2b respectively. In the ideal case, where noise and intersymbol interference do not disturb the system, the sampled detected signal d is either zero ($d=0$) when permanent "0" is transmitted or one ($d=1$) when permanent "1" is transmitted.

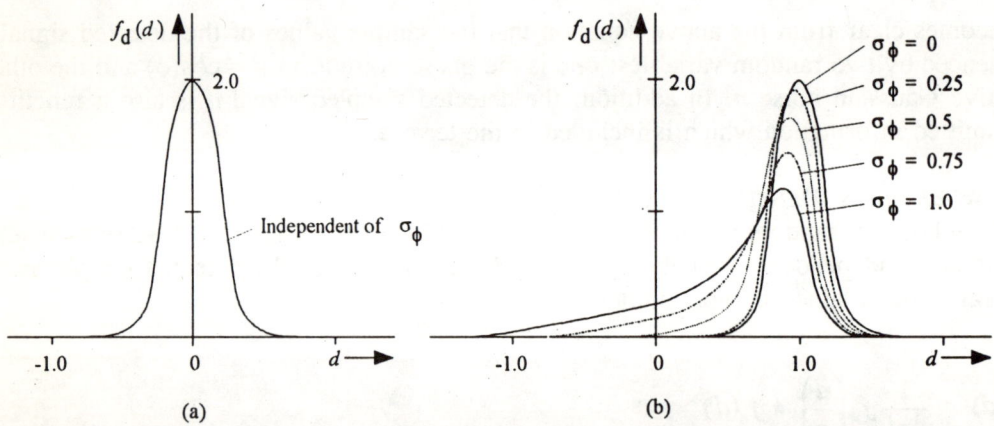

Fig. 5.2: Probability density function $f_d(d)$ of normalized sample values d of the detected signal $d(t)$ in ASK homodyne receiver. The normalized standard deviation of the additive Gaussian receiver noise is fixed at $\sigma=0.2$. The transmitted information is: (a) permanent "0" and (b) permanent "1".

Let us now consider the Fig. 5.2a which shows the pdf $f_d(d)$ when a sequence of only "0" (permanent "0") is transmitted. In this case, the transmitted binary symbol sequence $<q_\nu>$ is given by the special sequence $<q_\nu> = \cdots 000 \cdots$. As a result, noiseless sample value a equals zero ($a=0$) and the phase noise term w is permanently multiplied by zero. It means that laser phase noise does not impair the system. Thus, the detected signal d is only disturbed by the additive Gaussian noise and, consequently, the pdf $f_d(d)$ is Gaussian.

Consider next Fig. 5.2b which gives $f_d(d)$ for permanent "1". As the transmitted symbol sequence is $<q_\nu> = \cdots 111 \cdots$, it represents a simple DC signal. Let us first assume that the laser phase noise does not impair the system (i.e., $\sigma_\phi=0$). In such a case, pdf $f_d(d)$ of the detected signal d is again Gaussian, but now shifted to the expected value $d = 1$. However, as soon as laser phase noise influence the system in addition (i.e., $\sigma_\phi \neq 0$), the pdf $f_d(d)$ changes its shape and remains no longer Gaussian. As this influence increases, the pdf $f_d(d)$ becomes more and more flat and broad, but only in direction to the lower levels of d. However, in the opposite direction pdf $f_d(d)$ changes much less. This most important result is a typical feature of ASK homodyne and ASK heterodyne systems, which results in some decisive effects; for example, a required shift of the threshold in the decision circuit. As the effects of phase noise are of fundamental importance, these will be discussed in more detail now.

The coherent optical ASK homodyne systems are based on an *unipolar modulation scheme*, where the signal s_v is either zero ($s_v = 0$) or one ($s_v = 1$). The pdfs of the detected signal for the symbols "0" and "1" are usually asymmetrical. This characteristic feature of ASK systems becomes clear in Fig. 5.2. Due to phase noise, symbol "1" is always disturbed more than the symbol "0". The mathematical reason for this feature is given by the phase noise term $w = \cos(\phi)$ which is a multiplier to the information signal (Eq. 5.7). In contrast to shot noise and also thermal noise which are *additive* sources of noise, phase noise is a *multiplicative* source of noise. As a result, influence of phase noise rapidly increases with the increase in the amplitude of the information signal or the optical power at the input to the receiver. On the other hand, impact of phase noise is low when the incoming optical power is low. Here, performance degradation due to additive Gaussian receiver noise is dominant. For this reason, ASK homodyne system belongs to the group of optical communication systems which are characterized by *signal-dependent noise*. It should be remembered that conventional optical direct detection system is also included in this group.

A well-suited measure to assess the different influence of phase noise and additive Gaussian noise is given by the standard deviation σ_d or variance σ_d^2 and expected value η_d of the sample value d of the detected signal. These important measures can be determined as follows:

Expected value:

$$\eta_d = a\, e^{-\frac{\sigma_\phi^2}{2}} \tag{5.10a}$$

Variance:

$$\sigma_d^2 = \sigma^2 + \frac{a^2}{2}\left[1 - e^{-\sigma_\phi^2}\right]^2 \approx \sigma^2 + \frac{a^2}{2}\sigma_\phi^4 \tag{5.10b}$$

As the additive Gaussian receiver noise is a random process with zero mean, expected value η_d is determined by the phase noise process only. Without any phase noise (i.e., $\sigma_\phi=0$), expected value η_d is same as a which includes the information (Eq. 5.7). To achieve a low bit error rate, η_d should be as large as possible for transmitted symbol "1" and as small as possible for transmitted symbol "0". Since phase noise always decreases η_d (independent of transmitted information), quality of transmission is always deteriorated in case of symbol "1" (large a), but it is improved in case of symbol "0" (small a)! This surprising feature of ASK homodyne system becomes more clear when the eye pattern of the detected signal is considered (Chapter seven).

The variance σ_d^2 of the sampled detected signal d is determined by the variance of the additive Gaussian noise and phase noise. However, the influence of phase noise is much more serious as σ_d^2 depends on the square of the phase noise variance σ_ϕ^2 in contrast to direct relationship with variance σ^2 of additive circuit noise (Eq. 5.10b). The approximation employed in the above expression makes use of $\exp(x) \approx 1 + x$, which is valid if x is much less than 1.

(iii) INTERSYMBOL INTERFERENCE AND BIT ERROR RATE (BER)

So far we have considered how the pdf $f_d(d)$ of the detected sample values d is influenced by phase noise and additive Gaussian noise. However, the pdf $f_d(d)$ is also influenced by the transmitted symbol or bit sequence $<q_v>$ which is included in the noiseless part of d i.e., in a (see Eq. (5.7) above). The probability of an errorless detection of a transmitted symbol "1" or "0" is essentially determined by the previous and following symbols.

As explained in Section 2.5.1, mutual distortion of neighbouring symbols is caused by the restricted bandwidth of the system; for example, filters. Each restriction of bandwidth, which, of course, is always required to reduce the influence of shot and thermal noise, yields pulse broadening. It give rise to intersymbol interference (ISI). As shown in Section 2.5.1, influence of ISI can easily be assessed from the eye pattern measured by means of an oscilloscope. As an example, Fig. 5.3 shows a typical eye pattern.

In order to distinguish the possible symbol sequences of a transmitter, an index i is used. Hence, the totality of all symbol sequences is $<q_v>_i = \cdots q_{-2}, q_{-1}, q_0, q_1, q_2 \cdots$ where $i = 0, 1, \cdots \infty$.

For determining BER, a second distinction with respect to the present symbol q_0 at the current sampling point t_0 is required. Since this symbol can either be "1" or "0", the following distinction can be made with respect to sampled detected signal and symbol sequence:

$$d = \begin{cases} d_{1i} = a_{1i} \cos(\phi) + n & \text{if} \quad <q_v>_{1i} = <\cdots q_{-2}, q_{-1}, \mathbf{1}, q_1, q_2 \cdots> \\ d_{0i} = a_{0i} \cos(\phi) + n & \text{if} \quad <q_v>_{0i} = <\cdots \bar{q}_{-2}, \bar{q}_{-1}, \mathbf{0}, \bar{q}_1, \bar{q}_2 \cdots> \end{cases} \qquad (5.11)$$

It may be mentioned that symbol pattern $<q_v>_{1i}$ and $<q_v>_{0i}$ are complementary to each other. Thus, a symbol $q_v=$"1" of sequence $<q_v>_{1i}$ is related to a symbol $q_v=$"0" in the sequence $<q_v>_{0i}$.

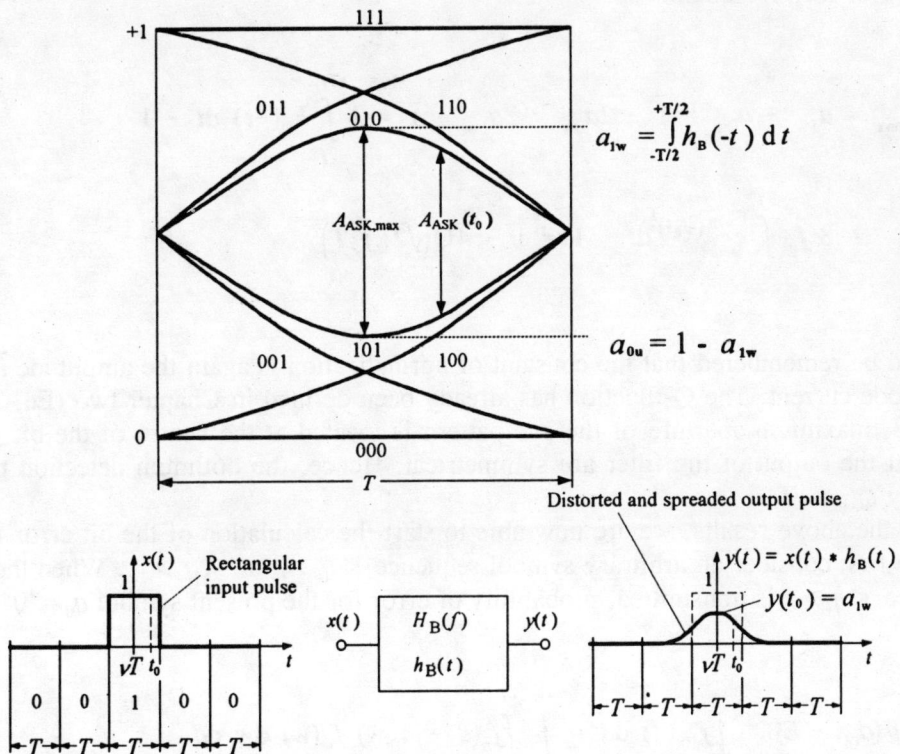

$$a_{1w} = \int_{-T/2}^{+T/2} h_B(-t)\, dt$$

$$a_{0u} = 1 - a_{1w}$$

Fig. 5.3: Eye pattern and aperture in ASK homodyne receiver with Gaussian baseband filter in absence of noise

As far as the totality of all possible symbol sequences are concerned, the worst-case pattern $<q_v>_w$ are of primary importance since these pattern directly affect the aperture of the eye pattern and, thus, the worst-case BER as shown in Section 2.5.1.

With a *Gaussian baseband filter* (Eq. 2.77), the worst-case pattern are given by the symbol sequences "single one" and "single zero" i.e.,

$$<q_v>_{1w} = <\cdots 0\ 0\ 0\ 1\ 0\ 0\ 0\ \cdots> \;\to\; a_{1w} = 1 - 2Q\!\left(\sqrt{2\pi} f_g T\right)$$

$$<q_v>_{0w} = <\cdots 1\ 1\ 1\ 0\ 1\ 1\ 1\ \cdots> \;\to\; a_{0w} = 1 - a_{1w}$$

(5.11b)

Here, a_{1w} and a_{0w} represent the worst-case sample values corresponding to the worst-case pattern $<q_v>_{1w}$ and $<q_v>_{0w}$ in the absence of noise (Eq. 5.7). It can be seen that both sample values a_{1w} and a_{0w} are only determined by bit rate $1/T$ and cut-off frequency f_g of the Gaussian baseband filter (Eq. 2.77). With a_{0w} and a_{1w}, maximum normalized eye aperture $A_{ASK,max}$ of an ASK homodyne system can be calculated as a well-suited measure to assess the influence of ISI. We obtain

$$A_{ASK,max} = a_{1w} - a_{0w} = 1 - 2a_{0w} = 2a_{1w} - 1 = 2 \int_{-T/2}^{+T/2} h_B(-t)\, dt - 1$$

(5.12)

$$= 8 f_g \int_0^{+T/2} e^{-\pi (2f_g t)^2} dt - 1 = 1 - 4Q\left(\sqrt{2\pi} f_g T\right)$$

It should be remembered that the constant of normalization is again the amplitude $\hat{\imath}_{PD}$ of the photodiode current. The Q-function has already been defined in Chapter two (Eq. 2.93). In Fig. 5.3, maximum aperture of the eye pattern is located at the centre of the bit provided pulses at the output of the filter are symmetrical. Hence, the optimum detection times are given by $t_{0,opt} = vT$.

With the above results, we are now able to start the calculation of the bit error rate. For this, we first consider an arbitrary symbol sequence $<q_v>_{0i}$ or $<q_v>_{1i}$. When the symbol sequence $<q_v>_{0i}$ is transmitted, probability of error for the present symbol $q_0 = $"0" is given by

$$p_{0i} = p(d_{0i} > E) = \int_E^{+\infty} f_{d0i}(d)\, dd = \int_E^{+\infty} \int_{-1}^{+1} f_n(d - a_{0i}w) f_w(w)\, dw\, dd$$

(5.13)

which has been derived using Eq. (5.8). In this equation, E represents the threshold level normalized to amplitude $\hat{\imath}_{PD}$ of the photodiode current and $f_{d0i}(d)$ the pdf of the sample value d_{0i}. Hence, this pdf provides that $q_0 = $"0" or $<q_v>_i = <q_v>_{0i}$. By means of the Q-function or the erfc function

$$Q(x) = \frac{1}{\sqrt{2\pi}} \int_x^{+\infty} e^{-\frac{u^2}{2}} du = \frac{1}{2} \operatorname{erfc}\left(\frac{x}{\sqrt{2}}\right)$$

(5.14)

Eq. (5.13) can be simplified as follows [61]:

$$p_{0i} = \int\limits_{-1}^{+1} Q\left(\frac{E - a_{0i}\,w}{\sigma}\right) f_w(w)\; dw = \int\limits_{-1}^{+1} \tilde{p}_{0i}(w)\, f_w(w)\; dw \tag{5.15}$$

It becomes clear from the above equation that the calculation of error probability has to be performed in two steps: First, we have to determine the probability of error $\tilde{p}_{0i}(w)$ by neglecting phase noise. Thereby, the random variable $w = \cos(\phi)$ is assumed to be a constant. To show that phase noise is neglected, the sign " $\sim$ " on the top of the probability $\tilde{p}_{0i}$ has been chosen. The result of the first step is same as obtained in the case of conventional digital communication systems i.e., the well-known dependence of bit error rate on Q-function or erfc function [e.g., 256, 269]. In the second step, random nature of $w = \cos(\phi)$ has to be considered. For this, pdf $f_w(w)$ determined in Section 3.3.4 will be taken into account. Finally, real probability of error is obtained by evaluating the expected value of $\tilde{p}_{0i}(w)$, which has been performed in the second part of Eq. (5.15).

Next we shall calculate the probability of error p_{1i} for the other symbol $q_0 = $"1" in the symbol sequence $<q_v>_{1i}$. This calculation can be performed in the same way as for symbol "0". Accordingly, we obtain

$$p_{1i} = p(d_{1i} < E) = \int\limits_{-\infty}^{E} f_{d1i}(d)\; dd = \int\limits_{-\infty}^{E} \int\limits_{-1}^{+1} f_n(d - a_{1i}w)\, f_w(w)\; dw\; dd \tag{5.16}$$

This expression can also be simplified by means of the Q-function (Eq. 5.14). We obtain:

$$p_{1i} = \int\limits_{-1}^{+1} Q\left(\frac{a_{1i}\,w - E}{\sigma}\right) f_w(w)\; dw = \int\limits_{-1}^{+1} \tilde{p}_{1i}(w)\, f_w(w)\; dw \tag{5.17}$$

(a) AVERAGE PROBABILITY OF ERROR p_a

The most important measure to assess the quality of transmission and to determine the overall performance of a digital communication system is the average probability of error p_a. As already mentioned in Section 2.5.1, probability of error is frequently called as bit error rate (BER) although it is actually not a rate. When a communication system is already realized, BER is usually measured by means of an appropriate measurement equipment which includes at least a digital transmitter (normally a PN-generator) and a digital receiver with a bit error counter. For a new communication system to be designed, the probability of error in the system must be calculated or estimated first. For this, all possible bit pattern $<q_v>_i$ with $i = 1, 2, \cdots \infty$, and their related probabilities of error have to be considered. This is, however, a rather comprehensive task. Normally, a limited number of bits which are very close neighbours influence each other in practical systems. Therefore, only a limited number of different symbol sequences need to be considered. As explained in Section 2.5.1, number of symbol pattern which must actually be used for the BER calculation equals the number of different lines in the eye pattern of detected signal $d(t)$. If, for example, each bit is only disturbed by n previous bits and v following bits, then the number of distinguishable lines in the eye pattern equals 2^{n+v+1} (see Eq. 2.81). In that case, probability of error is given by

$$P_a = \sum_{i=1}^{2^{n+v}} p(<q_v>_i) \left(p_{0i} + p_{1i} \right) = 2^{-(n+v+1)} \sum_{i=1}^{2^{n+v}} \left(p_{0i} + p_{1i} \right) \tag{5.18}$$

where $p(<q_v>_i)$ represents the occurrence probability of the symbol sequence $<q_v>_i$ or line number i in the eye pattern. If all the symbols q_v are statistically independent and both the occurrence probabilities $p(q_v = 0)$ and $p(q_v = 1)$ are equal, then the occurrence probabilities $p(<q_v>_i)$ of the symbol sequences are also equal. Let us consider, for example, that only $2^{n+v+1} = 8$ different symbol pattern have to be taken into account. In that case, eye pattern exhibits $2^{n+v} = 4$ distinguishable lines with $q_v = 0$ and 4 distinguishable lines with $q_v = 1$. The occurrence probability $p(<q_v>_i)$ equals $2^{-(n+v+1)} = 1/8$.

(b) WORST-CASE PROBABILITY OF ERROR p_w

Even for a low and limited number of different symbol pattern (for example, $2^{n+v+1} = 8$), calculation of the average probability of error p_a becomes rather comprehensive. In comparison to this, worst-case probability of error p_w can be evaluated much faster. As mentioned previously, two worst-case pattern $<q_v>_{0w}$ and $<q_v>_{1w}$ are at least existent among all the possible symbol sequences $<q_v>_i$. For this reason, it is useful to define the probability

$$P_w = \frac{1}{2} \left(p_{0w} + p_{1w} \right) \tag{5.19}$$

as the worst-case probability of error or briefly the worst-case BER. In the above expression, p_{0w} and p_{1w} correspond to the worst-case symbol pattern $<q_v>_{0w}$ and $<q_v>_{1w}$ respectively. In the case of ASK homodyne systems, both probabilities p_{0w} and p_{1w} are normally different since an ASK system is disturbed by signal-dependent noise. Remembering that the influence of phase noise is different for symbol "1" and symbol "0". When both threshold level E and cut-off frequency f_g of the Gaussian baseband filter are optimized for a minimum BER (see Subsection iv below), then the following relationship is valid:

$$p_{0w} \leq p_{1w} \tag{5.20}$$

The equality sign is used for ASK homodyne system without any phase noise. It should be noted that p_{0w} and p_{1w} are always identical if an optimized PSK homodyne system instead of an ASK system is used (see Section 5.1.2).

In coherent optical digital communication systems, the product $2f_g T$ of filter bandwidth $2f_g$ and bit duration T is usually more than one i.e., $2f_g T > 1$. For this practical reason, only the adjacent symbols q_{-1} and q_{+1} normally influence the present symbol q_0 at the current sampling point t_0. Hence, the eye pattern exactly includes eight distinguishable lines (Fig. 5.3) which are related to eight distinguishable symbol sequences. When all these sequences occur with same probability $p(<q>_i)=1/8$, the following practical estimation can be given:

$$\frac{1}{4} p_w \leq p_a \leq p_w \tag{5.21}$$

In this important relationship, lower bound on the average probability p_a assumes that p_a is determined by both the worst-case pattern only. The contribution of all other (six) symbol sequences is neglected. Thus, their probability of error is assumed to be zero. In contrast to this, upper bound provides an equal and maximum probability of error (i.e., the worst-case BER of the worst-case pattern) for all the eight symbol sequences. Since accuracy of one order of magnitude is mostly sufficient in practice, above estimation represents a very useful tool for a fast and practically-based calculation of probability of error or BER. Taking into account Eq. (5.15) and Eq. (5.17), worst-case probability of error can now be expressed as:

$$p_w = \frac{1}{2} \int\limits_{-1}^{+1} \left[Q\left(\frac{E - a_{0w}\, w}{\sigma} \right) + Q\left(\frac{a_{1w}\, w - E}{\sigma} \right) \right] f_w(w)\ dw \tag{5.22}$$

Substituting the argument of Q-functions by using the maximum aperture $A_{ASK}:=A_{ASK,max}$ of the eye pattern given in Eq. (5.12), we obtain

$$p_w = \frac{1}{2} \int\limits_{-1}^{+1} \left[Q\left(\frac{2E - (1-A_{ASK})\, w}{2\sigma} \right) + Q\left(\frac{(1+A_{ASK})\, w - 2E}{2\sigma} \right) \right] f_w(w)\ dw \tag{5.23}$$

Finally, the pdf $f_w(w)$ of the random process $w = \cos(\phi)$ can be replaced by using Eq. (3.68). After some mathematical operations, following result is obtained:

$$P_w = \frac{1}{\sqrt{2\pi}\sigma_\phi} \int_0^{+\infty} \left[Q\left(\frac{2E - (1-A_{ASK})\cos(\phi)}{2\sigma} \right) + Q\left(\frac{(1+A_{ASK})\cos(\phi) - 2E}{2\sigma} \right) \right]$$

$$\cdot \exp\left(-\frac{\phi^2}{2\sigma_\phi^2} \right) \, d\phi \tag{5.24}$$

According to the above equation, worst-case BER p_w (and, of course, the average BER p_a also) is a function of several system parameters: threshold E, aperture $A_{ASK,max}(f_g, T, t_0)$ of the eye pattern, standard deviation $\sigma_\phi(\Delta f)$ of the OPLL rest-phase noise (Eq. 5.57), and standard deviation $\sigma(f_g, G_c)$ of the additive Gaussian receiver noise i.e., shot noise and thermal noise. It should be remembered that threshold E and standard deviation σ have been normalized with respect to the amplitude $\hat{i}_{PD} = \hat{i}_{PD}(P_r, P_l)$ of the photodiode current. Therefore, the worst-case BER p_w is also influenced by the optical power levels P_r and P_l of the received and the local laser light waves respectively. As these parameters are again functions of other system parameters, the worst-case BER also depends on symbol or bit duration T, cut-off frequency f_g of the Gaussian baseband filter, resulting laser linewidth Δf of both transmitter and local lasers and power spectral density G_c of the additive Gaussian receiver noise.

Finally, the sampling point t_0 is a system parameter of primary importance since this parameter determines the vertical aperture A_{ASK} of the eye pattern and, hence, the BER. In the above expressions, sampling point t_0 is not included since an optimum sampling point t_0 with maximum eye aperture A_{ASK} has already been taken.

(iv) Optimization

The purpose of this Subsection is to optimize the system parameters for the minimum (worst-case) BER. To simplify calculation, a Gaussian baseband filter with cut-off frequency f_g is considered (Eq. 2.77). The first step in the process of optimization is to decide which of the system parameters can be optimized and which parameters cannot be optimized.

The characteristic features of a *non-optimizable system parameter* is first a fixed value and second a virtual optimum. The BER could only be minimized when the numerical value of this parameter is chosen to be either infinite or zero and both cannot be realized normally. A typical non-optimizable system parameter is, for example, the emission linewidth Δf of the transmitter laser. This linewidth is a fixed parameter determined by the type of laser which has been selected. Moreover, the optimum linewidth is virtual since the optimum linewidth is zero ($\Delta f = 0$) and, thus, not realizable. Other non-optimizable system parameters are the received light power P_r (the virtual optimum is $P_r \to \infty$), the bit rate $1/T$ and the power spectral density G_c of the additive Gaussian receiver noise. In contrast to that, an *optimizable system parameter* exhibits a real, mathematical optimum value. Considering ASK coherent optical communication system, three optimizable system parameters are at least existent: threshold level E, cut-off frequency f_g of the Gaussian baseband filter and sampling point t_0. Since we have already selected a Gaussian baseband filter, optimization of this filter is restricted to the cut-off frequency only. In general, entire filter function $H_B(f)$ of the baseband filter has to be optimized. As it is well-known from conventional digital communication techniques, the result of this optimization is a so-called matched filter which is a very common technique in practice [260]. To avoid generality, we have decided to optimize the bandwidth $2f_g$ only and focus our interest on the principle.

(a) SAMPLING POINT t_0

By changing the sampling point t_0, effective aperture A_{ASK} of the eye pattern is changed simultaneously. If sampling point is optimized then the effective eye aperture reaches its maximum value. Considering Fig. 5.3, optimum sampling point $t_{0,opt} = vT$ is located at the centre of the bit (in the strict sense at the centre of each bit) since this eye pattern is determined on the basis of Gaussian baseband filter which exhibits a symmetrical impulse response $h_B(t) = h_B(-t)$. If a baseband filter with an asymmetrical impulse response $h_B(t) \neq h_B(t)$ is chosen, then the optimum sampling point is shifted from the centre; for example, to the end of each bit in case of a simple RC filter [256]. In this book, the discussion will be restricted to symmetrical impulse responses only.

(b) THRESHOLD E

The optimum and normalized threshold E_{opt} is located at the cross point of probability density functions $f_{d0w}(d)$ and $f_{d1w}(d)$. It should be remembered that $f_{d0w}(d)$ and $f_{d1w}(d)$ are related to both the symbols $q_v = "0"$ and $q_v = "1"$ in the worst-case pattern $<q_v>_{0w}$ and $<q_v>_{1w}$ respectively. Therefore, the equation for determining the optimum threshold level is given by

$$f_{d1w}(E_{opt}) = f_{d0w}(E_{opt}) \tag{5.25}$$

Proof

$$\frac{dp_w}{dE} = \frac{1}{2} \frac{d}{dE} \left[\int\limits_{E}^{+\infty} f_{d0w}(d) \; dd \; + \int\limits_{-\infty}^{E} f_{d1w}(d) \; dd \right]$$

$$= \frac{1}{2} \left[f_{d1w}(E) - f_{d0w}(E) \right] = 0 \; \rightarrow \; f_{d1w}(E) = f_{d0w}(E) \; \rightarrow \; E = E_{opt}$$

Unfortunately, an analytical solution of Eq. (5.25) is not possible for optimum threshold E_{opt}. Hence, a numerical method of iteration must be used on computer, which also yields the desired solution very fast. The result will be shown and discussed in Subsection (v) below.

The optical ASK homodyne system is disturbed by signal-dependent noise. As a higher signal level is more disturbed than a lower signal level, following inequality can be given for this system:

$$E_{opt} \le \frac{1}{2}(a_{0w} + a_{1w}) = 0.5 \qquad (5.26)$$

Here, the equality sign applies for an ASK homodyne system without any phase noise i.e., $\sigma_\phi = 0$. In this special case, both probability density functions $f_{d0w}(d)$ and $f_{d1w}(d)$ are Gaussian distributed and symmetrical with respect to the optimum and normalized threshold $E_{opt} = 0.5$. Thus, these probability density functions are determined only by the normalized standard deviation $\sigma = \sigma(f_g)$ of the additive Gaussian receiver noise (i.e., shot noise and thermal noise). As the optimum threshold E_{opt} is a function of the standard deviation $\sigma(f_g)$, optimum threshold indirectly is also a function of the optimizable cut-off frequency f_g of the Gaussian baseband filter.

(c) CUT-OFF FREQUENCY f_g OF THE LOW-PASS FILTER

Owing to the complexity of both the Eqs. (5.23) and (5.24), an analytical calculation of the optimum cut-off frequency $f_{g,opt}$ of the low-pass (baseband) filter is also not possible. Therefore, numerical methods of iteration are again to be used. The results are shown and discussed in the following Subsection (v). The accuracy of numerical calculation (termed as method one) strongly depends on the number of iterative steps and, therefore, on the computer processing time. To get a fast solution, appropriate approximation methods are more convenient.

A rather rough method of approximation (method two) can be obtained by neglecting the phase noise. In that case, optimum cut-off frequency $f_{g,opt}$ is obtained by

$$\frac{A_{ASK}(f_g)}{\sigma(f_g)} \rightarrow \text{max. provided that } \sigma_\phi = 0 \qquad (5.27)$$

By means of simple numerical evaluation, this operation of maximization yields

$$f_{g,opt} T \approx 0.79$$

where T is the bit period. This simple approximation is well-suited when the optical power level P_r at the input to the receiver is less than -50 dBm (Fig. 5.4b). Since ASK homodyne receiver exhibits a high sensitivity, this requirement is normally fulfilled. If a better approximation is needed, then the expected value

$$E\left\{\frac{A_{\text{ASK}}(f_g)}{\sqrt{\sigma_d^2(f_g)|_{\text{a0w}} + \sigma_d^2(f_g)|_{\text{a1w}}}}\right\} \quad \rightarrow \quad \max \quad \rightarrow \quad f_{g,\text{opt}}T \tag{5.28}$$

can be used to perform the maximization process and estimate the optimum product $f_{g,\text{opt}}T$. We call this method of approximation as method three. In contrast to approximation given in Eq. (5.27), expected value above also takes into account the phase noise and its different influence on both the symbols "0" and "1". As given in Eq. (5.10b), variances $\sigma_d^2(f_g)|_{\text{a0w}}$ and $\sigma_d^2(f_g)|_{\text{a1w}}$ of the detected samples d include the influence of additive Gaussian noise (described by $\sigma(f_g)$) as well as the influence of the phase noise (described by σ_ϕ). In the case of $\sigma_\phi = 0$, which means that phase noise does not disturb the system, both Eqs. (5.27) and (5.28) yield the same result for $f_{g,\text{opt}}T$.

All three methods discussed above, require a computer or at least an electronic calculator to perform the numerical iterations. While method one allows to determine $f_{g,\text{opt}}T$ exactly or at least with desired accuracy (only depending on the number of iterative steps), methods two and three are approximations. Each single iterative step in method one requires the numerical evaluation of an integral which finally requires a large computer processing time. In comparison to that, methods two and three can be performed much faster since no evaluation of integral is involved. Even when a nongaussian baseband filter is provided, methods two and three represent a fast and well-suited tool to determine the optimum value of $f_{g,\text{opt}}T$.

(v) EVALUATION AND DISCUSSION

The purpose of this last Subsection is to evaluate the results obtained in previous Subsections to achieve a better insight of partial complex interrelations in ASK coherent optical communication system.

To begin with, Fig. 5.4a shows the normalized and optimized threshold level E_{opt} as a function of the received optical power P_r expressed in dBm. Another parameter in Fig. 5.4a is the standard deviation σ_ϕ of the OPLL rest-phase noise as given in Eq. (5.57). Evaluation of optimum threshold E_{opt} requires the cut-off frequency f_g of the baseband filter. As optimum value of f_g is a function of the received optical power P_r also i.e., $f_{g,\text{opt}}=f_{g,\text{opt}}(P_r)$, cut-off frequency f_g must be optimized afresh for each P_r to obtain Fig. 5.4a. Thus, each change in P_r requires an optimization of both f_g and E again. For this, method one has been used for calculation (see optimization of cut-off frequency f_g). All other system parameters required to obtain Fig. 5.4a are summarized in Table 7.1 of Chapter seven.

In the absence of phase noise (i.e., $\sigma_\phi = 0$), normalized and optimized threshold always equals $E_{\text{opt}} = 0.5$ irrespective of received optical power level P_r. If phase noise impairs the system, then the optimum threshold E_{opt} becomes a function of P_r. Since phase noise always deteriorates a transmitted symbol "1", but improves the detection of a transmitted symbol "0", normalized optimum threshold E_{opt} is always less than 0.5 as shown in Fig. 5.4a. Thereby, the deviation from $E_{\text{opt}} = 0.5$ increases with a growing distortion of symbol "1".

For this reason, deviation is large when standard deviation σ_ϕ of the OPLL rest-phase noise is large. This deviation also increases with the increase in the optical power P_r at the receiver input. It can be explained as follows: high optical signal power P_r decreases the influence of the additive Gaussian noise and indirectly increases the influence of phase noise factor $\cos(\phi)$. Therefore, it leads to increase in the deviation.

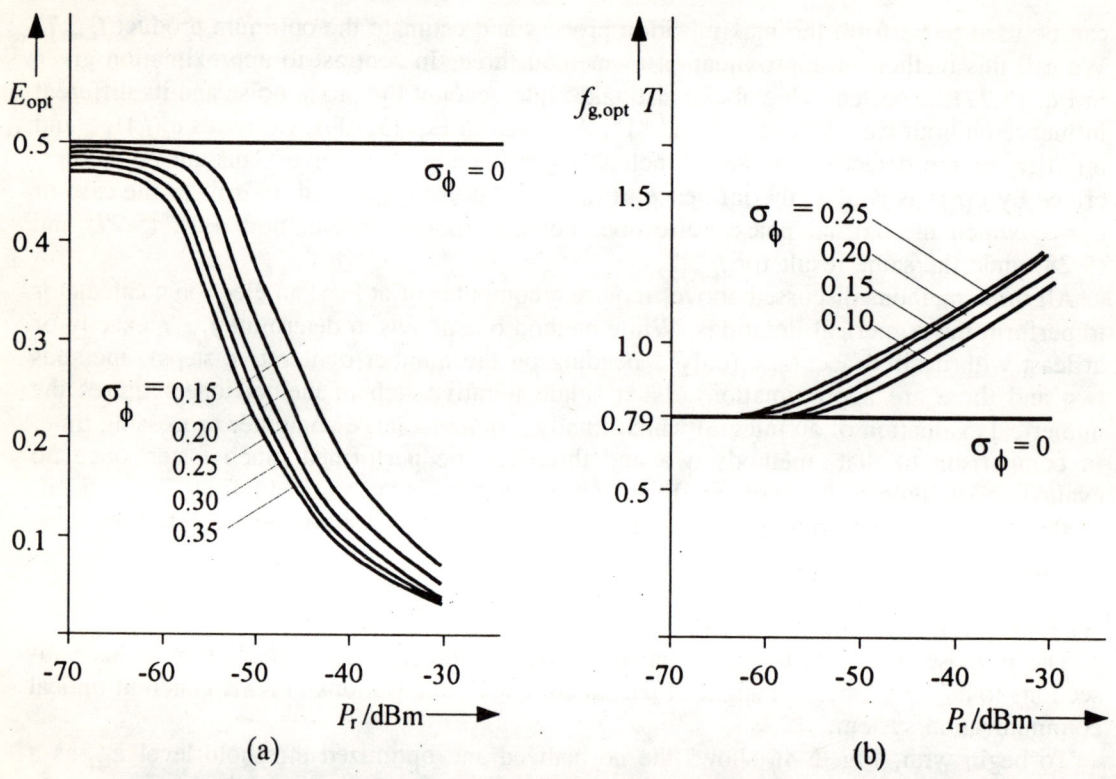

Fig. 5.4: (a) Normalized and optimized threshold E_{opt} and (b) normalized and optimized cut-off frequency $f_{g,opt}T$ of low-pass filter in ASK homodyne receiver

The optimum cut-off frequency $f_{g,opt}$ as a function of the received optical power P_r and standard deviation σ_ϕ of the phase noise is shown in Fig. 5.4b. Analogous to Fig. 5.4a, this figure is also based on method one (exact calculation). Thereby, the threshold E has been optimized afresh for each new value of P_r. It becomes clear from Fig. 5.4b that the bandwidth $2f_g$ of the baseband filter must be increased when phase noise disturbs the system. The physical reason for this is a broaden signal spectrum (due to phase noise) which requires a broader filter to give appropriate high energy at the filter output.

The most important measure to assess the system performance is given by the probability of error which should, of course, be as low as possible. The worst-case probability of error p_w as a function of received optical power P_r at the input to the receiver and standard deviation σ_ϕ of the OPLL rest-phase noise is shown in Fig. 5.5. This figure provides that the cut-off frequency f_g of the low-pass filter is already optimized. The curves shown in Fig. 5.5 are briefly called the *BER curves* or the *performance curves*. It can be seen that these curves can be divided into three characteristic regions:

First region: In the this region where received optical power P_r is low, all the curves show a rather steep slope since the additive Gaussian receiver noise dominates the phase noise. This part of the BER curves is in agreement with the well-known BER curves of an optical transmission system with direct detection (see Fig. 7.2) as well as all standard electrical digital transmission systems [256, 260]. Here, even a small increase in the optical power yields a considerable improvement in BER.

Second region: As optical power level P_r increases, influence of the additive Gaussian noise decreases while the influence of the phase noise indirectly increases. As a result, the BER curves exhibit a bend.

Third region: When the received optical power P_r is very high, the system is practically disturbed by phase noise alone. In this case, influence of the additive Gaussian noise is negligibly small. It may be mentioned that the influence of an additive noise can always be reduced by raising the optical power P_r, while the influence of phase noise remains unchanged. For this reason, BER curves reach a saturation level called the *error rate floor* or the *BER floor*.

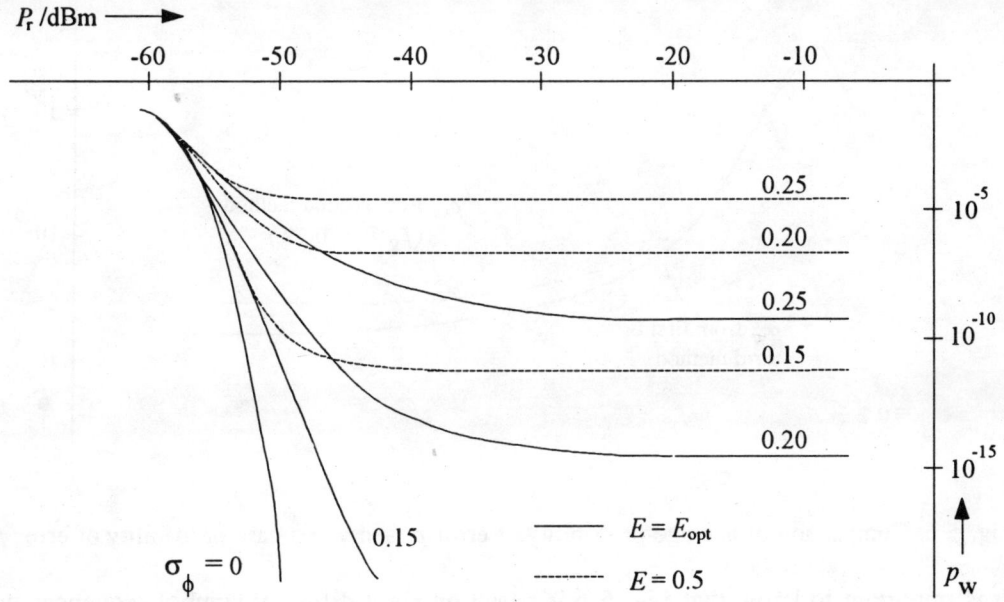

Fig. 5.5: Worst-case probability of error p_w in ASK homodyne receiver

It becomes evident from the above figure that a desired probability of bit error, for example; $p_w = 10^{-10}$ cannot be achieved when standard deviation σ_ϕ of phase noise exceeds a certain value (for example; $\sigma_\phi \approx 0.25$). Therefore, the error rate floor is of great practical interest. As the maximum allowed standard deviation is related to maximum allowed laser linewidth Δf_{max} (Section 5.1.3), Fig. 5.5 directly enables to decide whether a selected laser satisfy the communication system requirements or not. It should be noted that an error rate floor normally exists in all types of coherent optical communication systems independent of modulation scheme used.

In order to emphasize the influence of a non-optimized threshold E, Fig. 5.5 additionally shows the BER curves for a threshold $E=0.5$. This normalized threshold only represents optimum threshold for ASK homodyne system without any phase noise. It becomes clear from Fig. 5.5 that a small change in threshold yields a considerable change in system performance.

As mentioned in the previous Sections, worst-case bit error rate is a well-suited and very powerful measure to approximate the average error rate which is the real error rate of a digital communication system. To verify the accuracy of this worst-case approximation, Fig. 5.6 shows both worst-case probability of error p_w and average probability of error p_a as a function of the received optical power P_r. It becomes evident from this figure that the both probabilities of error do agree very well.

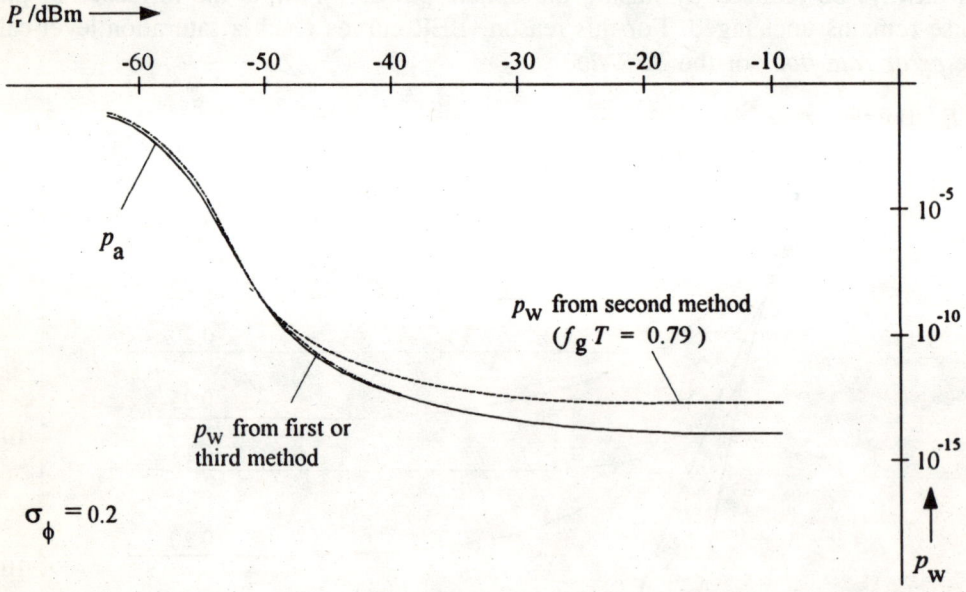

Fig. 5.6: Comparison of average probability of error p_a and worst-case probability of error p_w

It is important to know that Fig. 5.6 is based on eight different symbol sequences which have been taken into account to calculate p_a. Thus, one previous and one following neighbouring symbols have been considered. As mentioned earlier, only eight symbol sequences are sufficient in case of Gaussian baseband filter with optimum bandwidth $2f_{g,opt}$. This bandwidth must always be optimized afresh for each optical power level P_r for both minimum worst-case probability of error p_w and minimum average probability of error p_a.

As the optimum bandwidth increases with the increase in the optical power P_r (Fig. 5.4b), influence of ISI simultaneously decreases. As a result, both probabilities of error are equal at high levels of P_r provided first or second method has been chosen to calculate p_w. It becomes clear from Fig. 5.6 that second and third methods represent very good approximations to predict the average probability of error which can be measured in optical ASK homodyne system. Moreover, both these methods require very less computer run time in comparison to first method.

In the design of coherent optical communication systems, maximum standard deviation σ_ϕ of the OPLL rest-phase noise to achieve a demanded probability of error is an important parameter and is required to be known. For this, let us consider that the system is neither disturbed by additive Gaussian receiver noise ($\sigma=0$) nor by ISI. Thus, the system is only disturbed by phase noise of both transmitter and local lasers. In this case, worst-case probability of error can be easily calculated by

$$p_w = \frac{1}{2} \int_{-1}^{0} f_w(w) \, dw \approx Q\left(\frac{\pi}{2\sigma_\phi}\right) \quad \text{provided } \sigma = 0 \tag{5.29}$$

The approximation on the right-hand side of above equation is valid for $\sigma_\phi < 1$. As the additive Gaussian noise is neglected and phase noise only deteriorates (not improves) the detection of symbol "1", optimum threshold level is very close to zero i.e., $E_{opt} = 0 + \epsilon$. To derive this equation, a threshold $E_{opt} = 0$ was chosen. Numerical results computed from it are shown in Fig. 5.7. In the first part of this figure, small increase in standard deviation σ_ϕ of the phase noise causes a very steep rise in probability of error. This part is quite important since the most demanded error rates ($p_w < 10^{-6}$) are located in this region. For a worst-case probability of error $p_w = 10^{-10}$, standard deviation σ_ϕ of the phase noise must always be less than 0.25. Hence, the *maximum allowed standard deviation* is

$$\sigma_{\phi,max} \approx 0.25 \quad \text{for } p_w = 10^{-10} \tag{5.30}$$

This important relationship is frequently used in coherent optical homodyne systems. As will be discussed in Section 5.1.3, influence of rest-phase noise primarily depends on the architecture of optical phase-locked loop (OPLL) and, of course, on the laser linewidth of both transmitter laser and local laser. By using Eq. (5.57) discussed in Section 5.1.3, maximum allowed laser linewidth $\Delta f_{max} = \Delta f_{T,max} + \Delta f_{L,max}$ can be calculated. This linewidth represents a very important parameter for every coherent optical communication system. A summarized discussion on the phase noise problem and the linewidth requirements is given in Chapter seven.

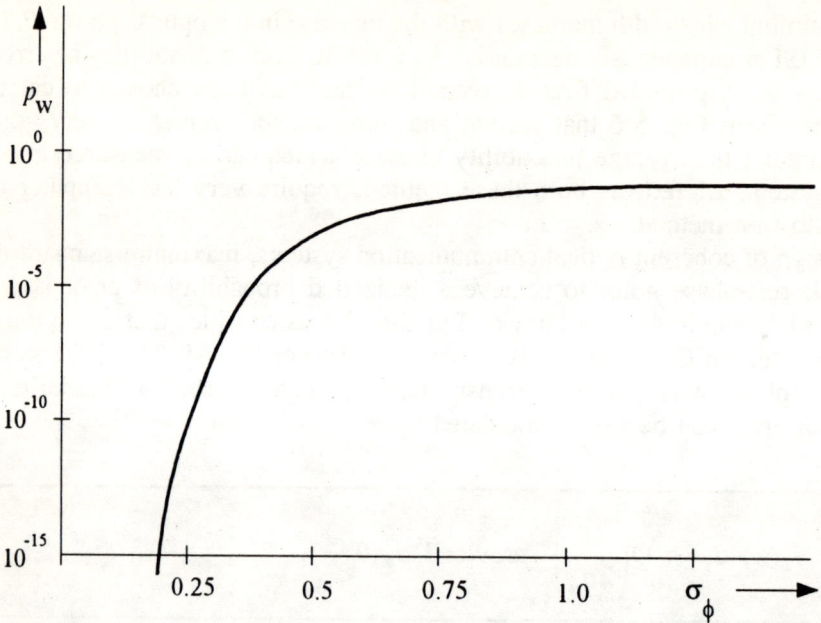

Fig. 5.7: Probability of error for ASK homodyne system when only phase noise disturbs the system

Some important results obtained in this Section on coherent optical ASK homodyne systems are summarized below.

- ASK homodyne systems are very sensitive to the laser phase noise of both transmitter and local lasers.

- Performance degradation caused by laser phase noise is different for symbols "0" and "1".

- ASK homodyne systems require an optical phase-locked loop (Section 5.1.3).

- Owing to phase noise, BER exhibits an error rate floor.

- Optimum threshold level in presence of laser phase noise is always less than the optimum level when phase noise is absent.

5.1.2 PSK HOMODYNE SYSTEM

(i) DETECTED SIGNAL $d(t)$ and DETECTED SAMPLE VALUES $d(\nu T + t_0)$

Like in ASK homodyne system, photodiode current in the signal branch of optical PSK homodyne receiver (Fig. 5.1) is given by

$$i_{PD}(t) = \hat{i}_{PD}\, s(t)\, e^{j\phi(t)}$$

$$= \hat{i}_{PD}\, \exp\!\left(j \sum_{\nu=-\infty}^{+\infty} \pi\,(1-s_\nu)\, \text{rect}\!\left(\frac{t-\nu T}{T}\right) + j\phi(t)\right) \tag{5.31a}$$

where $s_\nu \in \{0, 1\}$ again represents the information. The amplitude $\hat{i}_{PD}$ of photodiode current and phase noise process $\phi(t)$ are given by Eqs. (2.63) and (5.2) respectively. With the substitution $\tilde{s}_\nu = 2s_\nu - 1$, above equation can be written as

$$i_{PD}(t) = \hat{i}_{PD} \sum_{\nu=-\infty}^{\infty} \tilde{s}_\nu\, \text{rect}\!\left(\frac{t-\nu T}{T}\right) e^{j\phi(t)} \tag{5.31}$$

Comparing the photodiode currents for ASK system (Eq. 5.1) and PSK system, close relationship of both homodyne systems becomes clear immediately. In ASK system, s_ν is either 0 or 1, whereas in PSK system $\tilde{s}_\nu$ is either -1 or +1. Thus, a PSK homodyne system can be regarded as ASK homodyne system with a *bipolar modulation* i.e., $\tilde{s}_\nu \in \{-1, +1\}$. For this reason, these systems exhibit many similarities. To prevent repetitions, this Section is primarily focused on the differences only. For other description, Section 5.1.1 can be referred.

According to optical ASK homodyne system, statistical features of the sampled detected signal

$$d(t_0) = \hat{i}_{PD}\, a(t_0)\, \cos(\phi(t_0)) + n(t_0) \tag{5.32}$$

are of fundamental importance to assess the system performance. Difference in the sampled detected signals of both the homodyne systems is only given by the range of values of

$$a(t_0) = \int_{-\infty}^{+\infty} \underline{s}(\tau)\, h_B(t_0-\tau)\, d\tau \quad \text{with} \quad -1 \le a(t_0) \le +1 \tag{5.33}$$

It should be remembered that $a(t_0)$ equals the sample value $d(t_0)$ in absence of any noise i.e., $n(t_0)=0$ and $\phi(t_0)=0$ (Eq. 5.6). The range of values given in the above equation presumes a Gaussian baseband filter as given in Eq. (2.77). With the same normalizations and sub-stitutions as in the previous Section, normalized and sampled detected signal $d = d(t_0)/\hat{\imath}_{PD}$ of PSK homodyne system can be expressed as

$$d = a\cos(\phi) + n = aw + n \quad \text{with} \quad |w| \le 1 \quad \text{and} \quad -1 \le a \le +1 \tag{5.34}$$

Except the range of noiseless sample value a, this expression is same as Eq. (5.7) for ASK homodyne system. As Eq. (5.34) is of significant importance, physical meaning of various parameters is summarized below.

d: normalized detected sample value,

a: noiseless part of d disturbed by intersymbol interference (a includes the

             transmitted information),

$\cos(\phi)$:   phase noise factor,

n: additive Gaussian noise.

It becomes clear from the above equation that the sample values of the detected signal are again dependent on two random variables: one is the phase noise term $w=\cos(\phi)$ and the other is additive Gaussian receiver noise n.

(ii) PROBABILITY DENSITY FUNCTION $f_d(d)$

As the sampled detected signals in both ASK and PSK homodyne systems are formally equal, probability density functions can be calculated by using the same equation i.e., Eq. (5.9) derived in the previous Section. However, the shape of the probability density functions are different due to the different range of noiseless sample value a.

The pdf $f_d(d)$ in the special case of $a=-1$ and $a=+1$ which means $\bar{s}_r=-1$ and $\bar{s}_r=1$ respectively is shown in Fig. 5.8. In the first case, permanent "0" is transmitted whereas in the second case permanent "1". Comparing Figs. 5.8 and 5.2, an important difference between both the homodyne systems becomes clear. As compared to ASK homodyne system, pdf of the detected signal corresponding to symbol "0" and symbol "1" are now symmetrical with respect to $d=0$. The reason for this is the bipolar signal transmission scheme i.e., $\bar{s}_{\nu} \epsilon \{-1, +1\}$. Thus, both the binary symbols "0" and "1" are equally disturbed by phase noise and additive receiver Gaussian noise. The optimum threshold is always located at zero i.e., $E_{opt}=0$. In contrast to ASK homodyne system, PSK homodyne system belongs to the group of transmission systems which are characterized by *signal-independent noise*.

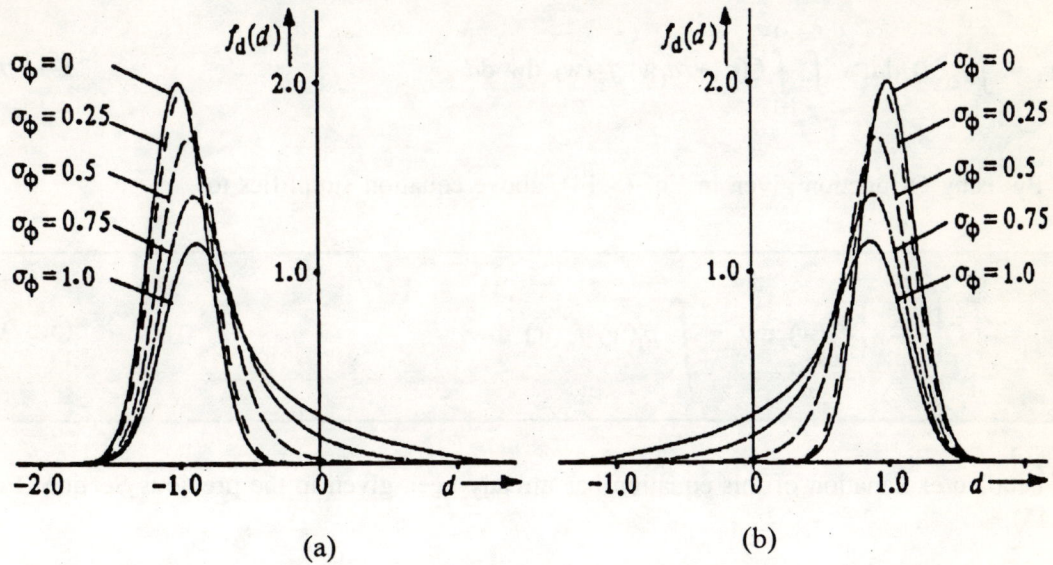

Fig. 5.8: Probability density function $f_d(d)$ of normalized sample value d of the detected signal $d(t)$ in PSK homodyne receiver. The normalized standard deviation of receiver additive Gaussian noise is fixed at $\sigma=0.2$. The transmitted information is: (a) permanent "0" and (b) permanent "1".

The pdf of detected signal d is again mainly determined by its expected value $E\{d\}=\eta_d$ and standard deviation σ_d as given in Eqs. (5.10a) and (5.10b) respectively. As both these quantities are functions of the noiseless sample value a, new range of a must be taken into account.

(iii) PROBABILITY OF ERROR OR BIT ERROR RATE (BER)

As the symbols "0" and symbol "1" are equally disturbed in PSK homodyne system, probabilities of error p_{0i} and p_{1i} for symbol pattern $<q_{\nu}>_{0i}$ and $<q_{\nu}>_{1i}$ are same. This is in contrast to ASK homodyne system where both probabilities of error are different. Hence, no distinction between the symbol pattern $<q_{\nu}>_{0i}$ and $<q_{\nu}>_{1i}$ is required to calculate the probabilities of error (see Section 2.5.3) provided that optimum threshold level is used i.e., $E_{opt}=0$. Hence,

$$a_i := a_{1i} = -a_{0i} \quad \rightarrow \quad f_{di}(d) := f_{d1i}(d) = f_{d0i}(-d) \quad \rightarrow \quad p_i := p_{1i} = p_{0i} \tag{5.35}$$

Like ASK homodyne system, probability of error for PSK homodyne system can be calculated as follows (compare with Eqs. 5.13 and 5.16):

$$p_i = \int_{-\infty}^{0} f_{di}(d) \; dd = \int_{-\infty}^{0} \int_{-1}^{+1} f_n(d + a_i w) \, f_w(w) \; dw \; dd \tag{5.36}$$

By using Q-function given in Eq. (5.14), above equation simplifies to

$$p_i = \int_{-1}^{+1} Q\left(\frac{a_i \, w}{\sigma}\right) f_w(w) \; dw = \int_{-1}^{+1} \tilde{p}_i(w) \, f_w(w) \; dw \tag{5.37}$$

A detailed explanation of this equation has already been given in the previous Section (Eq. 5.15).

(a) AVERAGE PROBABILITY OF ERROR p_a

Due to symmetrical properties of PSK homodyne system, average probability of error can be calculated somewhat easier than for ASK homodyne system (Eq. 5.18):

$$p_a = \sum_{i=1}^{2^{n+v+1}} p(<q_u>_i) \, p_i = 2^{-(n+v+1)} \sum_{i=1}^{2^{n+v+1}} p_i \tag{5.38}$$

It is seen from this equation that a distinction between both the probabilities p_{0i} and p_{1i} is not required since p_{0i} and p_{1i} are equal for PSK homodyne system i.e., $p_{0i}=p_{1i}=p_i$.

(b) WORST-CASE PROBABILITY OF ERROR p_w

In ASK homodyne system, calculation of the worst-case probability of error p_w was made by dividing p_w into two components p_{0w} and p_{1w}. This step was required because ASK system is based on unipolar signal transmission with $s_r \in \{0, 1\}$. In contrast, this division is not required when PSK homodyne system with bipolar signal transmission is considered i.e., $\tilde{s}_r \in \{-1, +1\}$. In this case, we maintain that

$$p_{0w} = p_{1w} = p_w \tag{5.39}$$

taking into account Eq. (5.37), we obtain

$$p_w = \int_{-1}^{+1} Q\left(\frac{a_w\, w}{\sigma}\right) f_w(w)\ dw = \int_{-1}^{+1} \tilde{p}_w(w)\, f_w(w)\ dw \tag{5.40}$$

where a_w equals the sampled detected signal $d_w = a_w \cos(\phi) + n$ when there is no noise (i.e., $\sigma = 0$ and $\sigma_\phi = 0$) and under worst-case pattern. Assuming a Gaussian baseband filter (Eq. 2.77), normalized worst-case sample value a_w can be calculated by using the equation

$$a_w = 1 - 4Q\left(\sqrt{2\pi}f_g T\right) = a_{1w} = -a_{0w} \tag{5.41}$$

As mentioned above, $a_{1w} = +a_w$ corresponds to the worst-case pattern

$$<q_v>_{1w}\ =\ <\cdots 0\ 0\ 0\ 1\ 0\ 0\ 0\ \cdots>$$

whereas the inverse pattern

$$<q_v>_{0w}\ =\ <\cdots 1\ 1\ 1\ 0\ 1\ 1\ 1\ \cdots>$$

is related to the normalized and noiseless sample value $a_{0w} = -a_w$. Both quantities a_{1w} and a_{0w} determine the *normalized aperture of the eye pattern*

$$A_{PSK} = a_{1w} - a_{0w} = 2a_w = 2\left[1 - 4Q\left(\sqrt{2\pi}f_g T\right)\right] = 2A_{ASK} \tag{5.42}$$

Like in ASK homodyne system, eye aperture again represents a very important measure to determine the system performance, particularly the probability of error. From Eqs. (5.40) and (5.42), we obtain:

$$P_w = \int_{-1}^{+1} Q\left(\frac{A_{PSK}\, w}{2\sigma}\right) f_w(w)\ dw \quad \text{provided} \quad E = E_{opt} = 0 \tag{5.43}$$

The eye aperture A_{PSK} of PSK homodyne system is twice the eye aperture A_{ASK} of ASK homodyne system. For this reason, PSK homodyne system offers a *6 dB higher sensitivity* than the ASK homodyne system provided phase noise does not disturb the system (i.e., $\sigma_\phi = 0$). This results in a lower probability of error. With the same error rate in both the systems, PSK homodyne system requires a *6 dB lower optical power* at the input to the receiver.

Substituting the pdf $f_w(w)$ of random variable $w = \cos(\phi)$ in terms of the pdf $f_\phi(\phi)$ of phase noise, Eq. (5.43) can be written as follows:

$$P_w = \frac{2}{\sqrt{2\pi}\sigma_\phi} \int_{0}^{+\infty} Q\left(\frac{A_{PSK}\, \cos(\phi)}{2\sigma}\right) \cdot \exp\left(-\frac{\phi^2}{2\sigma_\phi^2}\right) d\phi \tag{5.44}$$

(iv) OPTIMIZATION

The purpose of this Subsection is to optimize the system parameters for a minimum (worst-case) BER. Since the optimization of coherent optical PSK homodyne system is very similar to that of ASK homodyne system, this Subsection is restricted on the differences of both systems and principles. A more detailed discussion on the optimization of homodyne systems can be found in the previous Section 5.1.1 and also in the following Section 5.1.3, wherein the optical phase-locked loop (OPLL) is considered in more detail.

(a) THRESHOLD E

Due to the symmetrical features of optical PSK homodyne systems, optimum threshold level is

$$E_{opt} = 0 \tag{5.45}$$

This optimum threshold has already been used to derive the probability of error Eqs. (5.37), (5.43) and (5.44) given above.

(b) CUT-OFF FREQUENCY f_g OF THE LOW-PASS FILTER

As the error rate formulas (5.43) and (5.44) for PSK homodyne system are much simpler as compared to ASK homodyne system, optimization of the cut-off frequency f_g becomes easier. It becomes clear from Eq. (5.44) above that the worst-case probability of error reaches its minimum value if the argument of the Q-function is maximum. Therefore, the task of optimization is to maximize the expression

$$\frac{A_{\text{PSK}}(f_g)\,\cos(\phi)}{2\sigma(f_g)} = \frac{\left[1 - 4Q\left(\sqrt{2\pi}f_g T\right)\right]\cos(\phi)}{\sigma(f_g)} \;\to\; \max \tag{5.46}$$

with respect to f_g or $f_g T$. It should be noted that the first part of this expression is valid in general, whereas the second part is restricted on a Gaussian baseband filter. As the random variable $\cos(\phi)$ is not a function of cut-off frequency f_g, this variable (which is actually a constant here) does not influence the result. Hence, this term can be ignored during the maximization process. The normalized and optimized cut-off frequency $f_{g,\text{opt}}T$ depends only on the ratio of eye aperture to standard deviation of the additive Gaussian noise, which is same in ordinary digital communication systems [256, 260, 269].

By using appropriate iteration methods, maximization process yields again the following result

$$f_{g,\text{opt}}\,T \approx 0.79$$

It must be remembered that this numerical value presumes a Gaussian baseband filter. In contrast to ASK homodyne system, optimum cut-off frequency $f_{g,\text{opt}}$ is now independent of the phase noise.

(v) EVALUATION AND DISCUSSION

The worst-case probability of error p_w versus received optical power P_r at the receiver input for different standard deviation σ_ϕ of the OPLL rest-phase noise is shown in Fig. 5.9. Threshold E and cut-off frequency f_g of the Gaussian low-pass filter are already optimized. All other system parameters are summarized in Table 7.1 of Chapter seven.

Except a shift of about 6 dB to the lower optical powers P_r, BER curves shown in Fig. 5.9 exhibit the same characteristic features as for ASK homodyne system shown in Fig. 5.5. Therefore, the detailed discussion below the Fig. 5.5 in the previous Section is valid here also. The shift of about 6 dB caused by the improved sensitivity of PSK homodyne system in comparison to ASK homodyne system represents a significant advantage.

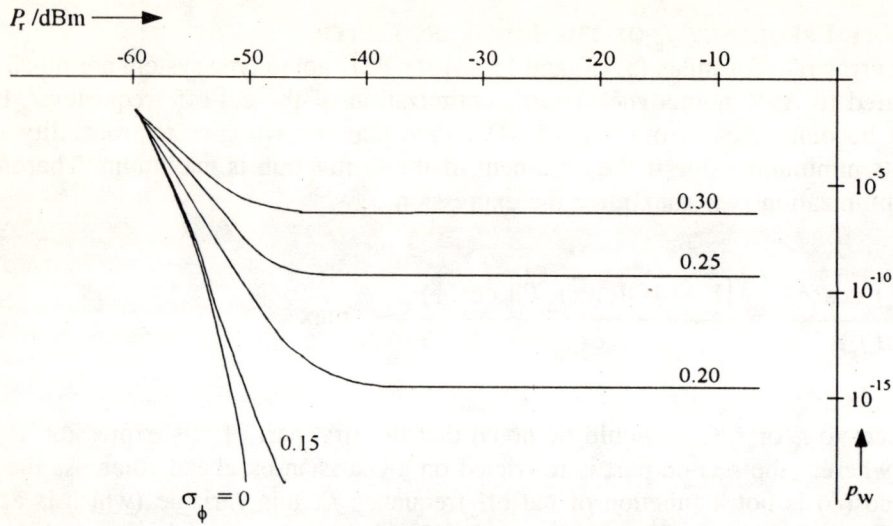

Fig. 5.9: Worst-case probability of error p_w for PSK homodyne system

Finally, let us consider again the probability of error in absence of ISI and additive Gaussian noise i.e., $\sigma=0$. Analogous to Eq. (5.29) in the previous Section, we obtain

$$p_w = \int\limits_{-1}^{0} f_w(w)\ dw \approx 2Q\left(\frac{\pi}{2\sigma_\phi}\right) \quad \text{provided } \sigma=0 \tag{5.47}$$

Comparing Eqs. (5.47) and (5.29), it becomes evident that the probability of error is now twice of ASK homodyne system. The reason for this surprising result is that *both* symbol $q_v="0"$ ($s_v=-1$) and symbol $q_v="1"$ ($s_v=1$) are disturbed by phase noise in PSK system, while symbol $q_v=0$ ($s_v=0$) remains undisturbed in ASK system. However, this is a rather unrealistic case ($\sigma=0$) since a sensitivity gain is normally achieved by PSK homodyne system when $\sigma\neq0$. We have to remember that Eq. (5.47) is indeed a very useful tool to estimate the maximum allowed standard deviation σ_ϕ, but it is not usable to evaluate the true error rate which can be measured in the receiver. For this, Eq. (5.44) must be employed. Assuming, for example, a probability of bit error $p_w=10^{-10}$, σ_ϕ should not exceed $\sigma_{\phi,max}\approx0.24$.

5.1.3 PHASE CONTROL IN HOMODYNE SYSTEMS

In coherent optical communication systems with homodyne detection, phase control and phase tracking is indispensable. Without any phase control, standard deviation σ_ϕ of the phase noise would increase more and more since the phase noise of transmitter and local lasers are nonstationary random processes. Finally, standard deviation σ_ϕ would reach infinity and probability of error would become very high and unacceptable i.e., 50 % in PSK and at least 25 % in ASK homodyning [61].

The most common architecture of phase-control circuit is a loop. The task of this feedback loop is to lock the phase of both local laser light wave and received optical light wave. Therefore, this loop is called a phase-locked loop (PLL). Since both phases are related to optical waves in optical homodyne system, this loop is called *optical phase-locked loop* (OPLL).

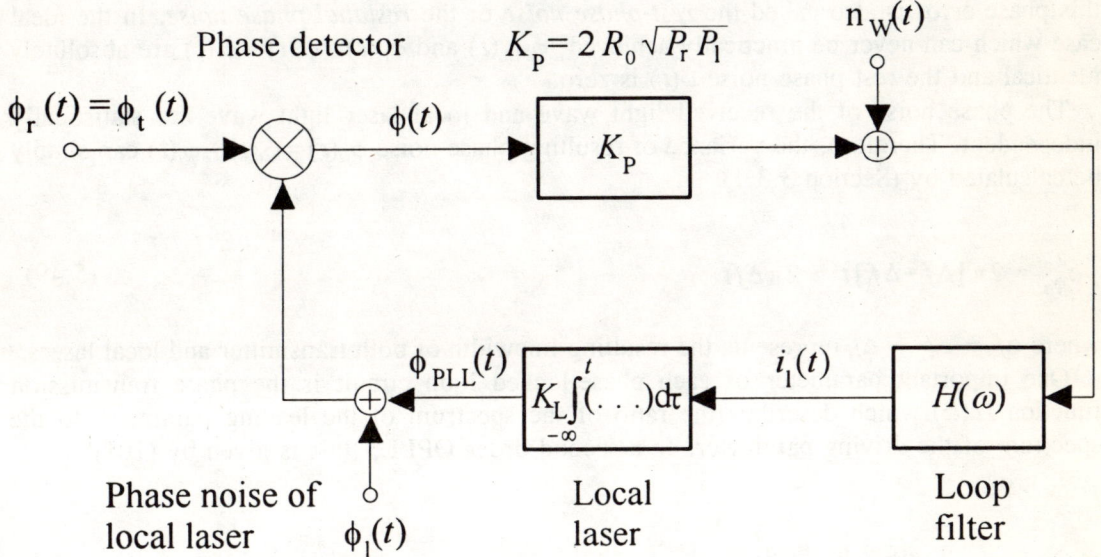

Fig. 5.10: Linearized model of optical phase-locked loop (OPLL)

In order to focus our discussion on the principle, we take advantage of the simplest configuration of optical phase-locked-loop circuit as already shown in Fig. 5.1. A more detailed discussion of phase control circuits is presented, for example, in [22, 23, 72, 103]. Linearized model of OPLL is shown in Fig. 5.10 [23, 103, 245, 246]. We talk about a linearized model, because we take advantage of the approximation $\sin(\phi) \approx \phi$. In contrast to Fig. 5.1, this model provides a balanced receiver with two photodiodes, which means one photodiode at each of the outputs of an optical 3 dB coupler (Chapter two). In this case, phase detector shown in Fig. 5.10 includes the optical coupler as well as both the photodiodes.

The phase to be controlled in the OPLL, frequently called the leading parameter or the leading phase, is the phase noise of the received optical light wave which equals the phase noise of the transmitter laser i.e., $\phi_r(t) = \phi_t(t)$. Driving parameter is the phase $\phi_{PLL}(t)$ of the local laser located in the OPLL circuit. This phase must continuously be tracked to keep the *phase error* $\phi(t)$ of the OPLL low i.e., the difference between driving and leading phases.

As shown in Fig. 5.10, operation of OPLL is always disturbed by the phase noise $\phi_l(t)$ of the local laser and receiver white Gaussian noise $n_w(t)$ i.e., shot noise of photodiodes and thermal noise. As mentioned above, fundamental operation of OPLL is to keep the phase error

$$\phi(t) = [\phi_t(t) - \phi_l(t)] - \phi_{PLL}(t) = \phi_{tl}(t) - \phi_{PLL}(t) \qquad (5.48)$$

low. As the time-varying phase error represents the phase noise which remains uncontrolled, this phase error is also called the *rest-phase noise* or the *residual phase noise*. In the ideal case which can never be practically achieved, $\phi_{PLL}(t)$ and $\phi_{tl}(t) = \phi_t(t) - \phi_l(t)$ are absolutely identical and the rest-phase noise $\phi(t)$ is zero.

The phase noise of the received light wave and local laser light wave are statistically independent. Therefore, the variance of resulting phase noise $\phi_{tl}(t) = \phi_t(t) - \phi_l(t)$ can simply be calculated by (Section 3.3.1)

$$\sigma^2_{\phi_{tl}} = 2\pi [\Delta f_t + \Delta f_l]t = 2\pi \Delta f t \qquad (5.49)$$

where $\Delta f = \Delta f_t + \Delta f_l$ represents the resulting linewidth of both transmitter and local lasers.

One important parameter of each phase-locked loop circuit is the phase transmission function $H(\omega)$ which describes the ratio of the spectrum of the leading parameter to the spectrum of the driving parameter. In a second order OPLL, this is given by [101]

$$H(\omega) = \frac{2j\xi\omega_n\omega + \omega_n^2}{(j\omega)^2 + 2j\xi\omega_n\omega + \omega_n^2} \qquad (5.50)$$

The phase transmission function includes two characteristic parameters of significant importance: one is the damping constant ξ the other is natural frequency ω_n of the loop. For a PI-loop filter (proportional and integrating-loop filter) with a frequency response

$$H_{PI}(\omega) = \frac{1 + j\omega\tau_2}{j\omega\tau_1} = \frac{\tau_2}{\tau_1}\left[1 + \frac{1}{j\omega\tau_2}\right] \qquad (5.51)$$

parameters ξ and ω_n are given by

$$\omega_n = \sqrt{\frac{K}{\tau_1}} \qquad \xi = \frac{1}{2}\omega_n\tau_2 \qquad K = K_P K_L \qquad (5.52)$$

Here, τ_1 and τ_2 are the time constants for PI filter and K the loop gain. It should be noted that a PI-loop filter normally provides sufficient stability to the OPLL. It becomes clear from Fig. 5.10 that the constant K_P is expressed in unit of A (Ampere) and the local laser constant K_L is measured in Hz/A. Hence, the loop gain itself is dimensionless as expected.

Taking now into account the impulse response $h(t)$ $\circ\!\!-\!\!\bullet$ $H(\omega)$ of the phase transmission function, following equation for the rest-phase noise $\phi(t)$ can be derived directly from the Fig. 5.10:

$$\phi(t) = [\phi_t(t) - \phi_l(t)] - [\phi_t(t) - \phi_l(t)] \star h(t) - K_P^{-1} n_w(t) \star h(t)$$

$$= \phi_{tl}(t) \star [\underline{\delta(t) - h(t)}] - n_w(t) \star \underline{K_P^{-1} h(t)}$$

$$\qquad\qquad \updownarrow \qquad\qquad\qquad \updownarrow$$

$$\qquad\qquad \frac{1}{2\pi}[1 - H(\omega)] \qquad \frac{1}{2\pi} K_P^{-1} H(\omega) \qquad (5.53)$$

Here, the sign $\star$ represents a convolution. The second line of this equation makes use of the sampling properties of Dirac delta function $\delta(t)$ i.e., $\phi_{tl}(t) \star \delta(t) = \phi_{tl}(t)$. Physical meaning of Eq. (5.53) using well-known components of system theory is illustrated in Fig. 5.11.

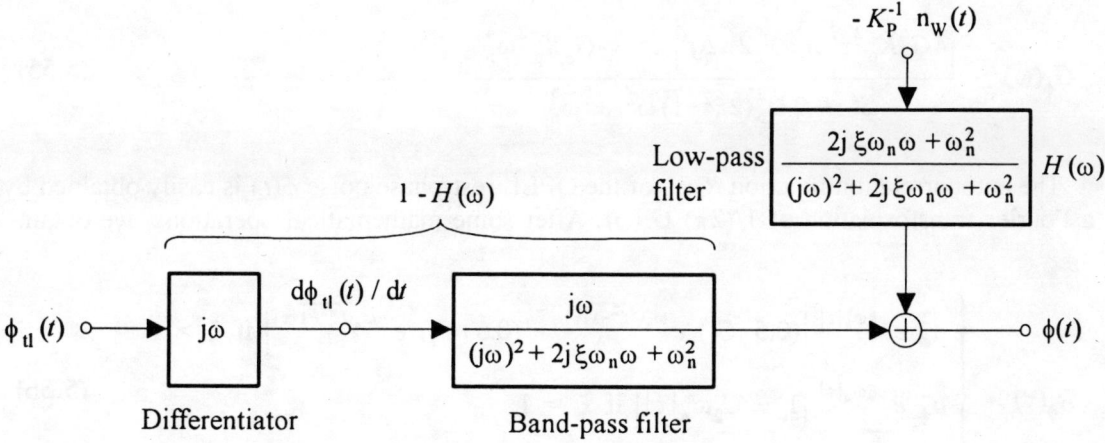

Fig. 5.11: Emergence of OPLL rest-phase noise $\phi(t)$ explained by means of system theory

As shown in this figure, resulting nonstationary laser phase noise $\phi_{tl}(t)$ is first fed to a system block which differentiates its input signal. In accordance with Section 3.3.3, output signal $d\phi_{tl}(t)/dt$ of this block represents a stationary white frequency noise with constant power spectral density $2\pi\Delta f = 2\pi[\Delta f_t + \Delta f_1]$. Thus, OPLL rest-phase noise $\phi(t)$ is composed of two different and independent white sources of noise: one is the white frequency noise $d\phi_{tl}(t)/dt$ and the other is receiver white Gaussian noise $n_w(t)$ i.e., shot and thermal noise. The first noise is filtered by a band-pass filter, whereas the second is filtered by a low-pass filter. As both sources of noise are stationary and Gaussian with zero mean, rest-phase noise exhibits the same statistical features. Therefore, rest-phase noise is a *stationary Gaussian random process with zero mean*.

For determining the BER, statistical features of the OPLL rest-phase noise $\phi(t)$ are of fundamental importance. For this reason, we now derive the following characteristic functions and parameters: noise power spectral density $G_\phi(f)$, autocorrelation function $R_\phi(\tau)$ and standard deviation σ_ϕ.

As both sources of noise are statistically independent, noise power spectral density $G_\phi(f)$ of the OPLL rest-phase noise equals the sum of both power spectral densities: filtered frequency noise and filtered shot and thermal noise. We obtain [101]

$$G_\phi(\omega) = 2\pi\Delta f \left|\frac{1-H(\omega)}{j\omega}\right|^2 + \frac{G_c}{K_P^2}|H(\omega)|^2 \tag{5.54}$$

It may be remembered that G_c represents the power spectral density of the receiver white Gaussian noise $n_w(t)$ (Chapter two). Replacing the phase transmission function $H(\omega)$ using Eq. (5.50), above equation finally takes the following form:

$$G_\phi(\omega) = \frac{\left[4 G_c K_P^{-2}\xi^2\omega_n^2 + 2\pi\Delta f\right]\omega^2 + G_c K_P^{-2}\omega_n^4}{\omega^4 + 2\omega_n^2(2\xi^2-1)\omega^2 + \omega_n^4} \tag{5.55}$$

The autocorrelation function $R_\phi(\tau)$ of the OPLL rest-phase noise $\phi(t)$ is easily obtained by a Fourier transformation of $1/(2\pi)\cdot G_\phi(\omega)$. After some mathematical operations, we obtain

$$R_\phi(\tau) = \begin{cases} \sigma_\phi^2\, e^{-\xi\omega_n|\tau|}\left[(0.5-C_1)\,e^{\omega_n\sqrt{\xi^2-1}|\tau|} + (0.5+C_1)\,e^{-\omega_n\sqrt{\xi^2-1}|\tau|}\right] & \text{if } \xi > 1 \\[2mm] \sigma_\phi^2\, e^{-\xi\omega_n|\tau|}\left[1 - C_2\omega_n|\tau|\right] & \text{if } \xi = 1 \\[2mm] \sigma_\phi^2\, e^{-\xi\omega_n|\tau|}\left[\cos\left(\omega_n\sqrt{1-\xi^2}|\tau|\right) - C_3\sin\left(\omega_n\sqrt{1-\xi^2}|\tau|\right)\right] & \text{if } \xi < 1 \end{cases} \tag{5.56}$$

where

$$\sigma_\phi^2 = R_\phi(0) = \frac{1}{2\pi} \int_{-\infty}^{+\infty} G_\phi(\omega) \, d\omega = \frac{\pi \Delta f}{2\xi \omega_n} + \frac{G_c \omega_n}{K_p^2} \left[\frac{1}{4\xi} + \xi \right] \qquad (5.57)$$

represents the variance. Mathematically, this variance is simply obtained by evaluating the area under the noise power spectral density $G_\phi(f)$. The amplitude coefficients C_1, C_2 and C_3 used in Eq. (5.56) are given by [54]

$$C_1 = \frac{1}{2} \frac{a + b \, (4\xi^2 - 1)}{a + b \, (4\xi^2 + 1)} \frac{\xi}{\sqrt{\xi^2 - 1}} \qquad (5.58)$$

$$C_2 = \frac{a + 3b}{a + 5b} \qquad (5.59)$$

$$C_3 = \frac{a + b \, (4\xi^2 - 1)}{a + b \, (4\xi^2 + 1)} \frac{\xi}{\sqrt{1 - \xi^2}} \qquad (5.60)$$

where

$$a = \frac{2\pi \Delta f}{\omega_n} \qquad (5.61)$$

and

$$b = \frac{G_c \omega_n}{K_p^2} \qquad (5.62)$$

are used as parameters. As explained in the previous Sections 5.1.1 (ASK homodyne) and 5.1.2 (PSK homodyne), σ_ϕ represents a very important system parameter which directly influences the BER and, therefore, the system performance. It was shown that the standard deviation σ_ϕ of the OPLL rest-phase noise should not exceed a certain value to achieve a desired BER; for example, $\sigma_\phi \leq 0.24$ for $p_w \leq 10^{-10}$ in PSK homodyne receiver. Considering Eq. (5.57), it can be seen that an optimum natural frequency ω_n must exist for a minimum standard deviation σ_ϕ i.e., a minimum influence of OPLL rest-phase noise. This optimum natural frequency usually called the OPLL-loop bandwidth is given by

$$\omega_{n,opt} = \sqrt{\frac{2\pi \Delta f K_p^2}{G_c(4\xi^2+1)}} \tag{5.63}$$

When $\omega_n < \omega_{n,opt}$, OPLL is no longer in a position to reduce the influence of the rest-phase noise sufficiently. As a result, influence of rest-phase noise $\phi(t)$ increases and system performance degrades. Under the condition $\omega_n > \omega_{n,opt}$, OPLL would more and more be disturbed by the receiver additive Gaussian noise $n_w(t)$. Again the influence of rest-phase noise $\phi(t)$ increases. The optimum loop frequency $\omega_{n,opt}$ given in Eq. (5.63) seems to be the best solution. However, this solution is only valid when the received optical light wave is unmodulated. In the case of modulated optical phase, phase-locked loop is not able to distinguish between the phase noise and the phase information. Therefore, great care has to be taken since a phase-locked loop circuit reduces the signal part of the phase in the same way as the noise part of the phase. For this reason, smaller loop bandwidth than given in Eq. (5.63) must be used to prevent signal cancellation. In this book, we consider that the loop bandwidth ω_n is fixed at $\omega_n = \omega_{n,max} = 2\pi \cdot 0.001/T$ where T represents the bit duration. This bandwidth provides a well-suited compromise between the reduction of influence of laser phase noise and maintaining the information part of the phase.

In the previous Section, we have derived that the standard deviation of the OPLL rest-phase noise σ_ϕ must be less than 0.24 to achieve a worst-case probability of error $p_w = 10^{-10}$ in optical PSK homodyne system. If, for example, an OPLL is used with $\xi = 1$ (non-periodical transient response), then the above requirements yield a *maximum allowed laser linewidth*

$$\Delta f_{max} = \begin{cases} \dfrac{K_p^2}{640\pi G_c} & \text{if} \quad \omega_n = \omega_{n,opt} \\[2em] \dfrac{1}{4}\dfrac{\omega_{n,max}}{2\pi}\left[1-20\dfrac{G_c\,\omega_{n,max}}{K_p^2}\right] & \text{if} \quad \omega_n = \omega_{n,max} \end{cases} \tag{5.64}$$

It should be noted that this linewidth derived using Eqs. (5.57) and (5.63) represents the resulting linewidth of both transmitter and local lasers. For simplicity $\sigma_\phi = 0.24$ has been taken as 1/4.

Example 5.1

Let the parameters of given system are: $P_r = -50$ dBm, $P_l = -10$ dBm, $R_0 = 1$ A/W, $G_c = 2.6 \cdot 10^{-23}$ A^2/Hz, $\xi = 1$ and $T = 6.43$ ns corresponding to a bit rate $R = 155.52$ Mbit/s in SDH (Synchronous Digital Hierarchy). In this case, Eq. (5.64) yields a maximum allowed laser linewidth of $\Delta f_{max} = \Delta f_t + \Delta f_l = 38.88$ kHz. This linewidth corresponds to 0.025% of the bit rate.

We have so far analyzed OPLL circuit which is characterized by a simple feedback loop such as shown in Figs. 5.1 and 5.10. Advantages and drawbacks of this special configuration are discussed below. Instead of simple feedback loop, some other important configurations are available for matching the phase and frequency of received and local laser waves. Now, we have to focus our interest on these approaches. In the frame of this book, a brief discussion is sufficient.

The most suitable phase control circuits for coherent detection can be classified into three categories:

- OPLL with pilot carrier,

- Costas loop circuit, and

- OPLL with synchbit control.

(i) OPLL WITH PILOT CARRIER

The optical phase-control circuit shown in Figs. 5.1 and 5.10 and discussed above belongs to this group. To maintain endless operation i.e., permanent phase tracking without any break, OPLL always requires a permanent signal at the optical carrier frequency f_t. However, an optical wave which is PSK modulated does not normally exhibit a carrier signal a priori. For this purpose, a pilot carrier which requires a certain part of the available transmitter power must be generated [101]. Usually, the pilot carrier is generated by a phase deviation less than $\pm 90°$. In the OPLL, phase error signal is separated from the information signal by a simple low-pass filter such as shown in Fig. 5.1 or Fig. 5.10. If ASK signal is applied to the system, then the OPLL requires a suppression of the amplitude modulation since a permanent DC signal (i.e., $s_\nu = 0$) would immediately cause a total break of the phase-control function.

An OPLL with pilot carrier needs very less number of components and the required circuit is very simple. Both features represent a significant advantage of using the pilot-carrier technique. Furthermore, this configuration is well-suited for optical integration. However, this technique exhibits some important drawbacks: First, a DC-coupled amplifier is required, which is difficult to realize when bit rate is high. Second, coupling ratio k of the optical coupler must be exactly 50% since each deviation pretends a phase error. Finally, the information signal should not have any spectral component in the OPLL bandwidth which, for example, can be achieved by coding. Since a decreased phase deviation reduces the available signal power, a sensitivity loss of about 0.5 dB can be observed.

(ii) COSTAS LOOP CIRCUIT

The most common OPLL technique is the Costas loop. Here, the phase error signal is obtained by detecting the inphase as well as the quadrature component of the received signal which are multiplied afterwards [102]. The inphase component is same as information signal (launched into the sample and hold circuit). The multiplied signal represents the phase error signal. As the quadrature component requires a certain part of the received optical power, a loss in sensitivity of about 0.5 dB is again effected.

The Costas loop technique avoids all the drawbacks of the pilot carrier technique. However, an 90° optical hybrid (90° coupler) is required to separate inphase and quadrature components, which is difficult to realize and integrate. The most convenient way to realize an optical hybrid is to model this device by an optical three-port fiber coupler. In contrast to pilot carrier technique, Costas loop requires at least two complete front ends, which represents a serious drawback.

(iii) OPLL WITH SYNCHBIT CONTROL

A third method of phase tracking is given by the synchbit control [280, 281]. This method combines the advantages and avoid the drawbacks of both pilot carrier technique and Costas loop. Requirements of somewhat more sophisticated digital signal processing is the only disadvantage. The synchbit approach is based on the principle of Costas loop. The Costas loop technique requires a permanent quadrature component to maintain phase tracking, whereas the synchbit technique switches between inphase and quadrature components. Most of the time, received signal is fed into the signal branch of the receiver (inphase component). Thereby, no signal loss occurs. To track the phase, it is periodically required to switch the received signal to the phase-control branch (quadrature component) for a short time. During this time of usually one bit duration, no information is sent. Instead, a synchronization bit is transmitted allowing to generate the phase error signal.

In order to maintain the information bit rate, channel bit rate (information and synchronization bits) must be increased. When one synchronization bit is always transmitted after eight information bits, the channel rate is 9/8 of the information bit rate. Since an increased channel bit rate requires an increased bandwidth (which increases the influence of the receiver additive Gaussian noise), the sensitivity is reduced by $10 \log(9/8)$ dB ≈ 0.5 dB. This loss in sensitivity is same as in Costas loop.

In the frame of this book, it is not required to discuss different configurations of phase-control circuits in more detail. Different techniques of PLL circuits are already well-known and well-described in many excellent books and papers e.g., [23, 101-103, 128, 129, 245, 246]. In addition, the principle function of a phase-control circuit remains the same irrespective of whether a conventional electrical PLL or an OPLL is considered. Therefore, the reader can see the theory and the design of phase-control circuits in one of the references mentioned above.

5.2   COHERENT HETERODYNE SYSTEMS

In a coherent optical heterodyne system, received optical light wave is first converted to an electrical intermediate frequency (IF) signal by means of an optical coupler, a local laser and one or two photodiodes. Thereby, the IF is equal to the difference of both transmitter laser and local laser frequencies (see Chapter two). Afterwards, IF signal is demodulated with respect to the modulation scheme applied in the transmitter. In order to realize the demodulator, all types of demodulation techniques which are well-known from ordinary digital communication can be employed.

In optical homodyne systems, received optical signal is always detected coherently (homodyne systems with noncoherent detection are impossible), whereas in optical heterodyne system either coherent or noncoherent detection of the modulated IF signal can be used in the receiver. This Section is focused on *coherent detection* or in other words *synchronous detection*.

As the principle function of heterodyne system with coherent detection is very much similar to homodyne system which already has been discussed in the previous Sections 5.1.1 (ASK homodyne) and 5.1.2 (PSK homodyne), discussion in this Section will be rather restricted. Further, heterodyne systems with coherent detection are of little practical interest as compared to heterodyne systems employing noncoherent detection (Section 5.3) or homodyne systems, in particular PSK homodyne system.

Due to the twofold meaning of the word "coherent" i.e., coherent laser and coherent detection (synchronous detection) the reader may read again the discussion at the beginning of this Chapter. Like homodyne systems, coherent heterodyne systems can also be realized by applying amplitude shift keying (ASK) or phase shift keying (PSK). In addition, frequency shift keying (FSK) can also be used. For this, a two-filter receiver with one synchronous detector in each filter branch is required. It should be remembered that FSK is impossible in homodyne systems (Section 5.1.1).

Synchronous detection of the received IF signal requires the generation of a separate IF carrier signal in the receiver, which must absolutely be synchronous (i.e., coherent) with the received IF carrier signal. Here, synchronous means that both signals must be equal in frequency *and* phase. For this, a phase-locked loop circuit is required. In contrast to homodyne system where an OPLL is needed, coherent heterodyne system only requires an electrical PLL which is usually based on standard voltage-controlled oscillator (VCO). The task of this PLL is to match the frequency and phase of the VCO and the received IF carrier signal to enable a real coherent detection.

The principle function of OPLL and electrical PLL is the same. Therefore, the model of the optical phase-control circuit shown in Fig. 5.10 (Section 5.1.3) can be used further to describe the electrical PLL in coherent heterodyne system provided some simple modifications are performed first. The modifications required are summarized in Table 5.1. In addition, following differences must also be taken into account:

- local laser operation is replaced by VCO,

- phase detector only includes electrical components instead of optical components such as photodiode and optical coupler,

- phase noise of IF signal is driving parameter of the electrical PLL. This is in contrast to OPLL, where this parameter is same as the phase noise of received light wave.

- source of receiver additive noise (shot and thermal noise) is located outside the PLL circuit i.e., at the input to electrical PLL. This is in contrast to OPLL, where this source is located inside the OPLL circuit. However, the OPLL model of Fig. 5.10 can be used further when the constant of proportionality K_P is modified appropriately (Table 5.1),

- realization of electrical PLL is much more convenient than the OPLL.

Table 5.1: Phase control in homodyne and coherent heterodyne systems

	Homodyne system	Heterodyne system	Remark
Phase detector	optical by means of an optical coupler and one or two photodiodes	electrical	
Local oscillator	local laser	VCO	
Leading parameter	$\phi_t(t)$	$\phi_t(t) - \phi_l(t)$	$\phi_t(t)$, $\phi_l(t)$: phase noise of transmitter and local lasers respectively
Additional sources of noise in PLL	$n_w(t)$ and $\phi_l(t)$	$n_w(t)$ and $\phi_{VCO}(t)$	$n_w(t)$: white noise (shot and thermal noise) $\phi_{VCO}(t)$: phase noise of VCO (negligible)
Constant K_P of proportionality	$R_0\sqrt{4P_rP_1}$	$R_0\sqrt{2P_rP_1}$	compare with [103]

As the analysis of coherent heterodyne system is similar to the analysis of homodyne system, an explicit system calculation will not be performed in this Section. For the BER calculation, same formulas as derived in Sections 5.1.1 and 5.1.2 can be used. We have only to take into account that the normalized standard deviations σ_ϕ and σ of the PLL rest-phase noise and the additive Gaussian noise have different numerical values in coherent heterodyne receiver (see Table 5.2).

In the absence of phase noise, homodyne system exhibits a 3 dB sensitivity gain in comparison to coherent heterodyne system provided that the same modulation scheme has been applied (ASK or PSK). Irrespective of whether coherent or noncoherent detection (Section 5.3) is used in the heterodyne receiver, limited bit rate in comparison to homodyne systems is a main disadvantage. A homodyne system can be regarded as a baseband system, whereas a heterodyne system operates in the IF domain. For this reason, the photodiodes as well as the amplifiers should be able to follow the high frequency IF signal. Assuming, for example, a bit rate of 1 Gbit/s, an IF of about 4 GHz and a cut-off frequency of about 5 GHz for the photodiodes and amplifiers are required. In contrast to this, both the components only require a cut-off frequency of about 1 GHz if an optical homodyne system is used. In Chapter seven (system comparison), this problem will be discussed in more detail.

Main differences of heterodyne systems with coherent detection and homodyne systems are summarized in Table 5.2. All equations in this table are based on the results of Chapter two. Thereby, a balanced receiver with two photodiodes ($K_B=2$) and an optical 3 dB coupler (i.e., $k=0.5$) are presumed.

An interesting result is highlighted in this table. In homodyne detection, baseband signal power S_B as well as the baseband noise power N_B are both always higher than that of heterodyne detection. However, homodyne detection offers a 3 dB improvement in the signal-to-noise ratio $(S/N)_B$. The reason for this is a fourfold increase in the signal power, but only a twofold increase in the noise power in comparison to coherent heterodyne system.

Like optical homodyne systems, performance of optical heterodyne systems with coherent detection is seriously influenced by the phase noise of both transmitter laser and local laser. Thus, requirements for the maximum permissible laser linewidths are very strong. For PSK heterodyne system, for example, a linewidth-bit duration product in the order of 10^{-4} is required (Chapter seven). This means that the laser linewidth of both the lasers must be less than 0.01 percent of the bit rate. In contrast to this, noncoherent heterodyne system such as DPSK heterodyne system with autocorrelation detection only requires a linewidth-bit duration product of about 10^{-2}. This corresponds to a maximum permissible laser linewidth of one percent of the bit rate. In noncoherent FSK and ASK heterodyne systems with envelope detection, even linewidths same as the bit rate are acceptable (Section 5.3 and Chapter seven).

Table 5.2: Characteristic features of coherent heterodyne and homodyne systems

	Homodyne system	Heterodyne system	Remark
Optical input wave $E(t)$	$[\underline{E}_r(t) + \underline{E}_1(t)]\cos(\omega_1 t)$	$\underline{E}_r(t)\cos(\omega_r t)$ $+ \underline{E}_1(t)\cos(\omega_1 t)$	$\omega_r = \omega_1$ (see Section 2.4.2)
IF signal $i_{IF}(t)$		$\hat{i}_{PD}\cos(\omega_{IF}t) + n_{IF}(t)$ where $n_{IF}(t) = x(t)\cos(\omega_{IF}t) +$ $y(t)\sin(\omega_{IF}t)$	$(\hat{i}_{PD})^2 = 4\,R_0^2 P_r P_1$ provided a 3 dB coupler and a balanced receiver are used (see Eq. 2.63) $n_{IF}(t)$: noise in IF band
Baseband signal $i_B(t)$	$\hat{i}_{PD} + n_B(t)$	$\tfrac{1}{2}\hat{i}_{PD} + \tfrac{1}{2}x(t)$	$(\hat{i}_{PD})^2 = R_0^2 P_r P_1$ (see remark above) $n_B(t)$: noise in baseband
Power budget in IF domain		$S_{IF} = 2\,R_0^2 P_r P_1$ $N_{IF} = 2\,G_c B_{IF}$ $= \sigma_x^2 = \sigma_y^2 = \sigma_{het}^2$ $\left(\dfrac{S}{N}\right)_{IF} = 1\dfrac{R_0^2 P_r P_1}{G_c B_{IF}}$	G_c: double-sided noise power spectral density as given in Eq. (2.66) B_{IF}: noise-equivalent bandwidth of IF filter
Power budget in baseband	$S_B = 4\,R_0^2 P_r P_1$ $N_B = 1\,G_c B_B = \sigma_{hom}^2$ $\left(\dfrac{S}{N}\right)_B = 4\dfrac{R_0^2 P_r P_1}{G_c B_B}$	$S_B = 1\,R_0^2 P_r P_1$ $N_B = \tfrac{1}{2}G_c B_B = \tfrac{1}{4}\sigma_x^2$ $= \tfrac{1}{4}\sigma_{het}^2$ $\left(\dfrac{S}{N}\right)_B = 2\dfrac{R_0^2 P_r P_1}{G_c B_B}$	B_B: double-sided noise-equivalent bandwidth of the baseband filter $B_B = B_{IF}$ $\boxed{\dfrac{(S/N)_{B,hom}}{(S/N)_{B,het}} = 2}$
Normalized variance σ^2 of receiver additive Gaussian noise	$\sigma^2 = \dfrac{\sigma_{hom}^2}{\hat{i}_{PD}^2}$ $= \dfrac{1}{4}\dfrac{G_c B_B}{R_0^2 P_r P_1}$	$\sigma^2 = \dfrac{\sigma_{het}^2}{\hat{i}_{PD}^2} = 2\dfrac{\sigma_{hom}^2}{\hat{i}_{PD}^2}$ $= \dfrac{1}{2}\dfrac{G_c B_B}{R_0^2 P_r P_1}$	see Eq. (2.78)
Variance σ_ϕ^2 of rest-phase noise	see Eq. (5.57)	Eq. (5.57) with K_P as given in Table 5.1	

5.3 NONCOHERENT HETERODYNE SYSTEMS

In comparison to systems with coherent detection (i.e., homodyne systems or coherent heterodyne systems), noncoherent heterodyne systems do not require a phase tracking circuit, which represents a fundamental advantage of employing noncoherent heterodyne systems. To stabilize the intermediate frequency (IF), standard automatic frequency control (AFC) is quite sufficient. The purpose of this Section is to analyze and optimize the following noncoherent heterodyne systems in detail:

- ASK heterodyne system with envelope detection (Section 5.3.1),

- FSK heterodyne system with one- or two-filter configuration and envelope detection (Section 5.3.2),

- DPSK heterodyne system with autocorrelation demodulator or two-filter detection (Section 5.3.3).

In the calculation and optimization of the systems listed above, we follow the same steps as in the previous Sections.

5.3.1 ASK HETERODYNE SYSTEM

This Section describes the system calculation and optimization of optical ASK heterodyne receiver with standard envelope detection. In particular, the intermediate frequency signal (IF signal) and the signal after envelope detection (called the envelope signal) are derived taking into consideration laser phase noise, receiver additive Gaussian noise, influence of the intermediate frequency filter (IF filter) on both sources of noise and intersymbol interference (ISI).

Based on the derived envelope signal which includes the transmitted information, probability density function and the bit error rate are calculated exactly and by approximations. For a minimum bit error rate, optimum IF filter bandwidth is evaluated as a function of line-widths of both transmitter laser and local laser. To highlight typical features of noncoherent ASK heterodyne systems with envelope detection, computer simulated eye pattern are finally presented.

As discussed in the previous Sections, optical homodyne systems (Section 5.1) and optical heterodyne systems with coherent detection (Section 5.2) are very sensitive to the laser phase noise of transmitter laser and local laser. The most sensitive system is coherent PSK homodyne, whereas the least sensitive system is noncoherent ASK heterodyne with envelope detection. As described in this Section, even in the latter system laser phase noise cannot be neglected.

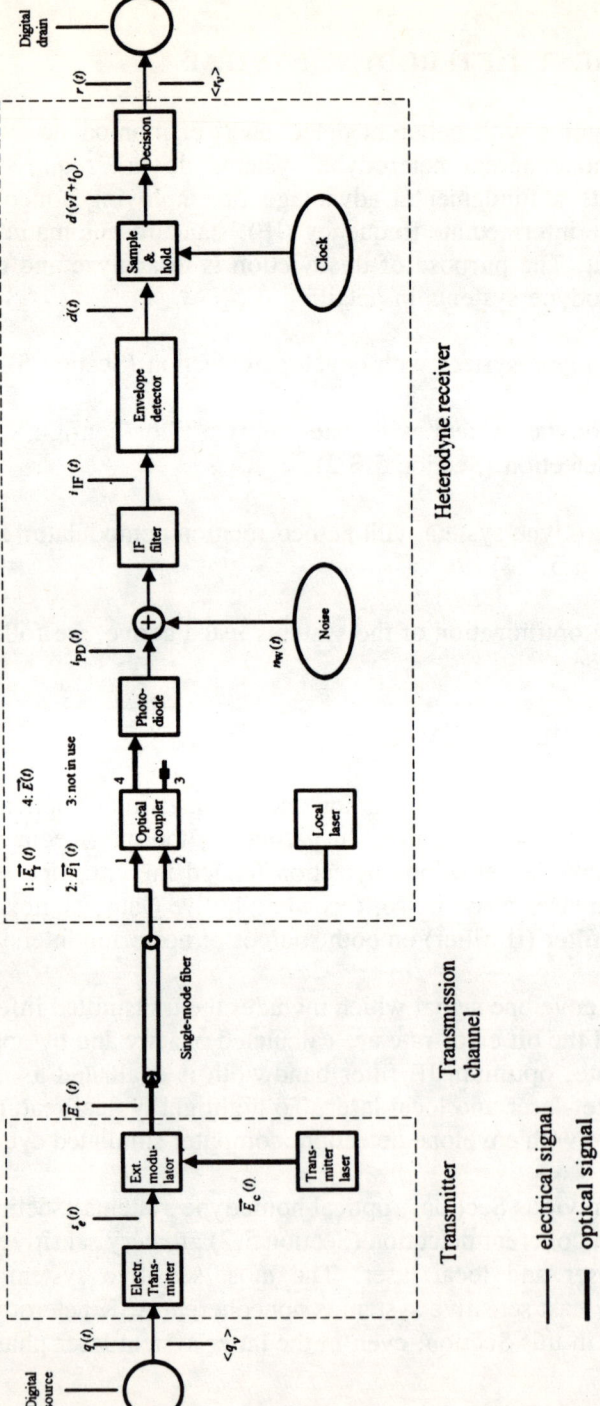

Fig. 5.12: Coherent optical communication system with noncoherent ASK heterodyne receiver

In optical ASK heterodyne receiver, phase noise can be regarded as a superposition of an undesired phase- or frequency modulation. In ideal ASK heterodyne system, this undesired phase- or frequency modulation does not disturb the desired amplitude modulation. In this case, disturbance to the envelope signal is given only by the shot noise of photodiode and thermal noise of amplifier. Both are additive to the IF signal. The same result can be obtained by neglecting the influence of IF filter on the phase noise. As the analysis of the IF filter influence is rather comprehensive when phase noise disturbs the system, this influence is usually neglected in the early system calculation [e.g., 60, 193, 287]. In that case, system calculation results in an envelope signal which is again independent of the laser phase noise (Subsection iii below). Therefore, the results obtained are only useful as a rough estimate.

In real ASK heterodyne receivers, required band-limiting of the IF filter causes a strong correlation between the undesired phase modulation and the desired envelope signal which contains the information. As a result, an additional disturbance to envelope and, hence, information is obtained due to the laser phase noise. This, of course, is not a particular feature of noncoherent optical ASK heterodyne systems. It is rather a generally valid and well-known result that any band-limiting of phase- or frequency modulated signal causes a phase-to-envelope conversion. Distortion of the envelope due to laser phase noise leads to an additional deterioration in the bit error rate (BER). Thus, an increase in the IF filter bandwidth decreases the BER due to phase noise, whereas the BER is increased due to additive Gaussian noise of receiver (Subsection vi). Therefore, an optimum IF filter bandwidth should exist (Subsection vii).

For determining the influence of laser noise on coherent optical communication systems, different methods of analysis are available. Therefore, Section 5.5 presents a brief comparison and discusses advantages and drawbacks of these methods.

In the following, we consider noncoherent optical ASK heterodyne system which is disturbed only by laser phase noise, additive Gaussian noise (shot noise of photodiodes and thermal noise of amplifiers) and intersymbol interference. Apart from this, transmission system is ideal. System components and their input and output signals have already been described and explained in Chapter two in detail. Again, the most important signal to assess transmission quality and system performance (in particular the BER) is the detected signal $d(t)$ and its sample values $d(vT + t_0)$.

The block diagram of a typical optical communication system employing ASK modulation scheme and noncoherent envelope detection is shown in Fig. 5.12.

(i) DETECTED SIGNAL $d(t)$ AND ITS SAMPLE VALUES $d(vT + t_0)$

As shown in Fig. 5.12, detected signal $d(t)$ at the input of sample and hold circuit equals the envelope $|\underline{i}_{\mathrm{IF}}(t)|$ of the intermediate frequency signal at the output of envelope demodulator. The intermediate frequency signal is given by

$$i_{IF}(t) = \int_{-\infty}^{+\infty} i_{PD}(\tau) h_{IF}(t-\tau) d\tau + \underline{n}(t)$$

(5.65)

$$= \hat{i}_{PD} \int_{-\infty}^{+\infty} s(\tau) e^{j\phi(\tau)} e^{j2\pi f_{IF} t} h_{IF}(t-\tau) d\tau + \underline{n}(t)$$

Photodiode current $i_{PD}(t)$ and normalized ASK information signal $s(t)$ have already been given and explained in Section 2.4.3. In order to analyze the influence of the filtered additive Gaussian receiver noise $\underline{n}(t)$, we again take the advantage of narrow-band representation (see Eq. 2.76)

$$\underline{n}(t) = x(t) e^{j2\pi f_{IF} t} - j y(t) e^{j2\pi f_{IF} t} = (x(t) - j y(t)) e^{j2\pi f_{IF} t}$$

(5.66)

This equation can be applied when the ratio of noise equivalent IF filter bandwidth B_{IF} to centre frequency f_{IF} of the IF filter is much less than 1.

As already explained in Chapter two, photodiode current $i_{PD}(t)$ at the input to the IF filter normally contains two baseband signal components: one is the DC component $kR_0 P_1$ and the other is baseband component which is proportional to the absolute square of the transmitted information signal $|s(t)|^2$ as given in Eq. (2.60). Both the undesired signal components can be eliminated either by the IF filter itself or by an appropriate balanced receiver configuration (Section 2.4.3).

In order to avoid spectral overlapping of IF spectrum and undesired baseband spectrum included in $i_{PD}(t)$, intermediate frequency f_{IF} must be chosen high enough. Because of the quadratic baseband term $|s(t)|^2$, undesired baseband spectrum exhibits a bandwidth of $2f_{s,max}$ (i.e., $0 \leq f \leq 2f_{s,max}$), where $f_{s,max}$ represents the maximum frequency of the information signal $s(t)$. For digital modulation, this frequency is approximately $f_{s,max} = 1/(2T)$. The IF spectrum is located in the range of $f_{IF} - f_{s,max}$ to $f_{IF} + f_{s,max}$. Hence, the intermediate frequency should fulfil the following relation:

$$f_{IF} > 3f_{s,max}$$

(5.67)

Next we assume an IF filter with frequency response $H_{IF}(f)$ which can be expressed in terms of an equivalent baseband filter with frequency response $H_B(f)$ i.e., $H_{IF}(f) = H_B(f - f_{IF}) + H_B(f + f_{IF})$. Taking into account the baseband filter impulse response $h_B(t)$ and the narrow-band representation (Eq. 5.66) of the additive noise, IF signal in Eq. (5.65) can be rewritten as follows:

$$i_{IF}(t) \approx \left[\hat{i}_{PD} \int_{-\infty}^{+\infty} s(\tau)\, e^{j\phi(\tau)}\, h_B(t-\tau)\, d\tau \; + \; x(t) \; - \; jy(t) \right] e^{j2\pi f_{IF} t}$$

$$= \left| i_{IF}(t) \right| e^{j\Psi_{IF}(t)}\, e^{j2\pi f_{IF} t}$$

(5.68)

It becomes clear from the above equation that the IF signal represents a phase- and amplitude modulated carrier signal with carrier frequency f_{IF}. Eq. (5.68) is an approximation in case of spectral overlapping of the IF filter components $H_B(f\text{-}f_{IF})$ and $H_B(f\text{+}f_{IF})$. However, this equation yields the exact result when there is no spectral overlapping (Section 2.4.3). Detected signal $d(t) = |i_{IF}(t)|$ and phase $\Psi_{IF}(t)$ are given by

$$d(t) = \left| i_{IF}(t) \right| \approx \sqrt{\mathrm{Re}^2\{i_{IF}(t)\} \; + \; \mathrm{Im}^2\{i_{IF}(t)\}}$$

$$= \left[\left(\hat{i}_{PD} \int_{-\infty}^{+\infty} s(\tau)\, \cos(\phi(\tau))\, h_B(t-\tau)\, d\tau \; + \; x(t) \right)^2 \right.$$

$$\left. + \left(\hat{i}_{PD} \int_{-\infty}^{+\infty} s(\tau)\, \sin(\phi(\tau))\, h_B(t-\tau)\, d\tau \; - \; y(t) \right)^2 \right]^{1/2}$$

(5.69)

and

$$\Psi_{IF}(t) = \arctan\left[\frac{\mathrm{Im}\{i_{IF}(t)\}}{\mathrm{Re}\{i_{IF}(t)\}} \right]$$

$$= \arctan\left[\frac{\hat{i}_{PD} \int_{-\infty}^{+\infty} s(\tau)\, \sin(\phi(\tau))\, h_B(t-\tau)\, d\tau \; - \; y(t)}{\hat{i}_{PD} \int_{-\infty}^{+\infty} s(\tau)\, \cos(\phi(\tau))\, h_B(t-\tau)\, d\tau \; + \; x(t)} \right]$$

(5.70)

The detected signal $d(t)$ is sampled at the centre of each bit. If the sample value $d(vT+t_0)$ is either above or below the threshold, symbol "1" or symbol "0" is detected respectively. The detected signal $d(t)$ can be regarded as a stationary random process. It means that the statistical properties do not change with time. Therefore, it will be sufficient to calculate the statistical properties at the sampling time $t = t_0$. In the strict sense, detected signal is a nonstationary random process since the random processes $\cos(\phi(t))$ and $\sin(\phi(t))$ are also nonstationary. However, both processes become stationary after a very short time $t \gg 1/(2\pi\Delta f)$ as discussed in Section 3.3.4.

In order to determine the statistical properties of the sample value $d(t_0)$, sign of the Gaussian random process $y(t)$ with zero mean can be chosen arbitrary without changing the statistical properties of $d(t_0)$. To obtain a clear description, positive sign will be used now onwards. Further, following substitutions and abbreviations will also be used to simplify our calculation:

$$d := \frac{d(t_0)}{\hat{i}_{PD}} \qquad n := \frac{n(t_0)}{\hat{i}_{PD}} \quad \rightarrow \quad \sigma := \frac{\sigma_n}{\hat{i}_{PD}} = \frac{\sigma_{het}}{\hat{i}_{PD}}$$

$$x := \frac{x(t_0)}{\hat{i}_{PD}} \qquad y := \frac{\dot{y}(t_0)}{\hat{i}_{PD}} \quad \rightarrow \quad \sigma_x = \sigma_y = \sigma$$

Due to above normalizations of system parameters "local laser power P_l" and "received optical power P_r" are now only included in the normalized standard deviation σ of the receiver additive Gaussian noise. The normalized detected signal d is given by

$$d = \sqrt{\left[\int_{-\infty}^{+\infty} s(\tau) \cos(\phi(\tau)) \, h_B(t_0 - \tau) d\tau + x \right]^2 + \left[\int_{-\infty}^{+\infty} s(\tau) \sin(\phi(\tau)) \, h_B(t_0 - \tau) d\tau + y \right]^2} \qquad (5.71)$$

As this expression is of prime importance for our further consideration, physical meaning of the signals included in Eq. (5.71) is briefly summarized below:

$s(t)$:   information signal,

$h_B(t)$:   impulse response of the equivalent baseband filter $H_B(f)$ corresponding to real IF filter $H_{IF}(f) = H_B(f - f_{IF}) + H_B(f + f_{IF})$

$\phi(t)$:   laser phase noise of both transmitter and local lasers,

$x(t)$:   inphase component of the receiver additive Gaussian noise,

$y(t)$:   quadrature component of the receiver additive Gaussian noise.

It becomes clear from the above equation that the normalized and sampled detected signal d is a function of the transmitted information signal $s(t)$ and impulse response $h_B(t)$ of the equivalent baseband filter. This signal is disturbed by three sources of noise: resulting laser phase noise $\phi(t)$, inphase component $x(t)$ and quadrature component $y(t)$ of the receiver additive Gaussian noise.

(ii) PROBABILITY DENSITY FUNCTION $f_d(d)$

The most difficult problem while calculating the probability density function (pdf) and the related BER is deriving the statistical properties of the interaction of phase noise and envelope due to restricted bandwidth of IF filter. In particular, the difficulties are due to the nongaussian behaviour of signals which are disturbed by laser phase noise (Section 3.5). It is well-known that an additive Gaussian process at a linear filter input always leads to another additive Gaussian process at the filter output. Thus, the shape of the pdf at the filter output remains unchanged when a Gaussian pdf at the filter input is assumed. Examples for such processes are the approximately Gaussian shot noise of the photodiode and the Gaussian thermal noise of the amplifier. However, the shape of the output pdf is different from the input pdf when the input process is nongaussian. In such a case, calculation of the pdf at the filter output is generally complicated. The analytical solution of this problem can only be found by comprehensive mathematical and statistical operations as explained in Section 3.5. In the following Subsections, this problem is solved exactly, approximately and by computer simulation.

In order to calculate the pdf $f_d(d)$ of the sampled and normalized envelope $d = |i_{IF}|$ given in Eq. (5.71), we represent the first integral in Eq. (5.71) by A and the second by B i.e.,

$$A = \int_{-\infty}^{+\infty} s(\tau)\,\cos(\phi(\tau))\,h_B(t_0-\tau)\,d\tau \tag{5.72}$$

and

$$B = \int_{-\infty}^{+\infty} s(\tau)\,\sin(\phi(\tau))\,h_B(t_0-\tau)\,d\tau \tag{5.73}$$

With these parameters, Eq. (5.71) can be written in the following simplified form:

$$d = \sqrt{(A+x)^2 + (B+y)^2} \tag{5.74}$$

It becomes evident from the above equation that the detected signal d is a function of four random variables, namely A, B, x, y. As mentioned in Chapter two, random variables x and y are statistically independent and each is defined by a Gaussian pdf with zero mean and normalized standard deviation σ. The new random variables A and B are, however, statistically dependent and, in general, nongaussian. The nongaussian pdf is due to nongaussian pdf of the random variables $\cos(\phi)$ and $\sin(\phi)$ as explained in Section 3.3.4. Considering first that the random variables A and B are constant, we evaluate the conditional probability density function $f_{d|A,B}(d\,|A,B)$ by means of the statistical theory of two-dimensional random variables (Section 3.5). We obtain

$$f_{d|A,B}(d,A,B) = \frac{d}{\sigma^2} e^{-\frac{d^2 + (A^2 + B^2)}{2\sigma^2}} J_0\left(\frac{d\sqrt{A^2 + B^2}}{\sigma^2}\right) \tag{5.75}$$

where J_0 is the Bessel function of order zero [1]. Substitution of

$$C = \sqrt{A^2 + B^2} = d(\sigma = 0) \tag{5.76}$$

in the above equation yields the well-known Rice probability density function [206]

$$f_{d|C}(d,C) = \frac{d}{\sigma^2} e^{-\frac{d^2 + c^2}{2\sigma^2}} J_0\left(\frac{Cd}{\sigma^2}\right) \approx \frac{1}{\sqrt{2\pi}\sigma} e^{-\frac{(d-C)^2}{2\sigma^2}} \tag{5.77}$$

For a sufficiently high C/σ-ratio corresponding to a high signal-to-noise ratio $(C/\sigma)^2$, the Rice density function can be approximated by a Gaussian density function. The new random variable C equals the normalized and sampled detected signal d provided variance $\sigma^2 = 0$ i.e., $x = y = 0$. Thus, $C = d(\sigma = 0)$ only includes the influence of phase noise. Taking into account the above considerations, random variable C can be described by one of the following three alternate forms:

$$C = \left| \int_{-\infty}^{+\infty} s(\tau) e^{j\phi(\tau)} h_B(t_0 - \tau) \, d\tau \right| \tag{5.78}$$

$$C = \sqrt{\left[\int_{-\infty}^{+\infty} s(\tau) \cos(\phi(\tau)) h_B(t_0 - \tau) \, d\tau\right]^2 + \left[\int_{-\infty}^{+\infty} s(\tau) \sin(\phi(\tau)) h_B(t_0 - \tau) \, d\tau\right]^2} \tag{5.79}$$

and

$$C = \sqrt{\int_{-\infty}^{+\infty}\int_{-\infty}^{+\infty} s(\tau) s(t) e^{j[\phi(\tau) - \phi(t)]} h_B(t_0 - \tau) h_B(t_0 - t) \, d\tau \, dt} \tag{5.80}$$

When the randomness of C is again considered by using its pdf $f_C(C)$, pdf $f_d(d)$ of the detected signal d is finally given by

$$f_d(d) = \frac{d}{\sigma^2} \int\limits_0^{+\infty} e^{-\frac{d^2 + c^2}{2\sigma^2}} J_0\left(\frac{Cd}{\sigma^2}\right) f_C(C) \, dC \tag{5.81}$$

A particular case of interest is obtained when the influence of IF filter on the phase noise is neglected. In that case, the term $\exp(j\phi(t))$ in Eq. (5.78) can be brought outside the convolution integral and, therefore, neglected since $|\exp(j\phi(t))| = 1$. This results in a non-random constant $C = C_0$ and its pdf $f_C(C) = \delta(C - C_0)$ which is a Dirac delta function. Therefore, Eq. (5.81) yields the following special forms

$$C_0 = 0: \quad f_d(d) = \frac{d}{\sigma^2} e^{-\frac{d^2}{2\sigma^2}} \qquad \text{Rayleigh distribution} \tag{5.82}$$

and

$$C_0 \neq 0: \quad f_d(d) = \frac{d}{\sigma^2} e^{-\frac{d^2 + c_0^2}{2\sigma^2}} J_0\left(\frac{C_0 d}{\sigma^2}\right) \quad \text{Rice distribution} \tag{5.83}$$

which are well-known from standard digital ASK communication theory [260]. It should be noted that a constant C is also obtained in an ideal system i.e., without any phase noise ($\sigma_\phi = 0$) or phase noise-to-amplitude conversion.

(iii) PROBABILITY DENSITY FUNCTION $f_c(C)$

This Subsection describes the exact calculation of the pdf $f_C(C)$ required to calculate the pdf $f_d(d)$ of the sampled detected signal d and also the BER. As the exact calculation is rather comprehensive, approximations would be much more convenient for practical system design. For this reason, next Subsection (iv) is dedicated on approximations. As the theoretical considerations of this Subsection are not directly required to understand Subsection (iv), the reader may skip the present Subsection without any discontinuity.

The random variable C defined in Eqs. (5.78) to (5.80) depends upon the random process $\phi(t)$, transmitted information signal $s(t)$ and impulse response $h_B(t)$ of equivalent baseband filter. To calculate the pdf $f_d(d)$ of sampled detected signal d, several analytical methods based on the sampling theorem exists (Section 3.5). One of the methods, namely the method of integral, will representatively be discussed in this Section in more detail. By using statistics of two-dimensional random variables [206], pdf $f_C(C)$ of the random variable C can be written as follows:

$$f_C(C) = C \int_0^{2\pi} f_{A,B} \ (C\cos(\phi), \ C\sin(\phi)) \ d\phi \qquad\qquad (5.84)$$

In the above equation, the most difficult problem is to evaluate the joint probability density function $f_{A,B}(A,B)$. As just mentioned above, a particular case is given by neglecting the influence of the IF filter on the phase noise. This results in a pdf $f_C(C)=\delta(C - C_0)$ corresponding to a parameter C_0 which only depends on the transmitted symbol pattern $<q_v>$ included the information signal $s(t)$. Thus, phase noise is completely extracted from the calculation and pdf $f_d(d)$ and BER becomes independent of laser phase noise.

In order to obtain a generally valid solution, we take the advantage of sampling theorem and describe the random variables A and B by the two sums

$$A = \sum_{n=-\infty}^{+\infty} s_n \cos(\phi_n)\, \alpha_n \qquad\qquad (5.85)$$

and

$$B = \sum_{n=-\infty}^{+\infty} s_n \sin(\phi_n)\, \alpha_n \qquad\qquad (5.86)$$

where s_n, α_n and the random variables ϕ_n represent the sample values $s(nT_s)$, $T_s h_B(t_0- nT_s)$ and $\phi(nT_s)$. Here, T_s is the sampling period. To fulfil the sampling theorem requirement, sampling period T_s must be chosen small enough. Only in that case, A and B are exactly determined by sums. The choice of T_s becomes clear when both frequency response $H_B(f)$ *and* frequency spectrum of the IF filter input signal $s(t)\exp(j\phi(t))$ are band-limited, whereas the choice is not clear when these frequency spectra are not band-limited. In such a case, we have to set up an arbitrary bandlimit depending upon the required accuracy of calculation.

For simplicity, let us presume an impulse response $h_B(t)$ which approaches zero within a specified time. In that case, it is sufficient to add up from n=-N to N, where N is finite and appropriately high enough.

Employing the statistical theory of multidimensional random variables (Section 3.5), we now have to design a set of (2N+1) independent equations, where (2N+1) equals the number of random variables included in Eqs. (5.85) and (5.86).

It should be noted that various possible sets of equations are actually existent, but not all lead to a straightforward calculation of $f_C(C)$. As explained in Section 3.5, it is most convenient to create a set of equations which can unambiguously be solved i.e., a single solution for the variables on the right hand side of this set. In this sense, the set given below represents a very useful set of equations for the calculation of pdf $f_C(C)$.

It can be shown that the calculation of pdf $f_C(C)$ can additionally be simplified, when two more equations are added to the set of (2N+1) equations. For this purpose, it is required to define the two new random variables u and v. These random variables are statistically independent and will be set to zero later. Taking into consideration the above discussion, we obtain

$$A = \sum_{n=-N}^{+N} s_n \cos(\Delta\phi_{-N} + \cdots + \Delta\phi_n)\alpha_n + u$$

$$B = \sum_{n=-N}^{+N} s_n \sin(\Delta\phi_{-N} + \cdots + \Delta\phi_n)\alpha_n + v$$

$$\phi_{-N} = 0 \quad + \Delta\phi_{-N}$$
$$\vdots \qquad \vdots \qquad \vdots$$
$$\phi_0 = \phi_{-1} \quad + \Delta\phi_0$$
$$\vdots \qquad \vdots \qquad \vdots$$
$$\phi_N = \phi_{N-1} \quad + \Delta\phi_N$$

(5.87)

Except the special phase ϕ_{-N} at the starting point $t=-NT_s$ all other sampled phases ϕ_n with $-N+1 < n < N$ are gradually circumscribed in Eq. (5.87) by real sampled phase deviations $\Delta\phi_n$ with respect to the starting phase ϕ_{-N}. These real phase deviations are statistically independent and each one is described by a simple Gaussian pdf with zero mean and variance

$$\sigma_{\Delta\phi}^2 = 2\pi[\Delta f_t + \Delta f_l]T_s \tag{5.88}$$

Here, Δf_t and Δf_l represent the laser linewidth of both transmitter and local lasers (see Eq. 3.41) and T_s the sampling period.

In order to simplify calculation, we further represent the sampled starting phase ϕ_{-N} by a virtual phase deviation $\Delta\phi_{-N}$. Because of the nonstationarity of Gaussian phase itself, variance of the random variable $\phi_{-N} = \Delta\phi_{-N}$ grows linearly with time and is in general infinite (Section 3.2). As far as the pdfs of both the nongaussian random variables $\cos(\phi_{-N})$ and $\sin(\phi_{-N})$ are concerned, it is immaterial whether the pdf of the starting phase ϕ_{-N} is Gaussian distributed with infinite variance or uniformly distributed between 0 and 2π or $-\pi$ and $+\pi$.

In the following, we consider the random variable ϕ_{-N} to be uniformly distributed. Taking into account the above considerations, multidimensional joint probability density function including all the $(2N+3)$ independent random variables of Eq. (5.87) can be written as

$$f_{A,B,\phi_{-N}\cdots\phi_N}(A, B, \phi_{-N} \cdots \phi_N) =$$

$$\frac{1}{2\pi} f_u\left[A - \sum_{n=-N}^{+N} s_n \cos(\phi_n) \alpha_n\right] \cdot f_v\left[B - \sum_{n=-N}^{+N} s_n \sin(\phi_n) \alpha_n\right] \cdot \prod_{n=-N+1}^{+N} f_{\Delta\phi}(\phi_n - \phi_{n-1}) \tag{5.89}$$

It should be noted that the generally valid equation for calculating multidimensional joint-density functions is normally much more comprehensive than given in Eq. (5.89) and has been described in Section 3.5. However, due to the proper design of our set of equations (Eq. 5.87), generally valid equation could be replaced by the much easier Eq. (5.89). This equation simply represents a multiple product of independent pdfs. It becomes clear from Eq. (5.89) that the required two-dimensional joint probability density function $f_{A,B}(A,B)$ is obtained by a multiple integration over the random variables ϕ_{-N} to ϕ_N (method of integral). Further, taking into account $u=v=0$ which corresponds to $f_u(u)=\delta(u)$ and $f_v(v)=\delta(v)$ and inserting $f_{A,B}(A,B)$ into Eq. (5.84), we finally obtain

$$f_C(C) = \frac{C}{(2\pi)^{N+1} \sigma_{\Delta\phi}^{2N}} \int\limits_{-\infty}^{+\infty} \cdots \int\limits_{-\infty}^{+\infty} \int\limits_{0}^{2\pi} \exp\left[-\frac{\sum_{n=-N+1}^{+N}(\phi_n - \phi_{n-1})^2}{2\sigma_{\Delta\phi}^2}\right] \cdot$$

$$\delta\left[C - \sqrt{\sum_{n=-N}^{+N}\sum_{m=-N}^{+N} s_n s_m \cos(\phi_n - \phi_m) \alpha_n \alpha_m}\right] d\phi_{-N} \cdots d\phi_N \tag{5.90}$$

Again, it should be remembered that $C=d(\sigma=0)$ equals the sampled and normalized detected signal d for a zero variance $\sigma=0$ i.e., $x=y=0$. This result represents an exact analytical solution for the pdf $f_C(C)$ required to calculate the probability of bit error exactly. However, Eq. (5.90) is only of theoretical interest since evaluation of this multi-integral solution would require too much computer run time. Obviously, the Dirac delta function included in Eq. (5.90) describes a well-defined curve in a $(2N+1)$ dimensional space. This curve is clearly described by setting the argument of the Dirac delta function to zero. For this reason, above equation can also be expressed in terms of a line integral. We obtain

$$f_C(C) = \oint\limits_s \exp\left[-\frac{\sum_{n=-N+1}^{+N}(\phi_n - \phi_{n-1})^2}{2\sigma_{\Delta\phi}^2}\right] \left|\,\text{grad}\left(\arg(C, \phi_{-N} \cdots \phi_N)\right)\right|^{-1} ds \tag{5.91}$$

Here, the exponential function pictorially describes a $(2N+1)$ dimensional mountain, s a certain path through this mountain which is defined by the argument of the Dirac delta function given in Eq. (5.90) and ds a small stretch of this path. Related to the above equation, each stretch has to be weighted by the inverse of absolute of the gradient grad$[\arg(C, \phi_{-N} \cdots \phi_N)]$. In comparison to Eq. (5.90), evaluation of the above equation does not require a multiple integration with respect to $(2N+1)$ dimensional space. Instead an integration along the path s is sufficient. However, this solution is also of theoretical interest only as it requires an unacceptable high computer run time.

For the design and realization of coherent optical communication systems with heterodyne detection, a more practical solution is required. For this, approximations and computer simulations represent well-suited and powerful tools. In the following, Gaussian approximation method and some simulation results are explained and discussed as a sample.

(iv) GAUSSIAN APPROXIMATION OF THE PROBABILITY DENSITY FUNCTION $f_C(C)$

To obtain a more practical solution for the pdf $f_C(C)$, we take the advantage of following simple Gaussian approximation. Some other important methods of approximations and a detailed comparison of all these methods will be presented in Section 5.5.

The Gaussian approximation

$$f_C(C) \approx \frac{1}{\sqrt{2\pi}\,\sigma_C} \, e^{-\frac{(c - \eta_C)^2}{2\sigma_C^2}} \tag{5.92}$$

is entirely described by its mean value η_C (first moment) and its variance σ_C^2. Due to the square root dependence of C, it is easy to calculate exactly the second moment $E\{C^2\}$, whereas the expected value $\eta_C = E\{C\}$ can only be found by approximations (E: expected value). The variance σ_C^2 is given by the well-known statistical relation

$$\sigma_C^2 = E\{C^2\} - \eta_C^2 \tag{5.93}$$

It may be noted that even a small error in η_C due to approximations could lead to a great error in the variance σ_C^2 when this variance is calculated by using the above equation.

Example 5.2

Let us assume an approximate mean value $\eta_C = 0.999$, a real mean value $\eta_C = 1.000$, and a real second moment $E\{C^2\} = 1.001$. In this particular case, relative error in η_C is only 0.1%, whereas it is 1100% in the variance σ_C^2! Thus, there is no use to calculate the variance σ_C^2 by the equation mentioned above when the second moment $E\{C^2\}$ is given exactly while the first moment η_C is given only approximately. This is valid, in particular, when the predicted variance σ_C^2 is very small.

In order to prevent this high error in variance, we make use of the following substitutions and approximations:

$$
C = \sqrt{D} \quad \rightarrow \quad
\begin{cases}
\eta_C \approx \sqrt{E\{D\}} = \sqrt{\eta_D} \\[2mm]
\sigma_C \approx \dfrac{1}{2\sqrt{\eta_D}}\sigma_D
\end{cases}
\tag{5.94}
$$

Here, evaluation of the statistical parameters of C is reduced to the evaluation of the corresponding parameters of $D = C^2 = A^2 + B^2$. Obviously, D equals the argument of the square root given in Eq. (5.80). Thus, D is same as the square of the normalized and sampled detected signal $d(\sigma=0)$ i.e., $D = C^2 = d^2(\sigma=0)$. Therefore, the quantity D physically represents a measure of the average power of d including the influence of phase noise. Using Eq. (5.80), we obtain:

$$
\eta_D = \int_{-\infty}^{+\infty}\int_{-\infty}^{+\infty} s(t_1)\, s(t_2)\, E\{e^{j[\phi(t_1)-\phi(t_2)]}\}\, h_B(t_0-t_1)\, h_B(t_0-t_2)\, dt_1\, dt_2
\tag{5.95}
$$

$$
E\{D^2\} = \int_{-\infty}^{+\infty}\int_{-\infty}^{+\infty}\int_{-\infty}^{+\infty}\int_{-\infty}^{+\infty} s(t_1)\, s(t_2)\, s(t_3)\, s(t_4)\, E\{e^{j[\phi(t_1)-\phi(t_2)]+j[\phi(t_3)-\phi(t_4)]}\}
\tag{5.96}
$$

$$
\cdot h_B(t_0-t_1)\, h_B(t_0-t_2)\, h_B(t_0-t_3)\, h_B(t_0-t_4)\, dt_1\, dt_2\, dt_3\, dt_4
$$

and

$$
\sigma_D^2 = E\{D^2\} - \eta_D^2
\tag{5.97}
$$

Computing the expected values required to evaluate the integrands of Eqs. (5.95) and (5.96), we have to consider the statistical properties of the Gaussian phase noise, which have already been derived in Section 3.3.4. By using the variables

$$
w(\tilde{t}) = e^{j\phi(\tilde{i})}
\tag{5.98}
$$

and

$$
H = 2\pi\Delta f\left(\min(\tilde{t}_4,\tilde{t}_2) - \min(\tilde{t}_4,\tilde{t}_1) + \min(\tilde{t}_3,\tilde{t}_1) - \min(\tilde{t}_3,\tilde{t}_2)\right)
\tag{5.99}
$$

in conjunction with Eq. (3.81), expected values required in Eqs. (5.95) and (5.96) can be calculated as follows:

$$E\{w(\tilde{t}_1)w^*(\tilde{t}_2)\} = E\{e^{j[\phi(\tilde{t}_1)-\phi(\tilde{t}_2)]}\} = R_w(\tilde{\tau}) = e^{-\pi\Delta f|\tilde{\tau}|} \tag{5.100}$$

and

$$E\{w(\tilde{t}_1)w^*(\tilde{t}_2)w(\tilde{t}_3)w^*(\tilde{t}_4)\} = e^{-\pi\Delta f\left[|\tilde{t}_2-\tilde{t}_1| + |\tilde{t}_4-\tilde{t}_3|\right]} e^H \tag{5.101}$$

In Eq. (5.100), $R_w(\tilde{\tau})=R_w(\tilde{t}_2-\tilde{t}_1)$ represents the autocorrelation function of the complex random process $w(\tilde{t})=\exp(j\phi(\tilde{t}))$. The variables of time $\tilde{t}_1$ to $\tilde{t}_4$ in Eqs. (5.99) to (5.101) are related to $t=-\infty$ when laser operation and, hence, phase noise process are assumed to have begun. Therefore, $\tilde{t}$ is formally determined by $\tilde{t}=t+T_\infty$ with $T_\infty\to\infty$ and phase noise variance is already infinite around $t=0$, which is in the relevant time interval of the integrals given in Eqs. (5.95) and (5.96). It should be noted that the integration which have to be performed in these equations approximately yields zero outside the pulse width Δt_B of the baseband filter impulse response $h_B(t)$.

An alternate equation for the expected value $\eta_D=E\{C^2\}$ is obtained by taking into account the Lorentzian power spectral density $G_w(f)$ of the random process $w(t)$ derived in Chapter three (Eq. 3.88). We obtain

$$\eta_D = \int_{-\infty}^{+\infty} G_w(f) \left| \int_{-\infty}^{+\infty} s(t)\, h_B(t_0-t)\, e^{-j2\pi f(t_0-t)} dt \right|^2 df \tag{5.102}$$

The squared expected value η_C^2 and the variance σ_C^2 as a function of the normalized IF filter bandwidth $B_{IF}T=2f_gT$ and the normalized laser linewidth ΔfT for the particular bit pattern $<q_v> = \cdots 0001000 \cdots$ (referred as single one) are shown in Figs. 5.13a and 5.13b. The IF filter is assumed to have a Gaussian frequency response $H_B(f)=\exp[-\pi(f/B_{IF})^2]$ in the baseband representation. Physically, the mean square value η_C^2 and the variance σ_C^2 represent the signal power (useful power) and the noise power caused by phase noise only. It must be remembered that C equals the sampled detected signal d in absence of receiver additive Gaussian noise i.e., $\sigma=0$.

As plotted in Fig. 5.13a, signal power $E\{C\}^2=\eta_C^2$ increases with the increase in the normalized IF filter bandwidth $B_{IF}T$ because the effect of intersymbol interference becomes smaller. On the other hand, signal power will get attenuated when bandwidth is decreased. Further, Fig. 5.13a clearly indicates that the finite linewidths of real lasers lead to a severe reduction of the signal power.

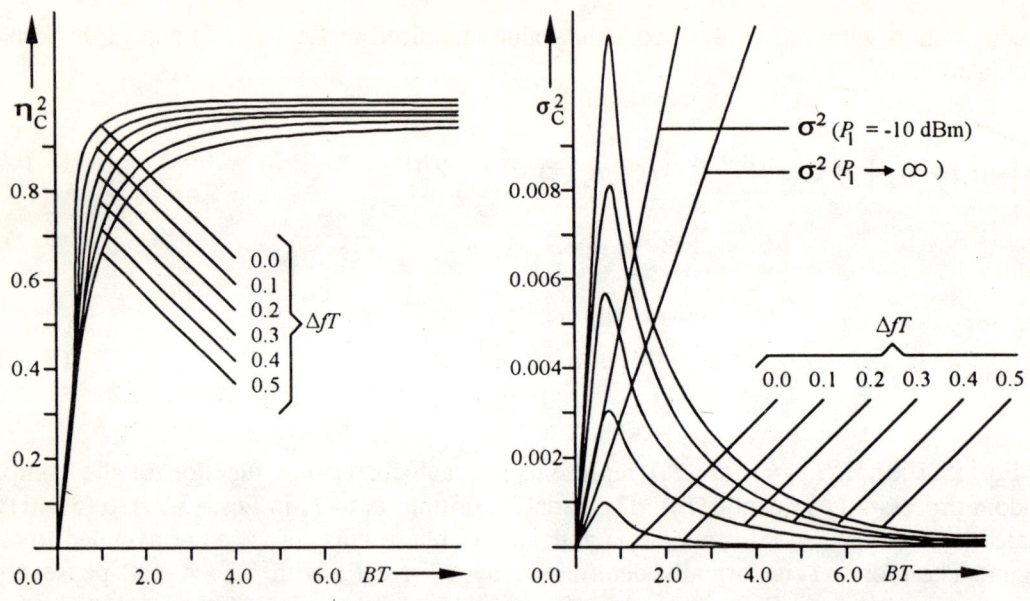

Fig. 5.13: (a) Square of expected value η_C representing the signal power and (b) variance σ_C^2 representing the noise power due to phase noise of random variable $C = d \, (\sigma = 0)$. As a comparison, straight lines in (b) represent the normalized variance σ^2 of additive Gaussian noise.

The normalized noise power σ_C^2 as shown in Fig. 5.13b is zero for $B_{IF}T = 0$ (i.e., the transmission path is broken) and also for $B_{IF}T \to \infty$ (i.e., no influence of the IF filter due to undesired phase noise-to-amplitude conversion). It can be seen from Fig. 5.13b that maximum noise power due to phase noise is existent. As a comparison, the straight line illustrates the influence of additive Gaussian noise whose normalized noise power σ^2 increases proportional to the IF filter bandwidth B_{IF}. Finally, it should be noted that the analytical results shown in Figs. 5.13a and 5.13b do agree very well with the respective values obtained by simulation experiments.

(v) Computer Simulation of the Probability Density Function $f_c(C)$

Besides analytical solutions, the computer simulation represents a very useful and very powerful alternative. Moreover, simulation gives a clear insight into the typical behaviour of noncoherent ASK heterodyne systems. In particular, the effects of a change in system parameters, for example in bit rate can quickly be evaluated and assessed.

The simulated probability density functions $f_C(C)$ and $f_d(d)$ shown in Figs. 5.14a and 5.14b are the typical results of a simulation experiment. These results are based on 100000 (one hundred thousand) computer simulated random values for each of the random variables C and d. Variable parameter in Fig. 5.14 is the normalized laser linewidth ΔfT. It can be regarded as a direct measure to assess the strength and effect of laser phase noise. In addition, both figures are based on the following system parameters:

- IF filter with a Gaussian frequency response $H_B(f) = \exp[-\pi(f/B_{IF})^2]$ in the baseband representation, $B_{IF}T = 2f_gT = 1$,
- $<q_v>_{1i} = \cdots 0001000 \cdots$ (single "1") and $<q_v>_{0i} = \cdots 1110111 \cdots$ (single "0").
- receiver additive Gaussian noise with normalized variance $\sigma^2 = 0$ in Fig. 5.14a and $\sigma^2 = 0.01$ in Fig. 5.14b respectively.

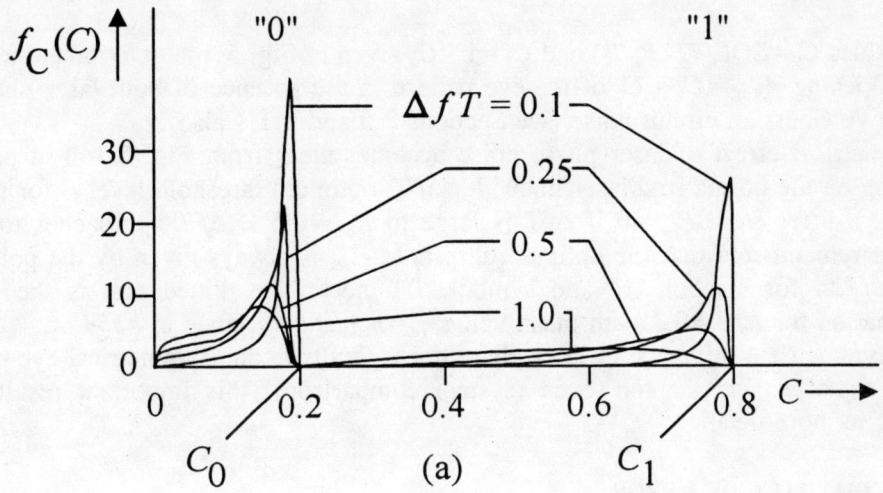

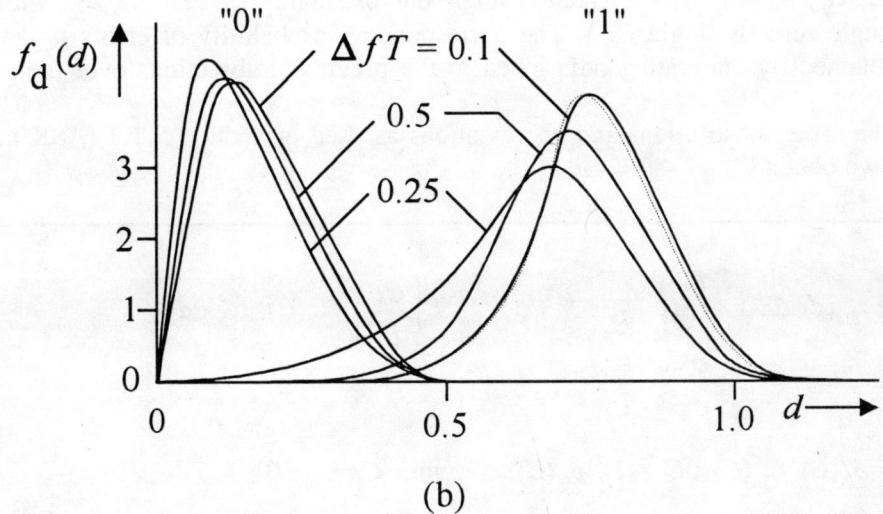

Fig. 5.14: (a) Computer simulated probability density functions $f_C(C)$ and (b) $f_d(d)$ of random variables $C = d$ ($\sigma = 0$) and d (detected signal). The dotted plot makes use of the Gaussian approximation given in Eq. (5.92) and $\Delta fT = 0.1$.

It becomes evident from Fig. 5.14a that the laser phase noise gives rise to a distortion for transmitted symbol "1", whereas it leads to an improvement for symbol "0"! As already explained in Chapter three, the reason for this very surprising result is that laser phase noise acts as a multiplier with the information signal in ASK heterodyne and ASK homodyne receivers. However, the deterioration is always dominant as compared to improvement. Thus, laser phase noise always yields an overall degradation of system performance as expected.

The values $C_0 = 2Q(\sqrt{\pi/2}B_{IF}T)$ and $C_1 = 1 - C_0$ given in Fig. 5.14a determine the normalized eye opening $A_{ASK} = C_1 - C_0$ of the eye pattern in the absence of both laser phase noise and additive Gaussian circuit noise (see Sections 2.5 and 5.1.1 also).

Asymmetrical effect of laser phase noise becomes clear from Fig. 5.14b in particular. Depending on the normalized laser linewidth ΔfT, optimum threshold level E for minimum BER could differ from $E_{opt} = 0$ if ΔfT is large to $E_{opt} = 0.5$ if ΔfT approaches to zero. It should be remembered that the optimum threshold E_{opt} is always given by the point where the pdfs $f_d(d)$ for symbol "1" and symbol "0" meet. The dotted plot is the Gaussian approximation for $\Delta fT = 0.1$ with mean value $\eta_d = \eta_{C|1}$ and variance $\sigma_d^2 = \sigma_C^2 + \sigma^2$.

The asymmetrical influence of laser phase noise results in an asymmetrical eye pattern of detected signal d. In Chapter seven (system comparison), this important result will be discussed in more detail.

(vi) PROBABILITY OF ERROR

For the calculation of probability of error or BER, we have to distinguish between the bit sequences $<q_v>_{1i}$ with $q_0 = $ "1" (called single one or single "1") and $<q_v>_{0i}$ with $q_0 = $ "0" (called single zero or single "0"). The corresponding probability of errors p_{0i} and p_{1i} are simply obtained by integrating pdfs given in the previous Subsections over the respective areas.

With the same substitutions and abbreviations as used in Section 5.1.1 (ASK homodyne system), we obtain

$$p_{0i} = \int_{E}^{+\infty} f_{d0i}(d) \; dd = \int_{E}^{+\infty} \int_{0}^{+\infty} \frac{d}{\sigma^2} e^{-(d^2+C^2)/2\sigma^2} J_0\left(\frac{Cd}{\sigma^2}\right) f_{C0i}(C) \; dC \; dd$$

$$(5.103)$$

$$= \int_{0}^{+\infty} \tilde{p}_{0i}(C) f_{C0i}(C) \; dC = E\{\tilde{p}_{0i}(C)\} \qquad \text{with } C = C_{0i} \geq 0$$

and

$$
p_{1i} = \int_0^E f_{d1i}(d)\, dd = \int_0^E \int_0^{+\infty} \frac{d}{\sigma^2}\, e^{-\frac{d^2+c^2}{2\sigma^2}}\, J_0\!\left(\frac{Cd}{\sigma^2}\right) f_{C1i}(C)\, dC\, dd
$$

$$
= \int_0^{+\infty} \tilde{p}_{1i}(C)\, f_{C1i}(C)\, dC = E\{\tilde{p}_{1i}(C)\} \quad \text{with } C = C_{1i} \geq 0
$$

(5.104)

In the above equations, $C_{0i} = d_{0i}(\sigma=0)$ and $C_{1i} = d_{1i}(\sigma=0)$ are the sample values of the envelope $|i_{IF}(t)| = d(t)$ for the symbol sequences $<q_v>_{0i}$ and $<q_v>_{1i}$ respectively provided additive noise does not disturb the system ($\sigma=0$). Because $C = |i_{IF}(t)| \geq 0$, it is sufficient to integrate from 0 instead of from $-\infty$. The last term in Eqs. (5.103) and (5.104) follows by interchanging the order of integration of the second term and by representing the new inner integral with $\tilde{p}_{0i}(C_{0i})$ and $\tilde{p}_{1i}(C_{1i})$ respectively. These newly defined probabilities of error depend on the amplitudes C_{0i} and C_{1i} as well as on the threshold level E and the normalized standard deviation σ of the additive Gaussian noise, but independent of phase noise. Assuming that C_{0i} and C_{1i} are constant, $\tilde{p}_{0i}$ and $\tilde{p}_{1i}$ become the well-known probabilities of error for the conventional ASK transmission systems with envelope detection.

It becomes clear from above equations that the calculation of actual BER can be divided into two steps: First step is the well-known calculation of the BERs $\tilde{p}_{0i}$ and $\tilde{p}_{1i}$ neglecting laser phase noise [260]. In this step, C_{0i} and C_{1i} are constant. In the second step, we consider the randomness of C_{0i} and C_{1i} due to laser phase noise and evaluate the real BERs p_{0i} and p_{1i} by computing the expected values given in Eqs. (5.103) and (5.104). Again the problem is to determine the pdfs $f_{C0i}(C)$ and $f_{C1i}(C)$ as discussed in the previous Subsection.

(a) AVERAGE PROBABILITY OF ERROR

The average BER p_a is obtained by taking into account the occurrence probabilities of all possible symbol sequences $<q_v>_i$. Similar to the ASK homodyne system (Section 5.1.1), we obtain

$$
p_a = \sum_{i=1}^{2^{n+v}} p\!\left(<q_v>_i\right)\left(p_{0i} + p_{1i}\right) = 2^{-(n+v+1)} \sum_{i=1}^{2^{n+v}} \left(p_{0i} + p_{1i}\right)
$$

(5.105)

As mentioned in Section 5.1.1, evaluation of p_a is very comprehensive and not useful for a practical system design. Again, much more convenient is to calculate the worst-case probability of error as a very good approximation and as an upper bound.

(b) WORST-CASE PROBABILITY OF ERROR

For the calculation of the worst-case probability of error p_w, we have to consider the worst-case sequences $<q_v>_{0w}$ and $<q_v>_{1w}$. Assuming a Gaussian IF filter, worst-case pattern are given by the bit sequences

$$<q_v>_{0w} = \cdots 1110111 \cdots$$

and

$$<q_v>_{1w} = \cdots 0001000 \cdots .$$

In order to simplify calculation and to avoid generality, we take advantage of the following approximations and worst-case considerations:

- The pdf $f_{C1w}(C)$ should be calculated as given in Eq. (5.92).

- Instead of using the Rice distribution for calculating the conditional density function $f_{d|C}(d, C)$ with $d = d_{1w}$ and $C = C_{1w}$, we take Gaussian approximation given in Eq. (5.77). This approximation can always be employed when signal-to-noise ratio is high enough which is normally fulfilled when symbol "1" is transmitted.

- As mentioned above, laser phase noise always deteriorates the detection of symbol "1", whereas the detection of symbol "0" is always improved. Therefore, under worst-case consideration, influence of phase noise can be neglected for symbol "0". In that case, pdf of the detected signal d_{0w} is

$$f_{d0w}(d) = \frac{d}{\sigma^2} e^{-\frac{d^2 + c_{0w}^2}{\sigma^2}} J_0\left(\frac{C_{0w}d}{2\sigma^2}\right) \quad \text{with} \quad C_{0w} = 2Q\left(\sqrt{2\pi} f_g T\right) \tag{5.106}$$

Similar to standard ASK transmission systems with envelope detection, pdf $f_{d0w}(d)$ yields a Rice distribution if $C_{0w} \neq 0$ and a Rayleigh distribution if $C_{0w} = 0$.

With these three conditions, worst-case probability of error can now be calculated as

$$p_w = \frac{1}{2}\int_0^E f_{d1w}(d)\,dd + \frac{1}{2}\int_E^{+\infty} f_{d0w}(d)\,dd$$

$$= \frac{1}{2}Q\left(\frac{\eta_{C1w} - E}{\sqrt{\sigma_{C1w}^2 + \sigma^2}}\right) + \frac{1}{2}\int_E^{+\infty} \frac{d}{\sigma^2} e^{-\frac{d^2 + c_{0w}^2}{2\sigma^2}} J_0\left(\frac{C_{0w}d}{\sigma^2}\right)\,dd \tag{5.107}$$

In this equation, influence of laser phase noise is considered in terms of mean value η_{C1w} and the standard deviation σ_{C1w} which are determined from Eq. (5.94) under worst-case i.e., $C = C_{1w}$. As mentioned above, C_{1w} is equal to the sample value d_{1w} of the detected signal $d(t) = |i_{IF}(t)|$ under worst-case pattern $<q_v> = <q_v>_{1w}$ and in absence of additive Gaussian noise ($\sigma = 0$).

For a minimum BER, threshold level E and IF filter bandwidth $B_{IF} = 2f_g$ which is included in $\sigma = \sigma(f_g)$, $\eta_{C1w} = \eta_{C1w}(f_g)$ and $\sigma_{C1w} = \sigma_{C1w}(f_g)$ are optimizable system parameters. In contrast to that, laser linewidth Δf included in η_{C1w} and σ_{C1w} and constant power spectral density G_c included in σ are non-optimizable system parameters. Moreover, local laser power P_l and received light power P_r in the normalized standard deviation σ of the Gaussian circuit noise are, of course, also non-optimizable. It must be remembered that σ is normalized with respect to the amplitude $\hat{i}_{PD}$ of photodiode current which is also a function of P_l and P_r.

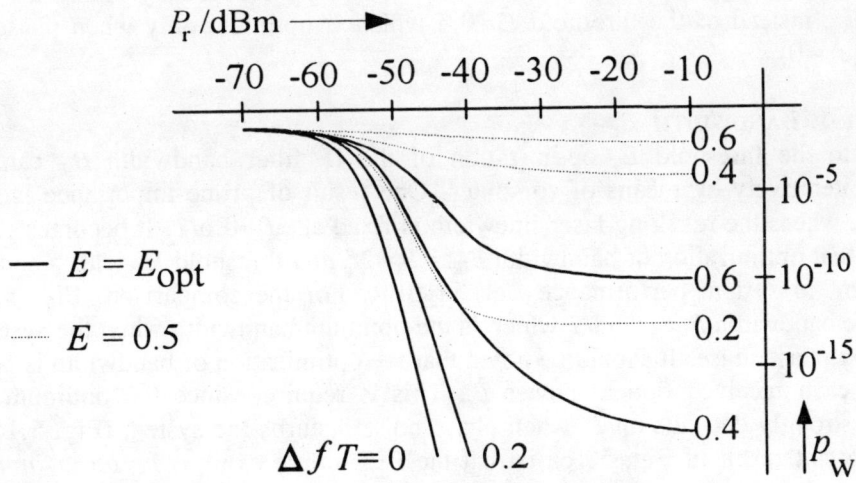

Fig. 5.15: Worst-case probability of error p_w for ASK heterodyne system with envelope detection with optimum and non-optimum thresholds

The worst-case BER as a function of the light power P_r at the input to the receiver and normalized resulting laser linewidth $\Delta f T$ is shown in Fig. 5.15. The IF filter bandwidth is fixed at $B_{IF} = 1.85/T$ which is the optimum bandwidth in case of $\sigma_\phi = 0$.

The trend of the BER curves shown in Fig. 5.15 are similar to that of coherent optical systems described in the previous Sections. If the received optical power P_r is low, then the additive Gaussian noise dominates the phase noise. Thus, the curves are rather steep. As the optical power P_r increases, influence of additive Gaussian noise decreases, while the influence of the laser phase noise increases. Consequently, BER curves exhibit a bend. As the optical power P_r is increased further beyond this point, performance curves finally reach a saturation called the bit error rate floor (Section 5.1.1).

(vii) OPTIMIZATION

(a) THRESHOLD E

Because of the complexity of Eq. (5.107), optimum threshold level for a minimum BER can only be evaluated by means of numerical methods using computer. Thereby, the relation

$$E_{opt} \leq 0.5 \qquad\qquad\qquad\qquad\qquad\qquad (5.108)$$

is always valid since both the binary symbols "0" and "1" are influenced differently by the laser phase noise as discussed above. It should be noted that the optimum threshold level E_{opt} given in Eq. (5.108) is also normalized with respect to the amplitude $\hat{i}_{PD}$ of the photodiode current given in Eq. (2.63).

It is seen from Fig. 5.15 that the BER is considerably reduced by taking the optimum threshold E_{opt} instead of the threshold $E=0.5$ which is optimum only when phase noise is neglected ($\sigma_\phi = 0$).

(b) IF FILTER BANDWIDTH B_{IF}

Similar to the threshold E, optimization of the IF filter bandwidth B_{IF} can only be performed iteratively by means of computer. One result of prime importance is shown in Fig. 5.16a, where the resulting laser linewidth is fixed at $\Delta f = 0.6/T$. It becomes clear from this figure that optimization of bandwidth $B_{IF} := B = 2f_g$ *and* threshold E yields a considerable improvement in system performance i.e., in BER. For the comparison, Fig. 5.16a also includes the bandwidth $B_{IF} = 1.58/T$ which is the optimum bandwidth when the system is not disturbed by phase noise. It should be noted that the optimization of bandwidth is performed afresh for each received optical power P_r. This is required since the optimum IF filter bandwidth strongly depends on P_r when phase noise disturbs the system (Fig. 5.16b).

An important result of Fig. 5.16a is that the BER curve exhibits *no bit error rate floor* when both bandwidth B_{IF} *and* threshold E are optimized. This means that a bit error rate of 10^{-10} is now generally achievable for any laser linewidth Δf. There exists no limit and no error rate floor in principle. This very surprising behaviour can easily be explained as follows: The system deterioration caused by large laser linewidth can be compensated by a respective broad IF filter bandwidth. This, however, causes a rise in additive Gaussian noise at the IF filter output. On the other hand, effect of this additional additive noise can always be reduced in principle by an appropriate high received light power P_r. The only limitation is given by the maximum available light power P_r and realizable intermediate frequencies. Nevertheless, to achieve a large repeater spacing, received optical P_r should be allowed to be as low as possible.

Finally, Fig. 5.16b shows the normalized optimum IF filter bandwidth $B_{opt}T$ as a function of the received optical power P_r at the input to receiver and the normalized resulting laser linewidth $\Delta f T$. Like in ASK homodyne receiver, a broader filter again reduces the influence of phase noise.

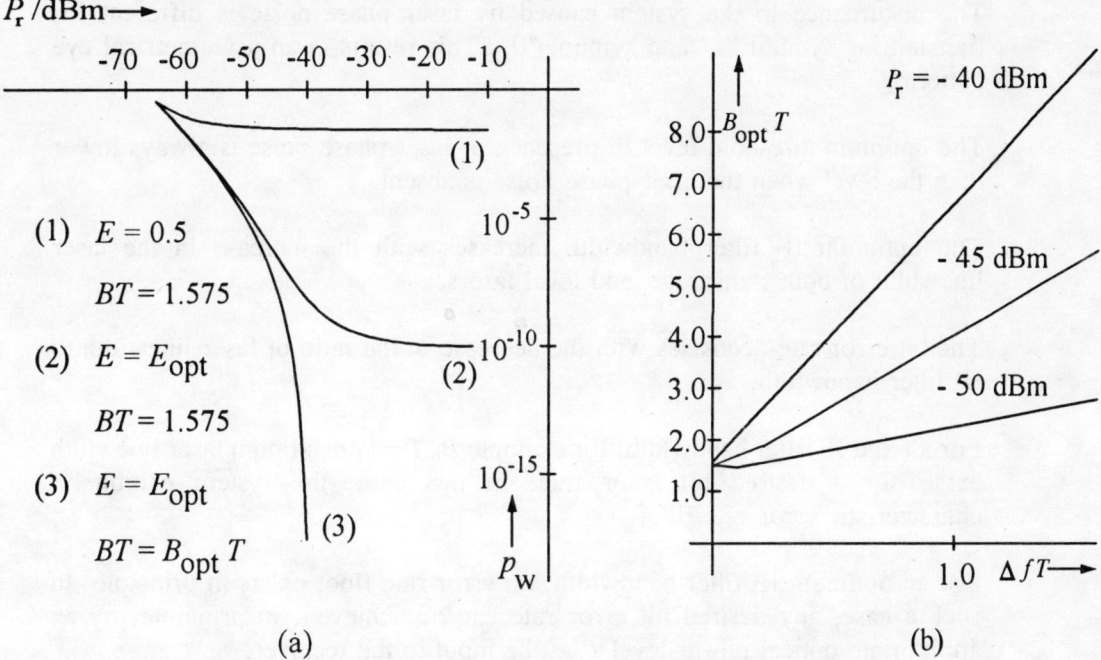

P_r/dBm⟶

-70 -60 -50 -40 -30 -20 -10

(1) $E = 0.5$

$BT = 1.575$

(2) $E = E_{opt}$

$BT = 1.575$

(3) $E = E_{opt}$

$BT = B_{opt} T$

(1)

10^{-5}

10^{-10}

(2)

10^{-15}

p_w

(3)

(a)

$B_{opt} T$

8.0

7.0

6.0

5.0

4.0

3.0

2.0

1.0

$P_r = -40$ dBm

-45 dBm

-50 dBm

1.0 $\Delta f T$⟶

(b)

Fig. 5.16: (a) Reduction of probability of error p_w by optimizing IF filter bandwidth B and threshold E for $\Delta f T = 0.6$ and (b) Optimum IF filter bandwidth as a function of normalized resulting laser linewidth $\Delta f T$ and received optical power P_r.

The present optimization corresponds to a minimum worst-case BER provided a Gaussian IF filter is used. To obtain a more general solution, an arbitrary IF filter frequency response must be taken into account. Moreover, the system becomes optimum when the system parameters mentioned above are optimized for a minimum average BER instead for a minimum worst-case BER. However, this optimization can only be carried out by very comprehensive and time intensive numerical computer calculation.

Owing to the fact that each bit sequence is related to its own BER, optimum receiver is actually obtained by changing the receiver configuration including only one IF filter to a configuration providing a matched IF filter for each possible bit sequence $<q_v>_i$. It will be the best receiver which can be theoretically obtained. However, it will not be of much practical use due to the very comprehensive design and realization.

(c) CHARACTERISTIC FEATURES OF NONCOHERENT ASK HETERODYNE SYSTEMS

Most important results of our considerations in the previous Subsections on noncoherent ASK heterodyne system with envelope detection are summarized below:

- The disturbance to the system caused by laser phase noise is different for transmitting symbol "1" and symbol "0". This results in an asymmetrical eye pattern.

- The optimum threshold level in presence of laser phase noise is always lower than the level when the laser phase noise is absent.

- The optimum IF filter bandwidth increases with the increase in the laser linewidth of both transmitter and local lasers.

- The bit error rate decreases with the decrease in the ratio of laser linewidth to IF filter bandwidth.

- For a fixed IF filter bandwidth; for example $B_{IF}T = 1$, maximum laser linewidth exists for a desired bit error rate. In this case, the system exhibits a characteristic error rate floor.

- For an optimum IF filter bandwidth, no error rate floor exists in principle. In such a case, any desired bit error rate can be achieved, in principle, by an appropriate optical power level P_r at the input to the receiver.

It should be noted that the first two items become evident in particular when the computer simulated eye pattern is considered (see Chapter seven).

An optical ASK heterodyne receiver with noncoherent detection (e.g., envelope detection) exhibits a 9 dB lower sensitivity than the most sensitive PSK homodyne receiver and a 3 dB lower sensitivity as compared to noncoherent dual-filter FSK heterodyne receiver (Section 5.3.2).

In noncoherent ASK heterodyne system, laser linewidths of 20% to 30% of the bit rate $1/T$ can be allowed when normalized bandwidth of Gaussian IF filter is fixed at $BT = 1.58$. This bandwidth represents the optimum in absence of phase noise. When bandwidth is optimized by taking into consideration also laser phase noise, linewidths in the order of bit rate are acceptable. This is in contrast to homodyne receivers, where the linewidth requirements are very strong (e.g., 0.01% of the bit rate).

As the linewidth requirements, costs and problems of realization are approximately the same in noncoherent ASK and FSK heterodyne, noncoherent ASK heterodyne receiver is of little practical importance. However, most of the results obtained in this Section can be used to analyze noncoherent FSK heterodyne system (Section 5.3.2) and noncoherent DPSK heterodyne system (Section 5.3.3).

5.3.2   FSK Heterodyne System

In conventional analog communication systems such as analog radio and TV broadcasting, frequency modulated (FM) carrier signal is normally demodulated by means of a *frequency discriminator*. In order to obtain a linear discriminator transfer function with respect to input frequency and output voltage, a proper match between both the discriminator filters must be performed. Each deviation in linearity directly results in nonlinear distortion.

In comparison to that, linear distortion is not as critical in a digital transmission system with FSK modulation. Here, the shape of the received signal is less important, whereas an errorless decision between both binary symbols "0" and "1" is most important. Thus, a linear discriminator is not absolutely required and the centre frequencies of both discriminator filters may be separated by a much larger frequency deviation than in case of a linear discriminator employed in analog FM systems. Since both filters are now no longer matched, this demodulator is usually called *dual-filter demodulator* or *two-filter demodulator*. Owing to the relatively large frequency deviation, these demodulators enable a proper separation and a clear detection of both binary symbols "0" and "1".

The block diagram of a noncoherent optical FSK heterodyne transmission system with a two-filter demodulator in the receiver is shown Fig. 5.17. The FSK demodulator contains two branches: one is the "1"-branch or the "1"-channel; other is the "0"-branch or the "0"-channel. Both symbols "0" and "1" are selected by two band-pass filters with centre frequencies f_{IF0} and f_{IF1} respectively. With a frequency deviation f_d and an intermediate frequency (IF) f_{IF}, both centre frequencies are given by

$$f_{IF0} = f_{IF} - f_d \quad \text{and} \quad f_{IF1} = f_{IF} + f_d \tag{5.109}$$

Hence, lower the frequency deviation f_d, lower is the frequency difference between both the centre frequencies. Thereby, an errorless detection of binary symbols "0" and "1" becomes more and more difficult. In the worst-case, $f_d = 0$ and distinction between binary symbols is, of course, no longer possible.

As indicated in Fig. 5.17, both the IF filter output signals $i_{IF0}(t)$ and $i_{IF1}(t)$ disturbed by laser phase noise and receiver additive Gaussian noise (i.e., shot noise and thermal noise) are fed to the input of an envelope detector. The detector output signals referred as the detected signals $d_0(t)$ and $d_1(t)$ are finally sampled after every T (i.e., symbol- or bit duration) and fed to a maximum decision circuit. Depending on which of the two branches has the highest sample value, decision circuit decides either the symbol "0" or symbol "1".

It becomes clear from Fig. 5.17 that noncoherent two-filter demodulator can be regarded as a simple parallel circuit of two noncoherent ASK envelope demodulators. Therefore, many results of the previous Section 5.3.1 can be used to analyze noncoherent FSK heterodyne system.

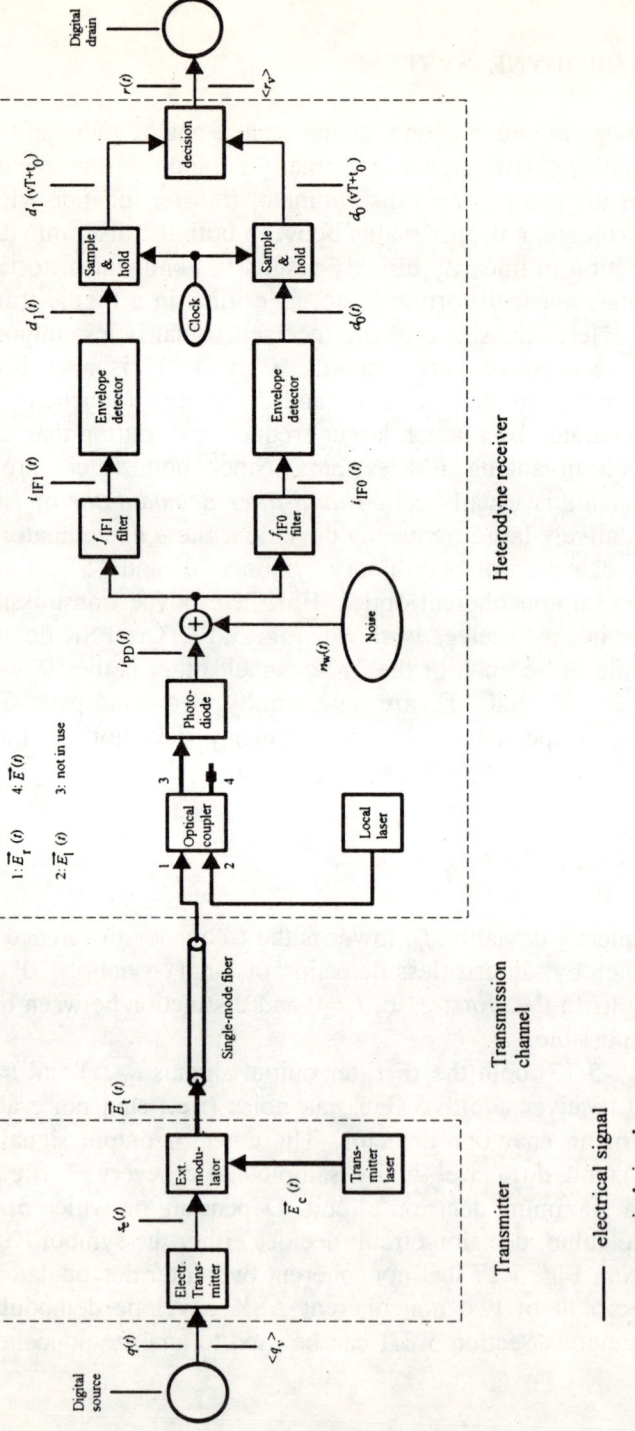

Fig. 5.17: Coherent optical communication system with noncoherent FSK heterodyne receiver

In order to realize an optical FSK system with a large frequency deviation f_d, which is required in presence of considerable system impairment, intermediate frequency f_{IF} must be chosen appropriately high enough. Otherwise, the filter with centre frequency f_{IF0} becomes unrealizable (Eq. 5.109). Of course, due to the limited bandwidths of photodiode, preamplifier and electronic components of the IF circuit, intermediate frequency f_{IF} cannot be increased indefinitely.

This problem is overcome in a *single-filter demodulator* also called a *one-filter demodulator* (Fig. 5.18). This demodulator is simply realized by leaving out the "0"-branch of the two-filter demodulator. Thereby, the maximum decision circuit can be replaced by a threshold decision circuit. Single-filter demodulator enables the realization of a relatively large frequency deviation f_d. However, overall available signal power is divided by two since one branch is missed. Therefore, *receiver sensitivity* is reduced by 3 dB as compared to FSK receiver with dual-filter detection.

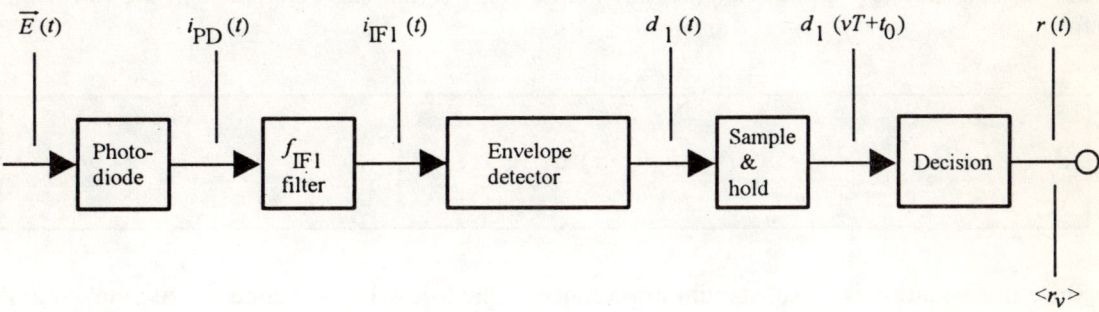

Fig. 5.18: Single-filter demodulator in noncoherent FSK heterodyne receiver

(i) DETECTED SIGNAL $d(t)$ AND SAMPLE VALUES $d(\upsilon T + t_0)$

Taking into account the results of the previous Section 5.3.1 (noncoherent ASK heterodyne system), detected signals at the output of both envelope detectors can be described by the common equation

$$
d(t) \approx \left| \hat{i}_{PD} \int_{-\infty}^{+\infty} \underline{s}(\tau)\, e^{j(2\pi f_{IF}\tau \,+\, \phi(\tau))}\, h_B(t-\tau)\, e^{j2\pi(f_{IF} \pm f_d)(t-\tau)}\, d\tau \;+\; \underline{n}(t) \right|
$$

$$
(5.110)
$$

$$
= \left| \hat{i}_{PD} \int_{-\infty}^{+\infty} \underline{s}(\tau)\, e^{\mp j2\pi f_d \tau}\, e^{j\phi(\tau)}\, h_B(t-\tau)\, d\tau \;+\; x(t)\; -\; jy(t) \right|
$$

Here, the upper sign in the exponent represents the "1"-branch with $d(t)=d_1(t)$ and the lower sign the "0"-branch with $d(t)=d_0(t)$. Like the analysis of noncoherent ASK heterodyne system (Section 5.3.1), above equation takes advantage of baseband representation of both the IF filters and also of the narrow-band condition given in Eq. (2.75). Again, $h_B(t)$ represents the impulse response of the equivalent baseband filter $H_B(f)$ as explained in Section 2.4.3. As the modulation scheme is FSK, information signal $\underline{s}(\tau)$ is given by (compare with Eq. 2.36)

$$\underline{s}(\tau) = \exp\left[j\int_{-\infty}^{\tau}\sum_{\upsilon=-\infty}^{+\infty} 2\pi f_d(2s_\upsilon - 1)\,\mathrm{rect}\left(\frac{t-\upsilon T}{T}\right)\,dt\right] = \exp\left[j\int_{-\infty}^{\tau}\omega_s(t)\,dt\right] = e^{j(\phi_s(\tau)\,+\,\phi_0)} \quad (5.111)$$

where $\omega_s(t)$ and $\phi_s(t)$ represent the modulated frequency and phase of $\underline{s}(\tau)$.

With the same normalizations and substitutions as used in the previous Section 5.3.1, normalized sample value $d = d(t_0)/\hat{\imath}_{PD}$ of the detected signal can be written in the following form:

$$d = \left|\int_{-\infty}^{+\infty} e^{j\phi_0}\,e^{j[\phi_s(\tau)\,\mp\,2\pi f_d\,\tau]}\,e^{j\phi(\tau)}\,h_B(t_0-\tau)\,d\tau\;+\;x\;-\;jy\right| \quad (5.112)$$

Since this equation is of substantial importance in the following system analysis, meaning of its physical quantities is summarized below:

ϕ_0:   arbitrary, but constant phase,
$\phi_s(t)$:   signal phase including the transmitted information,
f_d: frequency deviation,
$\phi(t)$:   laser phase noise of both transmitter and local lasers,
$h_B(t)$:   impulse response of the equivalent baseband filter, ·
$x(t)$:   inphase component of the receiver additive Gaussian noise,
$y(t)$:   quadrature component of the receiver additive Gaussian noise.

We have to distinguish "0"-branch and "1"-branch considering the term $\phi_s(\tau) \mp 2\pi f_d \tau$ given in the above equation. In the "1"-branch, this term is either a constant (if $q_\upsilon = $"1") or proportional to $4\pi f_d \tau$ (if $q_\upsilon = $"0"). In the "0"-branch, this relationship is reversed i.e., this term is constant for $q_\upsilon = $"0" and proportional to $4\pi f_d \tau$ for $q_\upsilon = $"1".

Assuming that no receiver and phase noise exist in the system i.e., $x = y = 0$ and $\phi = 0$, Eq. (5.112) simplifies to

$$d = \left| \int\limits_{-\infty}^{+\infty} e^{j\phi_0}\, e^{j[\phi_s(\tau)\, \mp\, 2\pi f_d \tau]}\, h_B(t_0 - \tau)\ d\tau \right| \tag{5.113}$$

An alternate equation is obtained by dividing the integral into real and imaginary parts:

$$d = \left[\left(\int\limits_{-\infty}^{+\infty} \cos(\phi_s(\tau)\, \mp\, 2\pi f_d \tau\, +\, \phi_0)\, h_B(t_0 - \tau)\ d\tau \right)^2 \right.$$

$$\left. + \left(\int\limits_{-\infty}^{+\infty} \sin(\phi_s(\tau)\, \mp\, 2\pi f_d \tau\, +\, \phi_0)\, h_B(t_0 - \tau)\ d\tau \right)^2 \right]^{1/2} \tag{5.114}$$

In case of an ideal envelope demodulator, where the detected signal at the output of demodulator equals the absolute of complex IF signal (Eqs. 5.110 and 5.113), a constant phase ϕ_0 does not influence the detection process. Hence, this phase can be either set to zero or chosen in such a way that the imaginary part of the convolution integral (i.e., the part with the sine function) becomes zero. However, this formal step is only possible in case of an even impulse response $h_B(t) = h_B(-t)$.

Generation of sample value $d = d_1$ in the "1"-branch of a dual-filter FSK demodulator is illustrated in Fig. 5.19. For this, worst-case pattern $\cdots 0001000 \cdots$ (single-"1") and $\cdots 1110111 \cdots$ (single-"0") are assumed to be the transmitted information. The upper curves in Fig. 5.19 first show the corresponding modulated frequency $\omega_s(t)$ and the modulated phase $\phi_s(t)$. In accordance with Eq. (5.114), we then have to subtract the linearly increasing phase $2\pi f_d t$. As a result, we obtain the phase $\phi_1(t) = \phi_s(t) - 2\pi f_d t$ which is also plotted in Fig. 5.19. Because of the fact that an additional constant phase ϕ_0 does not influence the result (i.e., the detected signal at the output of the FSK receiver), we may use ϕ_0 to simplify calculation. It is most convenient to choose ϕ_0 such that the new phase $\phi_2(t) = \phi_1(t) + \phi_0$ is an odd function i.e., $\phi_2(t) = -\phi_2(-t)$.

Next, we have to accomplish the required convolution between the impulse response $h_B(t)$ and signal $\cos(\phi_2(t)) = \cos(\phi_2(-t))$. Finally, the sample value d_1 is obtained by taking the absolute of the convolution result and setting $t_0 = 0$. Owing to our proper choice of the constant phase ϕ_0, imaginary part of the convolution integral yields zero since the sine function is an odd function i.e., $\sin(\phi_2(t)) = -\sin(\phi_2(-t))$.

Comparing ASK and FSK heterodyne receivers, it becomes clear that the sample value d_1 (provided single-"1" transmitted) is always higher (i.e., better) when FSK is employed. If a single-"0" is transmitted, then the FSK sample value d_0 can either be above or below the corresponding ASK sample value. Which of the two cases is actually existent depends on the frequency deviation f_d. If the normalized frequency deviation approaches infinity (i.e.,

$f_d T \to \infty$) or $f_d T = k$ where $k \in \mathbb{N}$, then sample value d_0 is same for both ASK and FSK hetero-dyne receivers (see also Fig. 5.20).

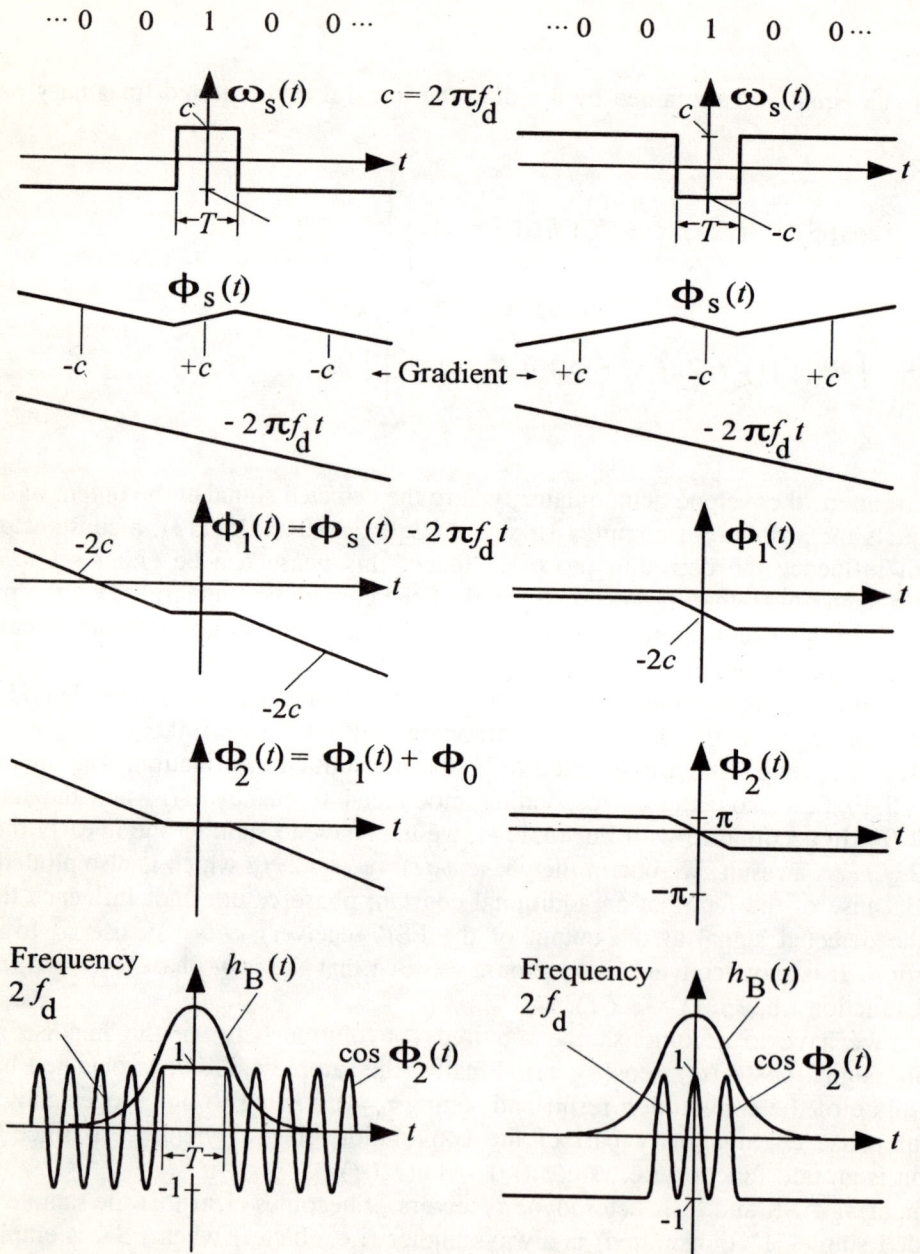

Fig. 5.19: Principle of generation of sample value d_1 in the "1"-branch of dual-filter FSK receiver

It is obvious that "0"-branch and "1"-branch are absolutely symmetrical. Thus, a symbol "1" in the "0"-branch always yields the same sample value as a symbol "0" in the "1"-branch. Therefore, Fig. 5.19 can also be used to illustrate how the detected signal d_0 is generated in the "0"-branch of a dual-filter FSK demodulator. For this, all binary symbols included in Fig. 5.19 are merely to be interchanged i.e., "0"→"1" and "1"→"0".

The sample value d_1 in the "1"-branch as a function of the normalized frequency deviation $f_d T$ in case of different symbol sequences $<q_v>$ is shown in Fig. 5.20. For a Gaussian filter with normalized bandwidth $B_{IF}T = 2f_g T = 1$ (Eq. 2.77), it is sufficient to consider only one previous and one following neigbouring symbol. In the unrealistic case of $f_d T = 0$, normalized sample value d_1 always equals 1 irrespective of which symbol sequence is actually transmitted. As expected, both binary symbols are now no longer distinguishable and, hence, detectable.

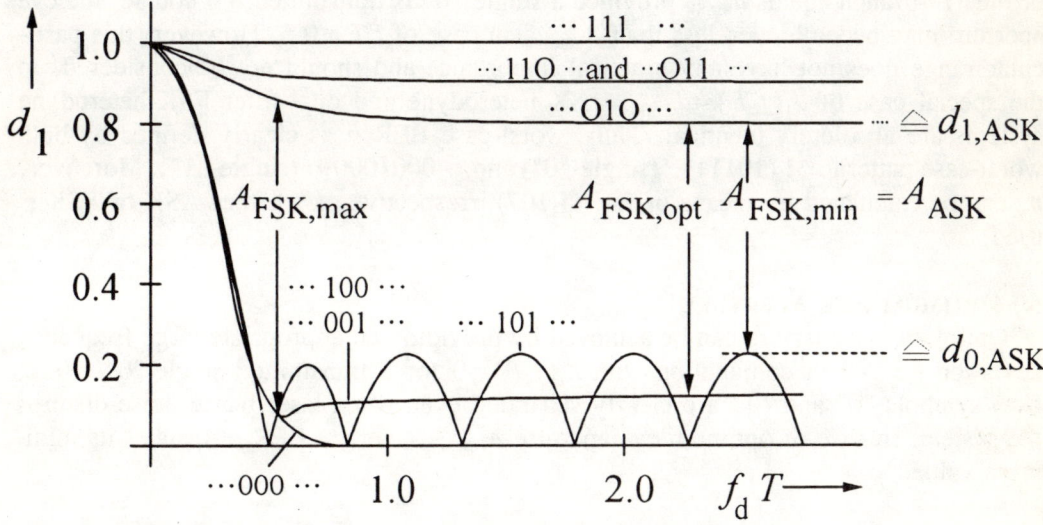

Fig. 5.20: Normalized sample value d_1 of "1"-branch in dual-filter FSK receiver as a function of normalized frequency deviation $f_d T$ for various symbol pattern $<q_v>$. Influence of receiver additive Gaussian noise is reduced by Gaussian filter with normalized bandwidth $BT = 2f_g T = 1$.

As normalized frequency deviation $f_d T$ increases, symbol detection improves. Depending on the number of periods in the symbol duration T (Fig. 5.19), a transmitted symbol sequence $\cdots 1110111 \cdots$ (single-"0") yields a sample value d_1 which is located in the range of $0 \le d_1 \le d_{0,ASK}$ provided frequency deviation f_d is high enough (e.g., $f_d T > 0.5$). Here, $d_{0,ASK}$ represents the sample value in case of ASK homodyne receiver provided $q_v = $"0". By interchanging symbol "0" and "1", Fig. 5.20 becomes valid for the sample value d_0 of the "0"-branch.

On the basis of Fig. 5.20, following three characteristic eye apertures can be defined:

(a) MAXIMUM EYE APERTURE $A_{FSK,max}$

This maximum eye aperture is given at a normalized frequency deviation $f_d T \approx 0.5$. When the phase noise does not disturb the system, this aperture is optimum for a minimum BER. However, at $f_d T \approx 0.5$ symbol "0" ($f_{IF} - f_d$) and symbol "1" ($f_{IF} + f_d$) are located closely. Hence, as soon as laser phase and frequency noise disturb the system in addition, both symbols can no longer be distinguished clearly. It should be noted that laser frequency noise permanently results in a random shift in frequency deviation f_d, which finally increases BER although the eye aperture is large.

(b) MINIMUM EYE APERTURE $A_{FSK,min}$

As shown in Fig. 5.20, eye aperture becomes minimum when sample value $d_1(f_d T)$ of the "1"-branch equals $d_{0,ASK}$ provided a single-"0" is transmitted. Of course, the eye aperture may become even less than $A_{FSK,min}$ in case of $f_d T \ll 0.5$. However, this particular range does not have any practical importance and should not be considered. In the special case of $d_1(f_d T) = d_{0,ASK}$, ASK heterodyne and dual-filter FSK heterodyne systems are absolutely identical. Thus, worst-case BER p_w is clearly defined by both worst-case pattern $\cdots 1110111 \cdots$ (single-"0") and $\cdots 0001000 \cdots$ (single-"1"). Moreover, p_w can be calculated by means of Eq. (5.107) irrespective of whether ASK or FSK is used.

(c) OPTIMUM EYE APERTURE $A_{FSK,opt}$

Optimum eye aperture can be achieved by providing an appropriate large frequency deviation $f_d T > 5$ in conjunction with $d_1(f_d T) = 0$ for a transmitted single-"0". Here, both symbols "0" and "1" are clearly separated even when laser phase noise disturbs the system. In case of optimum eye aperture $A_{FSK,opt}$, average BER p_a reaches its minimum value.

(ii) PROBABILITY OF ERROR
(a) AVERAGE PROBABILITY OF ERROR p_a

In accordance with our discussion in the previous Subsection, average probability of error p_a reaches minimum in case of an infinite frequency deviation provided $d_1(f_d T) = 0$ for a transmitted single-"0" (compare Fig. 5.20). Calculation of p_a is very comprehensive and can only be performed by means of appropriate numerical methods and fast computer program.

As the FSK sample values d are always "better" than the corresponding ASK values (i.e., a symbol "1" always results in a higher and a symbol "0" a lower sample value), average probability of error p_a is somewhat less when FSK is applied to the system. In terms of receiver sensitivity, this results in an improvement of about 1 dB in single-filter FSK receiver and about $(1+3)$ dB$=4$ dB in dual-filter FSK receiver.

(b) WORST-CASE PROBABILITY OF ERROR p_w

As mentioned earlier, noncoherent ASK and FSK systems exhibit same features with respect to the worst-case probability of error p_w provided frequency deviation is sufficiently large ($f_d T > 5$) and $d_1(f_d T) = d_{0,\text{ASK}}$ in case of a transmitted symbol sequence $\cdots 1110111\cdots$ (Fig. 5.20). Therefore, the same formulas can be used to evaluate p_w. Moreover, the BER curves of Fig. 5.15 (noncoherent ASK heterodyne system) are also valid for noncoherent FSK heterodyne system with single-filter detection.

In a dual-filter detection, maximum decision must be used instead of simple threshold decision. For this, probability density functions $f_{d1}(d)$ and $f_{d0}(d)$ of the worst-case sample values $d_1 > 0$ and $d_0 > 0$ are to be determined. We obtain

$$p_w = \frac{1}{2}\left[p\left(d_1 > d_0 | q_v = "0"\right) + p\left(d_0 > d_1 | q_0 = "1"\right)\right] = p\left(d_0 > d_1 | q_0 = "1"\right)$$

(5.115)

$$= \int\limits_{d_1=0}^{+\infty} \int\limits_{d_0=d_1}^{+\infty} f_{d1}(d_1)\, f_{d0}(d_0)\; \mathrm{d}d_0\, \mathrm{d}d_1$$

This equation first assumes same occurrence probabilities for both symbols "0" and "1" and second statistical independence of both sample values d_1 ("1"-branch) and d_0 ("0"-branch). Hence, $f_{d1,d0}(d_1, d_0) = f_{d1}(d_1) \cdot f_{d0}(d_0)$. As noise in "0"-branch and "1"-branch are characterized by different noise spectra, this condition is usually fulfilled provided discriminator filters do not exhibit spectral overlapping. In this case, dual-filter FSK receiver offers a sensitivity gain of 3 dB in comparison to single-filter FSK receiver. Evaluation of Eq. (5.115) requires the probability density functions $f_{d1}(d)$ and $f_{d0}(d)$, which can be obtained by applying one of the methods discussed in Section 5.3.1.

Using same approximations as given in Section 5.3.1 for special sequences viz., permanent-"0" and permanent-"1" (i.e., no intersymbol interference), Eq. (5.115) simplifies to

$$p_w = \frac{\sigma_0^2}{\sigma_1^2 + \sigma_0^2}\, \exp\left(-\frac{C_1^2}{2\left(\sigma_1^2 + \sigma_0^2\right)}\right) \quad \text{where } \sigma_0^2 = \sigma^2 \text{ and } \sigma_1^2 = \sigma^2 + \sigma_{C_1}^2$$

(5.116)

The physical quantities required in the evaluation of Eq. (5.116) can be derived as in Section 5.3.1 by taking permanent-"0" and permanent-"1". The influence of phase noise is included in the amplitude C_1 as well as in the standard deviation σ_{C1}. As described in Section 5.3.1, C_1 represents the sampled detected signal d_1 (i.e., $q_v = "1"$) provided receiver additive Gaussian noise is neglected ($\sigma = 0$). Assuming that $\sigma_1 = \sigma_0 = \sigma$ which represents a system without any laser phase noise, Eq. (5.115) is simply the well-known BER formula for electrical FSK transmission system.

(iii) FREQUENCY DEVIATION ESTIMATION

In the realization of noncoherent optical FSK heterodyne systems, one question of primary importance arises: What frequency deviation f_d is required to obtain best system performance? To answer this question, a simple estimation is presented and used in this Subsection. Two conditions are, however, required to use this powerful estimation: first we have to provide an ideal frequency discriminator characterized by a linear input-output transfer function (Fig. 5.21) and second we have to neglect receiver additive Gaussian noise i.e., $x=y=0$ or $\sigma=0$. Thus, the system is only disturbed by laser phase noise. Influence of additive noise may at least be reduced partly either by an amplitude limiter or appropriate large signal power level.

Transfer function of an ideal frequency discriminator including the corresponding frequencies for both binary symbols "0" and "1" is shown in Fig. 5.21.

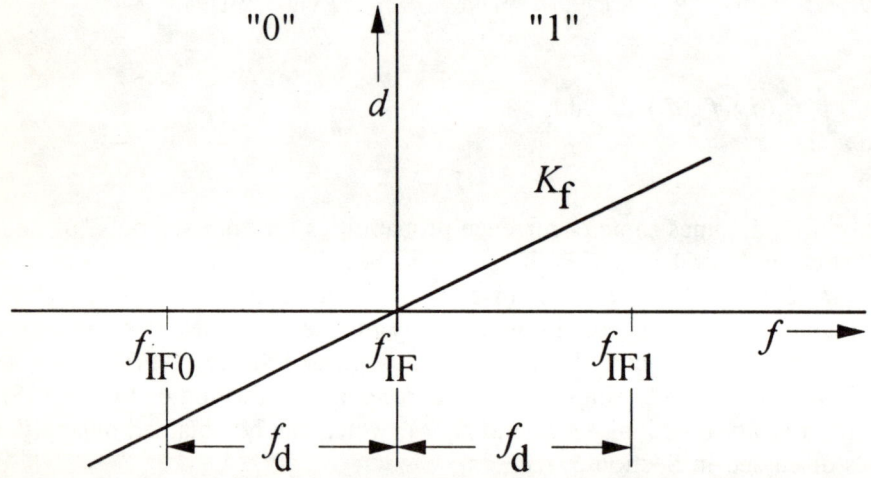

Fig. 5.21: Idealized transfer function of frequency discriminator

At the output of the frequency discriminator, detected signal

$$d(t) = K_\omega[\omega_s(t) + \omega(t)] = K_\omega\left[\sum_{\upsilon=-\infty}^{+\infty} 2\pi f_d\left(2s_\upsilon - 1\right)\text{rect}\left(\frac{t-\upsilon T}{T}\right) + \omega(t)\right] \qquad (5.117)$$

is obtained. In the above equation, $K_\omega=2\pi K_f$ represents the gradient of the discriminator transfer function and $\omega_s(t)$ and $\omega(t)$ the transmitted information and laser frequency noise respectively. As described in Section 3.3.3, laser frequency noise is a white random process with constant power spectral density

$$G_\phi(f) = G_\omega = 2\pi\Delta f = 2\pi(\Delta f_t + \Delta f_l) \tag{5.118}$$

Here, Δf_t and Δf_l represent the laser linewidth of both transmitter and local lasers. When the worst-case pattern $\cdots 0001000 \cdots$ and $\cdots 1110111 \cdots$ are transmitted and noise is reduced by a Gaussian filter, detected signal given in Eq. (5.117) results in a worst-case probability of error

$$p_w = Q\left[\frac{f_d T\left(1 - 4Q\left(\sqrt{2\pi} f_g T\right)\right)}{\sqrt{\Delta f T\sqrt{2} f_g T}}\right] \tag{5.119}$$

In the particular case of a normalized Gaussian filter bandwidth $2 \cdot f_{g,opt} T \approx 1.58$, worst-case probability of error p_w reaches to a minimum value. It is important to note that the optimum filter bandwidth neither depends on frequency deviation f_d nor on resulting laser linewidth Δf.

Let us consider, for example, a worst-case probability of error $p_w = 10^{-10}$ and an optimum filter bandwidth $B = 2 \cdot f_{g,opt}$. Eq. (5.119) yields the following result:

$$\frac{(f_d T)^2}{\Delta f T} \approx 55.30 \rightarrow f_d T \approx 7.44\sqrt{\Delta f T} \tag{5.120}$$

By using this simple, but very useful result, required frequency deviation f_d to obtain a worst-case probability of error $p_w = 10^{-10}$ can easily be estimated as a function of resulting laser linewidth Δf and bit rate $R = 1/T$.

Example 5.3

For a noncoherent optical FSK heterodyne system with bit rate $R = 622.08$ Mbit/s (SDH) and resulting laser linewidth $\Delta f = 56$ MHz, frequency deviation f_d should be at least $f_d = 2.23/T \approx 1.39$ GHz.

It must be noted that Eq. (5.120) has to be handled carefully. This equation suggests that each laser linewidth Δf can be accepted provided that frequency deviation f_d is chosen appropriately high enough. If laser phase noise becomes large, then the undesired interaction between phase noise and amplitude of the IF signal becomes more and more dominant. Thus, phase noise results in amplitude noise in addition.

In the particular case of an ideal frequency detection as considered above, amplitude fluctuations do not influence the detection process at all. If a real detection process is considered; for example, a dual-filter circuit with an envelope demodulator in each branch, amplitude (envelope) fluctuations directly disturb the detection process. This fact, however, is not considered in Eq. (5.120). Nevertheless, the above equation represents a very useful practical tool for a fast estimation of the required frequency deviation f_d, always keep in mind the underlying conditions and restrictions mentioned above.

When the real permissible laser linewidth is to be determined, BER curve $p_w(P_r, \Delta f)$ has to be computed as a function of received optical power level P_r and resulting laser linewidth Δf of both transmitter and local lasers. Assuming, for example, noncoherent FSK heterodyne system with a frequency deviation such as given in Eq. (5.120), BER curves are approximately same as the BER curves for noncoherent ASK heterodyne system (Fig. 5.22 and [213]).

(vi) EVALUATION AND DISCUSSION

Worst-case probability of error p_w as a function of received optical power P_r, resulting laser linewidth Δf and frequency deviation f_d is shown in Fig. 5.22. It should be mentioned that this figure is obtained by means of numerical methods and computer program [54].

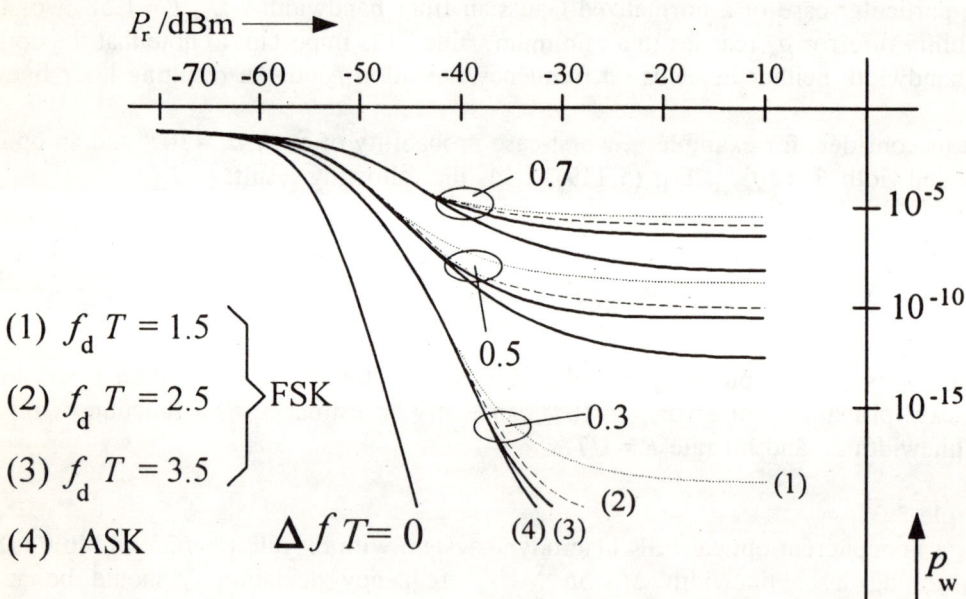

Fig. 5.22: Worst-case probability of error p_w in FSK heterodyne receiver with non-coherent single filter detection

It becomes clear from the above figure that ASK heterodyne (curve 4) and FSK heterodyne (curves 1 to 3) do agree very well provided frequency deviation f_d is chosen large enough (for example, $f_d T = 3.5$). In the special case of an infinite frequency deviation ($f_d \to \infty$) with $d_1(f_d T) = d_{0,ASK}$ (for a transmitted single-"0", see Fig. 5.20), both systems are absolutely identical as far as worst-case BER p_w is concerned.

5.3.3   DPSK HETERODYNE SYSTEM

Differential phase shift keying (DPSK) is a variant of PSK. In PSK system, information is located in the phase itself, whereas in DPSK system information is located in the phase difference of two neighbouring symbols. Thereby, symbol "0" is represented by a phase change of 180° and symbol "1" by a phase change of 0° i.e., phase remains unchanged. Each change in phase of 180° (i.e., π) reverse the polarity of received optical signal and, hence, IF signal also. Thus, the transmitted information is included in the change of sign of the received optical signal. Thereby, a change in sign represents a symbol "0", whereas no change in sign represents a symbol "1". Of course, the assignment of symbol and change in sign is arbitrary and can be chosen by the system designer. Therefore, the special assignment provided above may also be turned over i.e., "0" ≙ 0° and "1" ≙ 180°. In this Section, first assignment will only be used.

The block diagram of a typical DPSK transmission system is shown in Fig. 5.23. The main components and signals have already been discussed in Chapter two in detail. As shown in this figure, demodulator is based on a *delay line* (the required delay equals the symbol duration T) and *multiplication device* characterized by a multiplication constant C_M. Moreover, the demodulator also contains a *low-pass filter* suppressing the high frequency ($2f_{IF}$) signal components which are an additional undesired product of the multiplication process. Since this demodulator compares the input signal with a delayed version of itself, it is known as *autocorrelator demodulator*. If no phase change was sent (mark, q_v = "1"), intermediate frequency signal $i_{IF}(t)$ and delayed version $i_{IF}(t-T)$ are equal and low-pass filtered product, the detected signal $d(t)$, is positive. A phase change of 180°, which means that space was sent (q_v = "0"), results in a negative output.

It should be noted that the baseband signal part of the product signal $p(t)$ which includes the information passes through the low-pass filter unchanged. We again call the low-pass filter output signal as detected signal $d(t)$. Like in other digital systems, this signal is fed to the input of a sample and hold and decision circuits as shown in Fig. 5.23. There, the bit stream is regenerated by a simple threshold discriminator. Typical signals of the noncoherent optical DPSK heterodyne transmission system are shown in Fig. 5.24.

(i) DETECTED SIGNAL $d(t)$ AND SAMPLE VALUES $d(vT + t_o)$

In order to simplify calculation, we consider that the *IF filter* does not influence the useful part of the IF filter input signal. Thus, this filter only reduces the effect of receiver additive Gaussian noise. Moreover, the input signal passes through the filter without any change. Therefore, intersymbol interference do not occur and phase noise remains unchanged.

In case of noncoherent ASK heterodyning, these special conditions result in an envelope or detected signal $d(t)$ which is not disturbed by phase noise. Therefore, BER also become independent of laser phase noise (compare with Section 5.3.1 also).

In DPSK system, laser phase noise disturbs the detection process twice: First is the *direct distortion* of the signal phase due to additive phase noise and second is *indirect distortion* due to the coupling of phase and amplitude of the IF signal.

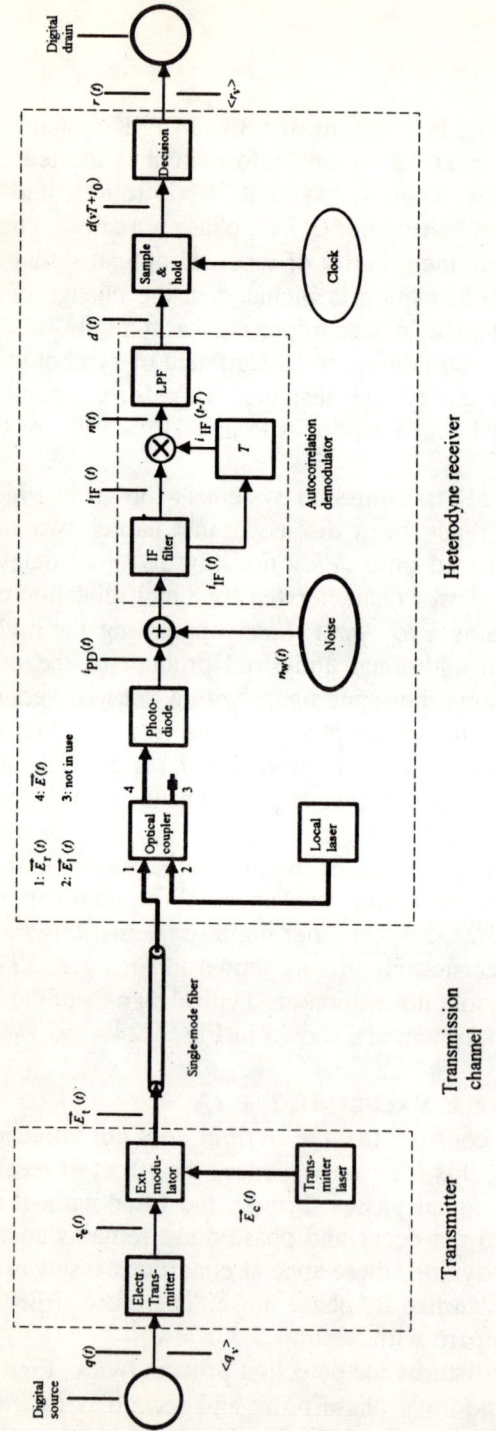

Fig. 5.23: Coherent optical communication system with heterodyne receiver employing autocorrelation detection

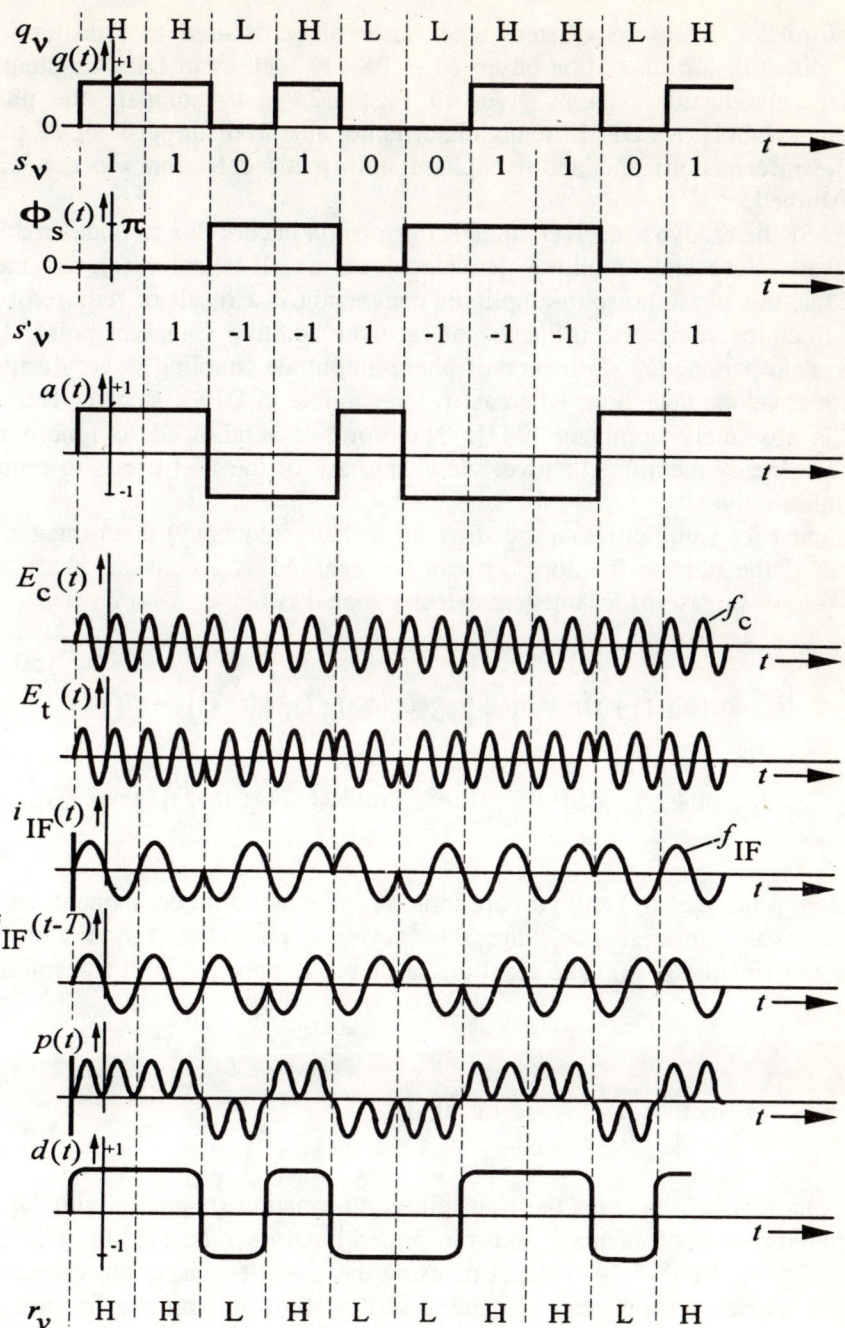

Fig. 5.24: Typical signals in noncoherent optical DPSK heterodyne system with autocorrelation detection

Direct distortion is always existent when laser phase is used to transmit information. Therefore, direct distortion can be observed in PSK as well as in DPSK systems. The task of a DPSK demodulator such as given in Fig. 5.23 is to compare the phase of two neighbouring symbols. As DPSK demodulator is not able to distinguish signal phase (which includes the information) and additive phase noise, this detection process is, therefore, directly disturbed.

Like in ASK heterodyne receiver, *indirect distortion* occurs due to undesired interactions between phase noise and amplitude (envelope) of the IF signal $i_{IF}(t)$. As mentioned in Section 5.3.1, this phase noise-to-amplitude conversion is a result of required bandlimiting of the IF filter to reduce the influence of receiver additive Gaussian noise. Considering practical system parameters, influence of phase-amplitude coupling is very critical in ASK system with envelope detection, whereas it is negligible in DPSK system. Here, the direct distortion is absolutely dominant [243]. Therefore, it is allowed to ignore the indirect distortion as already mentioned above. Now, the task of the IF filter is to reduce additive Gaussian noise only.

It is assumed for simplicity that the IF (intermediate frequency) is an integer multiple of the bit rate. If the ratio is fractional, it can be regarded as a constant phase deviation as discussed below. In case of an integer, detected signal is

$$d(t) = \frac{1}{2}C_M\left[\hat{i}_{PD}\cos\left(\phi_s(t)+\phi(t)\right) + x(t)\right]\left[\hat{i}_{PD}\cos\left(\phi_s(t-T)+\phi(t-T)\right) + x(t-T)\right]$$

$$+ \frac{1}{2}C_M\left[-\hat{i}_{PD}\sin\left(\phi_s(t)+\phi(t)\right) + y(t)\right]\left[-\hat{i}_{PD}\sin\left(\phi_s(t-T)+\phi(t-T)\right) + y(t-T)\right]$$

$$(5.121)$$

Here, random processes $x(t)$ and $y(t)$ are inphase and quadrature components of the receiver additive Gaussian process (i.e., the shot noise of photodiodes and thermal noise of amplifiers and resistors), $\phi(t)$ the resulting laser phase noise of both transmitter and local lasers and

$$\phi_s(t) = \pi\left(1 - s_v\right)\text{rect}\left[\frac{t-vT}{T}\right] \qquad s_v \in \{0, 1\} \qquad (5.122)$$

the signal phase which includes the transmitted information (compare with Eq. 2.35). The assignment between coefficients s_v and transmitted binary symbols $q_v \in \{"0", "1"\}$ becomes clear from Figs. 2.4 and 5.24. Instead of using the so-called phase representation given in Eq. (5.121), detected signal can also be described by applying the following amplitude representation:

$$d(t) = \frac{1}{2} C_M \left[\hat{i}_{PD}\, a(t)\, \cos(\phi(t)) + x(t) \right] \left[\hat{i}_{PD}\, a(t-T)\, \cos(\phi(t-T)) + x(t-T) \right]$$

$$+ \frac{1}{2} C_M \left[-\hat{i}_{PD}\, a(t)\, \sin(\phi(t)) + y(t) \right] \left[-\hat{i}_{PD}\, a(t-T)\, \sin(\phi(t-T)) + y(t-T) \right]$$

(5.123)

where

$$a(t) = \acute{s}_\upsilon\, \mathrm{rect}\!\left[\frac{t - \upsilon T}{T} \right] \qquad \acute{s}_\upsilon = 2s_\upsilon - 1 \qquad \acute{s}_\upsilon \in \{-1,\, 1\} \tag{5.124}$$

Without loss of generality, dimensionless product $C_M \hat{i}_{PD}$ can be taken as unity. If we use again same substitutions and abbreviations as in the previous Sections, then the normalized sample value $d := d(\upsilon T + t_0)/\hat{i}_{PD}$ is given by

$$d = \frac{1}{2} \left[a \cos(\phi) + x \right] \left[a_T \cos(\phi_T) + x_T \right]$$

$$+ \frac{1}{2} \left[-a \sin(\phi) + y \right] \left[-a_T \sin(\phi_T) + y_T \right] = \frac{1}{2} C_T C + \frac{1}{2} S_T S$$

(5.125)

Here, cosine term, delayed cosine term, sine term and delayed sine term are replaced by the abbreviations C, C_T, S and S_T respectively. The suffix T always represents the delayed version. Again, normalization is performed with respect to the amplitude $\hat{i}_{PD}$ of photodiode current. When, for example, all six sources of noise x, x_T, y, y_T, ϕ and ϕ_T are neglected, Eq. (5.125) simplifies to

$$d = \frac{1}{2} a\, a_T \tag{5.126}$$

The principle function of a DPSK demodulator i.e., the detection of change in sign of the sampled detected signal d becomes clear in particular from the above simple equation. The decision circuit always generates a received symbol $r_\upsilon = "1"$ when d is positive (no change in sign i.e., $a = a_T$) and a symbol $r_\upsilon = "0"$ when d is negative (change in sign i.e., $a = -a_T$). This operation in more detail is illustrated in Fig. 5.24 by mean of some typical signals.

Considering again Eq. (5.125), it becomes evident that the sampled detected signal d depends on six (!) sources of noise: x, x_T, y, y_T, ϕ and ϕ_T. Thus, the calculation of pdf

$f_d(d)$ requires the evaluation of a sixfold integral. Moreover, calculation of BER which is determined by a certain area under the pdf $f_d(d)$, will require the evaluation of a sevenfold integral! Even with a fast computer program, computer run time becomes extremely large.

Assuming, for example, that only 100 calculation steps are used for one integral (i.e., the effective area below the pdf is divided in 100 single area sections) and each integral may require a computer execution time of 0.01 s, overall computer run time will be $100^6 \cdot 0.01$ s $= 10^{10}$ s $= 317$ years. Hence, it becomes clear that the BER calculation must really be performed by means of alternate methods of calculation.

(ii) TWO-FILTER REPRESENTATION OF DPSK DEMODULATOR

A different, but functionally equivalent demodulator provided the linear filters are chosen properly is shown in Fig.5.25a. It consists of two linear filters followed by envelope detectors. The decision depends on which envelope detector output is high. A very similar system consisting of two band-pass filters is widely used for noncoherent FSK demodulation (Section 5.3.2). It is easier to build this system using current microwave technology since no microwave frequency multiplier is required.

In order to make the function of the two-filter representation more clear, we first modify Eq. (5.125) by means of some simple mathematical operations. We maintain that

$$d = \frac{1}{8}\left((C+C_T)^2 + (S+S_T)^2 - (C-C_T)^2 - (S-S_T)^2\right) = \frac{1}{8}\left(r_\Sigma^2 - r_\Delta^2\right) \qquad (5.127)$$

where

$$r_\Sigma = \sqrt{(C + C_T)^2 + (S + S_T)^2} \qquad (5.128)$$

and

$$r_\Delta = \sqrt{(C - C_T)^2 + (S - S_T)^2} \qquad (5.129)$$

are used as parameters. It should be noted that r_Σ and r_Δ physically represent the normalized and sampled envelopes of the IF sum signal $\Sigma(t) = i_{IF}(t) + i_{IF}(t-T)$ and the IF difference signal $\Delta(t) = i_{IF}(t) - i_{IF}(t-T)$ respectively. By using Eq. (5.127), threshold decision $d > 0$ or $d < 0$ is transferred to the maximum decision $r_\Sigma > r_\Delta$ (equal to $d > 0$) or $r_\Sigma < r_\Delta$ (equal to $d < 0$). Since sum term r_Σ and difference term r_Δ are Ricean distributed (due to its mathematical structure), calculation of the BER now becomes much easier as shown below. This is the only and main advantage of using Eq. (5.127).

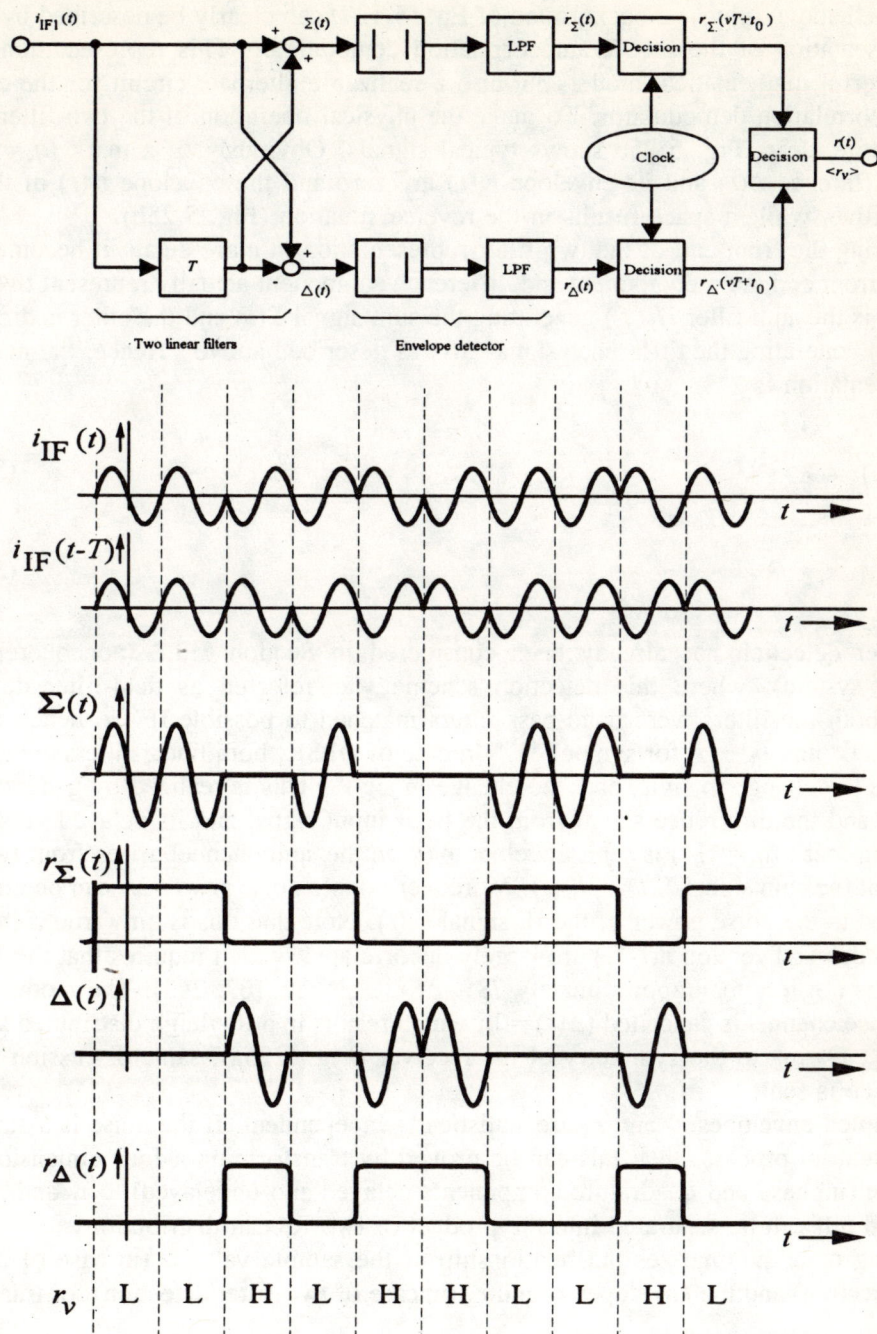

Fig. 5.25: Two-filter DPSK demodulator with typical signals

As mentioned above, physical background of Eq. (5.127) can clearly be described by a two-filter representation of the DPSK autocorrelation demodulator. This representation is not only a powerful mathematical model, but also a realizable alternate circuit for the conventional autocorrelation demodulator. To make the physical operation of the two-filter representation more clear, Fig. 5.25b shows typical signals. Obviously, if a mark (q_v="1") is sent, the difference $\Delta(t)$ and its envelope $r_\Delta(t)$ are zero and the envelope $r_\Sigma(t)$ of the sum $\Sigma(t)$ is positive, while a space results in the reverse situation (Fig. 5.25b).

Considering the front end of the two-filter representation in more detail, it becomes clear that in this front end, sum component and difference component actually represent two linear filters: one is the sum filter $H_\Sigma(f)$ generating the sum signal $\Sigma(t)$ and the other is difference filter $H_\Delta(f)$ generating the difference signal $\Delta(t)$ as described above. Hence, the mathematical representation is

$$H_\Sigma(f) = 1 + e^{-j2\pi fT} \tag{5.130}$$

$$H_\Delta(f) = 1 - e^{-j2\pi fT} \tag{5.131}$$

Two-filter detection has already been considered in Section 5.3.2 (noncoherent FSK heterodyne system), where this detection scheme was referred as dual-filter detection. However, both the filters were band-pass filters matched to possible IF frequencies $f_{IF} - f_d$ for symbol "0" and $f_{IF} + f_d$ for symbol "1." In case of DPSK, both linear filters are matched in principle to "change in sign" and "no change in sign". This is realized by generating the sum signal and the difference signal from the filter input signal and its delayed version.

Assuming mark (q_v="1") is sent, envelope $r_\Sigma(t)$ in the sum channel arises from twice the amplitude of the sum signal $\Sigma(t)=2i_{IF}(t)$. Moreover double noise power $2N$ can be measured as compared to the noise power of the IF signal $i_{IF}(t)$. Note that this is only true if the noise $n(t)$ and its delayed version $n(t-T)$ are largely uncorrelated, which requires that the IF filter bandwidth is not less than approximately 78% of the bit rate [67, 243]. The modulation in the difference channel is cancelled ($\Delta(t)=0$), which results in a Rayleigh distributed sampled envelope r_Δ. Owing to the symmetry of the receiver (Fig. 5.25a), same discussion is valid when a space is sent.

The sampled envelopes r_Σ and r_Δ are statistically independent. If the noise is a stationary Gaussian random process, then this can be proved by transforming a four dimensional pdf of the noise (inphase and quadrature components delayed and undelayed) to r_Σ and r_Δ. This transformed pdf can be separated into the product of two Ricean distributions.

Following table summarizes the relationship of the sample value d (in case of autocorrelation detection) and the envelopes r_Σ and r_Δ (in case of two-filter detection) for transmitted symbol q_v.

Table 5.3: Relationship of sample values d, r_Σ, r_Δ and transmitted symbol q_v

Transmitted symbol q_v	Autocorrelation detection with threshold decision	Two-filter detection with maximum decision
"1"	$d > 0$	$r_\Sigma > r_\Delta$
"0"	$d < 0$	$r_\Sigma < r_\Delta$

(iii) PROBABILITY OF ERROR

When a mark is sent ($q_v = $ "1"), bit error (i.e., $r_v \neq q_v$) will occur if the receiver additive Gaussian noise causes a greater envelope in the difference channel than in the sum channel i.e., $p = p(r_\Delta > r_\Sigma)$. Instead, if autocorrelation detection is used in the receiver, then a bit error occurs for $d > 0$. For space, calculation can be made in the same way and it yields the same result. If mark and space are sent with equal probability, it is sufficient to determine the BER for only one of them as given below. Neglecting intersymbol interference, probability of error for symbol q_v is independent of the neighbouring symbols. In this particular case, average probability of error is

$$P_a = \frac{1}{2}\left(p_0 + p_1\right) \tag{5.132}$$

Here, p_0 and p_1 represent the probabilities of wrongly detected symbol "0" and "1" respectively. It should be noted that above equation presumes same occurrence probabilities for both binary symbols "0" and "1". The probabilities of error p_0 and p_1 can be calculated as

$$p_0 = p\left(d > 0 | q_v = "0"\right) = p\left(r_\Sigma > r_\Delta | q_v = "0"\right) = \int_0^{+\infty} \int_{r_\Sigma = r_\Delta}^{+\infty} f_{r_\Sigma, r_\Delta}\left(r_\Sigma, r_\Delta\right) dr_\Sigma \, dr_\Delta \tag{5.133}$$

and

$$p_1 = p\left(d < 0 | q_v = "1"\right) = p\left(r_\Sigma < r_\Delta | q_v = "1"\right) = \int_0^{+\infty} \int_{r_\Delta = r_\Sigma}^{+\infty} f_{r_\Sigma, r_\Delta}\left(r_\Sigma, r_\Delta\right) dr_\Delta \, dr_\Sigma \tag{5.134}$$

The main problem in the calculation of p_0 and p_1 is to determine the two-dimensional joint density function $f_{r_\Sigma, r_\Delta}(r_\Sigma, r_\Delta)$. Since both the envelopes r_Σ and r_Δ are statistically dependent random variables, calculation of $f_{r_\Sigma, r_\Delta}(r_\Sigma, r_\Delta)$ becomes rather comprehensive. This problem, however, can be solved by using conditional probability density functions. For this, random phases ϕ and ϕ_T will be considered as constants first. In that case, both the random envelopes r_Σ and r_Δ given in Eqs. (5.128) and (5.129) are characterized by a Ricean distribution. This distribution is already known from Section 5.3.1 (noncoherent ASK heterodyne system), where the envelope of the ASK signal also had Ricean distribution. Like in Section 5.3.1, we obtain

$$f_{r_\Sigma \mid \Delta\phi}(r_\Sigma, \Delta\phi) = \frac{r_\Sigma}{2\sigma^2} \exp\left(-\frac{r_\Sigma^2 + 4\sin^2(\Delta\phi/2)}{4\sigma^2}\right) J_0\left(\frac{r_\Sigma \sin(\Delta\phi/2)}{\sigma^2}\right) \qquad (5.135)$$

and

$$f_{r_\Delta \mid \Delta\phi}(r_\Delta, \Delta\phi) = \frac{r_\Delta}{2\sigma^2} \exp\left(-\frac{r_\Delta^2 + 4\cos^2(\Delta\phi/2)}{4\sigma^2}\right) J_0\left(\frac{r_\Delta \cos(\Delta\phi/2)}{\sigma^2}\right) \qquad (5.136)$$

Here,

$$\sigma = \sigma_x = \sigma_y = \frac{\sigma_n}{\hat{i}_{PD}} = \frac{\sigma_{het}}{\hat{i}_{PD}} \qquad (5.137)$$

represents the normalized standard deviation of the Gaussian random processes x, x_T, y and y_T with zero mean and

$$\Delta\phi = \phi - \phi_T \qquad (5.138)$$

represents the phase difference between current phase ϕ and delayed phase ϕ_T. As $\Delta\phi$ is a function of the two random variables ϕ and ϕ_T, it is, of course, also a random variable itself (see also Section 3.3.2).

Considering the conditional probability density functions given in Eqs. (5.135) and (5.136) in more detail, it becomes clear that these functions only depend on the phase difference $\Delta\phi$ and not on the absolute phases ϕ and ϕ_T. It is of significant advantage in the calculation of BER. It should be noted that if there is a phase difference $\Delta\phi$ in addition to the $0/\pi$ modulation phase, sum envelope is not twice the transmission amplitude and the difference is no longer zero.

When the Gaussian distributed random variables x, x_T, y and y_T are uncorrelated and, hence, statistically independent, random envelopes r_Σ and r_Δ of sum and difference signals are also statistically independent [243, 260]. The conditions which are required to obtain statistical independence will be discussed later. In case of statistical independence, conditional joint density function $f_{r_\Sigma, r_\Delta|\Delta\phi}(r_\Sigma, r_\Delta, \Delta\phi)$ is simply obtained by the product of both the conditional density functions given in Eqs. (5.135) and (5.136). We get

$$f_{r_\Sigma, r_\Delta|\Delta\phi}(r_\Sigma, r_\Delta, \Delta\phi) = f_{r_\Sigma|\Delta\phi}(r_\Sigma, \Delta\phi)\, f_{r_\Delta|\Delta\phi}(r_\Delta, \Delta\phi) \tag{5.139}$$

Finally, we have to consider again the statistics of the phase difference $\Delta\phi$. As derived in Section 3.3.2, phase difference $\Delta\phi$ is stationary and Gaussian distributed with zero mean. By taking into account these characteristics, probabilities of error p_0 and p_1 are now given by

$$p_0 = \int_{-\infty}^{+\infty} \int_0^{+\infty} \int_{r_\Sigma = r_\Delta}^{+\infty} f_{r_\Sigma|\Delta\phi}(r_\Sigma, \Delta\phi)\, f_{r_\Delta|\Delta\phi}(r_\Delta, \Delta\phi)\, f_{\Delta\phi}(\Delta\phi)\; dr_\Sigma\; dr_\Delta\; d\Delta\phi$$

$$= \int_{-\infty}^{+\infty} \tilde{p}_0(\Delta\phi)\, f_{\Delta\phi}(\Delta\phi)\; d\Delta\phi \tag{5.140}$$

and

$$p_1 = \int_{-\infty}^{+\infty} \int_0^{+\infty} \int_{r_\Delta = r_\Sigma}^{+\infty} f_{r_\Sigma|\Delta\phi}(r_\Sigma, \Delta\phi)\, f_{r_\Delta|\Delta\phi}(r_\Delta, \Delta\phi)\, f_{\Delta\phi}(\Delta\phi)\; dr_\Delta\; dr_\Sigma\; d\Delta\phi$$

$$= \int_{-\infty}^{+\infty} \tilde{p}_1(\Delta\phi)\, f_{\Delta\phi}(\Delta\phi)\; d\Delta\phi \tag{5.141}$$

As in the previous Sections, sign " ~ " again represents the probability of error when phase noise is neglected. However, a constant phase distortion (i.e., an undesired phase shift) is allowed.

Due to the symmetry of two-filter demodulator, p_0 and p_1 are equal. Therefore, average probability of error in absence of intersymbol interference is given by

$$
\begin{aligned}
P_a &= \frac{1}{4\sqrt{2\pi}\sigma^4\sigma_{\Delta\phi}} \int\limits_{-\infty}^{+\infty} \int\limits_{0}^{+\infty} \int\limits_{r_\Delta=r_\Sigma}^{+\infty} r_\Sigma r_\Delta \, \exp\left(-\frac{r_\Sigma^2+r_\Delta^2+4}{\sigma^2}\right) \exp\left(\frac{\Delta\phi^2}{2\sigma_{\Delta\phi}^2}\right) \\
&\quad \cdot J_0\left(\frac{r_\Delta\cos(\Delta\phi/2)}{\sigma^2}\right) J_0\left(\frac{r_\Sigma\sin(\Delta\phi/2)}{\sigma^2}\right) dr_\Sigma \, dr_\Delta \, d\Delta\phi \\
&= \int\limits_{-\infty}^{+\infty} \tilde{p}_a(\Delta\phi) f_{\Delta\phi}(\Delta\phi) \, d\Delta\phi
\end{aligned}
\tag{5.142}
$$

In the derivation of above equation, Eqs. (5.132), (5.135), (5.136) and (5.140) have been used. To evaluate the average probability of error p_a, variance $\sigma_{\Delta\phi}^2$ of the random phase difference $\Delta\phi$ is required. From Section 3.3.2, we maintain

$$
\sigma_{\Delta\phi}^2 = 2\pi\Delta f T
\tag{5.143}
$$

where T is the bit duration and Δf the resulting laser linewidth of both transmitter and local lasers. It becomes clear from Eq. (5.142) that the calculation of probability of error is now reduced to a threefold integral instead of a sevenfold integral as described above. An additional simplification is, however, only possible if special conditions are given. Evaluating the threefold integral given in Eq. (5.142) by means of computer, run time of about $100^2\cdot0.01$ s$=100$ s is merely required. Thereby, the same numerical values as in Subsection (i) will be provided. In contrast to 317 years for sevenfold integral, this is truly a very large reduction in the computer run time.

In a system without any phase noise or phase distortion (i.e., $\Delta\phi=0$), Eq. (5.142) results in the well-known formula of conventional digital DPSK transmission system [260]:

$$
P_a\Big|_{\Delta\phi=0} = \tilde{p}_a(0) = \frac{1}{2} e^{-\frac{1}{2\sigma^2}}
\tag{5.144}
$$

It should be remembered that the standard deviation σ of the additive Gaussian noise is normalized to the amplitude $\hat{i}_{PD}$ of the photodiode current. Therefore, the variance σ^2 actually represents the reciprocal of signal-to-noise ratio.

Considering the conditional probability of error $\tilde{p}_a(\Delta\phi)$ in more detail, some important features of optical DPSK heterodyne systems become evident. For this reason, Fig. 5.26 shows $\tilde{p}_a(\Delta\phi)$ as a function of a constant phase shift $\Delta\phi$. When the normalized standard deviation σ of the receiver additive Gaussian noise is low (i.e., the signal-to-noise ratio is high), probability of error is also low provided phase shift $\Delta\phi$ is small enough. As the undesired phase shift $\Delta\phi$ increases, probability of error also increases. In the particular case of $\Delta\phi=\pi/2$ (i.e., 90°), probability of error reaches $\tilde{p}_a(\pi/2)=0.5$. Thereby, the signal components of two-filter demodulator are completely extinguished and only noise is given to the input of maximum decision circuit. If phase shift is above $\pi/2$, but less than π, then probability of error increases further since both the binary symbols "0" and "1" are interchanged now. In the special case of $\Delta\phi=\pi$ (i.e., 180°), probability of error reaches maximum i.e., $\tilde{p}_a(\pi)=1$. As each symbol is wrongly detected now, a simple symbol converter ("0"→"1", "1"→"0") yields again an errorless detection with $\tilde{p}_a(\pi)=0$. Hence, the actual maximum probability of error is 0.5 instead of 1. If $\Delta\phi>\pi$, then $\tilde{p}_a(\Delta\phi)$ again decreases.

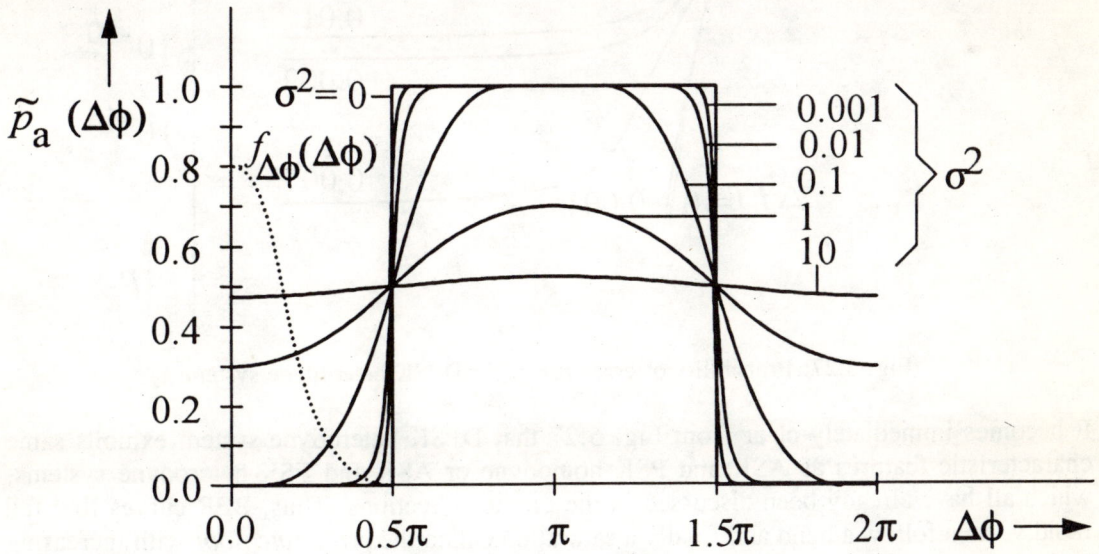

Fig. 5.26: Probability of error $\tilde{p}_a(\Delta\phi)$ in DPSK heterodyne receiver as a function of constant phase shift $\Delta\phi$

It becomes clear from Fig. 5.26 that $\tilde{p}_a(\Delta\phi)$ is symmetrical about $\Delta\phi=0$ (or multiple of π) and periodical with period 2π. Symmetry and periodicity are caused by both the periodic functions $\sin(\Delta\phi)$ and $\cos(\Delta\phi)$ which are included in the Eq. (5.142). Thereby, we have to take into account that the Bessel function $J_0(x)$ is an even function.

In the following, we reconsider the statistics of the phase difference $\Delta\phi$ which actually is a random variable due to laser phase noise. This random variable is completely described by a Gaussian pdf $f_{\Delta\phi}(\Delta\phi)$ shown in Fig. 5.26 (dotted line). Corresponding to Eq. (5.142), conditional probability of error $\bar{p}_a(\Delta\phi)$ must be weighted by $f_{\Delta\phi}(\Delta\phi)$ to obtain the probability of error p_a by integration. In contrast to $\bar{p}_a(\Delta\phi)$, probability of error p_a now also contains the statistics and, hence, the influence of phase noise. It should be noted that the pdf $f_{\Delta\phi}(\Delta\phi)$ plotted in Fig. 5.26 has been chosen unrealistic broad to obtain a clear description. Actually, this pdf is much narrower, especially if a probability of error $p_a = 10^{-10}$ is demanded and would nearly appear as Dirac delta function at $\Delta\phi = 0$.

The probability of error p_a versus received optical power P_r with normalized laser line-width $\Delta f T$ as a parameter is shown in Fig. 5.27.

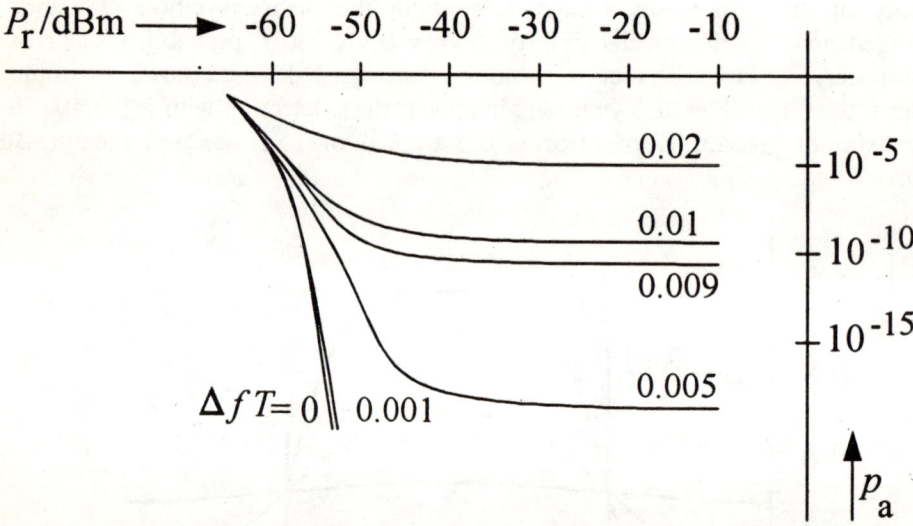

Fig. 5.27: Probability of error $p_a \approx p_w$ in DPSK heterodyne system

It becomes immediately clear from Fig. 5.27 that DPSK heterodyne system exhibits same characteristic features as ASK and PSK homodyne or ASK and FSK heterodyne systems, which all have already been discussed in the previous Sections. Thus, BER curves first fall rapidly, then follow a bend and finally a saturation called the *error rate floor* with increasing optical power levels. Physical description of this typical behaviour of all coherent optical communication systems has been given in Section 5.1.1 (see discussion on Fig. 5.5).

An error rate floor can always be observed when optical input power P_r is high or when standard deviation σ of the additive Gaussian receiver noise is low. If the laser phase noise is low ($\sigma_{\Delta\phi} < 0.5$), which is usually fulfilled in realized systems, then the probability of bit error at the error rate floor becomes

$$p_a = 2 \int_{\pi/2}^{+\infty} f_{\Delta\phi}(\Delta\phi) \, d\Delta\phi \quad \text{provided } \sigma = 0 \tag{5.145}$$

By using' Q-function given in Eq. (5.14) and Eq. (5.143), above equation yields

$$p_a = 2Q\left(\frac{\pi}{2\sigma_{\Delta\phi}}\right) = 2Q\left(\frac{1}{2}\sqrt{\frac{\pi}{2\Delta f T}}\right) \qquad \sigma = 0 \tag{5.146}$$

This important equation is formally same as Eq. (5.47) which was derived in Section 5.1.2 (PSK homodyne system). The only difference is given by the numerical value of the phase noise standard deviation.

Example 5.4

For a probability of error $p_a = 10^{-10}$, Eq. (5.146) yields a maximum permissible phase noise standard deviation $\sigma_{\Delta\phi} = 0.0077\pi = 0.24$. Assuming a bit rate of $1/T = 560$ Mbit/s, this result corresponds to a maximum allowed resulting laser linewidth of $\Delta f = 5.26$ MHz which is approximately 1% of the bit rate.

As mentioned above, all equations derived in this Section provide that the four random variables x, x_T, y and y_T are uncorrelated. Since x and y are inphase and quadrature components of the additive Gaussian noise, x and y are a priori uncorrelated and statistically independent. This, of course, is also valid for the delayed processes x_T and y_T. Whether the random processes x and x_T (or y and y_T) are uncorrelated or not can only be estimated by the correlation coefficients $\rho_x = \rho_y = \rho_n$. For an IF filter with Gaussian shape and bandwidth $B = 2f_g$, we obtain

$$\rho_x = \rho_y = \frac{E\{xx_T\} - E\{x\}E\{x_T\}}{\sigma_x \sigma_{x_T}} = \frac{E\{xx_T\}}{\sigma^2} = \frac{R_x(T)}{\sigma^2} = e^{-\frac{\pi}{2}(BT)^2} \tag{5.147}$$

where $R_x(\tau) = R_y(\tau) = R_n(\tau)$ represents the autocorrelation function (acf) of the filtered Gaussian noise $n(t)$ at the output of IF filter. After evaluating the above expression, following numerical values are obtained (Table 5.4). It becomes clear that the Gaussian distributed random variables x and x_T (or y and y_T) can be regarded as practically uncorrelated and statistically independent for normalized bandwidths $BT > 1.5$. This result verifies the BER calculation above, where statistical independence was a main precondition.

Table 5.4:Correlation coefficients ρ_x and ρ_y for several normalized IF filter bandwidths BT

BT	$\rho_x = \rho_y$	Remark
0	1.000	$x = x_T$ and $y = y_T$
0.5	0.675	
1.0	0.208	
1.5	0.029	$BT \approx B_{opt}T$
2.0	0.002	
5.0	≈ 0	uncorrelated

(iv) EFFECTS OF DEVIATION IN INTERMEDIATE FREQUENCY f_{IF} AND TIME DELAY T_d

The phase of undisturbed (i.e., no phase noise and no additive Gaussian noise) intermediate frequency signal $i_{IF}(t)$ is $2\pi f_{IF}t + \phi_s(t)$. The nominal phase difference to the delayed version $i_{IF}(t-T_d)$ is $2\pi f_{IF}T_d + \phi_s(t) - \phi_s(t-T_d)$, where $\phi_s(t) - \phi_s(t-T_d)$ is either 0 or π. If phase difference is equal to an integer multiple of 2π, the BER approaches minimum. Therefore, $f_{IF}T_d$ should be an integer. If f_{IF} is not chosen properly or implemented delay T_d differs from bit duration T, then the phase error

$$\phi_e = 2\pi f_{IF}T_d - k2\pi \qquad k \in \{\cdots -1, 0, +1, \cdots\} \tag{5.148}$$

which is additive to the phase difference $\Delta\phi$ cannot be neglected. Its effect is to shift the centre of the pdf $f_{\Delta\phi}(\Delta\phi)$ and, hence, increase the probability of error or the BER. Since the probability of error $\tilde{p}(\phi_e)$ is periodical with 2π and, in addition, independent of the sign of ϕ_e, following alternate equation for the phase error can be used:

$$\phi_e = (2\pi f_{IF}T_d) \bmod(2\pi) \quad \text{with} \quad 0 \le \phi_e \le 2\pi \tag{5.149}$$

Here, the function $a \bmod b = a - \text{Int}(a/b) \cdot b$ decreases a by an integer multiple (namely $\text{Int}(a/b)$) of b. It can easily be seen from Fig. 5.26 that a rise in ϕ_e increases the BER even to values greater than 0.5 if ϕ_e approaches π. Reversing the assignment of mark and space could again lower the BER, but this is not the solution if the deviation of f_{IF} or T_d is not known. Thus, it is essential to have an integer ratio of f_{IF} and $f_b = 1/T$ which is easily achieved if f_{IF}/f_b is not too large (moderate requirements on the relatively high accuracy).

As the constant phase error ϕ_e can be regarded as the mean of random variable $\Delta\phi$, the probability of error can easily be calculated by

$$P_e = \int\limits_{-\infty}^{+\infty} \tilde{p}_e(\Delta\phi)\, f_{\Delta\phi}(\Delta\phi - \phi_e)\, \mathrm{d}\Delta\phi \qquad (5.150)$$

By setting $\phi_e = 0$, Eq. (5.150) becomes same as Eq. (5.142).

For the realization of an appropriate intermediate frequency (IF), some important points should be kept in mind: First, IF should be high enough since a low IF results in spectral overlapping of IF and baseband spectrum. Second, IF should not be too large since realization problems increase with increasing IF. Moreover, realizing an integer ratio f_{IF}/f_b is more convenient when f_{IF}/f_b is not too large as already mentioned above. In addition, we have to keep in mind that photodiodes as well as preamplifiers must be able to follow the high frequency IF signal. For this reason, bandwidth of these components should be approximately $0 < f < f_{IF} + f_{s,\max}$, where $f_{s,\max}$ is the maximum frequency of the information signal.

(v)   INFLUENCE OF IF FILTER ON SYSTEM PERFORMANCE

Considering coherent optical heterodyne system, IF filter should first select the modulated IF signal and second reduce the influence of receiver additive Gaussian noise. For this reason, bandwidth of the IF filter should be as small as possible. On the other hand, a small bandwidth leads to intersymbol interference and, in addition, an undesired coupling of phase and amplitude of the IF signal. Hence, for the realization of an appropriate bandwidth, a compromise is always required. If bandwidth is optimum ($B_{IF,opt} = B_{opt}$), then overall distortion and BER are minimum.

In case of DPSK heterodyne receiver, numerical calculation of the optimum bandwidth B_{opt} is much more comprehensive than that of ASK or FSK heterodyne receiver. This is due to the comprehensive error rate formula given in Eq. (1.542) which requires the evaluation of a threefold integral even if the influence of the IF filter is neglected. An analytical calculation of B_{opt} is impossible. To determine B_{opt} approximately, we consider that phase noise is low. In that case, optimum bandwidth is approximately given by the optimum bandwidth of DPSK system which is not disturbed by phase noise i.e., $B_{opt}T = 2f_{g,opt}T \approx 1.58$ for a Gaussian IF filter. As phase-to-amplitude conversion decreases with increasing bandwidth, real optimum bandwidth is only insignificantly higher. Influence of the IF filter can be separated in four parts: Firstly, as the filter bandwidth is reduced, there is an useful proportional reduction of the additive noise power. Secondly, if the cut-off frequency is reduced below the bit rate, signal becomes increasingly distorted by intersymbol interference. This influence is described by the eye pattern shown in Fig. 5.28, which closes if the bandwidth is reduced. Thirdly, the phase noise is converted into additional amplitude modulation, which increases the additive noise. For typical linewidths of 1% of the bit rate, however, this effect (called the indirect distortion) is rather small. The last effect is the reduction of phase noise. It was seen by simulation experiments that the effective linewidth is reduced somewhat by the IF filter, for example $\approx 80\%$ for $B_{opt}T = 1.58$ [243].

Eye pattern can be measured (which is normally the way in practice) or analytically derived. In DPSK heterodyne system, analytical method is, however, much more difficult than in all other heterodyne or homodyne systems. This is primarily due to the multiplication process in the DPSK autocorrelation demodulator. In accordance with Eq. (5.126), normalized detected signal $d(t)$ at the output of the demodulator is generated by the weighted product (weight factor 0.5) of both the signals $a(t) \star h_B(t)$ and $a(t-T) \star h_B(t-T)$, which are already disturbed by intersymbol interference. It should be remembered that $h_B(t)$ represents the impulse response of the equivalent baseband filter representation of the IF filter (Section 2.4.3). Physically, convolutional signal $a(t) \star h_B(t)$ represents the envelope of modulated IF filter output signal disturbed by intersymbol interference. To make it more clear, Fig. 5.28a shows the eye pattern of the normalized envelope $a(t) \star h_B(t)$ of IF signal at the input of demodulator and Fig. 5.28b of detected signal $d(t) = 0.5[a(t) \star h_B(t)] \cdot [a(t-T) \star h_B(t-T)]$ at the output of demodulator. Both eye pattern are based on an IF filter with Gaussian frequency response and normalized bandwidth $B = 2f_gT = 1.58$. Phase noise and additive Gaussian receiver noise are neglected ($\sigma_\phi = 0$, $\sigma = 0$).

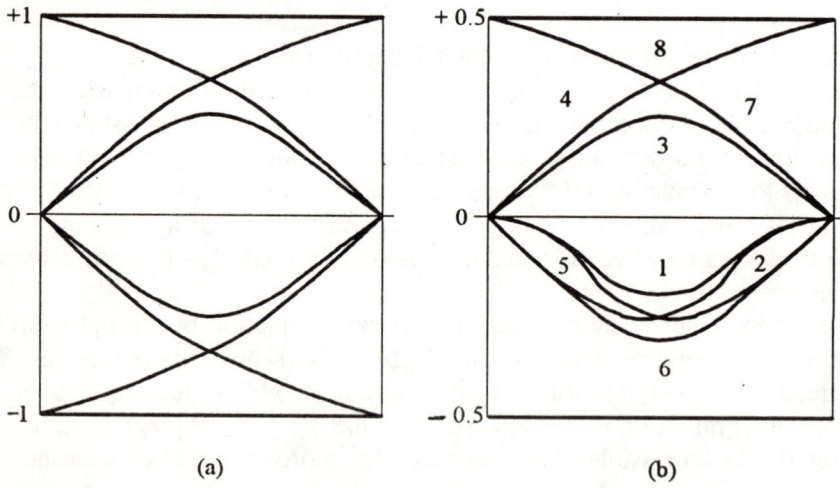

(a)   (b)

Fig. 5.28: Normalized eye pattern of the (a) envelope $a(t) \star h_B(t)$ of IF signal at the input of DPSK demodulator and (b) detected signal $d(t)$ at the output of demodulator. The numerical values given in (b) are explained in Table 5.5

The eye pattern of the envelope $a(t) \star h_B(t)$ shown in Fig. 5.28a is, of course, same as the eye pattern of the delayed envelope signal $a(t-T) \star h_B(t-T)$. It should be noted that the eye pattern of Fig. 5.28a is also same as the eye pattern of the detected signal in PSK homodyne system (see Fig. 7.4) and, in addition, of the detected signal in ASK heterodyne or homodyne system. For the underlying system parameters, eye pattern in Fig. 5.28a is characterized by eight eye lines only.

Keeping in view some special conditions (discussed below), eye pattern of the DPSK detected signal shown in Fig. 5.28b can be regarded as a product of two eye pattern: one is the eye pattern of the envelope $a(t) \star h_B(t)$ and other is the eye pattern of the delayed envelope $a(t-T) \star h_B(t-T)$. Since both these eye pattern are equal, the eye pattern in Fig. 5.28b corresponds to the square of the eye pattern in Fig. 5.28a. As a result, eye pattern in Fig. 5.28b will be based on $8 \cdot 8 = 64$ eye lines. However, this is actually not true since some special conditions must also be considered. As explained in the following example, only lines which belong to the same symbol sequence $<q_v>_i$ are allowed to multiply.

Example 5.5

Let us consider a symbol sequence $<q_v> = \cdots q_{-2}, q_{-1}, q_0, q_1, q_2 \cdots = \cdots 01101 \cdots$. Here, it is allowed to multiply sequence part $q_{-1}, q_0, q_1 = 110$ which represents a certain line in the eye pattern of $a(t) \star h_B(t)$ and sequence part $q_{-2}, q_{-1}, q_0 = 011$ which corresponds to the delayed eye pattern of $a(t-T) \star h_B(t-T)$. In contrast to this, it is not allowed to multiply the line which belongs to $q_{-1}, q_0, q_1 = 110$ and the line which belongs to $q_{-2}, q_{-1}, q_0 = 010$. It is true that both lines are included in the eye pattern in Fig. 5.28a, but the product of these lines does not yield a line in the eye pattern in Fig. 5.28b since both the underlying sequence parts do not belong to the same temporal sequence $<q_v>$ given above.

Table 5.5: Permitted (1-8) and forbidden (-) combinations of two successive input eye pattern (compare with Fig. 5.28b)

		q_{-1}, q_0, q_1							
		000	001	010	011	100	101	110	111
q_{-1}, q_0, q_1	000	1	2	-	-	-	-	-	-
	001	-	-	3	4	-	-	-	-
	010	-	-	-	-	5	6	-	-
	011	-	-	-	-	-	-	7	8
	100	1	2	-	-	-	-	-	-
	101	-	-	3	4	-	-	-	-
	110	-	-	-	-	5	6	-	-
	111	-	-	-	-	-	-	7	8

It should be noted that $d(t)$ is not symmetrical with respect to time axis because of the non-linear behaviour of demodulator. The "continuous mark", for example, will result in a received signal without any phase change and a steady $d(t)$ i.e., line 8. In contrast, "continuous space" has a rectangular envelope which is shaped sine-like by the IF filter. This yields a negative, but not constant demodulated signal (i.e., line 1). This makes the decision of the optimal threshold difficult. No bit sequence $<q_v>$ can be found that leads to a constant normalized signal $d(t) = -0.5$, except for an infinite large IF filter bandwidth. Table 5.5 shows the permitted combinations of the input pattern (symbols q_v) of two successively transmitted bits. Not all 64 combinations are allowed and only 8 of the remaining 16 combinations are distinguishable because DPSK reception is polarity independent.

(vi) OPTIMIZATION

In comparison to the systems discussed in the previous Sections such as ASK and PSK homodyne and ASK and FSK heterodyne systems, optimization of DPSK system is more comprehensive. Here, optimization is restricted on approximations or computer simulations since an exact analytical optimization is too comprehensive and without practical use.

(a) IF FILTER BANDWIDTH B_{IF}

The optimum IF filter bandwidth for continuously sent spaces (i.e., $\cdots 00000 \cdots$) is $B_{IF} = B = 1.58 f_b$ (like PSK). In contrast, spectrum of "continuous mark" is a single line at the intermediate frequency that would pass through any filter bandwidth even approaching zero. It should be noted that continuously sent spaces represent the worst-case bit sequence in DPSK. This is in contrast to the systems discussed in the previous Sections, where worst-case pattern was given by single-"0" and single-"1". Therefore, no simple calculation for the real optimum bandwidth exist and the value obtained for "continuous space" can be used as a worst-case approximation. It must be remembered that $B = 1.58 f_b$ is valid for a Gaussian IF filter. The real optimum bandwidth is only somewhat above (not below) this approximation.

(b) THRESHOLD E

Due to the asymmetry of the detected signal (see eye pattern in Fig. 5.28b), optimum threshold is no longer located at $E = 0$. It becomes clear from the eye pattern in Fig. 5.28b that the optimum threshold E_{opt} is always located somewhat above zero. The real value of E_{opt} can only be found by numerical methods using a fast computer program [243]. For practical system parameters, however, the optimum threshold remains approximately at zero. Hence, $E_{opt} \approx 0$ can be used as a very good approximation.

(c) CONCLUSION

Optical DPSK heterodyne system with autocorrelation detection can achieve a bit error rate of $p_a = 10^{-10}$ if the resulting laser linewidth is less than $\approx 1\%$ of the bit rate. For such a small linewidth, phase noise is largely unaffected by the IF filter which can, therefore, be optimized considering only the additive noise. The result is $B = 1.58 f_b$ for a Gaussian IF filter. The linewidth can be increased due to this intermediate filter to $\approx 1.2\%$ [67, 243].

Similar to the other heterodyne and homodyne systems, DPSK heterodyne system also exhibits a bit error rate floor. Increasing the received light power P_r even to infinity cannot reduce the BER below a limit that is only determined by the normalized, resulting laser linewidth $\Delta f T = (\Delta f_t + \Delta f_i) T$.

As compared to other optical systems, DPSK heterodyne system represents a very good compromise. It has higher sensitivity than the noncoherent ASK and FSK heterodyne systems and does not require the extremely small linewidth, for example, as in PSK homodyne systems. Moreover, no optical phase-locked loop (OPLL) circuit is required.

5.4   DIRECT DETECTION SYSTEM

This Section briefly reviews the background required to study the conventional optical transmission system with intensity modulation and direct detection (IM/DD). This system will be used in Chapter seven as a well-suited reference for system comparison.

The principle operation of optical direct detection system has already been discussed in Chapter two. Here, we focus our interest on BER calculation and system optimization. For this purpose, we take into account the distortion by intersymbol interference and additive Gaussian noise i.e., shot noise of photodiode and thermal noise of the amplifiers, in particular of the front-end preamplifier. As already explained in Chapter two, intersymbol interference is caused by the limited bandwidth of the system components, primarily the baseband filter (see Fig. 2.1a). Similar to the analysis of coherent optical communication systems in the previous Sections 5.1 to 5.3, we again begin our discussion with the detected signal and its sample values.

(i) DETECTED SIGNAL $d(t)$ AND SAMPLE VALUES $d(vT+t_0)$

Considering an optical direct detection receiver, detected signal $d(t)$ equals the signal at the output of the baseband filter (Fig. 2.1a). Taking into account the components of the simple receiver shown in Fig. 2.1a, we maintain that

$$d(t) = MR_0 P_r \int_{-\infty}^{+\infty} s(\tau) h_B(t-\tau) d\tau + n(t) \tag{5.151}$$

where $h_B(t)$ represents the impulse response of baseband filter or low-pass filter, $n(t)$ the additive Gaussian noise at the output of low-pass filter and

$$s(t) = \sum_{v=-\infty}^{+\infty} s_v \, \text{rect}\left(\frac{t-vT}{T}\right) \quad \text{with} \quad s_v = \begin{cases} 0 & \text{if} \quad q_v = "0" \\ 1 & \text{if} \quad q_v = "1" \end{cases} \tag{5.152}$$

the transmitted binary information. The product $s(t)P_r$ included in Eq. (5.151) represents the *intensity modulation* which is usually applied to optical direct detection system.

For a low-pass filter with Gaussian frequency response and cut-off frequency f_g (Eq. 2.77), variance of the zero mean Gaussian noise $n(t)$ is

$$\sigma_n^2 = G_d \int_{-\infty}^{+\infty} |H_B(f)|^2 \, df = \sqrt{2} G_d f_g \tag{5.153}$$

Here, the noise power spectral density G_d is actually not a constant since the shot noise which represents a dominant part of G_d depends on the modulated optical power $s(t)P_r$ at the input to the receiver. Hence, optical transmission system with intensity modulation and direct detection belongs to the systems which are characterized by *signal-dependent noise*. For this reason, both the binary symbols "0" and "1" are normally disturbed differently. The transmitted symbol "0" is not or even less disturbed by the shot noise, whereas symbol "1" is always disturbed more.

As well-known from standard digital communication systems, detected signal $d(t)$ is applied to sample and hold and decision circuits. Owing to the additive receiver noise $n(t)$, sample values $d(vT+t_0)$ are random variables. As the statistical features of $n(t)$ are time-invariant, it is sufficient to determine the statistical parameters of the random variables $d(vT+t_0)$ at an arbitrary, but fixed point of time, for example, $t=t_0$. The results obtained are valid in general i.e., at each sampling point of time $vT+t_0$. Taking advantage of the following normalizations and substitutions,

$$ d := \frac{d(t_0)}{MR_0P_r} \qquad n := \frac{n(t_0)}{MR_0P_r} \quad \rightarrow \quad \sigma := \frac{\sigma_n}{MR_0P_r} $$

$$ a := a(t_0) = \int_{-\infty}^{+\infty} s(\tau)\, h_B(t_0-\tau)\, d\tau \qquad 0 \le a(t_0) \le 1 $$

(5.154)

the sampled and normalized detected signal d can be expressed as

$$ d = a + n $$

(5.155)

Here d, a and n have the following physical meanings:

d: normalized sample value of the detected signal,
a: noiseless part of d disturbed by intersymbol interference (including the transmitted information),
n: receiver additive Gaussian noise.

(ii) PROBABILITY DENSITY FUNCTION (pdf) $f_d(d)$

As the additive receiver noise is Gaussian distributed, sampled detected signal d is also Gaussian distributed. As seen from Eq. (5.154), expected value $\eta_d = a$ of this Gaussian pdf is a function of the impulse response $h_B(t)$ of the low-pass filter and transmitted information $s(t)$ i.e., transmitted symbol sequence $<q_v>$. For this reason, normalized standard deviation $\sigma_d = \sigma$ of the sample value d is also a function of the transmitted bit pattern. Depending on whether symbol "0" or symbol "1" is transmitted, following variances must be distinguished (compare with Eq. 2.3):

$$\sigma_0^2 = \sigma_d^2\big|_{q_v = "0"} = \left(e M^{2+x} I_{\text{dark}} + G_{\text{th}}\right) \sqrt{2}\, f_g \tag{5.156a}$$

and

$$\sigma_1^2 = \sigma_d^2\big|_{q_v = "1"} = \left(e M^{2+x} (R_0 P_r + I_{\text{dark}}) + G_{\text{th}}\right) \sqrt{2}\, f_g \tag{5.156b}$$

Therefore, the Gaussian pdf of sampled detected signal is

$$f_d(d) = \frac{1}{\sqrt{2\pi}\,\sigma_d} \exp\left(-\frac{(d-a)^2}{2\sigma_d^2}\right) \quad \text{where} \quad \begin{cases} \sigma_d = \sigma_0,\ a = a_{0i} & \text{if}\quad q_v = "0" \\ \sigma_d = \sigma_1,\ a = a_{1i} & \text{if}\quad q_v = "1" \end{cases} \tag{5.157}$$

Here, the suffix "i" distinguishes symbol pattern $<q_v>_{0i}$ with $q_{0i} = "0"$ and $<q_v>_{1i}$ with $q_{1i} = "1"$ and noiseless sample values a_{0i} and a_{1i}. In the strict sense, standard deviation σ_d of the pdf $f_d(d)$ is actually a function of transmitted symbol sequence $<q_v>_i$. As the signal-independent additive Gaussian noise usually dominates the signal-dependent shot noise, it is sufficient to distinguish σ_0 and σ_1. Thus, more accurate division into σ_{0i} and σ_{1i} on the basis of symbol pattern $<q_v>_{0i}$ and $<q_v>_{1i}$ is normally not required.

(iii) PROBABILITY OF ERROR

As mentioned in the previous Sections, worst-case probability of error p_w is normally used to estimate the average probability of error $p_a \le p_w$. Hence, we again focus our interest on the calculation of p_w.

(a) WORST-CASE PROBABILITY OF ERROR p_w

The worst-case probability of error briefly called the worst-case BER is completely determined by considering both the worst-case bit pattern $<q_v>_{0w}$ and $<q_v>_{1w}$. If a Gaussian low-pass filter is employed, then these special pattern are given by

$$<q_v>_{1w} = <\cdots 0\ 0\ 0\ 1\ 0\ 0\ 0\ \cdots> \quad \to \quad a_{1w} = 1 - 2Q\left(\sqrt{2\pi} f_g T\right)$$

$$<q_v>_{0w} = <\cdots 1\ 1\ 1\ 0\ 1\ 1\ 1\ \cdots> \quad \to \quad a_{0w} = 1 - a_{1w} \tag{5.158}$$

Like in heterodyne systems with coherent detection and homodyne systems, noiseless sample values a_{0w} and a_{1w} are determined by the Q-function given in Eq. (5.14). With a_{1w} and a_{0w}, normalized eye aperture A_d of direct detection system can be calculated as a well-suited and very powerful measure to assess the influence of intersymbol interference. We obtain

$$A_d = a_{1w} - a_{0w} = 1 - 2a_{0w} = 2a_{1w} - 1 = 1 - 4Q\left(\sqrt{2\pi}f_g T\right) \qquad (5.159)$$

It should be remembered that the constant of normalization is again the amplitude $\hat{\imath}_{PD}$ of the photodiode current.

Comparing the above equation and Eq. (5.12) of Section 5.1.1, it becomes immediately clear that the normalized eye aperture for coherent optical ASK homodyne system and noncoherent direct detection system are formally identical. However, comparing the eye apertures, a gain of about 16 dB can be observed i.e., eye aperture in ASK homodyne is about 6.3 times larger than that of direct detection system.

With the above results, we are now able to calculate the bit error rate. For this, we first have to determine the worst-case probability density functions $f_{d1w}(d)$ and $f_{d0w}(d)$. Considering equal occurrence probabilities for binary symbols "0" and "1", we obtain

$$P_w = \frac{1}{2}\left[\int_E^{+\infty} f_{d0w}(d)\ dd + \int_{-\infty}^{E} f_{d1w}(d)\ dd\right] \qquad (5.160)$$

Since both the probability density functions $f_{d1w}(d)$ and $f_{d0w}(d)$ are Gaussian, above expression can be rewritten as follows:

$$P_w = \frac{1}{2}\left[Q\left(\frac{E - a_{0w}}{\sigma_0}\right) + Q\left(\frac{a_{1w} - E}{\sigma_1}\right)\right] \qquad (5.161)$$

Finally, we use the eye aperture A_d given in Eq. (5.159) to modify Eq. (5.161). We obtain

$$p_w = \frac{1}{2}\left[Q\left(\frac{2E - (1-A_d)}{2\sigma_0}\right) + Q\left(\frac{(1+A_d) - 2E}{2\sigma_1}\right)\right] \qquad (5.162)$$

Considering this equation in more detail, it becomes clear that the BER can be minimized by optimizing the following *optimizable system parameters*:

- cut-off frequency f_g of the low-pass filter since $\sigma_0 = \sigma_0(f_g)$, $\sigma_1 = \sigma_1(f_g)$ and $A_d = A_d(f_g)$,

- threshold level E and

- avalanche gain M of the photodiode since $\sigma_0 = \sigma_0(M)$ and $\sigma_1 = \sigma_1(M)$.

It should be remembered that both standard deviations σ_0 and σ_1 as well as eye aperture A_d and threshold level E are normalized with respect to the amplitude $\hat{\imath}_{PD} = MR_0P_r$ of the photodiode current.

(iv) OPTIMIZATION

(a) THRESHOLD E

As already proved in Section 5.1.1, optimum threshold E_{opt} for discriminating symbol "0" and symbol "1" is located at the crosspoint where the probability density functions $f_{d0w}(d)$ and $f_{d1w}(d)$ meet. Therefore, the equation for determining the optimum threshold level is given by

$$f_{d1w}(E_{opt}) = f_{d0w}(E_{opt}) \qquad (5.163)$$

Since both the probability density functions $f_{d0w}(d)$ and $f_{d1w}(d)$ are Gaussian, Eq. (5.163) can be solved analytically. Applying some simple mathematical operations, we finally get

$$E_{opt} = a_{0w} + A_d \frac{1 - \dfrac{\sigma_1}{\sigma_0}\sqrt{1 + 2\dfrac{\sigma_1^2 - \sigma_0^2}{A_d^2}\ln\left(\dfrac{\sigma_1}{\sigma_0}\right)}}{\left(1 + \dfrac{\sigma_1}{\sigma_0}\right)\left(1 - \dfrac{\sigma_1}{\sigma_0}\right)} \qquad (5.164)$$

When the BER is minimized, following relationship between the probabilities of error $p_{0w}=p$(symbol "0" is detected wrong) and $p_{1w}=p$(symbol "1" is detected wrong) is always valid:

$$p_{0w} \leq p_{1w} \tag{5.165}$$

This inequality is caused by the asymmetrical and unequal distortion for binary symbol "0" and symbol "1". As mentioned earlier, symbol "1" is always disturbed by shot noise more than the transmitted symbol "0". It is important to note that the optimum threshold E_{opt} and, hence, the minimum BER is normally not obtained when p_{0w} and p_{1w} are chosen to be equal. The equality sign in the above equation is only valid for signal-independent noise i.e., for $\sigma_0 = \sigma_1$.

Next, let us consider the Eq. (5.164) for calculating the optimum threshold E_{opt} in more detail. It becomes clear that the ratio in the square root represents the reciprocal of signal-to-noise ratio. With practical system parameters (for example, $p_w \leq 10^{-9}$), second term in the square root becomes

$$2 \frac{\sigma_1^2 - \sigma_0^2}{A_d^2} \ln\left(\frac{\sigma_1}{\sigma_0}\right) \ll 1 \tag{5.166}$$

and Eq. (5.164) simplifies to

$$E_{opt} \approx a_{0w} + A_d \frac{\sigma_0}{\sigma_0 + \sigma_1} = \frac{a_{0w}\sigma_1 + a_{1w}\sigma_0}{\sigma_0 + \sigma_1} \tag{5.167}$$

Assuming further that shot noise is negligibly small as compared to the thermal noise of amplifier, optimum threshold in that case is exactly located at the centre of the noiseless sample values a_{0w} and a_{1w} as expected:

$$E_{opt} \approx \frac{a_{0w} + a_{1w}}{2} = a_{0w} + \frac{A_d}{2} = 0.5 \quad \text{provided } \sigma_0 \approx \sigma_1 \tag{5.168}$$

Again, it should be remembered that this threshold level is normalized with respect to the amplitude $\hat{i}_{PD} = MR_0 P_r$ of the photodiode current.

(b) Cut-Off Frequency f_g of the Low-Pass Filter

Considering Eq. (5.162), it becomes clear that minimization of BER requires maximization of arguments of both Q-functions. This task can be performed by maximizing both the signal-to-noise ratios A_d/σ_1 and A_d/σ_0. As variances σ_0^2 and σ_1^2 are directly proportional to the cut-off frequency f_g, both signal-to-noise ratios reach their maximum value at the same cut-off frequency $f_g = f_{g,opt}$. Therefore, the optimum normalized cut-off frequency $f_{g,opt}T$ is determined by the conditional equation

$$\frac{A_d^2(f_gT)}{f_gT} = \frac{\left[1 - 4Q\left(\sqrt{2\pi}f_gT\right)\right]^2}{f_gT} \;\to\; \max \;\to\; f_{g,opt}T \tag{5.169}$$

where T is the bit or symbol period and $1/T$ the bit or symbol rate. By means of a simple computer program, we again obtain

$$2f_{g,opt}T \approx 1.58$$

which represents the well-known characteristic numerical value when a Gaussian low-pass filter is used in the digital receiver. It should be noted that irrespective of filter function used, optimum rectangular-equivalent filter bandwidth is always in the range of

$$1.0 < 2f_gT < 2.0$$

(c) Avalanche Gain M

As compared to the optimization of f_g and E performed above, optimization of the avalanche gain M is much more comprehensive since both the Q-functions given in Eq. (5.162) do not reach their maximum value at the same avalanche gain. In practical systems, it is sufficient to determine the optimum avalanche gain from the equation

$$M_{opt} = \left(\frac{2G_{th}}{xe(R_0P_r + I_{dark})}\right)^{\frac{1}{2+x}} \tag{5.170}$$

which has already been derived in Chapter two. However, this equation only represents an approximation since this semi-optimum avalanche gain only maximizes the second Q-function in Eq. (5.162), but not the first Q-function. For practical system parameters, this is a very good and frequently used approximation of excellent accuracy.

(v) EVALUATION AND DISCUSSION

Summarizing the results of optimization obtained above, minimum worst-case probability of error p_w is

$$p_w = Q\left(\frac{A_d}{\sigma_0 + \sigma_1}\right) = Q\left(\frac{1}{2}\sqrt{\frac{S}{N}}\right) \tag{5.171}$$

This simple equation represents a well-suited practical tool to predict the BER of an optical transmission system with intensity modulation and direct detection. As $S = A_d^2$ is a measure of the signal power and $N = \sigma^2$ with $\sigma = (\sigma_0 + \sigma_1)/2$ the noise power, Eq. (5.171) is also a very useful tool to estimate the required signal-to-noise ratio.

Example 5.6

For a worst-case BER of $p_w = 10^{-9} \approx Q(6)$, we obtain by using Eq. (5.171): $S/N = 144$ or $10 \log(S/N) = 24.6$ dB.

The BER curve p_w as a function of received optical power P_r at the input to the direct detection receiver is shown in Fig. 5.29. This performance curve is based on the BER Eq. (5.171) given above. The underlying system parameters such as bit rate, noise power spectral density and responsivity, excess noise exponent and dark current of the avalanche photodiode are summarized in Table 7.1 of Chapter seven. The avalanche gain of the photodiode is chosen to be optimum (Eq. 5.170).

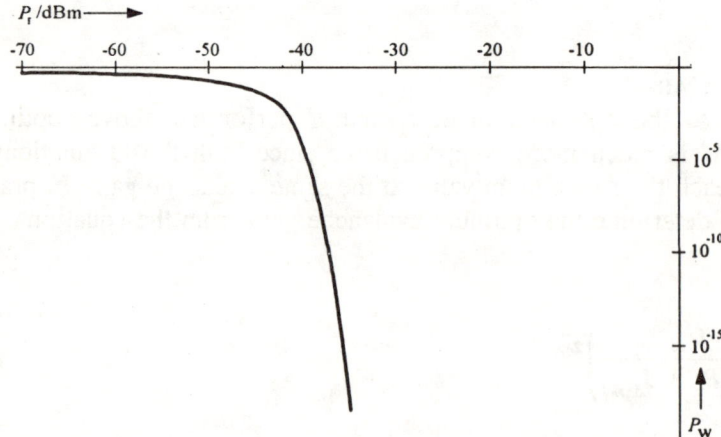

Fig. 5.29: Worst-case probability of error p_w as a function of received optical power P_r in optical direct detection system

It becomes clear from the above figure that the BER rapidly decreases with the increase in the received optical signal power P_r. Thereby, even a very small increase in the signal power P_r improves BER by some order of magnitudes. This means that direct detection system is characterized by a very steep BER curve, which is in agreement with standard electrical digital communication systems [256, 260, 269].

In comparison to optical heterodyne and homodyne systems, BER curve of direct detection system exhibits no saturation i.e., *no error rate floor*. However, the BER curve is shifted to the right, where received signal power is high. This becomes immediately clear by comparing Fig. 5.5 (ASK homodyne system) and Fig. 5.29 above. ASK homodyne system only requires an optical input power of about -52 dBm to achieve a BER of 10^{-10} (provided phase noise does not disturb the system), whereas the direct detection system requires about -39 dBm.

5.5 COMPARISON OF CALCULATION METHODS IN HETERODYNING

Last Section of this Chapter consider again noncoherent optical ASK and FSK heterodyning. Several different calculation methods are existing for the evaluation of BER and determining the performance of ASK or FSK heterodyne receivers with noncoherent demodulation. In section 5.3, only one of these methods has been used to analyze noncoherent heterodyne systems. This section compares these methods with regard to accuracy, complexity and required computer processing time.

5.5.1 WHY APPROXIMATION?

Scanning the literature, a large number of publications exist that theoretically examine the noncoherent optical heterodyne systems. A small representative sample of these papers is given by [58, 61, 64, 66, 75, 103, 110, 129-131, 138, 193, 201, 213, 231, 239, 262, 287]. The primary theoretical problems in analyzing optical heterodyne and homodyne systems are now solved. However, the available solutions are not always same and differ sometimes in accuracy, complexity and required computer processing time.

A typical example is given by the receiver performance evaluation of noncoherent ASK or FSK heterodyne systems including the effects of laser phase noise and also the influence of IF filter bandwidth. As noncoherent FSK heterodyne receiver with dual- or single-filter detection operates in nearly the same manner as noncoherent ASK heterodyne receiver, in this section we restrict our consideration to the latter receiver only.

As exact analytical calculation of the system performance is usually not possible, appropriate approximations must be derived. However, the problem is to assess the degree of accuracy of the different approximation methods since the exact solution is unknown. Of course, the most powerful verification of the accuracy is obtained by comparing the approximation results with the measurement results for a realized system; for example, eye pattern, BER and signal shape. On the other hand, an approximate system calculation is usually the first step in the design and realization of an optical communication system. Therefore, a comparison with realized system is not possible since the realized system is normally not available. In such a case, computer simulation of the system represents a very powerful tool to analyze the system and to verify the accuracy of approximations. As the accuracy of simulation results only depend on the available computer run time, it can theoretically be improved arbitrarily. Therefore, simulation results can usually be regarded as the exact solution.

In this Section, we make use of this substantial advantage of computer simulation to compare different approaches of approximate performance evaluation. It should be remembered that simulation results have already been employed and discussed in the previous Chapters several times.

5.5.2   PROBABILITY DENSITY FUNCTION AND BER

This Section first summarizes the basic formulas describing probability density function (pdf) and BER of optical ASK heterodyne receiver and the results of Section 5.3.1.

In the following, we consider noncoherent optical ASK heterodyne balanced receiver disturbed by laser phase noise, shot noise and thermal noise (Fig. 5.12). The signal of interest in this receiver is the envelope or detected signal $d(t) = |i_{IF}(t)|$ given in Eqs. (5.69) and (5.71). To calculate the BER, pdf $f_d(d)$ of the sampled (and normalized) detected signal d is required. As derived in Section 5.3.1, this pdf can be written as follow

$$f_d(d) = \frac{d}{\sigma^2} \int\limits_0^{+\infty} e^{-\frac{d^2 + C^2}{2\sigma^2}} J_0\left(\frac{Cd}{\sigma^2}\right) f_C(C) \, dC \tag{5.172}$$

It should be remembered that the random variable C equals the normalized and sampled envelope d provided standard deviation of the receiver additive Gaussian noise is zero i.e., $\sigma = \sigma_x = \sigma_y = 0$. Thus, shot noise and thermal noise of receiver are excluded.

In order to simplify BER calculation and focus our interest on the topic of this section, we neglect intersymbol interference. Thus, $C = 0$ if a symbol "0" is transmitted, whereas symbol "1" leads either to $C = 1$ when phase noise $\phi(t)$ is neglected or $C < 1$ when phase noise is included. As derived in Section 5.3.1, worst-case probability of error is given by

$$P_w = \frac{1}{2}\left(p_0 + p_1\right) = \frac{1}{2} \int\limits_E^{+\infty} f_{d0}(d) \, dd + \frac{1}{2} \int\limits_0^E f_{d1}(d) \, dd \tag{5.173}$$

$$= \int\limits_0^1 \tilde{p}(C) \, f_C(C) \, dC$$

where E represents the decision threshold and $\tilde{p}(C)$ the well-known BER formula of conventional ASK heterodyne systems in the absence of phase noise. It should be noted that occurrence probabilities for symbol "0" and "1" are assumed to be equal.

The Eq. (5.173) enables a nearly "exact" calculation of the BER for noncoherent ASK heterodyne systems. The only inherent approximations are the narrow-band consideration of the noisy IF signal, no intersymbol interference (which are considered in section 5.3.1) and the use of an ideal envelope detector with $d(t) = |i_{IF}(t)|$.

As explained in Section 5.3.1, main problem in solving Eq. (5.173) is the determination of pdf $f_C(C)$ of random envelope C which physically represents the envelope of IF signal in the absence of additive Gaussian shot and thermal noise. In principle, it is possible to obtain analytically the required pdf $f_C(C)$ as shown in Section 5.3.1. However, the numerical evaluation on computer is very time consuming, especially for a high laser linewidth-to-bandwidth ratio $\Delta f/B_{IF}$ [58, 64, 66]. For that reason, goal is to find useful methods of approximation for the pdf $f_C(C)$. Approximation problems will be discussed in more detail by considering three representative approaches in the following Sections.

5.5.3 METHODS OF APPROXIMATION

(i) APPROXIMATION METHOD 1 (QUASI-CONSTANT FREQUENCY APPROXIMATION)

This approximation method, first proposed by Garrett and Jacobsen [75, 110] and already introduced in Subsection 3.5 (vi) presumes that the intermediate frequency noise

$$\dot{\phi}(t) = \omega(t) = \omega_c \qquad (5.174)$$

is approximately constant during the time interval

$$\Delta t_B = \frac{1}{h_B(0)} \int_{-\infty}^{+\infty} h_B(t) \, dt = \frac{H_B(0)}{h_B(0)} = \frac{1}{B} \approx T \qquad (5.175)$$

which is in the order of bit duration T in practical systems. Physically, Δt_B represents the rectangular equivalent pulse width of the filter time response $h_B(t)$ and $B = (\Delta t_B)^{-1}$ the double-sided noise equivalent bandwidth of the filter frequency response $H_B(f)$. As explained in Chapter two, $H_B(f)$ is the baseband representation of the intermediate frequency filter i.e., $H_{IF}(f) = H_B(f-f_{IF}) + H_B(f+f_{IF})$. For example, a Gaussian baseband filter with frequency response $H_B(f) = \exp[-\pi(f/B)^2]$ and $B = 2f_g = 1/T$ yields $\Delta t_B = T$.

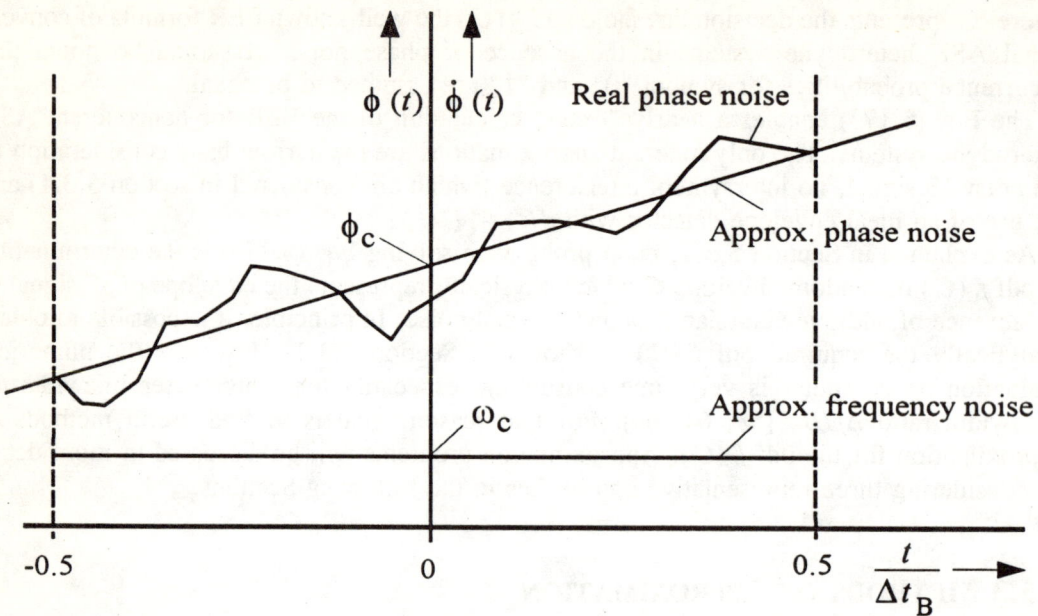

Fig. 5.30: Phase noise and frequency noise approximation related to approximation method 1

As shown in Fig. 5.30, a constant frequency ω_c during Δt_B is related to a linearly increasing or decreasing (ω_c can also be negative) phase

$$\phi(t) = \omega_c t + \phi_c \tag{5.176}$$

It may be mentioned that ω_c is actually a random variable which statistically changes its value from one time interval (or from one bit) to the following time interval (the next bit). The pdf of the zero mean Gaussian random frequency ω_c is given by

$$f_{\omega_c}(\omega_c) = \frac{1}{\sqrt{2\pi}\sigma_{\omega_c}} \exp\left(-\frac{\omega_c^2}{2\sigma_{\omega_c}^2}\right) \tag{5.177}$$

Its variance is

$$\sigma_{\omega_c}^2 = 2\pi\,\Delta f B \tag{5.178}$$

where Δf represents the resulting laser linewidth of both transmitter and local lasers.

Next, we reconsider the random envelope C for symbol "1" as given in Eq. (5.78). By substituting the phase noise $\phi(t)$ from Eq. (5.176), we obtain:

$$C = |H_B(\omega_c)| \tag{5.179}$$

The required pdf $f_c(C)$ can now easily be calculated by using the simple one-dimensional statistical transformation (see Section 3.5)

$$f_C(C) = f_{\omega_c}(\omega_c) \frac{1}{\left|\dfrac{dH(\omega_c)}{d\omega_c}\right|_{\omega_c = H^{-1}(C)}} \tag{5.180}$$

Here, H^{-1} represents the inverse function of H. As an example, we consider again a Gaussian filter with frequency response $H_B(f) = \exp[-\pi(f/B)^2]$. In this particular case, above equation yields

$$f_C(C) = \sqrt{\frac{B}{\Delta f}} \; C^{\frac{B}{\Delta f}-1} \; \frac{1}{\sqrt{-\pi \ln(C)}} \tag{5.181}$$

By substituting Eq. (5.181) in Eq. (5.173), we finally obtain

$$P_w = \frac{1}{2}\exp\left(-\frac{E^2}{2\sigma^2}\right) + \frac{1}{2}\sqrt{\frac{B}{\Delta f}} \int_0^1 Q\left(\frac{C-E}{\sigma}\right) \frac{C^{\frac{B}{\Delta f}-1}}{\sqrt{-\pi \ln(C)}} \, dC \tag{5.182}$$

This equation clearly indicates that the BER of noncoherent optical heterodyne system is primarily determined by three parameters: First, the normalized standard deviation σ of the additive Gaussian shot and thermal noise; second, the filter bandwidth-to-laser linewidth ratio $B/\Delta f$ and third, the threshold level E. Here, E and $B/\Delta f$ can be optimized to minimize the BER. Eq. (5.182) includes only one integral which, however, cannot be solved analytically. Nevertheless, Eq. (5.182) represents a relatively fast and also accurate method (Section 5.5.4) for calculating the BER for noncoherent optical ASK heterodyne system.

(ii) APPROXIMATION METHOD 2 (GAUSSIAN APPROXIMATION)

This method, already proposed and used in Section 5.3.1, takes advantage of the simple Gaussian approximation

$$f_C(C) \approx \frac{1}{\sqrt{2\pi}\,\sigma_C}\, e^{-\frac{(c-\eta_C)^2}{2\sigma_C^2}} \tag{5.183}$$

of the pdf $f_C(C)$. The remaining problem is to calculate the mean η_C (first moment) and the standard deviation σ. Owing to the square-root dependence of C (see Eq. 5.78 to 5.80), it is easy to calculate exactly the second moment $E\{C^2\} = \sigma^2 + \eta^2$, whereas the mean value $\eta_C = E\{C\}$ can only be found by approximations (E = expected value). Calculation of η_C and σ_C have been carried out in Section 5.3.1 (noncoherent ASK heterodyne system). If an ideal square-law detector with $d(t) = |i_{IF}(t)|^2$ instead of an ideal envelope detector with $d(t) = |i_{IF}(t)|$ is used, then the required moments η_D and σ_D with $D = C^2$ can be calculated exactly. With the above Gaussian approximation, calculation of the BER becomes very easy. We obtain

$$p_w = \frac{1}{2}\exp\left(-\frac{E^2}{2\sigma^2}\right) + \frac{1}{2}\,Q\left(\frac{\eta_C - E}{\sqrt{\sigma^2 + \sigma_C^2}}\right) \tag{5.184}$$

This simple formula includes no integral provided the Q-function is installed in the computer. Thus, the BER calculation can be performed very fast and with a relatively good accuracy (Section 5.5.4). However, a large computer execution time is necessary to evaluate the statistical values η_C and σ_C or η_D and σ_D. As these values only depend on the linewidth Δf and type of filter used in the receiver, it is possible to store these values in computer data files as a function of Δf for some typical filters normally used in practice.

(iii) APPROXIMATION METHOD 3 (POWER-ACCOMMODATION APPROXIMATION)

This approximation method proposed by Kazovsky [131] is entirely different from the methods discussed in Subsections (i) and (ii) as it is not based on foregoing BER calculation. This method takes into account the fact that the IF filter bandwidth $B_{IF} = \Delta f_M$ (the index M stands for modulation) must be in the order of three times the bit rate $R = 1/T$ to accommodate 95% of the ASK signal power [131]. Considering phase noise also, IF-filter bandwidth must now be increased to [131]

$$B_{IF} = \sqrt{\Delta f_M^2 + \Delta f_{95}^2} = \sqrt{(3R)^2 + \Delta f_{95}^2} \tag{5.185}$$

where Δf_{95} represents the 95% bandwidth of the photodetector current due to laser phase noise. It is assumed that Δf_{95} and linewidth $\Delta f/2$ of laser are related by $\Delta f_{95} = 12.7 \Delta f$. Provided that a

$$penalty = \frac{B_{IF}(\text{with phase noise})}{B_{IF}(\text{without phase noise})} = \sqrt{1 + \left(\frac{1}{3}\Delta f_{95}T\right)^2} \tag{5.186}$$

of 1 dB is permissible, following allowable linewidth-bit duration product can be obtained:

$$\Delta f T \leq 0.09 \tag{5.187}$$

Obviously, this simple relationship enables a very fast (no computer is required), but only a rough estimation of the maximum permissible laser linewidth. Because of the fact that a penalty of 1 dB does not necessarily lead to a total system collapse, a larger Δf than predicted by Eq. (5.187) can be used. A practical example for this is given in [250].

The Eq. (5.187) does not give any information about BER. It only imply that the unknown BER will increase if the penalty is above 1 dB and decrease otherwise. The BER, however, remains unknown. Clearly, this is the drawback of power-accommodation approximation. However, the permissible linewidth obtained from Eq. (5.187) is in good agreement with those obtained by the much more comprehensive methods 1 and 2 explained above.

5.5.4 COMPARISON

The objective of this Section is to compare the different approximation methods discussed above. The underlying comparison consider the following: accuracy, complexity and required computer processing time of the calculation.

The problem of comparing accuracy is that no exact analytical solution (for example for the BER) exists which can be used as a reference. However, a very good alternative reference is given by computer simulated numerical solutions which has already been used in Section 5.3.1. The accuracy of this solution depends only on the computer run time and can, therefore, theoretically be as high as required. For this reason, the results obtained by a computer simulation can be regarded as the exact solution.

Mean value η_C and standard deviation σ_C obtained by computer simulation and by approximation methods 1 and 2 are compared in Fig. 5.31. It may be noted that $C = d(\sigma = 0)$ physically represents the envelope of IF signal corrupted by phase noise when shot and thermal noise have been neglected ($\sigma=0$). The corresponding nongaussian pdf $f_C(C)$ of the random envelope C is required for an accurate calculation of BER from Eq. (5.173). The mean value η_C and standard deviation σ_C essentially determine location and width of $f_C(C)$.

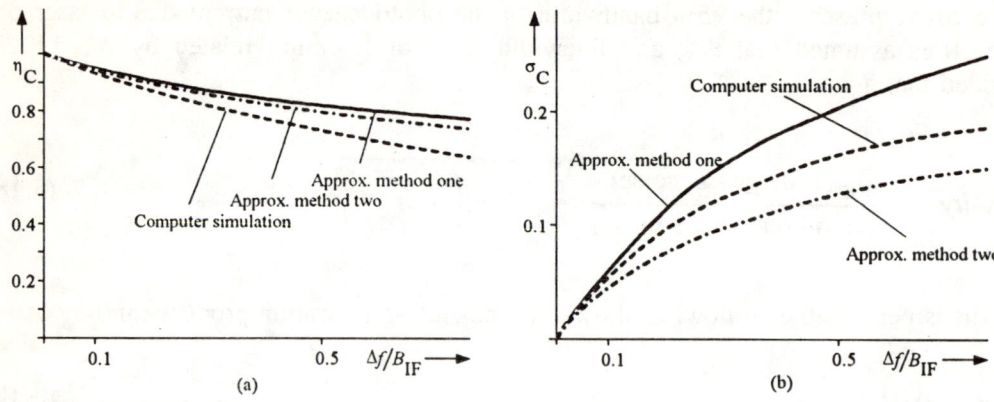

Fig. 5.31: (a) Mean value η_C and (b) standard deviation σ_C of the random envelope C which physically represents the envelope of IF signal $i_{IF}(t)$ corrupted by phase noise (additive Gaussian noise is neglected).

We can see from Fig. 5.31 that there is some deviation between both approximate solutions and the computer simulated (exact) solution. This deviation increases with the increase in the laser linewidth-to-bandwidth ratio $\Delta f/B_{IF}$. For $\Delta f/B_{IF}$ ratio less than about 0.2, deviation in η_C becomes negligibly small, while the deviation in σ_C still remains considerable. As shown in Fig. 5.31a, exact (computer simulated) solution for η_C is always below the both approximate solutions, whereas the exact solution for σ_C lies between these solutions (Fig. 5.31b).

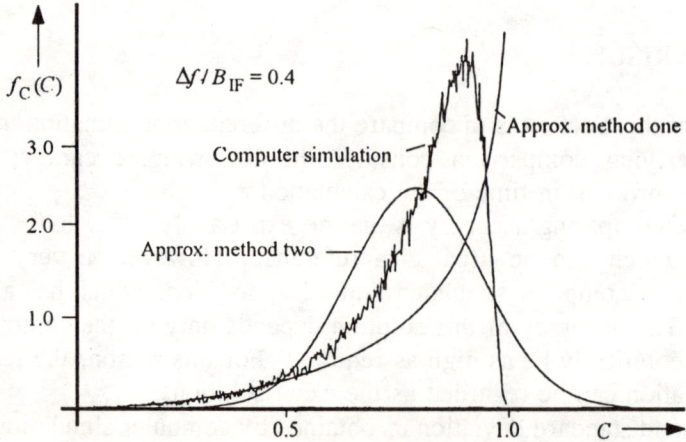

Fig. 5.32: Probability density function $f_C(C)$ of the random envelope C

The pdf $f_C(C)$ at $\Delta f/B = 0.4$ is shown in Fig. 5.32. It can be seen that $f_C(C)$ calculated by using approximation method 2 is slightly more close to the computer simulated plot than the approximation method 1. However, the main part of $f_C(C)$ for the BER calculation is the part for $C < 0.5$. The reason for this is that the normalized optimum threshold E is always less than 0.5 in ASK heterodyne receiver [66]. It becomes clear from Fig. 5.32 that in the range of $C < 0.5$, approximation method 1 represents the more accurate solution.

The probability p_1 of wrongly detected symbol "1" (as a part of the overall BER) is given by the area under the pdf $f_d(d)$ between $d = 0$ and $d = E < 0.5$. From Fig. 5.33, it becomes clear that the deviation between approximation method 1 and 2 is less when $\Delta f/B$ decreases. As $\Delta f/B$ increases, approximation method 1 again represents the more accurate solution. Approximation method 1 always yields a higher BER than method 2. The difference between BERs can reach several orders of magnitude [51]. In practice, where required input light power P_r for a desired BER and a given laser linewidth is of interest, difference between both approximation methods is less (see Example 5.7).

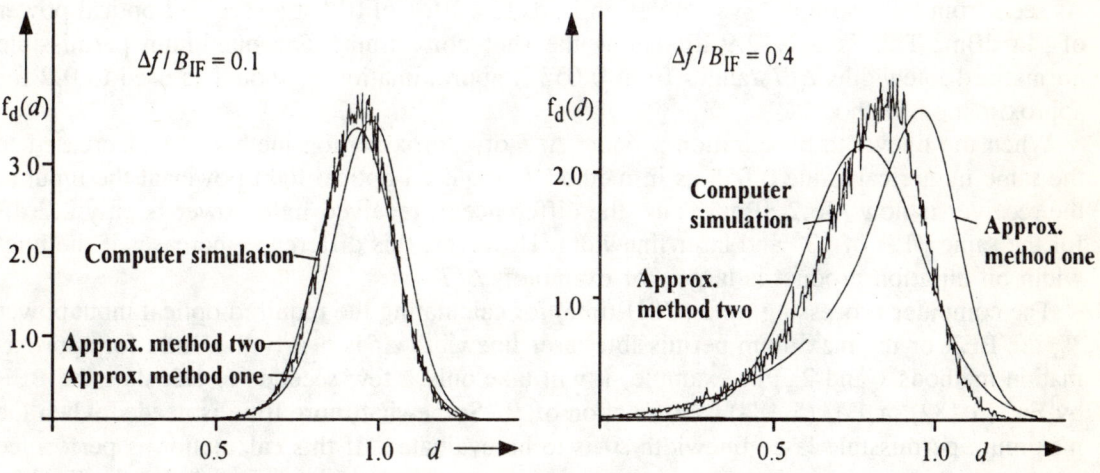

Fig. 5.33: Probability density function $f_d(d)$ of the sampled and normalized detected signal d

For the maximum permissible laser linewidth Δf, approximation method 3 can also be used. Based on typical system parameters, following numerical example shows the maximum allowed linewidth-bit duration product ΔfT obtained by approximation methods 1-3.

Example 5.5

Let us consider the following system parameters:

BER:	$p = 10^{-9}$,
Local laser power:	$P_l = 0$ dBm,
Responsivity:	$R_0 = 1$ A/W,

Coupling efficiency: $k = 0.5$ (3 dB coupler),
Receiver noise: $G_c = 10^{-24}$ A²/Hz,
Bit period: $T = 1$ ns (bit rate: 1 Gbit/s),
IF filter bandwidth: $B = 1.58/T = 1.58$ GHz (optimum if phase noise is neglected),
Threshold level: $E = E_{opt}$.

When the maximum permissible resulting laser linewidth Δf for optical input power levels $P_r = -44$ dBm and -46.2 dBm are calculated, the following results are obtained:

P_r	$(\Delta f T)_1$	$(\Delta f T)_2$	$(\Delta f T)_3$	Difference
- 44.0 dBm	0.055	0.20	0.09	2.2 dB
-46.2 dBm	-	0.055	-	

As seen from this example, system design leads to a BER of 10^{-9} at a received optical power of -44 dBm. This is only 2.9 dB below the shot noise limit. The maximum permissible normalized linewidths $\Delta f T$ ranges from 0.055 if approximation method 1 is used to 0.2 for approximation method 2.

When the linewidth-bit duration product $\Delta f T$ of approximation method 2 is decreased to the same numerical value 0.055 as in method 1, required optical light power at the input to the receiver is now -46.2 dBm. Thus, the difference in received light power is only 2.2 dB for the same BER of 10^{-9} and laser linewidth. However, this difference increases if the linewidth-bit duration product is large; for example, $\Delta f T = 1$.

The computer processing time (CPU-time) for calculating the required optical input power P_r, the BER or the maximum permissible laser linewidth Δf is nearly the same for approximation methods 1 and 2. For example, it will take only a few seconds to calculate the BER by Eq. (5.182) or Eq. (5.184) as a function of P_r. Somewhat more time is needed when the maximum permissible laser linewidth Δf is to be evaluated. If this calculation is performed in connection with a simultaneous optimization of the threshold level E and also the IF filter bandwidth $B_{IF} = B$, then the required computer run time can raise from a few minutes to some hours (depending on the computer). The computing time will continuously increase if intersymbol interference are also taken into account [64].

In contrast to this, no computer run time is needed to calculate the permissible laser linewidth Δf by approximation method 3, because Eq. (5.187) inherently contains the desired solution.

Finally, we will briefly consider the complexity of the theoretical derivations of approximation methods 1-3. The derivation of Eq. (5.187) as the result of approximation method 3 is straightforward and easy. No difficult calculation steps are required. In contrast to that, the derivations of approximation methods 1 and 2 are much more complex and involve some non-trivial calculation steps. Nevertheless, the derivations are also conceptually straightforward.

To conclude, Table 5.5 summarizes the results of our comparison performed in this Section. Again, this table considers the accuracy of calculation, required computer processing time and complexity of the underlying mathematical derivations as the most important criteria to compare the approximate calculation methods for noncoherent optical heterodyne receivers as discussed above.

Table 5.5 Features of three approximate calculation methods for evaluating the performance of noncoherent optical heterodyne systems

	Approximation method 1 (quasi-constant frequency)	Approximation method 2 (Gaussian approx.)	Approximation method 3 (power-accommodation)
Accuracy	very good	good	sufficient
CPU-time	medium	medium low, if η_c and σ_c are stored in data files	zero
Complexity	low	low	very low
Remarks	methods 1 and 2 are similar		only maximum permissible laser linewidth is approximated (not the BER)

It becomes clear from this table and our above discussion that the choice of an approximate calculation method depends on its application. If, for example, a fast estimation of the maximum permissible laser linewidth for a given bit rate is required, then approximation method 3 represents the best solution. Unfortunately, no information about the BER can be obtained by this method, which represents a main drawback of this simple method. If a more accurate calculation is required, then both approximation method 1 or method 2 can be used. Here, the required computer execution time is approximately the same. As method 1 is somewhat more accurate, especially for linewidth-to-bandwidth ratios greater than one, it gives the better result. If a solution is required that gives an overall and clear insight into the complex functional behaviour of the system including the interactions of all fundamental system parameters (i.e., BER, bit rate, laser linewidth, variance of shot and thermal noise, threshold, filter bandwidth and optical input power), computer simulation is the best method.

6 OPTICAL AMPLIFIERS AND SYSTEM ASPECTS

One of the most exciting development in the field of optics has been the discovery of optical amplifier. It is probably the most significant development since the discovery of single-mode fiber. The optical amplifiers will play an important role in optical communication systems as they enable the direct amplification of light with a minimum of electronics and thus eliminate the electronic bottleneck associated with many current systems. In a long-haul optical networks, system upgradation (in terms of increased data rate or improvement in performance or link length) can be achieved without the modification of hardware. For local area networks, optical amplifiers can be used to overcome the splitting loss associated with passive networks and increase the fan out capability.

In recent years, optical amplifiers have received increasing attention and very fast developments have taken place in terms of device performance and practical demonstrations. Initial field trials and subsequent use in operational links are under way. Indeed for many of the optical communication systems being proposed, optical amplifiers are essential components. In this Chapter, different types of optical amplifiers have been described and their comparative study has been made. Further, various applications of optical amplifiers in the optical communication systems have been discussed.

6.1 TYPES OF OPTICAL AMPLIFIERS

Optical amplifiers can be classified on the basis of device characteristics i.e., whether it is based on linear characteristic (semiconductor laser amplifiers and rare-earth doped fiber amplifiers) or nonlinear characteristic (Raman amplifiers and Brillouin amplifiers). Optical amplifiers can also be classified on the basis of structure i.e., whether semiconductor based or optical fiber based. In this classification, semiconductor laser amplifiers (SLAs) will fall in the first category, while the other three i.e., rare-earth doped fiber amplifiers, Raman and Brillouin scattering amplifiers in the second category.

Different rare-earth ions, such as erbium, holmium, neodymium, samarium, thulium and ytterbium can be used to realize fiber amplifiers operating at different wavelengths covering visible to infrared region (up to 2.8 μm). Erbium-doped fiber amplifiers (EDFAs) have attracted the most attention among them simply because they operate near 1550 nm, the wavelength region in which fiber attenuation is minimum. Raman and Brillouin amplifiers are based on the scattering of a photon to a lower energy photon such that the energy difference appears in the form of a phonon. The main difference between the two is that optical phonons participate in the Raman scattering, whereas acoustic phonons in the Brillouin scattering.

The SLAs are more developed at present, as they are based on existing laser structure with anti-reflection (AR) coating applied to the facets. The recent developments have resulted in devices exhibiting high gain, low polarization sensitivity and high saturation power. Progress in fiber amplifier technology during last few years has also been very rapid and EDFAs have been demonstrated in many system experiments. The fiber amplifiers have a particular advantage that they can be spliced into the system fiber with very low loss and this is in contrast to large coupling loss associated with the SLAs and fiber coupling. Therefore, the fiber amplifiers exhibit higher gain than the SLAs [33].

In the following Section, different types of optical amplifiers have been discussed.

6.1.1 SEMICONDUCTOR LASER AMPLIFIERS

The structure of SLA is similar to that of semiconductor laser source used in optical communication systems. The basic difference is that in SLA, resonant cavity is essentially eliminated by substantially reducing the facet reflectivities down to levels of 10^{-3} to 10^{-4} or even less [252]. This, in turn, will reduce the optical feedback. An optical amplifier without feedback is referred as travelling wave amplifier (TWA). The optical gain for such an amplifier depends not only on the frequency (wavelength) of the incident signal, but also on the intensity of the signal at any point inside the amplifier. The relationship of gain with the frequency and intensity depends upon the amplifier medium. When the gain is modelled as homogeneously broaden two-level system, gain coefficient of such a medium is given by [6]

$$g(f) = \frac{g_0}{1 + 4\pi^2(f-f_0)^2 T_2^2 + \dfrac{P_i}{P_s}} \tag{6.1}$$

Here, g_0 represents the peak value of gain determined by the pumping level of amplifier, f and P_i are the optical frequency and power of the incident signal respectively and f_0 the atomic transition frequency. P_s is the saturation power level (it depends on the gain medium) and T_2 is known as dipole relaxation time (typically varies from 0.1 ps to 1 ns). Depending on P_i/P_s ratio, there are two cases:

(i) UNSATURATED REGION

In this region $P_i/P_s \ll 1$ and $g(f)$ from Eq. (6.1) is approximately given by

$$g(f) = \frac{g_0}{1 + 4\pi^2(f-f_0)^2 T_2^2} \tag{6.2}$$

It is seen from this equation that the gain coefficient is maximum when $f=f_0$. Let us define the gain/amplification factor of the amplifier as

$$G = \frac{P_o}{P_i} \tag{6.3}$$

where P_o is the output power and P_i as defined earlier is the input power of the amplifier. This amplification factor for an amplifier of length L is given by [6]

$$G(f) = e^{g(f)L} \tag{6.4}$$

It is seen from the above equation that the $G(f)$ and $g(f)$ are maximum when $f=f_0$ and decreases when $f \neq f_0$. The maximum unsaturated gain G_0 of the amplifier will be

$$G_0 = e^{g_0 L} \tag{6.5}$$

The amplifier bandwidth defined as full width at half maximum (FWHM) is given by [6]

$$(\Delta f)_{TWA} = \left(\frac{1}{\pi T_2}\right) \cdot \left(\frac{\ln(2)}{g_0 L - \ln(2)}\right) \tag{6.6}$$

It is seen from Eqs. (6.5) and (6.6) that both G_o and $(\Delta f)_{TWA}$ depend on g_0 and L.

(ii) SATURATED REGION

In this region $P_i/P_s \gg 1$ and $g(f)$ from Eq. (6.1) will be relatively lower and hence G also. For simplicity, let the frequency of incoming optical signal is exactly tuned to the atomic transition frequency i.e., $f=f_0$. Effect of detuning $(f \neq f_0)$ on the gain can be easily incorporated. The large-signal amplifier gain is given by [78]

$$G = G_0 e^{(1-G)\frac{P_i}{P_s}} \tag{6.7}$$

This nonlinear equation is to be solved for G, once G_0, P_i and P_s are known or specified. The above equation shows that G decreases from its unsaturated value G_0 when P_i becomes comparable with P_s. The gain saturation not only reduces the amplifier gain, but also leads to nonlinear effects that can increase the crosstalk in a multichannel transmission system.

In semiconductor lasers, relatively large feedback is present due to reflections occurring at the cleaved facets (reflectivity $\sim 32\%$). Such lasers can also be used as amplifiers when biased below threshold and are referred as Fabry-Perot amplifiers (FPAs). The amplification factor for such amplifiers is given by [6, 286]

$$G_{FPA}(f) = \frac{(1-R_1)(1-R_2)G(f)}{\left(1 - G(f)\sqrt{R_1 R_2}\right)^2 + 4G(f)\sqrt{R_1 R_2}\ \sin^2\phi} \tag{6.8a}$$

where

$$\phi = \frac{\pi(f-f_r)}{\delta f} \tag{6.8b}$$

Here, R_1 and R_2 are the facet reflectivities, f_r represents the cavity resonance frequencies and δf ($=c/2nL$, n is the refraction index and c the velocity of light) the longitudinal mode spacing. The parameter G is the single pass amplification factor and corresponds to that of TWA (refer to Eq. 6.4). The above equation shows that the amplifier transmission characteristics contain resonant peaks whose absolute wavelength and spacing depends on the cavity dimensions. Typical transmission characteristics of a FPA with $R_1=R_2=0.3$ and a single pass gain of 25 dB is shown in Fig. 6.1.

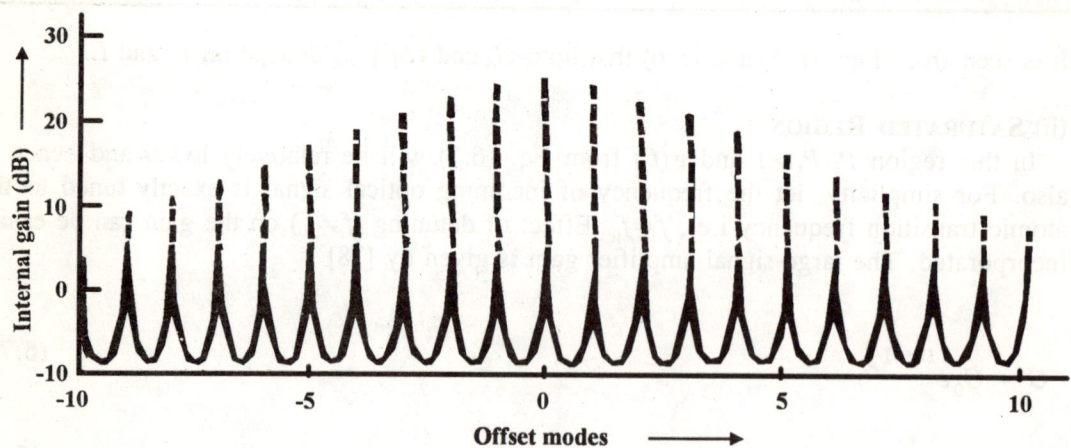

Fig. 6.1: FP amplifier passband, mode 0 corresponds to peak gain wavelength; mode spacing $\lambda^2/(2nL)=1.5$ nm for $L=200$ μm [204]

It is possible to determine the amplifier bandwidth by calculating the detuning $f-f_r$ for which G_{FPA} drops by 3 dB from its peak value. It is given by [233]

$$(\Delta f)_{FPA} = \frac{2\delta f}{\pi} \sin^{-1}\left[\frac{1 - G\sqrt{R_1 R_2}}{\left(4G\sqrt{R_1 R_2}\right)^{1/2}}\right] \qquad (6.9)$$

For large amplification factor, $G\sqrt{R_1 R_2}$ should be close to unity. (refer to Eq. 6.8a). Under this condition, amplifier bandwidth from Eq. (6.9) will be quite small (< 10 GHz). Such a small bandwidth makes the FPAs unsuitable for light wave communication systems and they are primarily used for signal processing applications. The FPAs are characterized by narrow bandwidth (in the order of GHz) and a high sensitivity to temperature, bias current and signal polarization fluctuations. For wide bandwidth applications, amplifier bandwidth can be increased by suppressing the reflection feedback from the end facets. This is normally achieved by applying AR coating to the facets. When the reflectivity is sufficiently reduced (less than 1%), it will become a TWA. The bandwidth of TWA is three order of magnitudes larger than that of FPA. Nevertheless, passband in the TWA comprises peaks and troughs whose relative amplitudes are determined by the residual facet reflectivities, single pass gain (and hence the applied bias current) and the input signal power level. Variations of amplifier gain with signal wavelength for a SLA whose facets are anti-reflection coated to reduce reflectivity to about 0.04% (SLA is operating in a nearly travelling wave mode) is shown in Fig. 6.2 [232].

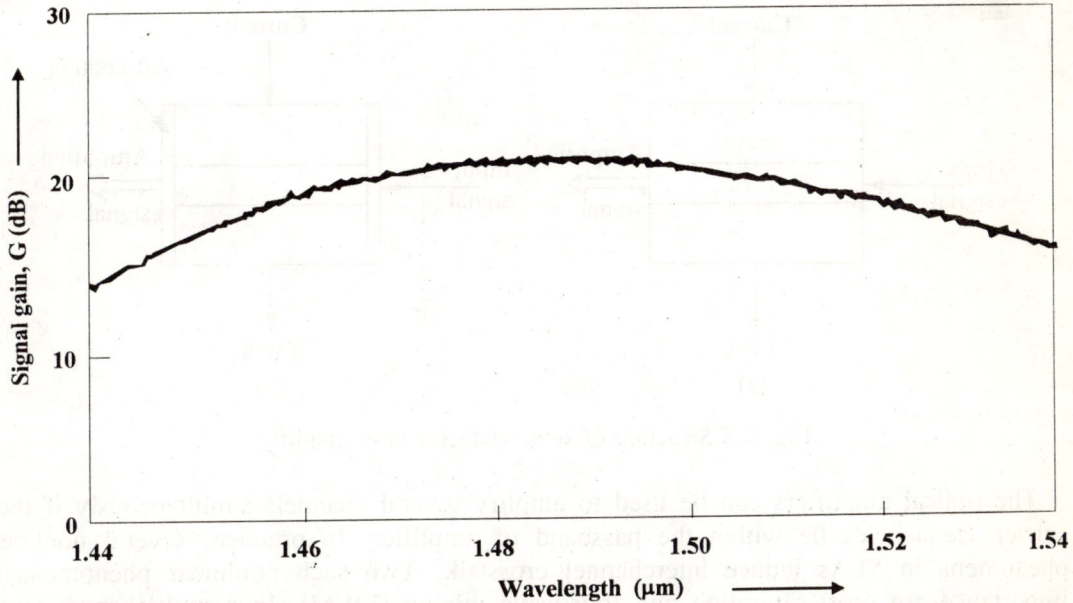

Fig. 6.2: Amplifier gain versus signal wavelength for semiconductor laser amplifier for transverse electric (TE) mode whose facets are anti-reflection coated to produce reflectivity to about 0.04% [232]

The gain undulation or peak-trough ratio of the ripple (ΔG) is given by [233]

$$\Delta G = \frac{1 + \sqrt{R_1 R_2}\, G}{1 - \sqrt{R_1 R_2}\, G} \qquad\qquad (6.10)$$

For wide-band operation, ΔG must be small and is considered to be less than 3 dB (i.e., $\Delta G < 2$) for TWAs over their signal gain spectrum. A gain ripple of zero imply an ideal TWA. Hence, an amplifier whose gain ripple significantly exceeds 3 dB is usually categorized as a FPA.

The structures of FPA and TWA are shown in Fig. 6.3. The FPAs operating around a central wavelength of 1300 nm or 1500 nm with a facet-facet gain of about 30 dB are presently available [252]. Their gain bandwidth product is about 40 nm (more than 4000 GHz). Devices based on special structure viz., multiple quantum well (MQW) have shown gain bandwidth of more than 200 nm. Noise figure (NF) of the SLAs is typically in the order of 7 dB [34]. In practical systems, it is necessary to couple the light to and from the amplifier and there are associated coupling losses. Typically, these losses are about 3.5 dB/facet. It means fiber-fiber gain will be 7 dB lower than the facet-facet gain. For this reason, maximum realistic gain of a wide-band amplifier is in the order of 20 dB.

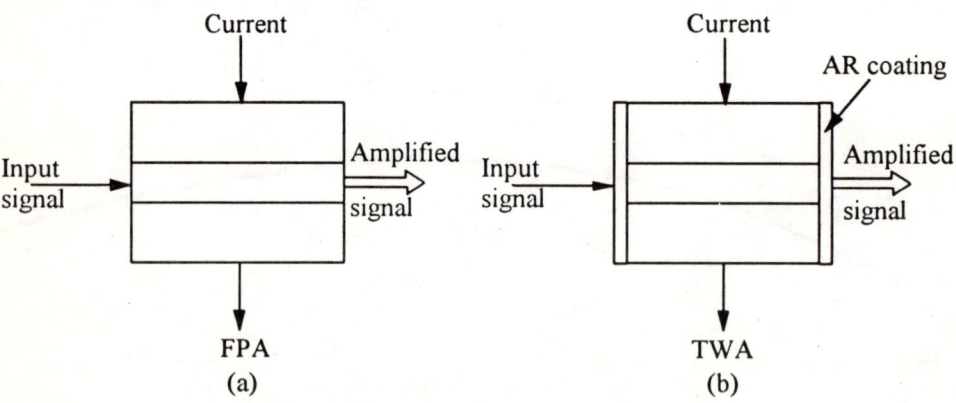

Fig. 6.3 Structure of semiconductor laser amplifiers

The optical amplifiers can be used to amplify several channels simultaneously if their carrier frequencies lie within the passband of amplifier. In practice, several nonlinear phenomena in SLAs induce interchannel crosstalk. Two such nonlinear phenomena of importance are cross-saturation and four-wave mixing (FWM). In a multichannel transmission system, total input optical power P_i is given by. [6]

$$P_i = \sum_{j=1}^{N} P_j + \sum_{j=1}^{N} \sum_{k=1}^{N} \sqrt{P_j P_k} \cos\left(2\pi(f_j - f_k)t + \phi_j - \phi_k\right) \qquad j \neq k \qquad (6.11)$$

Here, f_j is the frequency, ϕ_j the phase of 'j'th channel signal and N the total number of channels. The power as given by Eq. (6.11) cause the carrier population in the amplifier medium to oscillate at the beat frequency $f_j - f_k$. As a consequence, gain and refraction index are also modulated at the same frequency. Thus, multichannel signal creates the gain and index gratings. These gratings induce interchannel crosstalk by scattering a part of signal from one channel to another. This phenomena is referred as FWM [6]. If three optical fields with carrier frequencies f_1, f_2 and f_3 copropagate inside the amplifier simultaneously, new fields with frequencies $f_1 \pm f_2 \pm f_3$ are generated in principle due to FWM. In practice, most of these combinations are not build up due to phase-matching requirements for any FWM process. The frequency combination of the form $f_4 = f_1 + f_2 - f_3$ is most troublesome for multichannel communication systems especially when the channel spacing is relatively small ($\approx$1GHz) and optical power per channel is high. The phase-matching condition is nearly satisfied in this case. It may lead to power transfer among various channels.

As evident from Eq. (6.11), amplifier gain for a particular channel is saturated not only by its own power, but also by the power of neighbouring channels. When the input light intensity is low, signal gain becomes constant and crosstalk does not occur. On the other hand, if the input intensity is high, output intensity is not proportional to the input intensity i.e., amplifier is gain saturated. In that case, signal gain is dependent upon the total light intensity. In a multichannel transmission system, total light intensity varies randomly since each channel light is independently modulated. Thus, signal output of one channel varies according to signal gain fluctuations induced by the modulation of other channels, even when this signal input is constant. This is the crosstalk induced by gain saturation in the amplifier. This crosstalk occurs regardless of the extent of channel spacing. It is possible to avoid this crosstalk by operating SLAs in the unsaturated regime. It can also be avoided by using phase shift keying (PSK) or frequency shift keying (FSK) signalling schemes. For these signalling schemes, power in each channel and hence the total power remains constant with time. Best conventional devices have shown saturation power level of about 10 mW, while MQW devices have reached 110 mW saturation power level [34].

The interchannel crosstalk induced by FWM can occur for all multichannel communication systems irrespective of the modulation scheme used. In the case of equispaced channels, new frequencies coincide with the existing frequencies and FWM leads to power transfer among various channels. When channels are not equispaced, new frequencies fall in between the channels and interfere with the detection process. In both the cases, system performance degrades because of a loss in the channel power and noise introduced by the interchannel interference. Impact of FWM on system performance can be reduced by decreasing the channel power or by increasing the channel spacing. It becomes negligibly small when the channel spacing exceeds 10 GHz [4, 6, 40, 91, 103].

While cascading a number of SLAs, one has to take into account the fact that the amplifier noise keeps on accumulating in the chain. It limits the number of cascadable devices because of saturation of latter amplifiers. Further, any undesirable reflections along the line may cause feedback into the amplifier cavity and generate oscillations and instabilities. In order to prevent the oscillations, optical isolators are required which are quite expensive and bulky devices [34].

6.1.2 ERBIUM DOPED FIBER AMPLIFIER

Rare-earth doped fiber amplifier, particularly EDFA has recently emerged as a big competitor to SLAs [276]. It is based on glass fibers doped with an appropriate amount of erbium. The energy for amplification is provided in the form of optical radiation at an appropriate wavelength around 1550 nm and power level tens to hundreds of milliwatts. A schematic view of EDFA is shown in Fig. 6.4.

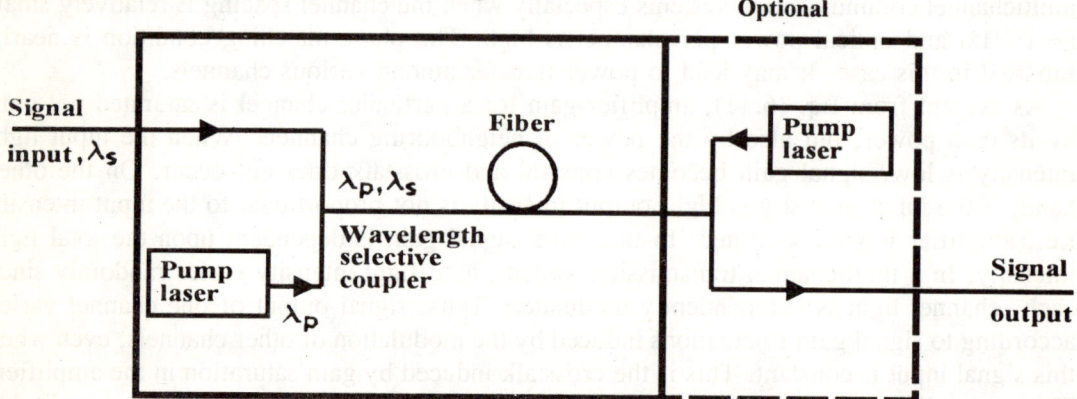

Fig. 6.4 Schematic of optical fiber amplifier

As shown in the figure above, EDFA consists of three basic components: length of suitable erbium-doped fiber, pump laser and wavelength selective coupler to combine the signal and pump wavelengths. In some applications, it may be required to remove any unused pump light. For this purpose, an optical filter can be placed at the opposite end of amplifier. When the pump enters the amplifier at the same end as the signal, beams are copropagating. An alternate geometry is possible where the pump laser is introduced at the opposite end of fiber amplifier and pump light propagates in the opposite direction from that of the signal. The beams are then counterpropagating. The use of insufficient pump power or too long fiber can lead to part of the amplifier providing loss rather than gain. Alternatively, increasing pump power arbitrarily or using too short a fiber leads to insufficient use of pump power. There is, therefore, an optimum fiber length that has a strong dependence on pump power, input signal power, amount of rare-earth doping, pumping wavelength and typically spans from few metres to around 100 metres. In the extreme case of very low level doping, kilometre lengths may be used. In that case, amplifier will provide a kind of continuous amplification along the fiber. It may be quite useful for local loop distribution or sustaining nonlinear propagation for soliton pulse transmission.

Various wavelengths are accessible for pumping, but the most useful are those that allow the use of semiconductor lasers as pumping source. These sources are the best choice because of size, ruggedness, power consumption and reliability. Use of lasers at 820 nm would have been the better choice, but it is not among the best from the technical point of view. The 980 nm wavelength has proved to be the best one in terms of efficiency (more than 10 dB gain per mW pump power) and better noise performance. The 1480 nm wavelength may be preferred because of higher source reliability as well as lower attenuation of communication fibers in this spectral range [34].

Generally, forward pumping leads to smaller NF and lower pumping light to signal light conversion efficiency η, than the backward pumping. Typically NF lies between 4-5 dB and η between 40-50% with forward pumping and the equivalent figures for backward pumping are 6-7 dB and 60-70% assuming 1480 nm pumping light is used. In the case of bidirectional pumping, NF and η values lie between those of forward and backward pumping [293].

The EDFAs give similar performance as SLAs in terms of gain, saturation power (can be increased with pump power) and available bandwidth. In addition, they have following additional advantages over SLAs: (i) Compatibility of the structure with optical fiber. This allows to retain practically all the available gain unlike SLAs, (ii) Insensitivity to polarization of the optical signal, (iii) High reproducibility of amplifier characteristics as they are essentially determined by the atomic structure and not by the geometry, (iv) Greater stability against environmental changes, (v) Lower NF than SLA, partly because of the intrinsic mechanism, but mainly because of the reduced coupling loss at the input, (vi) The lifetime associated with the process are such that crosstalk in presence of number of wavelengths is significantly reduced and the amplifier is an almost true travelling wave device, (vii) Saturation power in EDFA may be much higher than the power available with SLAs provided sufficient pump power is available and (viii) Possibility of remote optical pumping as the pump signal can be transmitted along the same fiber used for signal transmission.

On the negative side, their weak points are: large size and difficulty of integration with the other optical components.

6.1.3   RAMAN FIBER AMPLIFIER

In Raman fiber amplifier, amplification mechanism is the stimulated Raman scattering (SRS) in the fiber which causes an energy transfer from the pump to the signal. The vibrational spectra of core material defines the Raman shift. If the wavelength of signal laser beam is known, then the optimum wavelength for pump signal can be calculated i.e., Raman amplification can occur at any wavelength as long as appropriate pump laser is available. The structure of this amplifier is very much similar to that of EDFA. There are again three basic components: pump laser, wavelength selective coupler and fiber (Fig. 6.4). The major difference between these two amplifiers is that in the Raman amplifier standard single-mode optical fiber can be used, although for certain applications it may be desirable to use special fiber designs to enhance the amplifier performance. Gains of Raman amplifier for standard single-mode fiber of length 100 km in the 1550 nm transmission window and for 100 mW pump power are in the order of 2.5 dB for step index fiber and 5 dB for dispersion shifted fiber [59, 259]. The difference in gains between the two fibers arises mainly because of more germanium content in the fiber core of the dispersion shifted fiber. In both the cases, 95 % of the gain is obtained in the first 50 km of fiber. Pump power of 100 mW at the appropriate wavelength from each end of above 100 km length of the dispersion shifted fiber reduce the loss of link from 20 dB to only 10 dB at 1550 nm. This imply an effective gain of 15 dB with dispersion shifted fiber. A Raman power amplifier can be made using a short (few km) length of specific fiber of optimum length for a given pump power and then it can be incorporated in the optical network based on standard optical fibers.

The main features of the Raman amplification are that it may be realized as a continuous amplification along the fiber (which could be an ordinary telecommunication fiber) thus never letting the signal to become too low; it is bidirectional in nature; it offers more stability and insensitivity to reflections; spectral range of gain can be chosen in a continuous way along all the optical spectrum of interest. The saturation optical power level is very high as it depends on the pump power. The only disadvantage which has greatly reduced interest in this kind of amplifier is the requirement of relatively high power pump laser (100 mW-200 mW) in comparison with SLAs and EDFAs.

6.1.4   BRILLOUIN FIBER AMPLIFIER

The operating principle of Brillouin fiber amplifier is essentially the same as for Raman fiber amplifier except that in this amplifier, optical gain is obtained by stimulated Brillouin scattering (SBS) instead of SRS. The Brillouin fiber amplifiers are also pumped optically and a part of the pump power is transmitted to the signal through SBS. Physically, each pump photon creates a signal photon and the remaining energy is used to excite an acoustic phonon. Classically, the pump beam get scattered from an acoustic wave moving through the medium at the speed of sound. Despite a formal similarity between SBS and SRS, SBS differs from SRS in three important aspects: (i) Amplification occurs only when the signal beam propagates in direction opposite to that of the pump beam (backward pumping), (ii) Stokes shift i.e., difference in the signal and pump frequency for SBS (≈ 10 GHz) is smaller by three order of magnitudes compared with that of SRS and (iii) Brillouin gain spectrum is extremely narrow with a bandwidth < 100 MHz. Gains in the order of 25 dB have been obtained for pump power of only 5 mW [11]. The bandwidth available in Brillouin amplifier arises from the thermal distribution of the phonons in the fiber core material and is less than 100 MHz. The most efficient use of this amplification mechanism requires the use of narrow-band lasers for signal and pumping (< 100 MHz), separated by about 10 GHz to an accuracy of only a few MHz. Only for a certain specialized application, this complexity is justified. Gain of Brillouin amplifier can be as high as 50 dB with a moderate pump power level. For a gain of about 20 dB, few mW pump source is good enough.

The narrow bandwidth of Brillouin fiber amplifiers makes them less suitable as power amplifier, preamplifier or in-line amplifier in light wave systems. The same feature can, however, be exploited for some other applications in coherent and multichannel communication systems.

6.2 COMPARISON BETWEEN SEMICONDUCTOR AND FIBER AMPLIFIERS

The typical characteristics and main features of optical amplifiers are given in Table 6.1. Further, a representative gain spectral curve is shown in Fig. 6.5.

Table 6.1: Comparison of optical amplifier characteristics (na: not applicable)

Property	Laser	Erbium fiber	Raman fiber	Brillouin fiber
Unsaturated device gain	> 20 dB	> 20 dB	5 dB - 15 dB	> 25 dB
Optical pump power	na	20 mW - 50 mW	100 mW - 200 mW	< 10 mW
Optical pump wavelength	na	820 nm, 980 nm, 1400 nm-1500 nm	Stokes shift below signal	
Electrical bias current	50 mA	> 100 mA	> 500 mA	< 50 mA
Wavelength of operation	any	1525 nm-1565 nm	any, but subject to pump	
Bandwidth	20 nm - 50 nm	10 nm - 40 nm	20 nm - 40 nm	0.001 nm
Coupling loss	5 dB - 6 dB	< 1 dB	< 1 dB	< 1 dB
Polarization sensitivity	less than few dB	0 dB	0 dB	0 dB
Saturated output	less than few mW	few mW	limited only by pump power	
Directions	bidirectional	bidirectional	bidirectional	unidirectional
Noise	low	low	very low	very low

Some of the additional points to be mentioned in connection with the optical amplifiers are following: (i) For SLAs and EDFAs, typical unsaturated gain are greater than 20 dB, while for Raman fiber amplifiers, gain is restricted to lower values by the stringent pump power requirement. If coupling losses of the amplifiers are also considered, gain of SLAs will get reduced by about 7 dB, (ii) The SLAs need an electrical bias supply at levels of around 50 mA, while the supply requirement is much more stringent in erbium and Raman fiber amplifiers because of the high power pump laser requirement, (iii) In SLAs, gain spectrum is dependent upon the band gap of material used (which broadly defines the wavelength at which maximum gain is obtained), bias current supplied to the device (which defines the broad envelope of the gain) and residual reflectivity of the facets (which governs the fine structure on the overall envelope). The wavelength of optimum gain can be selected by the growth of appropriate semiconductor material. In fiber amplifiers, pump laser and gain spectrum must be at or around specific wavelengths. The EDFAs require semiconductor pump laser at one of the 820 nm, 980 nm or 1400 nm-1500 nm wavelength and the gain spectrum can extend across 1525 nm-1565 nm by using alumina co-doped fibers. For Raman amplifiers, pump laser wavelength is usually 100 nm (Raman shift) below that of the signal. As the commercial high power lasers are available between 1470 nm and 1550 nm wave-length region, signal wavelength must be greater than 1570 nm to obtain optimized Raman gain from an amplifier. At this wavelength, fiber attenuation may not be minimum and therefore a part of the gain will be eaten away by the increased fiber attenuation.

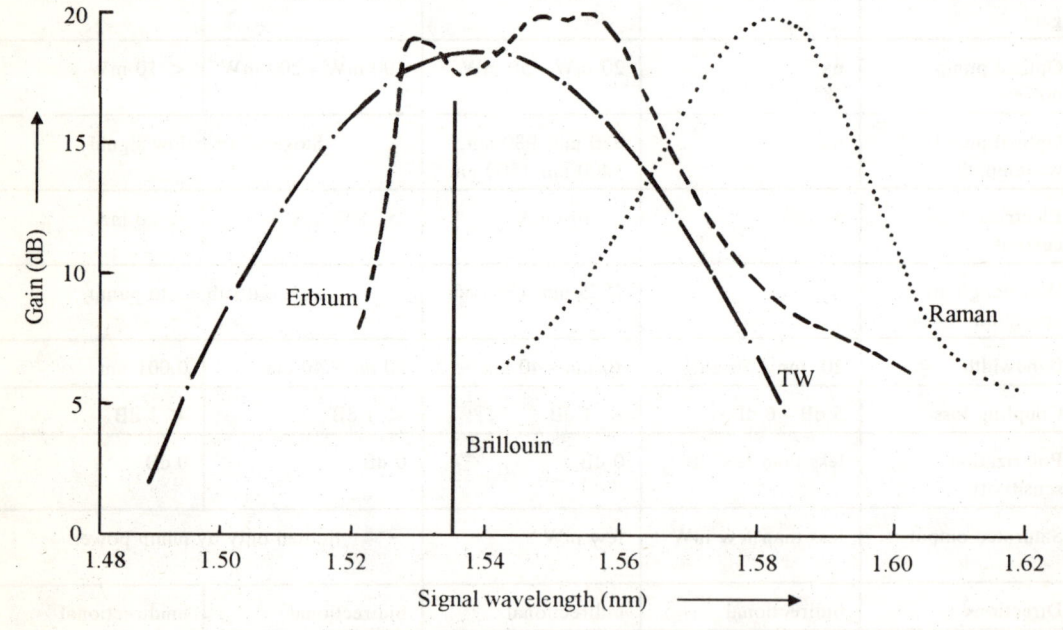

Fig. 6.5: Typical gain curves for various optical amplifiers

In Brillouin fiber amplifiers, problem is of matching the signal and pump source wave-lengths within 0.1 nm, (iv) The polarization dependence of SLAs can be reduced to levels of 1 dB or less by using special device structures. Although fiber amplification mechanism is polarization insensitive, care must be taken to ensure that couplers used do not introduce polarization dependence into the amplifiers, (v) For both SLAs and EDFAs, saturated output power is typically a few milliwatts. In Raman and Brillouin amplifiers, saturated output power could be much higher as it is limited only by the pump power and (vi) Intermodula-tion distortion (arise because of departure from linearity and is similar in nature to FWM distortion) and saturation induced crosstalk in WDM system is negligibly small in fiber amplifiers as compared to SLAs. However, if the power in any one signal channel becomes large (more than tens of milliwatts), then that signal can become a pump for Raman amplification of all other signals at longer wavelengths. This is known as Raman-induced crosstalk and puts a fundamental limit on the high power multichannel transmission. In contrast to Raman-induced crosstalk, Brillouin-induced crosstalk is easily avoided because of the narrow frequency range (≈ 100 MHz) over which such crosstalk occur [5].

6.3   APPLICATIONS OF OPTICAL AMPLIFIERS

All optical amplifiers (except the Brillouin type) may find applications in optical communication systems. However, due to specific characteristics, a particular amplifier may be better suited for certain applications rather than others. For instance, EDFA would be a better choice as in-line amplifier because of its compatibility and other characteristics. Similarly, Raman amplifier could be an excellent power amplifier in view of its high saturation power. If the integrated optic chip including optical amplifier has to be developed, SLA appears to be the best choice.

In this Section, possible applications of optical amplifiers in optical communication systems and suitability of various amplifiers for a particular application have been discussed.

(i) POWER AMPLIFICATION

The maximum loss capability of an unrepeatered optical communication system is set by the difference between transmitter laser launch power and receiver sensitivity. To design a system with a large loss capability, one has to increase launch power and use highly sensitive receiver. For a given requirements, receiver sensitivity may be significantly improved by the use of avalanche photodiode (APD) instead of PIN-photodiode in the optical receiver or coherent detection techniques if the extra complexity is justified. However, shot noise sets a fundamental limit beyond which the sensitivity will not be further improved.

The power launched from the source into the fiber can be increased by using an optical amplifier. The upper limit to the transmitter power is set by nonlinear optical effects in the fiber such as SRS and SBS. These effects can cause severe degradation in the system performance [35].

Though all types of amplifiers can be used as power amplifier, Raman fiber amplifier will be a better choice as the optical pump power required to drive this amplifier is the highest implying the highest saturated output power.

(ii) PREAMPLIFIER

When optical amplifier is used as preamplifier, it is placed directly in front of the receiver. Both SLA and EDFA can be used for this purpose. The SLA has a slight disadvantage that residual Fabry-Perot resonances can give small changes in the sensitivity of receiver for different signal wavelengths. Therefore, it will not be suited for wavelength division multiplex systems.

(iii) IN-LINE AMPLIFIER

Current medium- and long-haul optical communication systems use regenerative repeaters at regular intervals to compensate for both fiber loss and dispersion. When optical amplifier is used as a direct replacement for optoelectronic repeater in the system, regeneration of optical signal is not required. For long-haul systems (e.g., transatlantic optical fiber communication system), while a chain of amplifiers could compensate for the loss in the system, but the accumulation of spontaneous emission from amplifiers will set an upper limit on the number of amplifiers to be cascaded. The regenerative repeaters may still be required to prevent this build up of noise and also to restore the optical pulse distortion arising from the fiber dispersion. The Raman gain, which requires many kilometres of fiber, is most likely to be used to provide distributed gain in the transmission fiber itself [59]. However, SLA (< 1 mm long) and EDFA (tens of metres long) can both be used as discrete amplifiers.

(iv) NONLINEAR PROCESSING

In optical communications, there is a need to have an optical device which can be used for a variety of nonlinear applications including pulse reshaping allowing the prospect of all-optical regeneration. The semiconductor laser amplifiers, particularly FPA is more suitable for such applications. In FPA, change in the refraction index with input signal is sufficient to observe nonlinear behaviour [3]. This can be used, for example, to provide optical pulse shaping. The FWM in wide-band SLA is also of current interest. This phenomenon can be used for frequency translation, an important function for multiwavelength systems. The SLA can be used as phase modulator in coherent systems [66]. This type of modulator provides gain in contrast to the loss associated with lithium niobate modulator and has lower modulation voltage requirement.

(v) OPTICAL SWITCH

All optical amplifiers require the supply of electrical current to drive the amplification mechanism. Removal of this bias removes the optical gain. In some cases, optical amplifier may become opaque to the signal wavelengths. Therefore, it behaves as an electrically controlled optical switch.

The SLA and EDFA which become strongly absorbing without any bias are suitable for such an application. The SLA responds on nanosecond time scale (carrier recombination time ≈ 1 ns) and can be used as a high speed switch. The speed of EDFA is governed by the life time of the erbium ions (≈ 14 ms) and therefore it is only possible to make slow switches with this type of amplifier.

(vi) SOLITON TRANSMISSION

Soliton provides a means for balancing the pulse dispersion of silica with the nonlinear process of self-phase modulation [94, 169]. As a consequence, very short pulses can propagate undistorted along the fiber and ultrahigh bit rates (greater than 10 Gbit/s) can be achieved. For this, peak power of the pulse must be kept constant at a level determined by its temporal width. As the soliton pulse propagates along the fiber, it slowly loses energy and can fall below the soliton level. At this point, an in-line EDFA can amplify the pulse and bring it back to the soliton level. High bit rate and long distance soliton transmission has been achieved with lumped EDFA separated by a distance of 25-30 km [20].

6.4   NOISE IN OPTICAL AMPLIFIERS

An optical amplifier, in addition of amplifying the input signal, will add to it a signal due to spontaneous emission of light. A portion of the spontaneously emitted light is in the same direction as the signal and gets amplified along with the main signal. This added light, called the amplified spontaneous emission (ASE) noise is spread over a wide wavelength range (in the order of 50 nm) compared to signal as shown in Fig. 6.6.

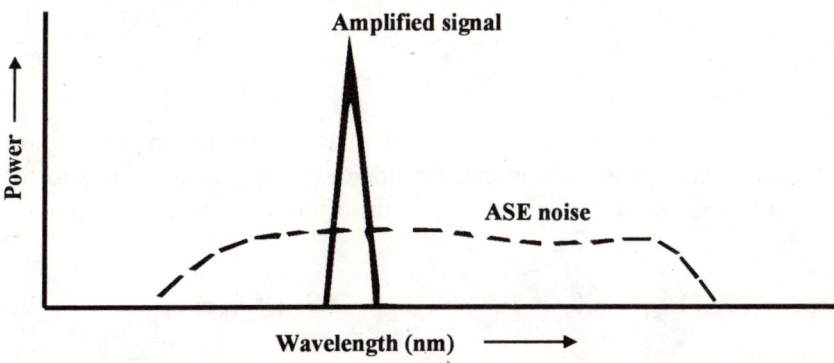

Fig. 6.6: Amplified optical signal and ASE noise at the amplifier output

The spectral density of ASE noise is nearly constant (white noise) in TWA and is given by [203]

$$\rho_{sp}(f) = (G-1)n_{sp}hf \tag{6.12}$$

where f is the optical frequency and n_{sp} the population inversion parameter. For amplifiers with complete population inversion (all atoms in the excite states), its value is minimum i.e., unity. For a two level system, it is given by

$$n_{sp} = \frac{N_2}{N_2 - N_1} \tag{6.13}$$

where N_1 and N_2 are the atomic populations in the ground and excited states respectively. Since EDFAs operate on the basis of a three-level pumping scheme, N_1 remains nonzero and

$n_{sp} > 1$. Further, N_1 and N_2 vary along the fiber length because of their dependence on the pump and signal power and n_{sp} should be averaged along the fiber length. As a consequence, n_{sp} is expected to depend on both the amplifier length and pump power [6].

The amplifier output-signal plus ASE when detected by a photodiode consists of desired signal component along with the beat and shot noise components. Signal-spontaneous beat noise is generated by mixing between the amplified signal and ASE components. Similarly, spontaneous-spontaneous beat noise is generated by mixing within the ASE components themselves. In addition to these two noise components, there will be shot noise components produced by the signal and the spontaneous emission.

The optical output (i.e., average number of photons) from an optical amplifier consists of an amplified optical signal and ASE noise. It is given by [201, 251]

$$<n_0> = G<n_i> + (G-1)n_{sp}b \qquad\qquad (6.14)$$

Here, $<n_i>$ is the average number of incident photons, G the amplifier gain and b the bandwidth of optical band-pass filter inserted at the output end of amplifier to limit the ASE noise. The variance (σ_{n0}^2) of the number of photons after amplification is given by

$$\sigma_{n0}^2 = G<n_i> + (G-1)n_{sp}b + 2<n_i>G(G-1)n_{sp} + (G-1)^2 n_{sp}^2 b$$
$$+ G^2\left(<n_i^2> - <n_i>^2 - <n_i>\right) \qquad\qquad (6.15)$$

In the above equation, first and second terms correspond to shot noise due to signal and ASE respectively. The third and fourth terms are beat noise originating from signal-ASE and ASE-ASE interference respectively. The fifth term depends on the state of incident electric field. With a coherent incident light, average and variance of the photon number are equal and, therefore, the last term vanishes [201].

6.4.1   NOISE FIGURE OF AMPLIFIER

Without an amplifier i.e., when $G = 1$, variance σ_{n0}^2 from Eq. (6.15) equals $<n_i>$. Thus, signal-to-noise (S/N) ratio becomes

$$\left(\frac{S}{N}\right)_i = \frac{<n_i^2>}{2B\sigma_{n0}^2} = <n_i>T \qquad\qquad (6.16)$$

where $B = 1/(2T)$ is the minimum receiver bandwidth and T the pulse duration. This means that the S/N ratio of input optical signal is equal to the number of photons within an optical pulse. It is the theoretical shot noise limit set by the quantum nature of light. On the other hand, when the amplified signal is detected by a photodiode, the S/N ratio is given by

$$\left(\frac{S}{N}\right)_o = \frac{G^2 <n_i>^2}{2\left[G<n_i> + (G-1)n_{sp}b + 2<n_i>G(G-1)n_{sp} + (G-1)^2 n_{sp}^2 b\right]B} \tag{6.17}$$

When G is high, shot noise terms (first and second) will be comparatively much smaller than the beat noise terms (third and fourth). Under this condition, above expression gets simplified to

$$\left(\frac{S}{N}\right)_o \simeq \frac{<n_i>^2}{2\left[2<n_i>n_{sp} + n_{sp}^2 b\right]B} \tag{6.18}$$

If b is so small that the ASE-ASE beat noise component is negligibly small as compared to signal-ASE beat noise component, then the second term in the denominator of above equation can be neglected. In that case, S/N ratio becomes

$$\left(\frac{S}{N}\right)_o = \frac{<n_i>T}{2n_{sp}} \tag{6.19}$$

The above S/N ratio is called the "beat noise" limited. By comparing Eqs. (6.16) and (6.19), noise figure (NF) of optical amplifier is defined as follows:

$$NF = \frac{\left(\frac{S}{N}\right)_i}{\left(\frac{S}{N}\right)_o} = 2n_{sp} \tag{6.20}$$

The best NF occurs when n_{sp} is unity. For an ideal TWA, n_{sp} is unity and therefore the NF will be 3 dB. For FPA, it ranges from 5 dB to 7 dB. EDFA with forward pumping has a NF of about 4.8 dB.

6.4.2   EVALUATION OF S/N RATIO IN DIRECT DETECTION SYSTEM

In this Section, evaluation of S/N ratio in direct detection system has been made when the optical amplifier is used as preamplifier, postamplifier and in-line amplifier (Fig. 6.7).

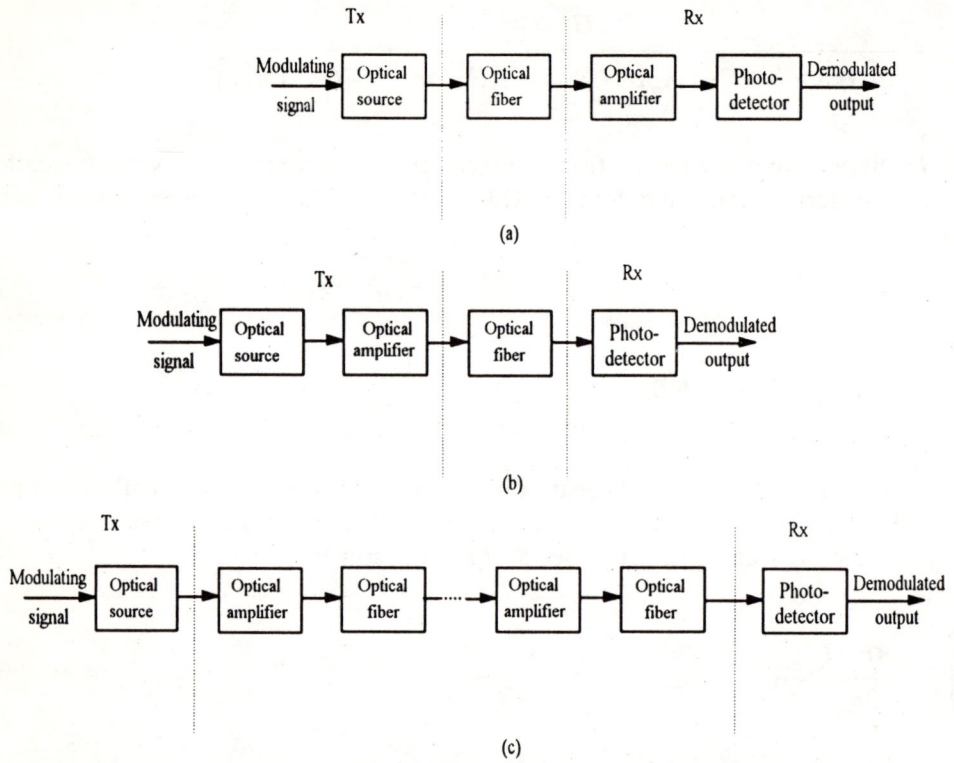

Fig. 6.7: Direct detection system with (a) optical preamplifier, (b) optical postamplifier and (c) cascade amplification using k optical in-line amplifiers

(i) OPTICAL PREAMPLIFIER

In this configuration, as shown in Fig. 6.7a, optical signal is first amplified and then detected by a photodetector. In order to include the effect of circuit noise, equivalent noise current due to circuit noise is taken as

$$\overline{i_c^2} = \frac{4k_B T_0 B}{R_1} \tag{6.21}$$

where k_B is the Boltzmann's constant, T_0 the noise temperature in degree Kelvin and R_1 the photodetector load resistance. Therefore, S/N ratio is given by [201]

$$\frac{S}{N} = \frac{e^2 G^2 <n_i>^2}{2e^2 \sigma_{n0}^2 B + 4k_B T_0 B/R_1} \tag{6.22}$$

With the increase in G, beat noise become predominant over the shot noise in the denominator of Eq. (6.22) as the latter noise has higher order dependence on G. Further, beat noise also become predominant over the circuit noise. If b is sufficiently small, S/N ratio approaches the beat noise limited value as given in Eq. (6.19), which under ideal conditions i.e., when $n_{sp}=1$ is 3 dB worse than the shot noise limited value.

(ii) POSTAMPLIFIER

In this application (refer to Fig.6.7b), input optical signal is first amplified and then made to propagate through the fiber. Finally, the signal is detected by a photodetector.

Let A be the attenuation factor by which the signal will get attenuated while propagating from one end of the fiber to other. At the photodetector output, average number of photons and their variance are given by [201]

$$<n_o> = AG<n_i> + A(G-1)n_{sp}b \tag{6.23a}$$

and

$$\sigma_p^2 = AG<n_i> + A(G-1)n_{sp}b + 2A^2<n_i>G(G-1)n_{sp} + A^2(G-1)^2 n_{sp}^2 b \tag{6.23b}$$

Therefore,

$$\frac{S}{N} = \frac{e^2 A^2 G^2 <n_i>^2}{2e^2 \sigma_p^2 B + 4k_B T_0 B/R_1} \tag{6.24}$$

Under the condition that the optical filter following the amplifier can reject the spontaneous emission sufficiently, S/N ratio is given by

$$\frac{S}{N} = \frac{e^2 A^2 G^2 <n_i>^2}{2e^2 \left[AG<n_i> + 2A^2<n_i>G(G-1)n_{sp}\right]B + 4k_B T_0 B/R_1} \tag{6.25}$$

Dependence of S/N ratio on the fiber attenuation factor (implying fiber length also) is as follows:

- Shot noise limited: When the first term in the denominator of Eq. (9.26) predominates, S/N ratio is proportional to A.

- Beat noise limited: When the second term predominates, S/N ratio is independent of A. Thus, under beat noise limited conditions, noise characteristics of amplifier are similar to an electrical system.

- Circuit noise limited: When the third term predominates, S/N ratio is proportional to A^2.

It is evident from Eqs. (6.22) and (6.25) that under shot noise and circuit noise limited conditions, postamplifier will give lower S/N ratio as compared to preamplifier.

(iii) IN-LINE AMPLIFIER

Let us consider a cascaded amplification system using k optical in-line amplifiers as intermediate repeaters (refer to Fig. 6.7c). For simplicity, let us presume that each amplifier has same characteristics and each section will reduce the signal by the same factor A. If the gain of amplifier is adjusted so that it compensates loss i.e., $GA=1$, average and variance of photon number at the output can be obtained from Eq. (6.23) by replacing n_{sp} by kn_{sp} and GA by unity and are given by

$$<n_o> = <n_i> + A(G-1)n_{sp}kb \tag{6.26a}$$

and

$$\sigma_c^2 = <n_i> + A(G-1)n_{sp}kb + 2A<n_i>(G-1)n_{sp}k + A^2(G-1)^2n_{sp}^2k^2b \tag{6.26b}$$

Therefore,

$$\frac{S}{N} = \frac{e^2<n_i>^2}{2e^2\sigma_c^2B + 4k_BT_0B/R_1} \tag{6.27}$$

Under beat noise limited condition

$$\frac{S}{N} = \frac{<n_i>^2}{2\left[2<n_i>n_{sp}k + n_{sp}^2k^2b\right]B} \tag{6.28}$$

A comparison of Eqs. (6.18) and (6.28) shows that the S/N ratio of the repeatered system is deteriorated at least by a factor of k and ASE-ASE beat noise becomes dominant due to multiplication by k^2. Consequently, in-line amplifiers used for cascaded amplification system should have smaller bandwidth filter. When the filter bandwidth b is sufficiently small, approximate S/N ratio from Eq. (6.28) is given by

$$\frac{S}{N} = \frac{<n_i>}{4n_{sp}kB} \tag{6.29}$$

If the amplifiers are not used in the system, S/N ratio in the shot noise limited condition is given by

$$\frac{S}{N} = \frac{A^k<n_i>}{2B} = \frac{<n_i>}{G^k2B} \tag{6.30}$$

It is seen from Eqs. (6.29) and (6.30) that S/N ratio gets improved by a factor of $G^k/2kn_{sp}$ with the use of k in-line amplifiers. This, of course, is possible only when ASE-ASE beat noise is sufficiently suppressed. Otherwise as the signal passes through the chain of amplifiers, ASE noise power grows and signal power decreases since amplifier saturation power level is constant for a given amplifier. Eventually, with the increase in the number of amplifiers, signal will get completely lost in the noise. In order to avoid this, either the number of amplifiers have to be restricted or repeaters at the appropriate location have to be used.

6.4.3 EVALUATION OF S/N RATIO IN COHERENT SYSTEMS

In coherent communication systems, local oscillator (LO) power is much higher than the input optical signal power. As a consequence, shot noise and beat noise will predominate over the circuit noise and the effect of latter noise can be neglected. This is in contrast to direct detection system, wherein the circuit noise may be comparable with the other sources of noise.

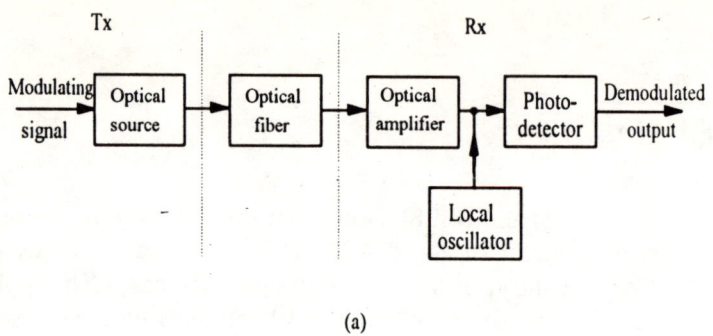

(a)

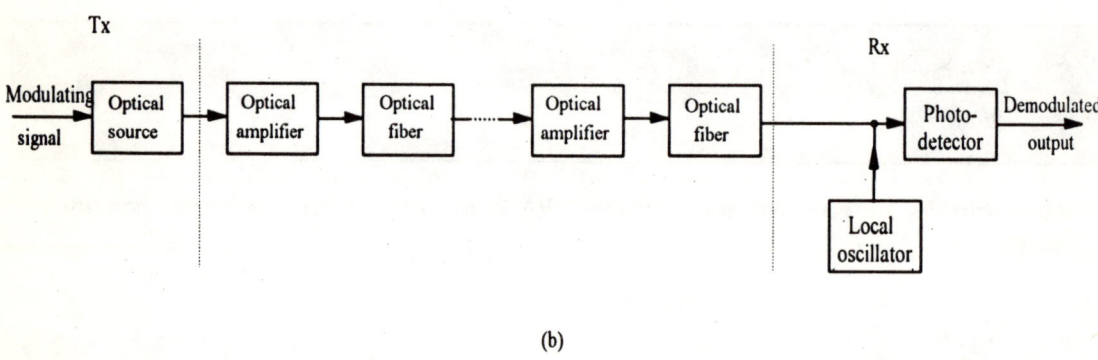

(b)

Fig. 6.8: Heterodyne system with (a) optical preamplifier and (b) cascade amplification using k optical in-line amplifiers

(i) PREAMPLIFIER

A laser preamplifier followed by heterodyne receiver is shown in Fig. 6.8a. The average output photon number and its variance are given by

$$<n_o> = G<n_i> + (G-1)n_{sp}b + <n_l> + 2\sqrt{G<n_i><n_l>}\cos(2\pi f_{IF}t) \qquad (6.31a)$$

and

$$\sigma_h^2 = G<n_i> + (G-1)n_{sp}b + <n_l> + 2<n_i>G(G-1)n_{sp}$$
$$+ 2<n_l>(G-1)n_{sp} + (G-1)^2 n_{sp}^2 b \qquad (6.31b)$$

In the above equation, $<n_l>$ is the average photon number from the LO and f_{IF} the intermediate frequency (IF). The fifth term in Eq. (6.31b) represents the ASE-LO beat noise. When the LO power is sufficiently high, this term becomes predominant over other terms. In that case, IF S/N ratio is given by

$$\frac{S}{N} = \frac{2e^2G<n_i><n_l>}{2e^2[2<n_l>(G-1)n_{sp}]2B} \tag{6.32}$$

For $G \gg 1$, it becomes

$$\frac{S}{N} = \frac{<n_i>}{4n_{sp}B} \tag{6.33}$$

In coherent system, beat noise limited SNR given by Eq. (6.33) can be achieved more easily than in direct detection system. Further, shot noise limited S/N ratio without an optical preamplifier from Eqs. (6.31a) and (6.31b) is given by

$$\left(\frac{S}{N}\right)_{shot} = \frac{<n_i>}{2B} \tag{6.34}$$

It is evident from Eqs. (6.33) and (6.34) that there is no advantage of using optical preamplifier with heterodyne receiver. In fact, it will degrade the performance by a factor of $2n_{sp}$. This, of course, is true only when the LO power is sufficiently high [115, 116].

(ii) POSTAMPLIFIER

Under the condition that LO-ASE beat noise predominates over the other noise, S/N ratio expression will be same as for preamplifier configuration.

(iii) IN-LINE AMPLIFIER

The block diagram of heterodyne system using k in-line amplifiers as repeaters is shown in Fig. 6.8b. In this case, average photon number and its variance at the output are given by

$$<n_o> = <n_i> + A(G-1)n_{sp}kb + <n_l> + 2\sqrt{<n_i><n_l>}\cos(2\pi f_{IF}t) \tag{6.35a}$$

and

$$\sigma_\mathbf{a}^2 = <n_\mathrm{i}> + A(G-1)n_\mathrm{sp}kb + <n_\mathrm{i}> + 2A<n_\mathrm{i}>(G-1)n_\mathrm{sp}k$$

$$+ 2A<n_\mathrm{i}>(G-1)n_\mathrm{sp}k + A^2(G-1)^2n_\mathrm{sp}^2k^2b$$

(6.35b)

Here, parameter A is as defined earlier. The above equations are obtained under the condition $GA=1$ i.e., gain of amplifier is adjusted so that it compensates loss. The fifth term in Eq. (6.35b) i.e., ASE-LO beat-noise will predominate over the other noise including the circuit noise if LO power is sufficiently large. In that case

$$\frac{S}{N} = \frac{<n_\mathrm{i}>}{4\,n_\mathrm{sp}kB}$$

(6.36)

This S/N ratio is same as obtained for the direct detection system (Eq. 6.29). In coherent systems also, transmission distance can be increased considerably by using cascaded amplifiers.

7 SYSTEM COMPARISON AND APPLICATIONS

The purpose of this Chapter is to compare the optical communication systems analyzed and optimized in Chapter five. First of all, we have to decide which criterion should be applied to make a fair system comparison. Number of different criteria exist depending on the required application. If, for example, an optical undersea communication link with a very large repeater spacing is required, then the highly sensitive optical PSK homodyne system represents a well-suited option. Here, the criterion of comparison is the repeater spacing or the receiver sensitivity achievable by the different systems. When a short distance and low-cost optical multichannel communication system is needed, noncoherent FSK heterodyne system is a good option. Now, the criterion of comparison is system cost and inherent problems of realization. Besides receiver sensitivity and special problems of realization, maximum permissible laser linewidth, eye pattern and maximum bit rate are other important criteria usable for fair comparison.

As an overview, this Chapter starts with a comparison under ideal conditions (Section 7.1). Subsequently, the comparison is made again by taking into account real conditions (Section 7.2). Applications and problems of realization are compared in Section 7.3. In the last Section 7.4, physical limits of optical communications are discussed.

7.1 SYSTEM COMPARISON UNDER IDEAL CONDITIONS

In this Section, the term *ideal conditions* means that only the receiver additive Gaussian noise frequently called the circuit noise i.e., shot noise of photodiodes and thermal noise of amplifiers disturb the systems. Moreover, the systems are considered to be ideal which means that laser phase noise and intersymbol interference are neglected. In this case, formulas for the BER calculation are very simple (see Table 7.2) and can, therefore, be evaluated very fast. For this, following physical quantities which have already been derived and discussed in Chapters two and five are required again:

$$\hat{i}_d = M R_0 P_r \tag{7.1}$$

$$\sigma_{d0}^2 = G_{d0} \int_{-\infty}^{+\infty} |H_B(f)|^2 \, df \quad \text{with} \quad G_{d0} = e M^{2+x} I_{dark} + G_{th} \tag{7.2}$$

$$\sigma_{d1}^2 = G_{d1} \int_{-\infty}^{+\infty} |H_B(f)|^2 \, df \quad \text{with} \quad G_{d1} = e M^{2+x} (R_0 P_r + I_{dark}) + G_{th} \tag{7.3}$$

$$\hat{i}_c = 2K_R\sqrt{k(1-k)}R_0\sqrt{P_rP_l} = \hat{i}_{PD} \quad \text{(compare with Eq. 2.63)} \tag{7.4}$$

and

$$\sigma_c^2 = \sigma_{het}^2 = 2G_c\int_{-\infty}^{+\infty}|H_B(f)|^2\mathrm{d}f \quad \text{with} \quad G_c = eK_R(R_0kP_r + I_{dark}) + G_{th} \tag{7.5}$$

Typical numerical values of different parameters subsequently required are given in the following table.

Table 7.1: Parameters for systems compared in Table 7.2

Parameters valid for all systems	
- Transmitted symbol sequence:	$<q_v> = \cdots111\cdots$ (permanent-"1") and $<q_v> = \cdots000\cdots$ (permanent-"0"),
- Symbol or bit rate:	$1/T = 560$ Mbit/s,
- Frequency of transmitter laser:	$f_t = 200$ THz,
- Quantum efficiency of photodiode:	$\eta = 0.83$,
- Responsivity of photodiode:	$R_0 = (e\eta)/(hf) = 1$ A/W,
- Dark current of photodiode:	$I_{dark} = 10^{-11}$ A,
- psd of thermal noise:	$G_{th} = 10^{-23}$ A²/Hz,
- Equivalent baseband filter:	$H_B(f) = 1$ for $|f| \le 1/(2T)$ and $= 0$ otherwise
- Threshold:	$E = E_{opt}$.
Additional direct detection system parameters	
- Avalanche gain of photodiode:	$M = M_{opt}$ (see Eq. 2.16),
- Excess noise exponent:	$x = 0.9$.
Additional coherent system parameters	
- PIN-photodiode:	$M = 1$,
- Local laser power:	$P_l \triangleq -10$ dBm,
- Balanced receiver:	$K_R = 2$,
- Coupling ratio of optical coupler:	$k = 0.5$ (i.e., 3 dB coupler).

A comparison of performance of various optical communication systems is given in the following table.

Table 7.2: Comparison of optical communication systems under ideal conditions

Communication system	Probability of error p	Required optical input power P_r	
		$P_1 = -10$ dBm, $\eta = 0.83$	$P_1 = \to \infty$, $\eta = 1$
Direct detection	$Q\left(\dfrac{\hat{i}_d}{\sigma_{d0} + \sigma_{d1}}\right)$	-39.4 dBm	-40.2 dBm (1287)
ASK heterodyne with envelope detection	$\dfrac{1}{2}\exp\left(-\dfrac{\hat{i}_c^2}{8\sigma_c^2}\right)$	-48.9 dBm	-51.8 dBm (89)
FSK heterodyne with envelope detection (single-filter)	$\dfrac{1}{2}\exp\left(-\dfrac{\hat{i}_c^2}{8\sigma_c^2}\right)$	-48.9 dBm	-51.8 dBm (89)
FSK heterodyne with synchronous detection (single-filter)	$Q\left(\dfrac{\hat{i}_c}{2\sigma_c}\right)$	-49.3 dBm	-52.2 dBm (81)
FSK heterodyne with envelope detection (dual-filter)	$\dfrac{1}{2}\exp\left(-\dfrac{\hat{i}_c^2}{4\sigma_c^2}\right)$	-51.9 dBm	-54.8 dBm (45)
FSK heterodyne with synchronous detection (dual-filter)	$Q\left(\dfrac{\hat{i}_c}{\sqrt{2}\sigma_c}\right)$	-52.3 dBm	-55.2 dBm (40)
ASK homodyne	$Q\left(\dfrac{\hat{i}_c}{\sqrt{2}\sigma_c}\right)$	-52.3 dBm	-55.2 dBm (40)
DPSK heterodyne	$\dfrac{1}{2}\exp\left(-\dfrac{\hat{i}_c^2}{2\sigma_c^2}\right)$	-54.9 dBm	-57.8 dBm (22)
PSK heterodyne with synchronous detection	$Q\left(\dfrac{\hat{i}_c}{\sigma_c}\right)$	-55.3 dBm	-58.2 dBm (20)
PSK homodyne	$Q\left(\dfrac{\sqrt{2}\hat{i}_c}{\sigma_c}\right)$	-58.3 dBm	-61.2 dBm (10)

This table, in addition to the approximate equations for the BER calculation (column 2), also gives the minimum received light power level P_r required to achieve a BER of 10^{-10}. The column 3 is based on real local laser power of -10 dBm and photodiode quantum efficiency of 83 %, whereas the last column presumes a shot noise limited receiver operation with infinite local laser power and 100 % quantum efficiency. It should be noted that in the table above, $(i_c/\sigma_c)^2$ represents the signal-to-noise ratio for coherent system and $(i_d/\sigma_d)^2$ for direct detection system. Here, σ_d is the average of σ_{d0} and σ_{d1} i.e., $\sigma_d = 0.5(\sigma_{d0}+\sigma_{d1})$.

It becomes clear from Table 7.2 that receiver sensitivity decreases from the top of this table to the bottom. Hence, direct detection system exhibits lowest sensitivity and PSK homodyne system is characterized by the highest sensitivity. A comparison of columns 3 and 4, shows that the shot noise limited receiver operation and operation at $P_l \triangleq -10$ dBm in terms of required input power only differ by about 3 dB.

While determining the sensitivity of optical communication system, two alternate measures can be employed: One is the required optical input power P_r and the other is number n_s of photons per bit. Both measures are related as

$$n_s = \frac{P_r T}{hf} \tag{7.6}$$

Here, the products $P_r T$ and hf represent the bit energy and the photon energy respectively. The column 4 of Table 7.2 shows both required number n_s of photons per bit and optical power P_r at the input to the receiver. Direct detection system with intensity modulation usually requires more than 1000 photons per bit, whereas coherent systems only require less than 100 photons per bit. With PSK homodyne system, even 10 photons are sufficient! Based on column 4, Fig. 7.1 indicates the relative sensitivity gains of different optical communication systems over the direct detection system.

The upper block in this figure represents the conventional direct detection system based on the intensity modulation (IM) of light. All the other blocks represent coherent systems. As shown in the figure, coherent systems can be separated in *homodyne and heterodyne systems*. In addition, coherent systems can also be separated on the basis of *coherent (synchronous) detection and noncoherent detection*; for example, envelope detection. In this classification, systems using coherent detection (demodulation) require phase-locking (PLL or OPLL), whereas it is not required in systems with noncoherent demodulation (see Chapter five).

Maximum sensitivity gain of coherent systems under ideal conditions is highlighted in the same figure. As mentioned above, ideal conditions mean no phase noise, no intersymbol interference and infinite local laser power which allows shot noise limited receiver operation. As shown in this figure, improvement in sensitivity increases from the upper blocks to the lower blocks and reaches to a maximum of about (12+3+3+3) dB=21 dB in the coherent PSK homodyne system. This large improvement in sensitivity can be attributed to a large gain of about 12 dB by changing from direct detection with APD detector to heterodyne detection and three steps of 3 dB each by changing the modulation and demodulation schemes.

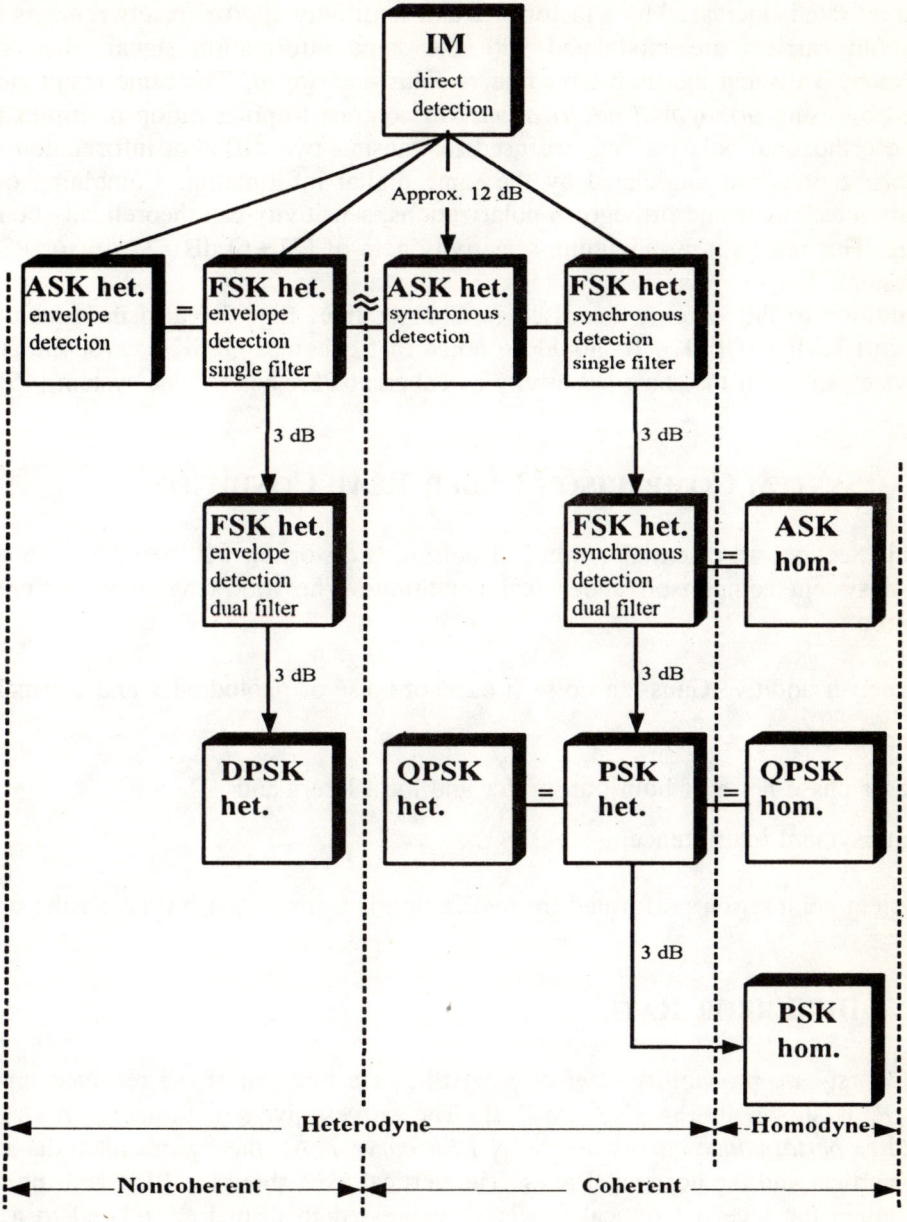

Fig. 7.1: Comparison of sensitivity gain of various optical communication systems under ideal conditions

Theoretically, sensitivity can further be improved by applying *quadrature modulation*. This technique is based on the orthogonality of sine and cosine carrier signals. If both optical carriers are modulated by different digital information signals, then the overall transmission bit rate is increased by a factor of 2 and sensitivity approximately remains the same. When both carriers are modulated with the same information signal, then sensitivity improves by 3 dB and the usable bit rate remains unchanged. The same result can also be obtained by using *orthogonal polarizations*. In contrast to polarization multiplexing where both the orthogonal polarizations are used to transmit two different information channels, both polarizations are modulated by the same digital information. Combining quadrature modulation technique and orthogonal polarizations, sensitivity can theoretically be improved by 6 dB. This results in a maximum sensitivity gain of (21+6) dB=27 dB for PSK homodyne system.

In addition to the systems discussed in Chapter five, Fig. 7.1 also includes quadrature phase shift keying (QPSK). It should be noted that coherent QPSK heterodyne and homodyne systems exhibit the same sensitivity as coherent PSK heterodyne system [45].

7.2   SYSTEM COMPARISON UNDER REAL CONDITIONS

In this Section, all essential sources of noise and distortion will be taken into account to obtain a system comparison under real conditions. The following imperfections will be considered:

- receiver additive Gaussian noise (i.e., shot noise of photodiodes and thermal noise of amplifiers),

- laser phase noise of both transmitter and local lasers and

- intersymbol interference.

The system comparison performed in this Section is again based on the results obtained in Chapter five.

7.2.1   BIT ERROR RATE

The worst-case probability of error p_w (BER) as a function of the received input optical power P_r is shown in Fig. 7.2a to 7.2f. The curves given in these figures are usually referred as *performance curves* or briefly *BER curves*. All the figures have the same scale on the vertical and the horizontal axes. The vertical axes show the BER and the horizontal axes contain the received optical power P_r expressed in dBm i.e., related to a reference power of 1 mW. Therefore, the optical power increases from the left side to right side.

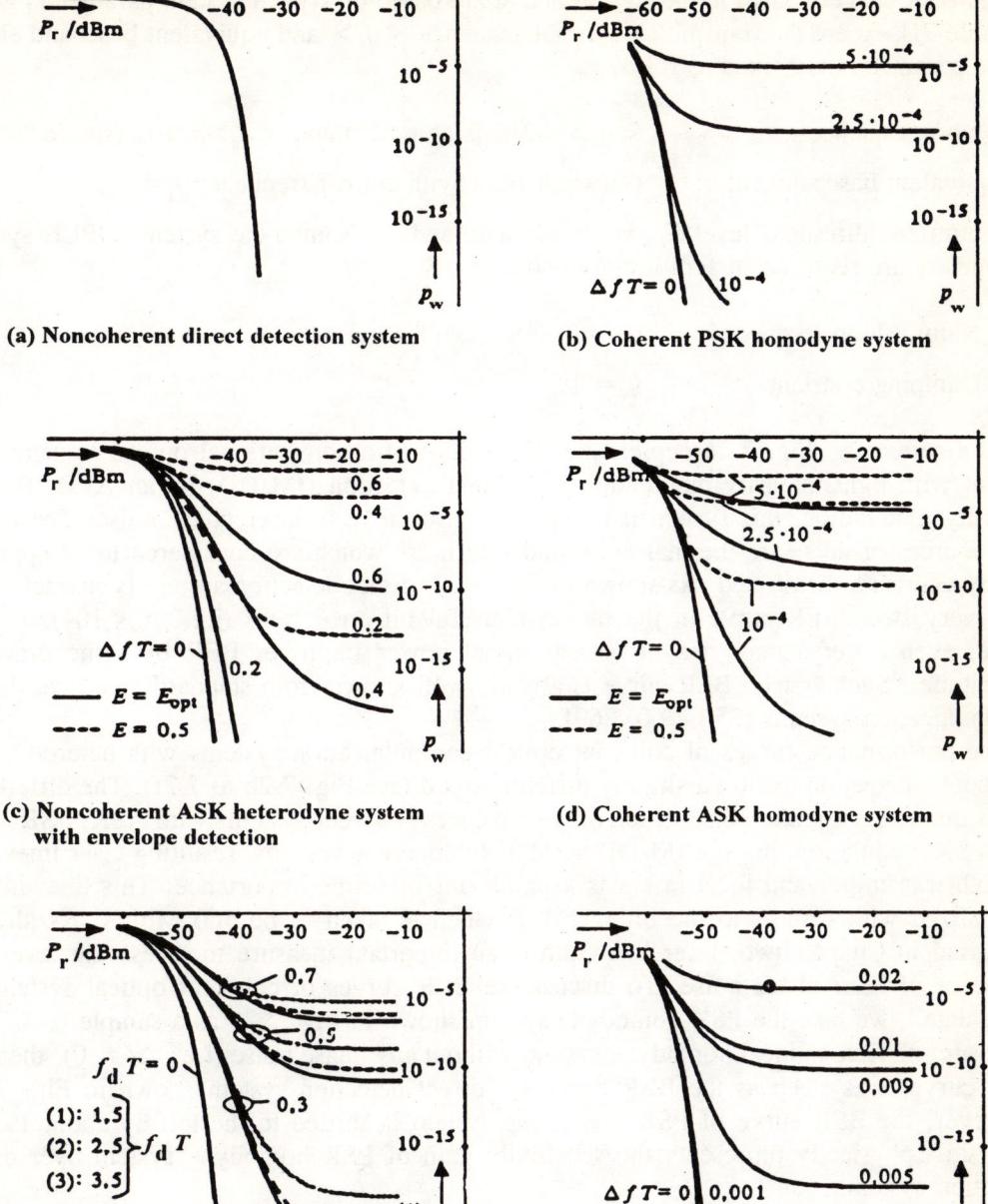

(a) Noncoherent direct detection system

(b) Coherent PSK homodyne system

(c) Noncoherent ASK heterodyne system with envelope detection

(d) Coherent ASK homodyne system

(e) Noncoherent FSK heterodyne system with envelope detection (single filter)

(f) Noncoherent DPSK heterodyne system

Fig. 7.2: BER curves for various optical communication systems

The BER curves shown in the figures above are based on typical system parameters given in Table 7.1 except the transmitted symbol sequence $<q_v>$ and equivalent baseband filter. These parameters are fixed as follows:

Symbol sequences: $<q_v>=010$ (single "1") and $<q_v>=101$ (single "0"),

Equivalent baseband filter: Gaussian filter with cut-off frequency $f_g=f_{g,opt}$.

An optimized threshold level E_{opt} is always assumed. In homodyne system, OPLL system parameters are required in addition. We consider:

- Natural loop frequency: $\omega_n = 0.001 \cdot 2 \cdot \pi \cdot (1/T)$ and

- Damping constant: $\xi = 1$.

As a reference, Fig. 7.2a depicts the BER curve of a conventional optical transmission system with light intensity modulation and direct detection (IM/DD) by an APD. Due to intensity modulation, this system is completely insensitive to laser phase noise. The dominant sources of noise are thermal noise and shot noise which are considered to be approximately Gaussian distributed. As shown in Fig. 7.2a, direct detection system is characterized by a very steep BER curve in the range of useful bit error rates (i.e., $p_w < 10^{-6}$). In this range, even a very small rise in optical signal power improves BER by some order of magnitudes. Such a steep BER curve is already well-known from standard electrical digital communication systems [256, 260, 269].

The performance curves of coherent optical communication systems with heterodyne or homodyne detection exhibit a slightly different trend (see Fig. 7.2b to 7.2f). The difference arises due to laser phase noise which only influences coherent systems with ASK, FSK, PSK or DPSK modulation, but not IM/DD system. In coherent systems, resulting laser linewidth of both transmitter and local lasers is a parameter of prime importance. This linewidth is normalized with respect to the bit rate $1/T$ which is taken to be 560 Mbit/s. As already discussed in Chapter two, laser linewidth is an important measure to assess the level and influence of laser phase noise. To discuss the BER curves of coherent optical systems in more detail, we take the PSK homodyne system shown in Fig. 7.2b as a sample.

Assuming first an ideal homodyne system without any phase noise (i.e., $\Delta f = 0$), then the BER curve is as steep as the BER curve for direct detection system shown in Fig. 7.2a. However, the BER curve of PSK homodyne system is shifted to the left by about 19 dB. This shift physically represents the sensitivity gain of PSK homodyne system over direct detection system.

Let us now turn our attention to a real homodyne system which is influenced by laser phase noise in addition. In this case, BER curves first remain very steep when the optical input power P_r is low. In this range, coherent system exhibits the same characteristic features as direct detection system or any other standard electrical digital communication system. The reason for this is that the influence of laser phase noise is negligibly small as compared to additive Gaussian noise of the receiver provided received optical power P_r is low enough. With the increase in P_r, influence of the phase noise increases, while the influence of the receiver additive noise decreases. As a result, a characteristic bend in the BER curve can be observed. When P_r is increased further, laser phase noise finally dominates the other noise. Influence of additive receiver noise can, in principle, be reduced by increasing P_r, whereas the influence of laser phase noise is independent of P_r. Therefore, the BER reaches a characteristic point of saturation termed as *error rate floor* or the *BER floor*.

The fundamental effect of the error rate floor is that certain probabilities of error (for example, $p_w = 10^{-10}$) cannot be achieved when the laser phase noise is too strong. In other words, every laser is not suitable to provide a BER of 10^{-10}. In homodyne systems, requirements for the maximum permissible laser linewidth are strong, in particular. This becomes clear by comparing the numerical values of the normalized linewidth $\Delta f T$ given in the Figs. 7.2b to 7.2f.

The BER curves of ASK heterodyne system with noncoherent detection and ASK homodyne system are shown in Figs. 7.2c and 7.2d. These curves are similar to the BER curve of PSK homodyne system in Fig. 7.2b i.e., first a steep course, then a bend and finally a BER floor. When normalized laser linewidths are compared in these figures, it becomes clear that a noncoherent heterodyne system is much less sensitive to phase noise than a homodyne system. Even lasers with a relatively poor quality in spectral emission are able to provide a BER of 10^{-10}. Further, above figures also illustrate the strong influence of threshold level. Even a small deviation in threshold level from the optimum threshold E_{opt} deteriorates the BER considerably.

The performance curves of noncoherent FSK heterodyne system with a single-filter detection is shown in Fig. 7.2e. For this system, normalized frequency deviation $f_d T$ is another parameter of importance. As explained in Section 5.3.2, this parameter is optimizable for a minimum BER. Finally, Fig. 7.2f shows the BER curves of DPSK heterodyne system with autocorrelation detection.

In conclusion, the following characteristic features of coherent optical communication systems are highlighted in Figs. 7.2a to 7.2f.

In *quality*, all the BER curves exhibit the same characteristic features irrespective of modulation scheme applied to the system. All BER curves show first a steep course, then a bend and finally a BER floor at high input optical power levels.

In *quantity*, the BER curves are very different. For example, a BER of 10^{-10} is obtained at rather different received optical power levels P_r and laser linewidths Δf. The quantitative shape of the BER curve is essentially determined by various other system parameters. The OPLL parameters (for example, the loop frequency) substantially influence the BER of homodyne system, whereas the BER of heterodyne system is fundamentally influenced by the features of IF filter (for example, the IF filter bandwidth) as already discussed in Chapter five.

7.2.2 LASER LINEWIDTH REQUIREMENTS

In the design and realization of coherent optical communication systems, two questions of primary importance are: How much maximum laser linewidth can be allowed to achieve a BER of 10^{-10}? What minimum optical power at the input to the receiver is required? The answers to these questions is available in Fig. 7.3.

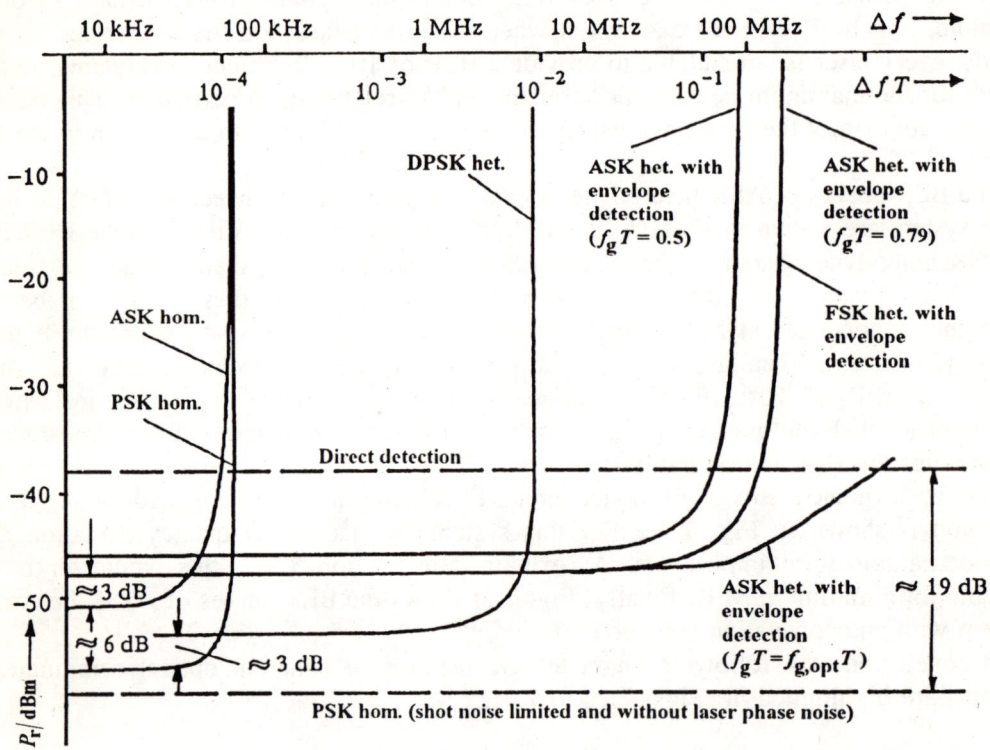

Fig. 7.3: Maximum permissible resulting laser linewidth Δf at a bit rate of $1/T = 560$ Mbit/s and bit error rate 10^{-10}

The horizontal axis in this figure shows the resulting laser linewidth $\Delta f = \Delta f_t + \Delta f_l$ of both transmitter and local lasers. Besides the normalized linewidth $\Delta f T$ (normalized to bit rate), this axis also shows the actual laser linewidth Δf for a bit rate of 560 Mbit/s. The required optical input power P_r expressed in dBm is given on the vertical axis. All other system parameters required to compute the results shown in this figure are taken from the Table 7.1.

As seen from the figure above, laser linewidth increases from left to right side, while the required optical power at the input to the receiver increases from bottom to top. Thus, optical communication system is more sensitive when it is placed at the bottom of this figure. In addition, influence of phase noise is less when the system is located more on the right. As a reference, direct detection system has been picked up in Fig. 7.3. This system is insensitive to laser phase noise (i.e., independent of laser linewidth Δf). Therefore, it appears as a horizontal straight line. In order to obtain best system performance, direct detection system has been optimized with respect to threshold level, avalanche gain of APD and bandwidth of low-pass filter (see Section 5.4). As a second reference, the same figure includes an optical PSK homodyne system under ideal conditions i.e., shot noise limited receiver operation without any laser phase noise. Therefore, it represents the theoretical optimum. Since phase noise is neglected in this ideal PSK homodyne system, it also appears as a horizontal straight line (i.e., independent of laser linewidth Δf).

On the basis of Fig. 7.3, coherent optical communication systems can be divided into three groups as far as laser linewidth requirements are concerned.

The *first group* contains the coherent optical homodyne systems which exhibit a high sensitivity (i.e., the lines are located far down). However, the requirements for maximum permissible laser linewidth are very strong (i.e., the lines are located on the extreme left side). Heterodyne systems with coherent (i.e., synchronous) detection not shown in the figure are also belong to this group. These systems have somewhat lesser sensitivity.

The *second group* includes only the DPSK heterodyne system based on noncoherent autocorrelation detection (see Section 5.3.3). This system represents a good compromise between the sensitivity gain and laser linewidth requirements. A DPSK heterodyne system can tolerate linewidths of about 1 % of the bit rate.

The *third group* contains all noncoherent ASK and FSK heterodyne systems. For these systems, laser linewidths of 20% to 30% of the bit rate $1/T$ can be allowed provided normalized bandwidth of the IF filter is $BT = 2f_gT = 1.58$. As discussed in Chapter five, this bandwidth is optimum only when there is a Gaussian IF filter with cut-off frequency f_g and no phase noise. If the bandwidth of IF filter is optimized by taking into consideration the influence of laser phase noise also, then lasers with linewidths in the range of bit rate $1/T$ and even more can be employed ($\Delta fT > 1$). As an example, Fig. 7.3 shows an optimized ASK heterodyne system with envelope detection. No error rate floor can be observed. As discussed in Section 5.3.1, optimum bandwidth of IF filter strongly depends on the received optical power P_r when laser phase noise is taken into account. It becomes clear again that heterodyne systems with noncoherent detection exhibit the lowest receiver sensitivity.

Comparing the required incoming power levels P_r in absence of laser phase noise (i.e., $\Delta fT = 0$), a difference of about 1 dB to 2 dB as compared to the optical power levels given in Table 7.2 can be recognized. The reason for this difference is that the Table 7.2 is based on the particular symbol sequences permanent "0" and permanent "1", whereas Fig. 7.3 is based on single "0" and single "1". Thus, Fig. 7.3 also includes the influence of intersymbol interference. This influence has been excluded in Table.7.2.

It may be mentioned that extent of realization and cost drastically increase for the systems located on the left side of this figure to those on the right side.

7.2.3   SENSITIVITY GAIN OF COHERENT SYSTEMS

The sensitivity gain of coherent optical communication systems has been shown in Fig. 7.1 under ideal conditions ($\Delta f = 0$). When phase noise disturbs the system, a similar, but modified graphic can be drawn. However, each change in laser linewidth would yield another constellation of the blocks. In this Section, procedure for obtaining the modified graphic will be discussed in brief.

In order to design a new graphic indicating the sensitivity gain of coherent systems corrupted by phase noise, results of Fig. 7.3 must be taken into account. With a fixed laser linewidth Δf, corresponding sensitivity gains can be taken directly from this figure by drawing a vertical straight line at the given linewidth Δf. The difference in received optical power levels P_r in dBm can either be positive or negative for the systems included in this figure. Hence, this difference represents the gain or the loss in receiver sensitivity. Depending on the laser linewidth, location of different blocks in Fig. 7.1 will change. Thereby, systems with a high sensitivity gain may change their position from the bottom to the top. For example, if laser linewidth is very broad, then PSK homodyne system may even exhibit a lower sensitivity than a conventional direct detection system with intensity modulation. In addition, when the laser linewidth exceeds a certain limit, underlying BER of $p = 10^{-10}$ is no longer achievable.

7.2.4   EYE PATTERN

As already mentioned in the previous Chapters, eye pattern measurement represents a very powerful and low-cost tool to assess system performance and to highlight characteristic features of electrical or optical digital communication systems. The eye pattern measurement technique is based on the detected signal at the input to the sample and hold and decision circuit. When this signal is displayed bitwise on the screen of oscilloscope (see Section 2.5), a pattern similar to an eye is created. As explained in Chapters two and five, quality of detected signal is of fundamental importance for the performance of overall transmission system, primarily for the BER.

Let us consider again the coherent optical communication systems discussed in Chapter five. These systems can in principle be classified into two groups:

- systems with a horizontal symmetrical eye pattern and
- systems with a horizontal asymmetrical eye pattern.

With the exception of both PSK homodyne and heterodyne systems (see Fig. 7.1), all other systems yield an asymmetrical eye pattern.

As an example, Fig. 7.4 displays the symmetrical and the asymmetrical eye pattern of coherent PSK and ASK systems respectively. Both the noisy eye pattern shown at the right side of Fig. 7.4 are simulated by means of fast computer program. In the simulation, receiver additive Gaussian noise (i.e., shot noise and thermal noise) have been neglected. This means that these eye pattern are obtained when the detected signal is disturbed only by laser phase noise. This figure also illustrates the corresponding noiseless eye pattern which are similar in shape, but different in amplitude level. All four eye pattern shown in Fig. 7.4 are based on a Gaussian filter of bandwidth $B = 2f_g = 1/T$ (see Eq. 2.77).

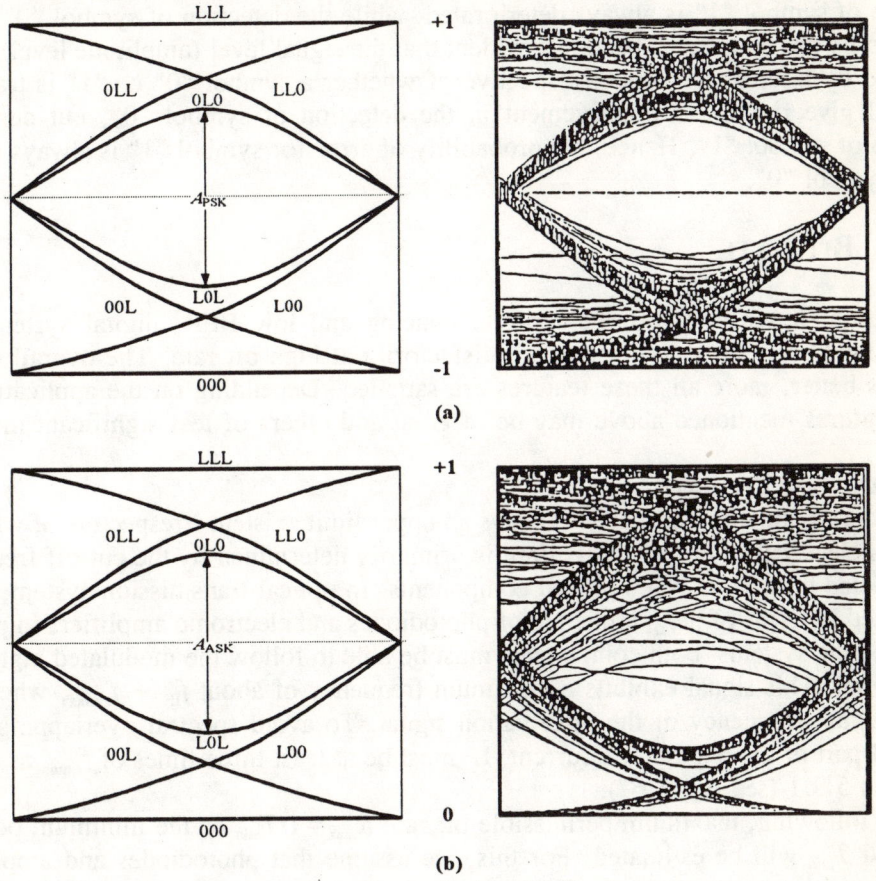

Fig 7.4: Comparison of eye pattern for coherent optical communication systems: (a) PSK system and (b) ASK system

Let us first consider the *eye pattern of PSK system* shown in Fig. 7.4a. This eye pattern is characterized by normalized amplitude levels +1 (symbol "1") and -1 (symbol "0"). Thus, PSK system is a bipolar digital transmission system and both the transmitted symbols "0" and "1" are likewise disturbed by phase noise (see Section 5.1.2). As a result, we observe a symmetrical eye pattern with respect to the horizontal axis at amplitude level 0. The optimum threshold level is always located at the centre line of the eye pattern i.e., $E_{opt} = 0$. In contrast to that, *eye pattern of ASK system* shown Fig. 7.4b is characterized by normalized amplitude levels +1 (symbol "1") and 0 (symbol "0"). Hence, ASK system is an unipolar digital transmission system (see Sections 5.1.1 and 5.3.1). It becomes clear from Fig. 7.4b that a transmitted symbol "1" is always disturbed by phase noise much more seriously than a transmitted symbol "0". This results in an asymmetrical eye pattern with an optimum threshold at $E_{opt} < 0.5$ which is always below the optimum threshold given in absence of laser phase noise (i.e., $E_{opt} = 0.5$). Further, Fig. 7.4b clearly indicates that the detection of symbol "1" is always deteriorated, while the detection of symbol "0" is always improved by phase noise! It becomes evident that the signal level (amplitude level) is always decreased by laser phase noise irrespective of whether a symbol "0" or "1" is transmitted. This will give rise to an improvement in the detection of symbol "0", but degrades the detection of symbol "1". Hence, the probability of error for symbol "1" is always more than that of symbol "0".

7.2.5 BIT RATE

Besides high sensitivity, large repeater spacing and low BER, digital system of high quality should also be able to operate satisfactorily at high bit rate. The overall quality of system is better, more all these features are satisfied. Depending on the application, some of the features mentioned above may be of great and others of less significant importance.

(i) MAXIMUM BIT RATE

Considering the bit rate, there is always an upper limit existent irrespective of whether the system is electrical or optical. This limit is primarily determined by the cut-off frequency or the restricted bandwidth of employed components. In optical transmission system, this limit is specified in terms of the parameters of photodiodes and electronic amplifiers in particular. In heterodyne systems, both components must be able to follow the modulated high frequency IF signal. This signal exhibits a maximum frequency of about $f_{IF} + f_{s,max}$, where $f_{s,max}$ is the maximum frequency of the information signal. To avoid spectral overlapping with the baseband part of the photodiode current, IF must be at least three times of $f_{s,max}$ as explained in Section 5.3.1 (see Eq. 5.67).

In the following, maximum permissible bit rate $R_{max} = 1/T_{min}$ or the minimum permissible bit period T_{min} will be estimated. For this, we assume that photodiodes and amplifiers are characterized by a common cut-off frequency f_{pa}. We also presume that information signal is band-limited to $0 < f_s < f_{s,max} = 1/(2T_{min})$. This means that only the sinusoidal wave with period $2T_{min}$ is transmitted when the periodic symbol sequence $\cdots 01010101 \cdots$ is applied to the system. It should be noted that this periodic symbol pattern represents the worst-case pattern with respect to maximum transmission speed and, therefore, maximum signal frequency $f_{s,max}$. Obviously, the sinusoidal signal still enables an errorless decision irrespective of whether symbol "0" or symbol "1" has been transmitted.

Summarizing the results of the above discussion, maximum permissible bit rate can be determined as

$$R_{max} = \frac{1}{T_{min}} = \frac{1}{2} f_{pa} \quad \text{for heterodyne systems}$$

$$\text{(7.7)}$$

$$R_{max} = \frac{1}{T_{min}} = 2 f_{pa} \quad \text{for homodyne systems}$$

Thus, in an optical homodyne system ($f_{IF}=0$), maximum allowable bit rate is at least four times the maximum bit rate of an optical heterodyne system ($f_{IF} \neq 0$) under ideal conditions. If real conditions are considered such as non-ideal filter with gradual instead of a sharp cut-off frequency response, then difference in bit rate is even somewhat higher.

The coherent optical communication systems are able to transmit very high bit rates. In the development of coherent systems, transmission speed has progressed from some Mbit/s to more than 10 Gbit/s. In combination with coherent multichannel communication (CMC) and optical time division multiplexing (OTDM), coherent optical communication systems even enable ultrahigh-speed optical transmission. In such system, each optical carrier (up to hundred and more) is modulated by a high bit rate signal of up to 100 Gbit/s by means of OTDM (see Section 7.3).

Example 7.1

For ten optical carriers (channels) at different optical wavelengths and a bit rate of 100 Gbit/s at each optical carrier obtained by OTDM, overall bit rate in the fiber link is 1000 Gbit/s = 1 Tbit/s.

It becomes clear by this simple example that coherent optical communication provides the basis for highly sophisticated Tbit/s transmission systems which offer many new interesting applications; for example, three-dimensional (holographic) multipoint-to-multipoint video transmission. Three-dimensional high-resolution video transmission is very promising for remote-controlled medical help or video conference with "telepresenting" speakers .

(ii) INTERRELATION BETWEEN BIT RATE AND LASER LINEWIDTH

For a fixed laser linewidth Δf, the allowable bit rate $R = 1/T$ is of significant importance in highly sophisticated transmission systems and networks; for example, data highways. As discussed in Chapter five, a high bit rate transmission system is relatively more tolerant to phase noise of the laser sources than a low bit rate system. Performance degradation due to laser phase noise is lower, better the following relationship is fulfilled:

$$\Delta f T < 1 \quad \text{or} \quad \Delta f < R \qquad \text{(7.8)}$$

Considering, for example, noncoherent DPSK heterodyne system, this feature of coherent optical communication systems becomes clear in particular. In this system, variance of phase noise (in the strict sense phase noise difference) increases proportional to the bit duration T

and laser linewidth Δf. As the bit rate increases (implying T decreases), variance of phase noise difference decreases and hence the influence of laser phase noise. Similar results can be observed by considering ASK and PSK homodyne or ASK, PSK and FSK heterodyne systems.

As explained in Section 5.3.1, each IF filter give rise to an undesired coupling of phase and amplitude. Hence, phase noise results in additional amplitude noise. As the IF filter bandwidth decreases, efficiency of the undesired phase noise-to-amplitude conversion increases. When the bit rate is high, distortion by phase-to-amplitude conversion is low since high bit rates always require large bandwidths. On the other hand, lower the bit rate, stronger is the influence of the undesired interaction of phase noise and amplitude. If bit rate reaches a certain lower limit, then system performance again improves by further decreasing the bit rate. This surprising result can be explained as follows: When the bit rate is very low and IF filter bandwidth is small, filtering of the phase noise becomes dominant (which yields an improvement in system performance) as compared to the undesired phase-to-amplitude conversion (which deteriorates the system performance). This fact, however, can only be observed when intersymbol interference is neglected. At a fixed bit rate and laser linewidth, optimum IF filter bandwidth exists for a minimum BER (see Fig. 5.16). To make it more realistic, effect of intersymbol interference must also be taken into account.

Performance of homodyne systems can also be improved by providing $\Delta f T \ll 1$. The reason for this is the operation principle of OPLL. As explained in Section 5.1.3, OPLL cannot distinguish between phase noise and phase information. Therefore, noise and information must be separated a priori in order to guarantee OPLL operation i.e., permanent phase tracking with low rest-phase error. Best separation is obtained when phase noise and information are characterized by non-overlapping frequency spectra. If the bit rate is low, then these spectra are located in the same frequency range and a spectral separation becomes nearly impossible. It should be remembered that the degrading effect of laser phase noise is more in the low frequency range (see Section 3.3.1). When bit rate is high enough, both spectra can be easily separated.

7.2.6 SENSITIVITY GAIN BY OPTICAL AMPLIFICATION

Optical amplifiers are used as booster amplifier in the transmitter, in-line amplifiers in the fiber link or as preamplifier in the receiver (see Chapter six). Optical preamplifier when employed to improve the sensitivity of optical receiver may provide performance close to shot noise limit. Hence, optical preamplifier is well-suited in combination with direct detection receiver which usually exhibits a difference of about 20 dB and more from the shot noise limit. A comparison between preamplified optical direct detection receiver and heterodyne receiver without an optical amplifier shows that both can have equivalent performance if the optical filters can be made with bandwidths comparable to electronic IF filters.

As heterodyne receiver already operates close to shot noise limit, sensitivity gain due to optical preamplification is negligible. In the strict sense, sensitivity even deteriorates by about 3 dB if the receiver was shot noise limited without preamplifier. The reason for this is given by the inherent noise of each optical amplifier due to spontaneous emission (see Chapter six).

The sensitivity gain of direct detection and heterodyne receivers as a function of amplifier gain G and local laser power P_1 is compared in Fig. 7.5. It becomes clear from this figure that sensitivity always reaches a saturation irrespective of whether direct or coherent detection is used. In the direct detection receiver, maximum sensitivity is achieved at the point of saturation. The coherent receiver may even exhibit a sensitivity loss (see curve for $P_1 = 10$ dBm in Fig 7.5b). As shown in this figure, performance of a coherent receiver can only be improved when local laser power is low, which means that a poor coherent receiver has been applied.

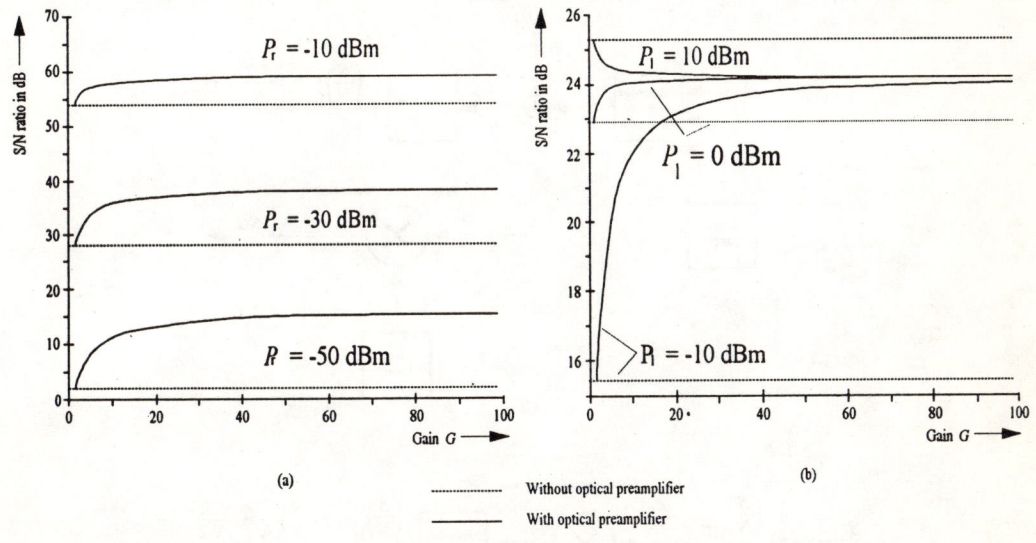

(a)

.... Without optical preamplifier

———— With optical preamplifier

(b)

Fig. 7.5: Sensitivity gain of (a) direct detection and (b) coherent detection receivers with optical preamplifier

In addition, Fig. 7.5 also illustrates that a preamplified direct detection receiver attains a sensitivity close to the high sensitivity of coherent receiver. For this reason, direct detection system with optical amplifier represent a well-suited, technical and economical alternative to high sensitivity coherent receivers.

7.2.7 MICROWAVE, OPTICAL HETERODYNING AND HOMODYNING

In this Section, we briefly discuss the difference between coherent microwave and coherent optical detection. When a microwave signal disturbed by additive Gaussian noise is detected by means of a heterodyne or homodyne receiver, sensitivity of heterodyne receiver normally equals the sensitivity of homodyne receiver. In comparison, a difference of 3 dB in sensitivity will exist when fiber-based optical heterodyne or homodyne systems are considered. Microwave heterodyne and homodyne systems gives nearly the same performance as optical heterodyne system, but their performance is 3 dB poorer as compared to optical homodyne system. This interesting result can be explained as follows: Let us consider microwave homodyne and optical homodyne systems. In microwave system, received signal is normally disturbed by additive noise while in optical system it is not so. Hence, the key difference between coherent microwave and coherent optical detection is the location of source of noise. As shown in Fig. 7.6, noise in optical receiver is primarily generated in the mixer i.e., within the photodiode. On the other hand, noise in microwave receiver is already existing in front of the mixer.

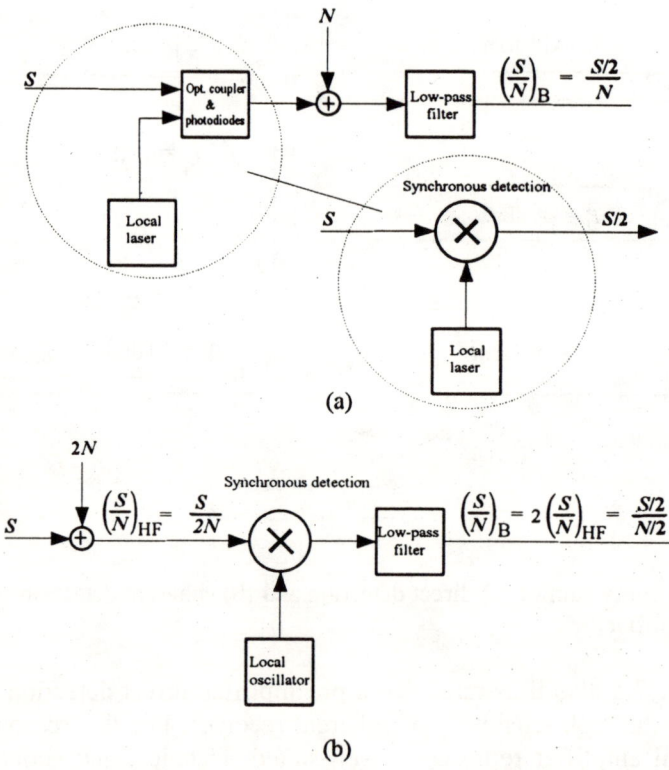

Fig. 7.6: Comparison of coherent (a) optical and (b) microwave detection receivers

It should be noted that in Fig. 7.6, same noise power spectral density in the microwave and optical receivers has been considered. In the microwave receiver, noise is in the HF range, while it is in the baseband frequency range in case of optical receiver. As a result, noise power in the former case will be twice of noise power in the latter case. This difference in microwave and optical systems is at least partly reduced when optical space communication is considered in Chapter eight. Due to background noise which is a fundamental source of noise in optical free-space applications, we now have one source of noise at the input to the optical receiver (background noise) and the other in the mixer (shot noise and amplifier noise).

7.3 APPLICATIONS AND SPECIAL PROBLEMS OF REALIZATION

We have so far compared optical systems by considering different criteria such as BER, receiver sensitivity, bit rate and laser linewidth requirements. In order to answer the question "which is really the best system", these features alone are not sufficient. The answer to this question primarily depends on the application. Each application is usually characterized by its own optimum solution (system), which represent a trade-off instead of a real optimum.

Besides the criteria mentioned above, the extent of realization as well as cost are another very important features for comparing optical systems. The purpose of this Section is to perform a system comparison on the basis of problems of realization and typical applications. For this, we shall focus our interest on principle problems since special problems of realization primarily depend on recent short-lived developments.

As far as extent of realization and cost are concerned, substantial differences can be observed between homodyne, coherent and noncoherent heterodyne as well as direct detection systems.

As compared to the realization of coherent optical communication systems, realization of *direct detection system* is very cost effective since local laser, optical coupler, PLL or OPLL, AFC and polarization control are not required. This represents the most important advantage of employing optical transmission system with intensity modulation and direct detection (IM/DD). Because of this, direct detection systems are used for various different applications. In particular, optical direct detection systems are most advantageous for single-wavelength data transmission over short and medium distances; for example, data transmission in local area networks (LANs), plants, cars, aircrafts, and so on. IM/DD systems are commercially available for data rates up to 10 Gbit/s and more. In combination with optical in-line amplifiers, booster amplifier or preamplifier, direct detection systems offer a performance which is comparable with coherent optical communication systems.

In *optical homodyne systems*, extent of realization and cost are maximum. The main reason for this is the OPLL which is required to match optical phase and frequency of both received and local laser light waves. In contrast to direct detection system, homodyne systems are, in addition, much more sensitive to change in the environmental conditions such as change in temperature or mechanical and acoustic vibrations. Moreover, optical homodyning requires high-quality single-mode laser sources with extremely narrow linewidths of about 0.01% of the bit rate. Therefore, solid-state lasers (for example, Nd:YAG laser) are frequently used to realize optical homodyne systems.

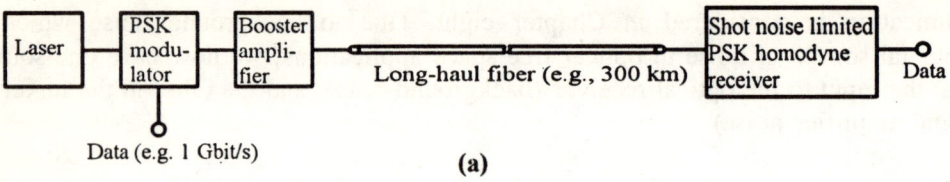

(a)

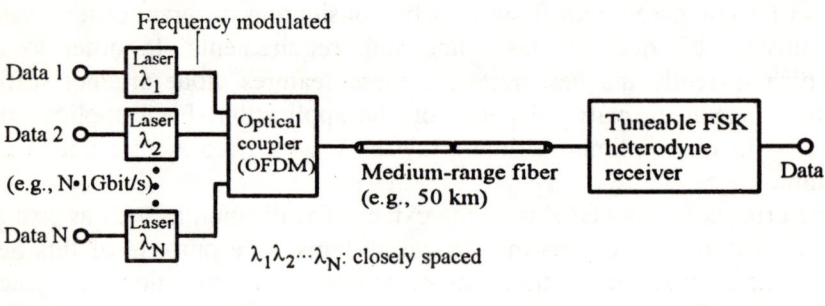

(b)

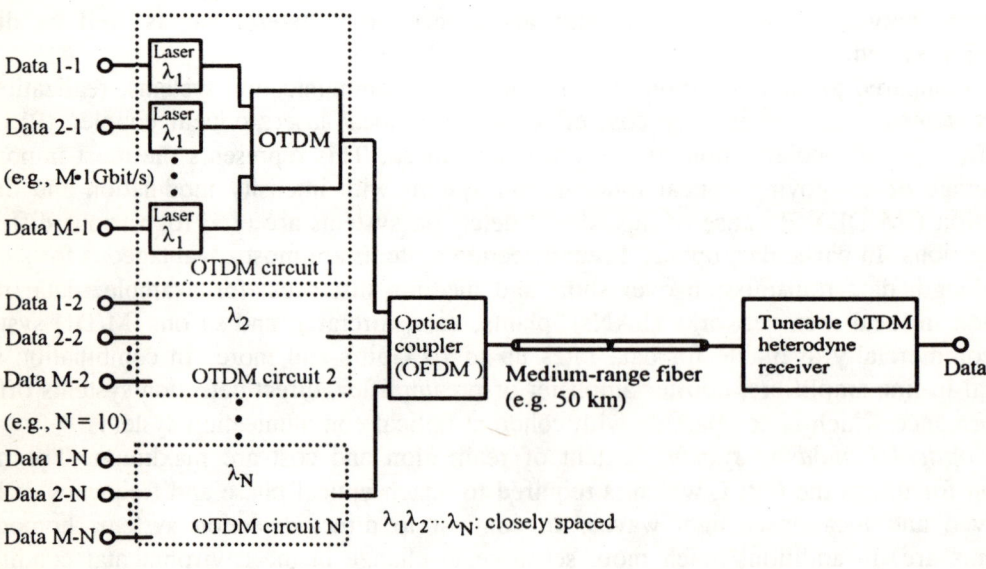

(c)

Fig. 7.7: Typical applications of coherent optical communication systems: (a) repeaterless, highly sensitive, long-haul point-to-point transmission using PSK, (b) coherent multichannel communication (CMC) with optical frequency division multiplexing (OFDM) and (c) multigiga-bit data transmission (e.g., 1 Tbit/s) with OFDM and optical time division multiplexing (OTDM)

Optical homodyne systems, because of their large extent of realization and high cost are useful for a very limited number of applications only. It should be remembered that these limitations have been overcome to some extent by using a synchbit controlled OPLL instead of a Costas loop or an OPLL employing a pilot carrier (see Section 5.1.3). Since PSK homodyne systems exhibit best receiver sensitivity, these systems are most suitable for *repeaterless long-haul point-to-point data transmission*; for example, transisland or trans-oceanic transmission links (Fig. 7.7a). Here, repeaterless means that *neither* an electric repeater *nor* an optical in-line amplifier is used and transmission channel is simply a passive fiber only i.e., without any active components.

The long-haul point-to-point transmissions links can also be realized by using direct detection systems employing a chain of optical in-line amplifiers such as erbium-doped fiber amplifiers (see Chapter six). Since each optical amplifier requires the supply of electric power to maintain amplification mechanism, this is actually an active instead of a passive fiber link.

In comparison to PSK homodyning, ASK homodyning is without any practical use since receiver sensitivity is less as compared to PSK, but extent of realization remains the same. When an *optical heterodyne system* instead of an optical homodyne system is used, extent of realization is reduced. Heterodyne systems with noncoherent detection represent a good compromise of receiver sensitivity and extent of realization. Neither an optical nor an electrical PLL is required. In order to match received and local laser frequencies and obtain a specified IF, standard AFC is sufficient.

A comparison of modulation schemes suitable for noncoherent heterodyning shows that FSK and DPSK offer best performance. FSK requires somewhat less critical components, whereas DPSK exhibits better sensitivity. It should be noted that DPSK receiver requires an electronic multiplication component for high frequency signals (IF signal $i_{IF}(t)$ and delayed IF signal $i_{IF}(t-T)$) and a signal shaping circuit when transmitter is modulated directly by modulating the injection current [186]. In the latter case, DPSK requires no external optical modulator as in PSK. The ASK system is less sensitive as compared to FSK system, but both the systems require same extent of realization. As a result, ASK system does not find much practical interest.

FSK and DPSK heterodyne systems are well-suited for applications which require coherent multichannel communication (CMC); for example, TV and HDTV distribution (Fig. 7.7b). Since coherent multichannel applications usually operate in the range of medium distances, receiver sensitivity is not critical parameter. Instead, receiver selectivity is most important to achieve best transmission capacity. Here, best transmission capacity means that the number of closely spaced optical channels and bit rate of each channel is as high as possible. Up to hundred and even more different optical carriers (channels) can simultaneously be transmitted in one single fiber by applying optical frequency division multiplexing (OFDM). In the coherent receiver, different optical channels can properly be selected by means of a tuneable local laser in conjunction with a high-quality, sharp IF filter at a fixed and specified IF. As the tuning range of a laser diode is limited, the channel spacing must be very close when number of optical channels is large.

Table 7.3: Comparison of optical communication systems in terms of bit rate, repeater spacing, extent of realization and typical applications

	Homodyne system	Coherent heterodyne system	Noncoherent heterodyne system	Direct system
Typical bit rate R	high e.g., 1 Gbit/s to 10 Gbit/s	relatively low e.g., 0.5 Gbit/s to 1 Gbit/s	relatively low e.g., 0.5 Gbit/s to 1 Gbit/s	high e.g., 1 Gbit/s to 10 Gbit/s
Typical repeater spacing at $R = 560$ Mbit/s and $\alpha = 0.2$ dB/km.	ASK: 212 km PSK: 242 km	ASK: 197 km FSK: 212 km PSK: 227 km	ASK: 195 km FSK: 210 km DPSK: 225 km	147 km
Extent of realization	very large	large	relatively small	small
Phase and frequency tracking	OPLL required	PLL required	no PLL, but AFC required	no PLL and AFC required
Typical applications	Long-range transmission Transoceanic transmission		Multichannel FDM BISDN HDTV distribution Multimedia services	Short range transmission e.g., LANs, remote control in plants, cars, aircrafts

Multichannel transmission can also be realized by using conventional direct detection systems with either subcarrier modulation (SCM) or wavelength division multiplexing (WDM). In SCM, an electric analog multichannel FDM signal (for example, an ordinary electric cable TV signal) modulates laser intensity by direct modulation of the laser injection current. As semiconductor laser can be modulated at frequencies in the multi-GHz range, very large number of analog TV channels and even large number of analog HDTV channels can be transmitted simultaneously by using the subcarrier technique. Since subcarrier transmission links are based on a single optical carrier wave only, fiber bandwidth is only inefficiently used. It may be noted that in the 1.5 μm region, an optical fiber exhibits a spectral bandwidth of about 15000 GHz!

In WDM, channel selection is performed by a set of optical filters or a single tuneable optical filter. Because the bandwidth of optical filter is very broad as compared to microwave filter, only a limited number in the order of 10 optical carriers with large channel spacing can be transmitted simultaneously. Therefore, WDM also does not enable to use the overall fiber bandwidth. In comparison to that, CMC systems with optical frequency division multiplexing are able to use the large fiber bandwidth most efficiently. In CMC systems, each optical carrier is again modulated by an analog microwave multichannel FDM signal or a digital signal of ultrahigh bit rate. By employing optical time division multiplexing in addition, even Tbit/s communication systems becomes possible (Fig. 7.7c).

In the realization of OFDM systems with a large number of optical channels, optical transmitter becomes very comprehensive since each optical channel requires a single-mode laser. To achieve a fixed frequency spacing, all lasers must be stabilized and synchronized. This can be achieved either absolutely or relatively by means of multichannel stabilization unit. For this, all frequencies present in the CMC transmitter comb are periodically measured, which can, for example, be performed by using a scanning heterodyne spectrometer.

Coherent multichannel communication (CMC) with FSK or DPSK modulation represent the primary application of coherent optical communications. Because of their flexibility, they are also well-suited for multipoint-to-multipoint networks (multiple-access networks), such as ultrahigh bit rate broadband ISDN (BISDN) or HDTV distribution networks including also video-on-demand. CMC systems will play a major role in highly sophisticated broadband networks based on information superhighways offering a broad spectrum of new services; for example interactive broadband multimedia services.

Fundamental differences and applications of various optical communication systems are summarized in Table 7.3 above. It becomes clear from the above discussion and Table 7.3 that only four optical communication systems are of significant practical interest:

- optical systems with intensity modulation and direct detection,

- optical FSK heterodyne systems with noncoherent detection,

- optical DPSK heterodyne systems with noncoherent detection and

- optical PSK homodyne systems.

With the realization of low-cost coherent optical communication systems, application field of direct detection systems will be more and more taken over by coherent systems. However, the technical and economical breakthrough of coherent optical communication systems is closely related with the development of *integrated optics*.

Table 7.4: Comparison of optoelectronic and conventional microelectronic integrated circuit

	Microelectronic integrated circuit (IC)	Optoelectronic integrated circuit (OEIC)
Wavelength	electrical	electrical *and* optical
Guide	metallic	metallic *and* optical waveguide
Components	only electrical	electrical, optical, and optoelectronic
Length-side ratio	about 1 : 1	about 1 : 1000 (e.g., diode laser)
Technology	planar	epitaxial
Material	Si, GaAs	GaAs, InP
Remark	many similar components (many thousand transistors)	less, but very different components

Optoelectronic integrated circuits (OEICs) represent the key components of low-cost economical and commercial coherent optical multichannel transmission systems. OEIC allows to integrate all required lasers along with the synchronization circuit and modulation electronics on one single substrate in the transmitter of optical multichannel transmission system. Considering the optical receiver, OEIC combines optical components (for example, the optical coupler), optoelectronic components (tuneable local laser and photodiodes) as well as the electrical components (low noise front-end amplifier) on one single substrate. In the realization of OEIC, know-how of realizing standard electronic integrated circuits (IC) can be used; for example, the clean-room technology. However, OEIC and ordinary electronic IC exhibit some essential differences which are summarized in Table 7.4 above.

In the following some important practical requirements are briefly discussed. We focus our interest on comparison only.

(a) TRANSMITTER

The main requirements for the transmitter laser are narrow linewidth, uniform FM response in case of direct modulation and high output power. Moreover, frequency stabilization is of primary importance. In CMC transmission systems, in addition of frequency stabilization, synchronization of all optical carriers must also be performed.

(b) CHANNEL

In order to achieve long-haul high bit rate transmission, single-mode fiber should have a low attenuation of less than 0.2 dB/km and a low dispersion, which can be achieved by dispersion equalization at the optical level or by dispersion shifted fiber. Free-space channels should have low background noise in addition.

(c) COHERENT RECEIVER

Irrespective of whether a homodyne receiver or heterodyne receiver is used, states of polarization of the optical incoming signal and the local laser must be matched. As explained in Chapter four, various polarization handling schemes exist which primarily differ in loss and speed. For the local laser, narrow linewidth and high output power are again the most important requirements. In addition, wavelength tuning is especially required when a coherent multichannel OFDM system has to be realized.

(d) HETERODYNE RECEIVER

In coherent optical communication systems with heterodyne detection, stable IF is of fundamental importance since temperature, mechanical and electromagnetic disturbances as well as low-frequency laser phase noise may cause fluctuations. The IF fluctuations are usually compensated with an AFC which tracks the frequency of the local laser with respect to the frequency of the received input signal. These fluctuations have to be kept to less than ± 10 MHz to keep the sensitivity loss below 0.1 dB. Imperfect demodulation circuit and non-ideal IF signal processing are primarily responsible for a deviation from shot noise limited operation in practical receivers.

(e) HOMODYNE RECEIVER

The most serious problem in the realization of homodyne receiver is the OPLL. Here, great care must be taken to separate phase noise and phase information and to keep the residual phase error low.

(f) INTRADYNE RECEIVER

An optical intradyne receiver is characterized by an intermediate frequency that is very close to zero. Like in homodyne receiver, received optical signal is converted into a base-band signal. It offers the possibility of transmitting very high bit rates. On the other hand, no optical phase matching (i.e., no OPLL) is required. Instead, this receiver requires either a symmetrical six-port coupler or an optical 90^0 hybrid as critical component. In addition, the electronic part of receiver is more comprehensive as compared to heterodyne and homodyne receivers.

7.4 COMPARISON OF PHYSICAL LIMITS FOR OPTICAL COMMUNICATION

In optical communications, following question frequently arises: How many photons per bit are at least required to achieve a certain BER; for example, 10^{-10}? To answer this question, physical limits for optical communication systems have to be considered. In practice, different physical limits are defined to determine the maximum possible performance of optical communication systems. The purpose of this Section is to discuss the physical meaning of these limits and present their comparison. The following limits will be discussed in details:

- Quantum limit,

- Shot noise limit and

- Shannon limit.

7.4.1   QUANTUM LIMIT

The quantum limit is related to an ideal direct detection receiver which is a simple photon counter. It is well-known from conventional optical communications that the photons of the incoming lightwave are Poisson distributed. The random nature of photons by means of a typical intensity modulated optical signal at the input to the photon counter is illustrated in Fig. 7.8. It becomes clear from this figure that the number of photons varies from bit to bit when symbol "1" is transmitted, whereas the number of photons is always zero when the transmitted symbol is "0".

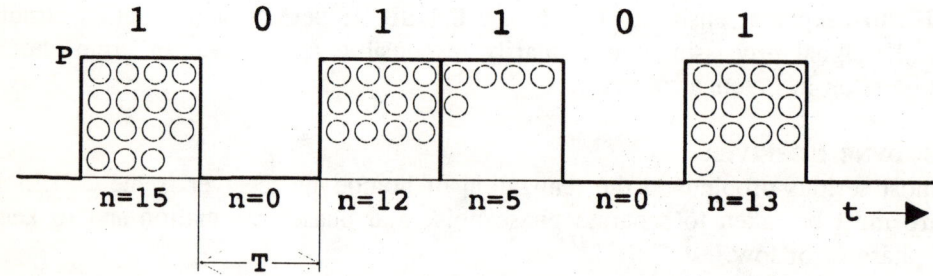

Fig. 7.8 : Poisson distributed photons at the input to optical receiver

The Poisson distribution is described by the following fundamental expression

$$f(n) = \frac{n_s^n e^{-n_s}}{n!}$$ (7.10)

where n_s represents the average number of photons for bit "1" (see Eq. 7.6). A bit error occurs when absolutely no photon is counted during the transmitted symbol "1". On the other hand, transmitted symbol "0" is always detected correctly. Hence, the average probability of bit error is

$$p = \frac{1}{2} f(n=0) = \frac{1}{2} e^{-n_s} \tag{7.11}$$

Here, occurrence probabilities for symbol "0" and symbol "1" are assumed to be equal. For a BER of $p = 10^{-9}$, on an average $n_s = 21$ photons are required from Eq. (7.11). In conclusion:

> To obtain a BER of 10^{-9}, an ideal optical receiver (photon counter) requires
>
> 21 photons/bit!

The question now arise: Is this result really the physical limit valid for all type of optical communication systems or is it possible to improve system performance even above this limit? This question will be discussed in the following Section.

7.4.2 SHOT NOISE LIMIT

Shot noise limited receiver operation can either be obtained by direct detection or coherent detection. The shot noise limit can practically be achieved with coherent detection provided local laser power is high enough (see Chapter two), whereas this limit is only of theoretical interest in direct detection.

(i) DIRECT DETECTION RECEIVER

As explained in Chapter two, signal-to-noise ratio of an optical direct detection receiver with an APD is given by

$$\left(\frac{S}{N}\right)_d = \frac{(M \, R_0 P_r)^2}{\left(e M^{2+x} R_0 P_r + e I_{dark} + G_{th}\right) B} \tag{7.12}$$

Shot noise limited detection can only be achieved with an infinite or at least an unrealistic high input light power P_r to the receiver. It is due to this reason that shot noise limit is of theoretical interest only in direct detection receiver. It must be noted that this statement is not valid if an optical preamplifier is used in addition. With a sufficient high power P_r, we obtain from Eq. (7.12)

$$\left(\frac{S}{N}\right)_d = \frac{R_0 P_r}{e M^x B} = \frac{\eta P_r}{hf M^x B} \tag{7.13}$$

This shot noise limited signal-to-noise ratio reaches its maximum value

$$\left(\frac{S}{N}\right)_{d,\max} = \frac{P_r T}{hf} = n_s \tag{7.14}$$

for $M^x = 1$ (i.e., PIN or $x = 0$), $\eta = 1$, and $B = 1/T$ (i.e., the double-sided noise equivalent bandwidth of a matched filter). As explained in Section 5.4, the BER is given by

$$p = Q\left(\frac{1}{2}\sqrt{\left(\frac{S}{N}\right)_d}\right) = Q\left(\frac{A_d}{2\sigma_d}\right) \tag{7.15}$$

where A_d represents the eye aperture and $\sigma_d = 0.5(\sigma_0 + \sigma_1)$ the average standard deviation of the additive Gaussian noise. Taking again a BER of $p = 10^{-9}$, on an average $n_s = 144$ photons per bit are required from Eq. (7.15). It should be remembered that $Q(6) \approx 10^{-9}$. In conclusion:

> To obtain a BER of 10^{-9}, shot noise limited
> optical direct detection receiver requires
>
> 144 photons/bit!

Main features of a shot noise limited direct detection receiver are summarized below:

- Infinite optical power P_r at the input to the receiver,

- Optimum photodiode quantum efficiency $\eta = 1$,

- PIN-photodiode or APD with $x = 0$,

- Matched filter for rectangular input signals with a double-sided noise equivalent bandwidth $B = 1/T$.

(ii) COHERENT RECEIVER

In coherent optical receiver, signal-to-noise ratio in the IF domain is given by (see Chapters two and five)

$$\left(\frac{S}{N}\right)_c = \frac{2R_0^2 P_r P_1}{2(eR_0 P_1 + eI_{dark} + G_{th})B} \tag{7.16}$$

In contrast to direct detection receiver, shot noise limited operation can easily be obtained by sufficiently high local laser light power P_1 (compare with Table 5.2). We obtain

$$\left(\frac{S}{N}\right)_{c,IF} := \frac{S}{N} = \frac{P_r T}{hf} = n_s \quad \text{in IF domain} \tag{7.17}$$

and

$$\left(\frac{S}{N}\right)_{c,B} = 2\frac{P_r T}{hf} = 2n_s \quad \text{in baseband frequency domain} \tag{7.18}$$

The bit error rate of coherent optical receiver also depends on the modulation scheme applied to the system. For the most important modulations schemes used in practice, bit error rates are given below (compare with the formulas given in Table 7.1):

$$p = Q\left(2\sqrt{\frac{S}{N}}\right) \quad \text{PSK homodyne} \tag{7.19}$$

$$p = \exp\left(-\frac{S}{N}\right) \quad \text{DPSK heterodyne} \tag{7.20}$$

and

$$p = \exp\left(-\frac{1}{2}\frac{S}{N}\right) \quad \text{FSK heterodyne with noncoherent dual-filter detection} \tag{7.21}$$

For a BER $= 10^{-9}$, we obtain the following result from Eqs. (7.19), (7.20) and (7.21):

> To obtain a BER of 10^{-9}, shot noise limited ideal heterodyne/homodyne receiver requires
>
> 41 photons/bit   for FSK heterodyne,
>
> 21 photons/bit   for DPSK heterodyne,
>
> 9 photons/bit   for PSK homodyne!

Characteristic features of ideal shot noise limited coherent receivers are summarized below:

- Infinite local laser power P_l,

- PIN-photodiode,

- Maximum photodiode quantum efficiency $\eta = 1$,

- Matched filter for rectangular input signals with a double-sided noise equivalent bandwidth $B = 1/T$.

It becomes clear from the above results that PSK homodyne system is actually better (i.e., 9 photons versus 21 photons or about 3.5 dB) than the quantum limit! This very surprising and important result, however, can easily be explained as follows: In photon counter system, photons are only transmitted in case of symbol "1", whereas in coherent homodyne system photons are transmitted for symbol "0" also.

Now, the same question arises again: What is the minimum number of photons required for a demanded BER or what is the real physical limit for optical communications? Is it possible to obtain a BER of 10^{-9} with less than 9 photons/bit?

7.4.3 SHANNON LIMIT

Given a digital source generating a bit rate R and a channel of capacity C. If

$$R \leq C \tag{7.22}$$

there exists a coding technique such that the output of the source may be transmitted over the channel with a probability of error which may be made arbitrarily small (Shannon's theorem). The channel capacity of a white, band-limited Gaussian channel is given by

$$C = B \log_2\left(1 + \frac{S}{N}\right) \text{ bit/s} \tag{7.23}$$

where B is the single-sided channel bandwidth, S the signal power and N the overall total noise power in the channel bandwidth [85, 93, 119, 164, 264].

For a fixed signal power and in presence of additive white Gaussian noise, channel capacity approaches an upper limit (termed as the Shannon limit) with increasing bandwidth:

$$C_{\max} = \frac{1}{\ln(2)} \frac{S}{G_d} \tag{7.24}$$

In optical PSK homodyne system, signal power is $S = 4R_0^2 P_r P_1$ and single-sided power density of white shot noise is $G_d = 2eR_0P_1$. Therefore, with $\eta = 1$ we obtain

$$R_{\max}T = \frac{1}{\ln(2)} \frac{2P_rT}{hf} = \frac{2n_s}{\ln(2)} = 1 \tag{7.25}$$

This yields

$$n_s = n_{s,\min} = \frac{\ln(2)}{2} = 0.347 \text{ photons !} \tag{7.26}$$

This important result actually represents the real physical limit of an optical communication system. Thus, only 0.347 photons per bit are required on an average to obtain errorless data transmission.

In may be mentioned that realization of this limit requires advanced highly sophisticated coding techniques. In direct detection system, an optical preamplifier is needed in addition since coding gain is restricted to some dB only. Sensitivities better than nine photons per bit for PSK homodyne system has already been demonstrated [281], where only five photons per bit are required for a BER of 10^{-9}.

8 OPTICAL FIBER LINK DESIGN

Over the past several years, optical fiber communication systems are extensively used all over the world for the telecommunication and data transmission purposes. The design of optical links to meet a given requirements has become an important issue. With the communication trend being towards digital, more and more digital based optical communication links are being designed and installed for commercial applications. In view of this, greater emphasis has been placed on the digital system though the same concepts are applicable for analog systems as well.

One of the motivations for developing optical fiber communication is its application to long-haul point-to-point communication systems and optical networks. The point-to-point feature implies a direct.fiber route between the sender and receiver with no intervening user stations. In long-haul systems, repeaters are often installed if the distance involved is sufficiently large. Applications like transfer of high resolution graphics, development of super computer networks and broadband integrated service digital networks (BISDNs) demand data transfer rates of several gigabits per second. Conventional networks cannot provide the required bandwidth, while the optical offer the most appropriate solutions.

This Chapter describes design procedure of a digital point-to-point link. Various types of optical networks and design issues involved in their implementation have been discussed.

8.1 LONG-HAUL SYSTEMS

The simplest transmission link is a point-to-point having transmitter on one end and receiver on the other. The link length can vary from less than a kilometre to thousands of kilometres depending on the type of application. It can be a digital computer data link connecting computer and terminals in the same building or between two buildings. It can also be an undersea light wave communication system for intercontinental communication over a distance of more than thousands of kilometres. In this application, low attenuation and large bandwidth of the fiber are more important, while in the former other properties of fiber like immunity to electromagnetic interference, low weight, small diameter are of prime importance [6].

In long-haul systems, communication links involved are point-to-point. As such the design of an optical link is a very involved process since it involves many interrelated variables among the fiber, source and photodetector characteristics. Link design is an iterative process and sometimes it may be necessary to relax the system requirements if these cannot be met with the existing technology or are too expensive. The main objective in the link design is not only to meet the system requirements, but it should also be cost effective.

In long distance communication links, where it is not feasible to transmit signals directly from the point of origin to the final destination, the total distance to be covered is divided into several shorter paths in tandem so that the transmission of adequate quality can be achieved. Each shorter path is connected to the next either through an optical amplifier or a regenerative repeater. In designing a multilink system, total system specifications must be reduced to one specification for each link e.g., if the overall system is required to have an error rate of less than 10^{-9} at a given data rate, then the individual link must have better error performance which is directly proportional to the number of links and other link parameters [284].

The key system requirements needed in the link design are: (i) data or bit rate/bandwidth, (ii) bit error rate (BER)/signal-to-noise (S/N) ratio and (iii) transmission distance or link length [112, 113]. Two basic issues involved in the link design are: (i) attenuation which determines the power available at the photodetector input for a given source power and (ii) dispersion which determines the limiting data rate or usable bandwidth. These are often referred as link power and time budgets respectively. Generally, first the link power budget is made and if found unsatisfactory, then some components might be changed. Once established, the designer can prepare time budget to ensure that the desired overall system performance is achieved.

On the basis of bit rate R and link length L, one can tentatively choose the operating wavelength and the source. The RL product of typical light emitting diodes (LEDs) and lasers at short (850 nm) and long wavelengths (1300 nm) are given in Table 8.1 [217]. The final choice, of course, is to be made on the basis of link power and time budgets. When the RL product is required to be more than 25 Gbit/s·km, one can operate near 1550 nm wavelength to take advantage of the lowest fiber loss occurring in this wavelength region. However, fiber dispersion becomes a major problem for such systems. If the dispersion shifted fibers are used, then both dispersion and loss become minimum near 1550 nm. As far as system cost is concerned, it is lowest near 850 nm and increases as the operating wavelength moves to 1300 nm-1600 nm. Further, the system robustness at 850 nm is more than the systems at higher wavelengths.

Table 8.1 : Data rate and distance product of typical LEDs and lasers

	Short λ (850 nm)	Long λ (1300 nm)
LED	< 150 Mbit/s·km	< 1500 Mbit/s·km
Laser	< 2.5 Gbit/s·km	< 25 Gbit/s·km

For optical fiber, choice is to be made between single-mode (SM) and multi-mode (MM). The MM fiber could be step-index (SI) or graded-index (GI). The choice depends on the amount of dispersion that can be tolerated and the power to be coupled into the fiber. LEDs are generally used with MM fibers, although edge emitting LEDs can launch sufficient power into SM fiber at data rates up to 560 Mbit/s over a distance of several kilometres. With a laser source, both SM and MM fibers can be used.

In making a choice of photodetector, minimum optical power required to satisfy the BER requirement at the specified data rate and receiver complexity are to be considered. A PIN-photodiode receiver is simpler and more stable with changes in temperature than an avalanche photodiode (APD) receiver. However, the sensitivity of an APD receiver is much higher (more than 10 dB to 15 dB). Therefore, it is to be used when the received optical power level is quite low. On the basis of above considerations, source, fiber and detector are tentatively decided. Then following the procedure outlined in Fig. 8.1 (discussed in details latter), design of the optical fiber link is made. In the iteration, various combinations in the order of increasing performance, complexity and cost are to be tried for the minimization of the system cost. Such combinations are: (i) LED-MM-PIN, (ii) LED-SM-PIN, (iii) LED-MM-APD, (iv) LED-SM-APD, (v) Laser-MM-PIN, (vi) Laser-SM-PIN, (vii) Laser-MM-APD and (viii) Laser-SM-APD. In the course of design, link power and time budgets are required to be made. These are described in the following Sections.

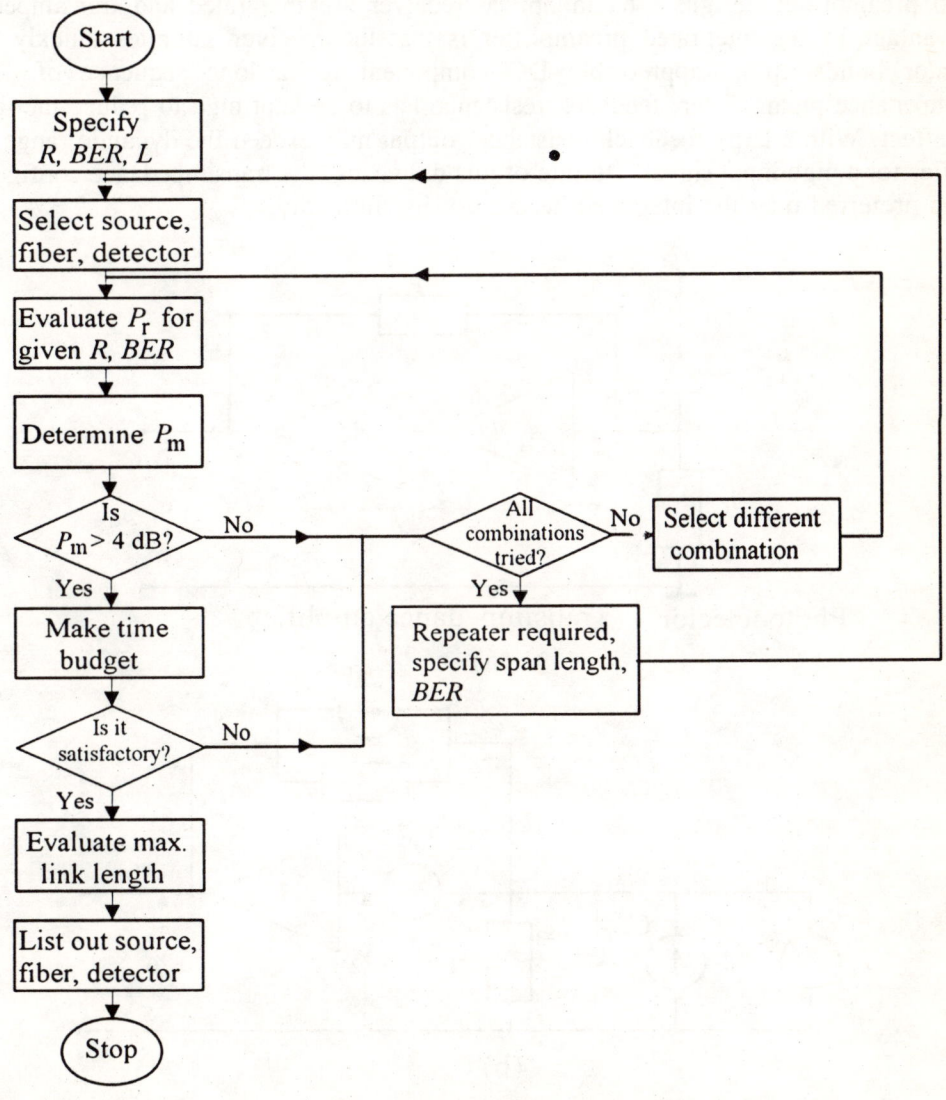

Fig. 8.1: Flow chart for link design

8.2   POWER BUDGET

In the preparation of link power budget, certain parameters like required optical power level P_r at the receiver to meet the system requirements, coupling losses etc. are required. These are discussed below:

8.2.1   MINIMUM REQUIRED POWER LEVEL

Two preamplifier designs used in optical receiver are integrated and transimpedance. Disadvantage of the integrated preamplifier is that the receiver saturates quickly as the integrator builds up an appreciable DC component for a long sequence of "1". In transimpedance preamplifier, feedback resistance has to be kept high to reduce the thermal noise effect. With a large feedback resistance, output may exceed the dynamic range of the amplifier for a high input signal. Despite of this drawback, the transimpedance configuration is often preferred over the integrated because of its simplicity.

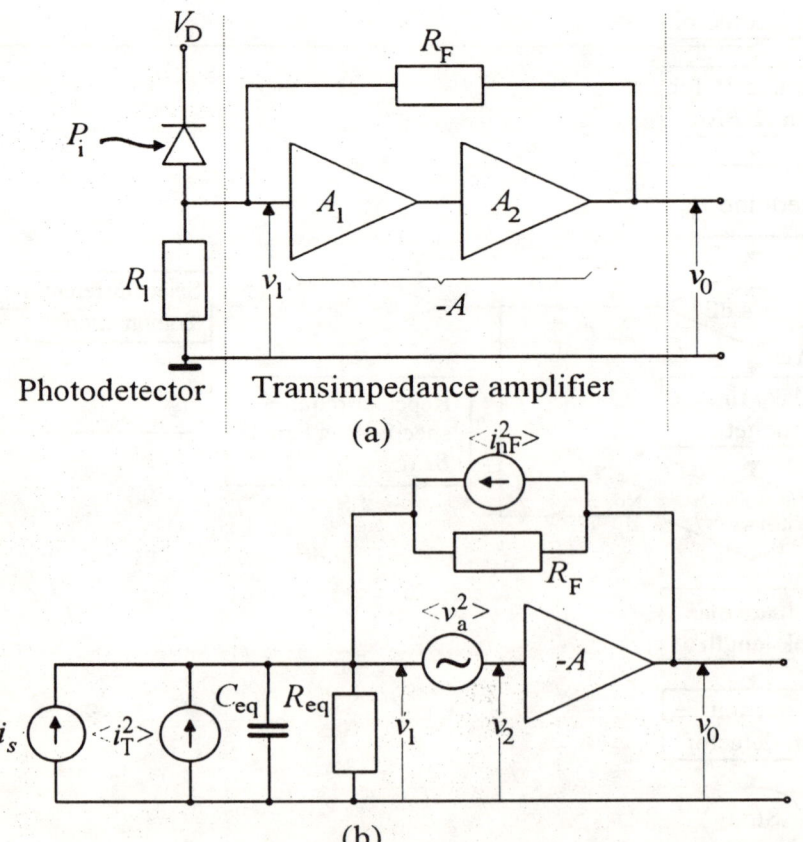

Fig. 8.2: (a) Circuit diagram of receiver with transimpedance preamplifier and (b) equivalent circuit ($<i_T^2> = 4k_B T_0/R_{eq}$ and $<i_{nF}^2> = 4k_B T_0/R_F$)

The S/N ratio of PIN-photodiode receiver with FET transimpedance preamplifier (Fig. 8.2) for an on-off signalling scheme is given by [284]

$$\frac{S}{N} = \frac{R_0^2 P_r^2}{\left\{ 2e(R_0 P_r + I_{dark}) + 4k_B T_0 \left(\frac{1}{R_{eq}} + \frac{1}{R_F} \right) + \{i_a\}^2 + \{v_a\}^2 \left[\left(\frac{1}{R_{eq}} + \frac{1}{R_F} \right)^2 + \frac{(2\pi B C_{eq})^2}{3} \right] \right\} B} \quad (8.1a)$$

where

$$R_{eq} = R_l \| R_d \| R_a \qquad\qquad (8.1b)$$

and

$$C_{eq} = C_d + C_s + C_a^{\cdot} \qquad\qquad (8.1c)$$

Here, R_l and R_F are the load and feedback resistances respectively; R_d and C_d the equivalent resistance and capacitance of the photodiode respectively; R_a and C_a the input resistance and capacitance of the amplifier respectively; C_s the miscellaneous parasitic capacitance; I_{dark} the dark current; $\{i_a^2\}$ and $\{v_a^2\}$ the noise spectral densities due to amplifier noise current and voltage respectively; B the receiver bandwidth ($=1/R$ where R is the data rate) and the remaining parameters have their usual meanings. We can determine the required S/N ratio for a given BER. It is used to determine required power P_r after solving the quadratic Eq. (8.1a).

When APD is used instead of PIN-photodiode in the receiver, the expression for S/N ratio gets modified to

$$\frac{S}{N} = \frac{R_0^2 P_r^2}{\left\{ 2e(R_0 P_r + I_{dark})F + \frac{4k_B T_0}{M^2} \left(\frac{1}{R_{eq}} + \frac{1}{R_F} \right) + \frac{\{i_a\}^2}{M^2} + \frac{\{v_a\}^2}{M^2} \left[\left(\frac{1}{R_{eq}} + \frac{1}{R_F} \right)^2 + \frac{(2\pi B C_{eq})^2}{3} \right] \right\} B} \quad (8.2)$$

where $F\, (=M^x)$ is the excess noise factor and M the gain of APD. From the above equation, we can determine the optimum value of gain M_{opt} which will maximize the S/N ratio. This M_{opt} will come in terms of P_r and other parameters. After the substitution of M_{opt} in Eq. (8.2), S/N ratio becomes a quadratic function of P_r and is used to determine the required P_r for a given S/N ratio (implying BER also). APD receiver will require about 10-15 dB less power as compared to PIN-photodiode receiver. The precise value, of course, depends on the characteristics of the photodiodes and other receiver parameters.

The coherent receivers generally operate under the shot noise limited conditions. Therefore, evaluation of S/N ratio or the required P_r for a given R and BER is much easier.

8.2.2 Coupling Loss Evaluation

Next step in the link design is to estimate losses in the link. These losses depend upon the source, fiber and detector. Power budget for the link starts from the receiver and proceeds towards the transmitter. Source power should at least be the sum of all losses and minimum optical power required along with some safety allowance P_m, which is normally taken as 4 dB. If this condition is not satisfied, then either source/fiber/detector or their combination has to be upgraded. The various losses encountered in an optical fiber link are briefly described as follows:

(i) Source-to-fiber Coupling Loss, L_{sf}

Most of the sources used in fiber optic application are available with a short length of optical fiber cable called pigtail connected in the optimum power coupling configuration. Manufacture usually specifies optical power from the source in the pigtail (ranges from microwatts to few milliwatts). In such cases, problem of coupling a source with fiber reduces to that of a fiber-to-fiber coupling.

When a source is to be coupled to fiber by external means, substantial amount of power gets lost at the interface. Loss in power depends on several parameters like radiation pattern, radii of the source and fiber, type of the fiber (SI or GI), refraction index (RI) profile for GI fiber etc. If the source is modelled as Lambertian with symmetric radiation pattern, then the coupling efficiency η_L into a SM fiber (i.e., the fraction of the source power coupled into the fiber) is given by [134]

$$\eta_L = \begin{cases} (NA)^2 & r_s \leq a \\[3mm] (NA)^2 \left(\dfrac{a}{r_s}\right)^2 & r_s > a \end{cases} \tag{8.3a}$$

The corresponding expression for GI fiber is [134, 217]

$$\eta_L = \begin{cases} (NA)^2 \left[1 - \left(\dfrac{2}{g+2}\right)\left(\dfrac{r_s}{a}\right)^g\right] & r_s \leq a \\[4mm] (NA)^2 \left(\dfrac{a}{r_s}\right)^2 \left(\dfrac{g}{g+2}\right) & r_s > a \end{cases} \tag{8.3b}$$

where the parameter r_s represents the source radius, a the core radius and g defines shape of the RI profile.

When there is no matching of the refraction index at the light source and core interface, an additional Fresnel reflection loss will occur. The power transmittance (TR) at the interface assuming perpendicular incidence is given by [134, 217]

$$TR = 1 - \left(\frac{n_1 - n_0}{n_1 + n_0}\right)^2 \tag{8.4}$$

where n_0 is the refraction index of the medium outside the fiber core and n_1 the refraction index of the core.

It is possible to improve the source directivity and hence source-to-fiber coupling efficiency by the use of a microlens if the size of source is smaller than the size of fiber [217]. The microlens will optically match the source with fiber. In that case, directivity of the source gets improved by the magnification factor i.e., a/r_s. Such techniques do not exist if the size of the source is larger than the size of fiber.

Coupling efficiency of LED depends on whether it is surface or edge emitting. Surface emitting LED produces radiation pattern quite close to an ideal Lambertian source. But edge emitting diode and also the laser diode have asymmetric radiation pattern (the pattern for laser diode is much narrower than the LED). In these cases, graded index lens (also called GRIN lens) and cylindrical lens have also been tried to improve the coupling efficiency, but the improvement is found to be very marginal. Though use of lenses improves the coupling efficiency, it also makes it sensitive to lateral misalignment errors. Further, presence of back reflections from the lens may affect the stability of laser source.

The coupling loss when determined lie in the range 1-2 dB. To be on safer side, we can presume it to be 2 dB.

(ii) FIBER-TO-FIBER COUPLING LOSS, L_{ff}

Fiber-to-fiber coupling loss L_{ff}, also referred as splice loss, has two components: intrinsic and extrinsic. Intrinsic loss arises due to difference in the waveguide properties of the two fibers. User does not have any control over it. The coupling efficiency (implying loss also) resulting from core diameter, numerical aperture and core RI profile mismatch can be easily determined.

(a) FIBER RADIUS EFFECTS

When the radii a_e and a_r are not equal, but the axial NA and RI profiles are equal ($NA_e(o) = NA_r(o)$ and $g_e = g_r$), the coupling efficiency η_a is given by [134, 217]

$$\eta_a = \begin{cases} (a_r/a_e)^2 & \text{for } a_r < a_e \\ \\ 1 & \text{otherwise} \end{cases} \tag{8.5}$$

where the subscript "e" and "r" stand for emitting and receiving fibers respectively.

(b) NUMERICAL APERTURE EFFECTS

When the radii and the index profiles of two coupled fibers are identical ($a_e = a_r$ and $g_e = g_r$), the coupling efficiency η_{NA} is [134, 217]

$$\eta_{NA} = \begin{cases} \left(\dfrac{NA_r(0)}{NA_e(0)}\right)^2 & \text{if } NA_r(0) < NA_e(0) \\ \\ 1 & \text{otherwise} \end{cases} \tag{8.6}$$

(c) INDEX PROFILE EFFECTS

When the radii and the axial numerical apertures are same ($a_e = a_r$ and $NA_e(0) = NA_r(0)$, the coupling efficiency η_g is given by [134, 217]

$$\eta_g = \begin{cases} \dfrac{g_r(g_e + 2)}{g_e(g_r + 2)} & \text{if } g_r < g_e \\ \\ 1 & \text{otherwise} \end{cases} \tag{8.7}$$

The above intrinsic efficiencies are used to determine the overall intrinsic efficiency η_{total}

$$\eta_{total} \propto \eta_a \eta_{NA} \eta_g \tag{8.8}$$

The total combined loss in dB is given by

$$L_{total}(\text{dB}) = -10\log(\eta_{total}) = L_a + L_{NA} + L_g \tag{8.9}$$

where L_a, L_{NA} and L_g are the intrinsic losses in dB due to mismatch in the radii, NAs and RI profiles respectively.

In addition, there may be other effects like ellipticity of the core, concentricity of the core within the cladding, variations in RI profile within the core etc. which may also produce coupling losses. But their quantitative evaluation is difficult. Out of all the effects, losses due to difference in the core radii and NAs are dominant [217].

The extrinsic loss arises due to mechanical misalignments introduced while splicing. These can be reduced by carefully aligning the prepared flat and fine end faces of the fibers. Basically there are three types of misalignments: lateral, longitudinal and angular (Fig. 8.3). In lateral misalignment (or axial misalignment), lateral axes of the two fibers are separated by a distance d. Longitudinal misalignment occurs when the fibers have same axis with a gap s between the end faces. In angular misalignment, fiber end faces are not parallel and fiber axes form an angle θ.

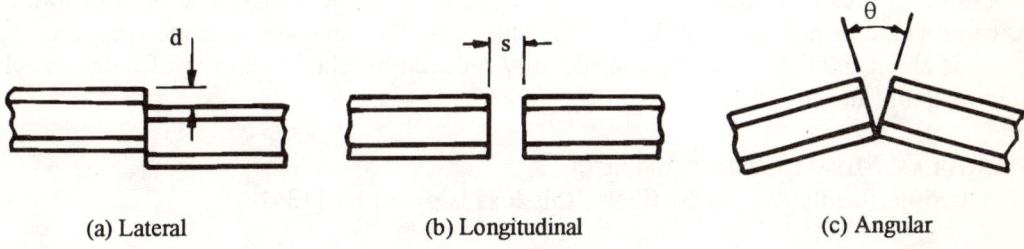

| (a) Lateral | (b) Longitudinal | (c) Angular |

Fig. 8.3: Mechanical misalignments between two fibers: (a) lateral, (b) longitudinal and (c) angular

(d) LATERAL MISALIGNMENT EFFECTS

It contributes maximum to the splicing loss among the three misalignments. Its value is different for SI and GI fibers.

For SI fiber [134]

$$L_{ds} = -10\log\left[\frac{2}{\pi}\cos^{-1}\left(\frac{d}{2a}\right) - \frac{d}{\pi a}\sqrt{1-\left(\frac{d}{2a}\right)^2}\right] \tag{8.10}$$

For GI fibers (g=2) [134]

$$L_{dg} = -10\log\left[\frac{2}{\pi}\cos^{-1}\left(\frac{d}{2a}\right) - \sqrt{1-\left(\frac{d}{2a}\right)^2}\,\frac{d}{6a}\left(5-\frac{d^2}{2a^2}\right)\right] \tag{8.11}$$

where d represents the lateral separation.

(e) LONGITUDINAL MISALIGNMENT EFFECTS

The loss due to longitudinal misalignment (also known as end separation loss) for two identical SI fibers with uniform power distribution is given by [134, 217]

$$L_{ss} = -10\log\left(\frac{a}{a + s\tan(\theta_c)}\right)^2 \tag{8.12}$$

where θ_c is the critical angle of acceptance ($=\sin^{-1}$ NA). For GI fiber, no such formula for the loss is available.

When there is a longitudinal misalignment (implying a gap between the fibers), there will be loss due to Fresnel reflection in addition to the misalignment loss. The Fresnel reflection loss can be determined by using Eq. (8.4). This loss at an air-glass interface ($\eta_{glass}=1.5$ and $\eta_{air}=1$) is about 0.2 dB. It can be considerably reduced by placing a drop of index matching liquid or gel at the interface.

(f) ANGULAR MISALIGNMENT EFFECTS

For two uniformly excited SI fibers, this loss is given by [134]

$$L_{\theta} = -10\log\left(\cos(\theta)\left\{\frac{1}{2} - \frac{1}{\pi}p\sqrt{1-p^2} - \frac{1}{\pi}\sin^{-1}(p)\right.\right. \tag{8.13a}$$

$$\left.\left. -q\left[\frac{1}{\pi}y\sqrt{1-y^2} + \frac{1}{\pi}\sin^{-1}(y) + \frac{1}{2}\right]\right\}\right)$$

where

$$p = \frac{\cos(\theta_c)\,(1-\cos(\theta))}{\sin(\theta_c)\,\sin(\theta)} \tag{8.13b}$$

$$q = \frac{\cos^3(\theta_c)}{\left(\cos^2(\theta_c) - \sin^2(\theta)\right)^{3/2}} \tag{8.13c}$$

and

$$y = \frac{\cos^2(\theta_c)\,(1-\cos(\theta)) - \sin^2(\theta)}{\sin(\theta_c)\,\cos(\theta_c)\,\sin(\theta)} \tag{8.13d}$$

In the evaluation of L_{ff}, uniform illumination in the fiber that excites all modes uniformly has been presumed [134, 217].

(iii) FIBER-TO-DETECTOR COUPLING LOSS, L_{fd}

Fiber-to-detector coupling does not pose that much problem as from source-to-fiber. Generally, active area of detector is much larger (10 times is quite common) than that of fiber end. Therefore, area mismatch loss does not exist. Also, transverse offset will not be a problem. The only significant loss need to be considered is the Fresnel reflection loss.

In any practical design, an allowance has to be made for the degradation of components with ageing, replacement, variations due to temperature fluctuations, manufacturing spreads, imperfect repeatability on reconnection, field repairs, maintenance, variations in drive conditions and so on. The designer has to keep a safe margin to take care of these factors [48, 209]. The loss model for a point-to-point optical fiber link is shown in Fig. 8.4. After computing various losses and fixing safety margin, power budget of the link is made as follows [113]:

Table 8.2 : Calculation of power budget
  * L_0 is the factory unit length of fiber and $[L/L_o]$ represents the integer part of L/L_o.
  $ $ \alpha $ is the attenuation coefficient of fiber in dB/km.

Source output power, P_t	.. dBm
Minimum receiver power, P_r (min)	.. dBm
Source-to-fiber coupling loss, L_{sf}	.. dB
Splice loss:	
- Number of splices, $N = [L/L_0]^*$	
- Fiber-to-fiber coupling loss, L_{ff} .. dB	
- Total splice loss, NL_{ff}	.. dB
Fiber loss$^\$$, αL	.. dB
Fiber-to-detector coupling loss, L_{fd}	.. dB

$$\text{Power margin } P_m = \left(P_t - P_r(\min) - L_{sf} - NL_{ff} - \alpha L - L_{fd} \right) \text{ dB} \qquad (8.14)$$

$P_m \geq 4$ dB is acceptable otherwise some components need to be upgraded. With $P_m < 4$ dB, system will become less reliable. The typical values of L_{sf}, L_{ff} and L_{fd} are about 2 dB, 0.5 dB and 0.2 dB respectively.

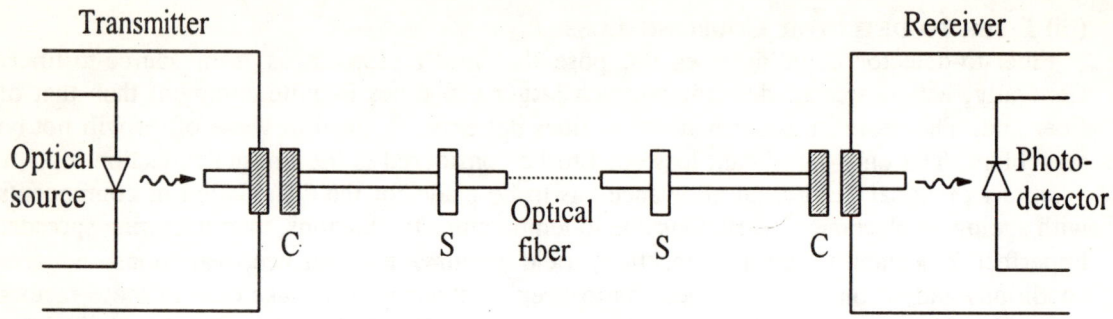

Fig. 8.4: Loss model of point-to-point optical fiber link (C: connector and S: splice)

8.3 TIME BUDGET

Objective of time budget is to ensure that the system is able to operate properly at the desired data rate. Sometimes, it may happen that the total system is not able to operate at the desired data rate even if the bandwidth of the individual subsystem exceeds the data rate. If the transmitter, fiber and receiver are considered to have a Gaussian impulse response with rms pulse widths σ_{tx}, σ_f and σ_{rx} respectively, the overall system response will also be Gaussian with a rms pulse width [284]

$$\sigma_{sys} = \sqrt{\sigma_{tx}^2 + \sigma_f^2 + \sigma_{rx}^2} \qquad (8.15)$$

As a safety margin, it is usually taken to be 10% higher than the above value. System time budget is considered to be satisfactory if σ_{sys} does not exceed $(1/4R)$ where R is the data rate. The rms pulse widths for different subsystems are determined as follows:

8.3.1 TRANSMITTER RISE TIME

Rise time of transmitter primarily depends on the electronic components of driving circuit and electrical parasitics associated with the optical source. It is few nanoseconds for LED-based transmitter, but can be as short as 0.1 ns for a laser-based transmitter [140]. Presuming an exponential rise and decay, σ_{tx} is nearly equal to half of the rise time [284].

8.3.2 RECEIVER RISE TIME

The receiver rise time t_{rx} is determined by 3 dB electrical bandwidth of the receiver front-end. If B is the receiver bandwidth in MHz and the receiver frontend is modelled by a first order low-pass filter having a step response $[1-\exp(2\pi \dot{B} t)]u(t)$ where $u(t)$ represents the unit step function, then t_{rx} in ns is approximately given by [134]

$$t_{rx} = \frac{350}{B} \tag{8.16}$$

As for the transmitter, σ_{rx} is determined from the rise time.

8.3.3   FIBER RMS DISPERSION

The rms pulse width of the fiber σ_f includes the contributions of both modal and material dispersion through the relation

$$\sigma_f = \sqrt{\sigma_{mod}^2 + \sigma_{mat}^2} \tag{8.17}$$

where σ_{mod} and σ_{mat} are the rms pulse widths due to modal and material dispersion of the fiber respectively. For SM fiber, modal dispersion does not contribute and therefore σ_f becomes same as σ_{mat}.

The optical fiber link may consist of several concatenated sections and the fiber in each section may have different dispersion characteristics. Further, there may be mode mixing at the splices and the connectors. As a consequence, propagation delay associated with different modes tends to average out. In the absence of mode mixing, σ_{mod} for SI fibers is [247]

$$\sigma_{mod} \approx \frac{n_1 \Delta L}{2\sqrt{3}\,c} \tag{8.18}$$

where c is the velocity of light and the parameter Δ depends on the core and cladding refraction indices ($\approx (n_1-n_2)/n_1$). For GI fibers, delay time is a function of g. The minimum intermodal rms pulse broadening with an optimum g is [86, 247]

$$\sigma_{mod} \approx \frac{n_1 \Delta^2 L}{20\sqrt{3}\,c} \tag{8.19}$$

In both the cases, effect of mode mixing can be included by changing the linear dependence on L by a sublinear dependence L^q, where q is a constant ranging from 0.5 to 1 and its typical value is 0.7.

The σ_{mat} is approximately given by

$$\sigma_{mat} \approx |D_{mat}|\sigma_\lambda L \tag{8.20}$$

where σ_λ is the rms spectral width of the source. The dispersion parameter D_{mat} may change along the fiber link due to different dispersion characteristics of fibers in the link. Therefore, average value of D_{mat} should be used in Eq. (8.20).

8.4 MAXIMUM LINK LENGTH CALCULATION

When a particular set of components meets the design requirements, one would like to know the maximum distance up to which these components could be used. Further, if the link length is quite large, it will help in determining the repeater location. The maximum link length is determined presuming that link is attenuation limited and there is no dispersion effect. And again when the link is dispersion limited and there is no attenuation effect. Minimum of the two is taken as the maximum practicable link length.

Under attenuation limited condition, total fiber attenuation should be at least equal to the difference between power at the input end of fiber and minimum power required at the output end of fiber to meet the receiver power requirement. If there are splices and/or connectors in-between, their attenuation is also to be considered (see Example 8.1). The maximum link length is the total attenuation divided by the fiber attenuation per unit length. For a dispersion limited link, maximum data rate R that can be transmitted over an optical fiber system is given by [248, 284]

$$R = \frac{1}{4\sigma_{sys}}$$

(8.21)

The maximum allowable fiber dispersion will be

$$\sigma_{al} = \sqrt{\left(\frac{1}{4R}\right)^2 - \sigma_{tx}^2 - \sigma_{rx}^2}$$

(8.22)

Using Eq. (8.17), we can determine σ_f/L i.e., fiber dispersion per unit length. Therefore, maximum link length under dispersion limited condition will be $L \cdot \sigma_{al}/\sigma_f$. The flow chart for link design is given in Fig. 8.1.

Example 8.1
Design an optical fiber link for a length 2.9 km and data rate 90 Mbit/s. The required BER is 10^{-9}. Available components are:

(a)   Laser at 850 nm; rms spectral width 1 nm; output power 0 dBm; transmitter rise time 2 ns,

(b)   LED at 850 nm; rms spectral width 50 nm; output power -13 dBm; transmitter rise time 15 ns,

(c)   Fiber GI type; $n_1=1.43$; $\Delta=0.015$; $\alpha=4$ dB/km; $D_{mat}=70$ ps/(nm·km) at 850 nm; unit factory length 1 km,

(d)   PIN-photodiode receiver of sensitivity presuming typical parameters is described by the equation

$$P_r = 200 \cdot 10^{-17} \text{Ws} \cdot R \left[1 + \sqrt{1 + 1.42 \cdot 10^{-8} \text{s} \cdot R + \frac{1.468 \cdot 10^9 \text{s}^{-1}}{R}}\right] \quad (8.23)$$

where R is the data rate.

(e)   APD receiver with sensitivity better than PIN receiver sensitivity by 11 dB at the same data rate and BER. Take receiver rise time with either PIN or APD to be 1 ns.

Procedure

With the given information, procedural steps involved in the link design are as follows:

(i)   Keeping in view the availability of components, region of operation is 800 nm-900 nm.

(ii)   Following Section 8.2.2, different coupling losses can be determined. Let these losses be $L_{sf} \approx 2$ dB, $L_{ff} \approx 0.5$ dB and $L_{fd} \approx 0.2$ dB.

(iii)   Let us consider the simple and economical LED-GI-PIN combination. The sensitivity of the receiver at $R=90$ Mbit/s from Eq. (8.23) will be -30.3 dBm. The power margin P_m from Eq. (8.14) is

$$P_m = (-13+30.3-2-1-11.6-0.2) \text{ dB} = 2.5 \text{ dB} \quad (8.24)$$

Since the power margin is less than 4 dB, this combination will not meet the power budget requirement.

(iv)   Now either the source or detector has to be upgraded. Let us replace the PIN-photodiode receiver by APD receiver. Sensitivity of the receiver will be -41.3 dBm. The power margin,

$$P_m = (-13+41.3-2-1-11.6-0.2) \text{ dB} = 13.5 \text{ dB} \quad (8.25)$$

This combination satisfies the power budget requirement. Let us now make the time budget. The modal and material dispersion rms pulse widths can be determined from Eqs. (8.19) and (8.20) respectively. Total system rise time from Eq. (8.15) with a safety margin of 10% is

$$\sigma_{sys} = 1.1\sqrt{(7.5)^2+(0.09)^2+(10.15)^2+(0.5)^2} \text{ ns} = 13.89 \text{ ns} \qquad (8.26)$$

Allowable rms pulse width ($=1/4R$) is 2.78 ns. Since σ_{sys} is more than the allowable rms value, time budget is not satisfied.

(v)   Now let us take Laser-GI-PIN combination. The power margin

$$P_m = (0+30.3-2-1-11.6-0.2) \text{ dB} = 15.5 \text{ dB} \qquad (8.27)$$

Thus, power budget requirement has been satisfied. The system rise time is

$$\sigma_{sys} = 1.1\sqrt{(1)^2+(0.09)^2+(0.2)^2+(0.5)^2} \text{ ns} = 1.25 \text{ ns} \qquad (8.28)$$

As σ_{sys} is less than the allowable rms value, time budget is also satisfied. Hence, this combination will meet the given system requirements.

(vi)   Maximum link length for the above combination under attenuation limited condition can be determined by using

$$L_a(max) = \frac{P_t-(P_r-11\,dB)-L_{sf}-L_{fd}-P_m}{\alpha+L_{ff}/L_o} \qquad (8.29)$$

In the above equation, P_t and P_r are in dBm; L_{sf}, L_{ff}, L_{fd} and P_m in dB; L_o in km and α in dB/km. Substituting the values of various parameters, we get

$$L_a(max) = \frac{-P_r+4.8}{4.5} \text{ km} \qquad (8.30)$$

P_r is computed using Eq. (8.23) for different R and then $L_a(max)$ is determined from the above equation. The results are shown in Fig. 8.5.
The maximum link length under dispersion limited condition for the above combination is given by

$$L_d(\text{max}) = \frac{\sqrt{\left(\dfrac{1}{4R}\right)^2 - \sigma_{tx}^2 - \sigma_{rx}^2}}{1.1\sqrt{\left(\dfrac{n_1\Delta^2}{20\sqrt{3}c}\right)^2 + \left(|D_{mat}|\,\sigma_\lambda\right)^2}} \qquad (8.31)$$

In the above equation, factor 1.1 in the denominator represents a safety margin of 10% in the fiber dispersion as P_m in the calculation of $L_a(\text{max})$. When values of various parameters are substituted in the above equation, it gives

$$L_d(\text{max}) = \frac{\sqrt{\left(\dfrac{1}{4R}\right)^2 - 1.25}}{0.084} \text{ km} \qquad (8.32)$$

where R is in Gbit/s. Numerical results computed from the above equation are also shown in Fig. 8.5. It is seen from the figure that at a data rates of 10 Mbit/s, channel is attenuation limited and the maximum repeater spacing is 9 km. At higher data rate, say 200 Mbit/s, channel becomes dispersion limited and the maximum repeater spacing is 6.6 km.

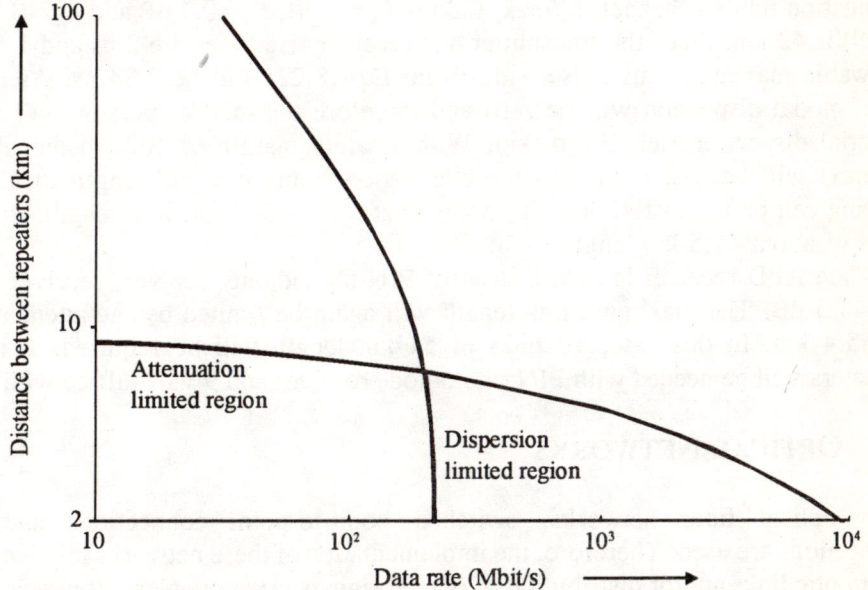

Fig. 8.5: Variations of maximum distance between repeaters with data rate under attenuation and dispersion limited conditions

Example 8.2

An optical communication link is to be designed for a data rate 100 Mbit/s, BER 10^{-9} and link length 540 km. The available components are:

(a)   Laser at 1300 nm; rms spectral width 3 nm; output power -5 dBm; transmitter rise time 2 ns.

(b)   Fiber SI-SM type, $n_1 = 1.43$; $\Delta = 0.0037$; $\alpha = 0.6$ dB/km; $D_{mat} = 3.2$ ps/(nm·km) at 1300 nm, $L_0 = 3$ km.

(c)   For signal detection either a PIN-photodiode or an APD receiver can be used. The PIN has the same specifications as APD ($M < 100$, $F(M) \approx M^{0.5}$). $C_d = 1$ pF; $C_s = 1$ pF; $C_a = 2$ pF; $R_F = 6.5$ kΩ, $R_{eq} \gg R_F$; $T_0 = 300$ K; $I_{dark} = 15$ nA; $\{i_a^2\} = 0.6 \times 10^{-24}$ A^2/Hz; $\{v_a^2\} = 0.04 \times 10^{-24}$ V^2/Hz; $R_0 = 0.8$ A/W (for PIN) and receiver rise time 1 ns.

Procedure

As the link length is quite large, one has to use repeaters in-between. As discussed in section 8.1, the BER of each link must be better than the overall system BER, say 10^{-10}. For this BER, an average S/N ratio of about 16 dB is required [284]. Let us now consider Laser-GI-PIN combination. The required optical power level at the receiver for the above S/N ratio can be determined by substituting the values of various parameters in Eq. (8.1a). It will be approximately -36.8 dBm [284]. With this sensitivity, maximum link length for the attenuation limited channel, L_a(max) (taking $L_{sf} = 1$ dB, $L_{ff} = 0.1$ dB and $L_{fd} = 0.2$ dB) from Eq. (8.29) is 42 km. Since the transmitter and receiver rise times are 2 ns and 1 ns respectively, allowable maximum rms pulse width from Eq. (8.22) will be 2.24 ns. As the fiber is SM type, modal dispersion will be zero and therefore the total dispersion will be same as the material dispersion i.e., 9.6 ps/km. With a safety margin of 10% in the fiber dispersion, L_d(max) will be 212.1 km. As the attenuation limits the link length to 42 km, repeater spacing can be at most 42 km. It implies that for the 540 km link length, there must be 13 links of about 41.5 km length each.

When APD receiver is used instead of PIN-photodiode receiver, receiver sensitivity will be -45.3 dB. The maximum link length will again be limited by the attenuation and it will be 55.4 km. In this case, 10 links of 54 km length will be required. It implies that 12 repeaters will be needed with PIN-photodiode receiver and 9 will suffice with APD receiver.

8.5   OPTICAL NETWORKS

In optical fiber networks, switched point-to-point connections and/or multipoint connections are used. Therefore, the implementation of these networks involves point-to-point fiber optic links and/or distribution of optical signals using couplers. It may also use the fiber optic repeaters when implemented in the multiple segments. Depending on the application, optical networks are divided into two categories: (i) Broadcast and distribution and (ii) Local area networks (LANs). Main characteristics and applications of such networks are described below.

(a) BROADCAST AND DISTRIBUTION NETWORKS

These networks are used to distribute wide range of services viz., telephone, facsimile, computer data, video signals. Transmission distances are relatively small (<50 km) and the data transmission rate as high as 10 Gbit/s. Topologies commonly used in such networks are hub and bus (Fig. 8.6). In hub topology, channel distribution takes place at the central locations (or hub). Role of the fiber is to carry high bandwidth inter-hub traffic just like in point-to-point links. Main problem associated with this topology is of reliability. Failure of a single link can affect the service to a large portion of the network. Telephone networks employ this topology for the distribution of audio channels.

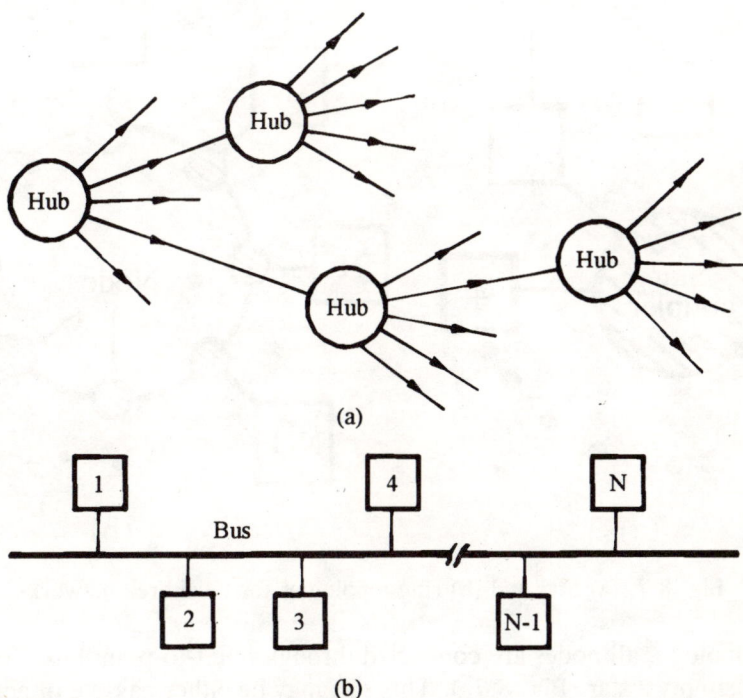

(a)

(b)

Fig. 8.6: (a) Hub and (b) bus topologies for distribution networks

In bus topology, a single fiber cable carries the optical signal and distribution is done by using optical taps. Large bandwidth available with the optical fiber permits the use of more channels. Disadvantage of this topology is that the signal loss in dB increases linearly with the number of taps. Therefore, it limits the number of subscribers served by a single bus (see

Section 8.5.1 for details). A simple application of this topology is the distribution of multiple video channels within a city.

(b) LOCAL AREA NETWORKS

In LANs, large number of users within a small area can access the network to transmit data to any other user. Transmission distance involved is less than 10 km. Therefore, main motivation behind the use of optical fiber is its large bandwidth rather than the low attenuation.

Several fiber optic local area network (FOLAN) topologies have been tried and implemented. These are the variations of three basic topologies viz., bus, star and ring. The bus topology is quite similar to that shown in Fig. 8.6b. But in LANs, each user has random access to the network. In these networks also, number of users are limited by the loss at each tap on the bus.

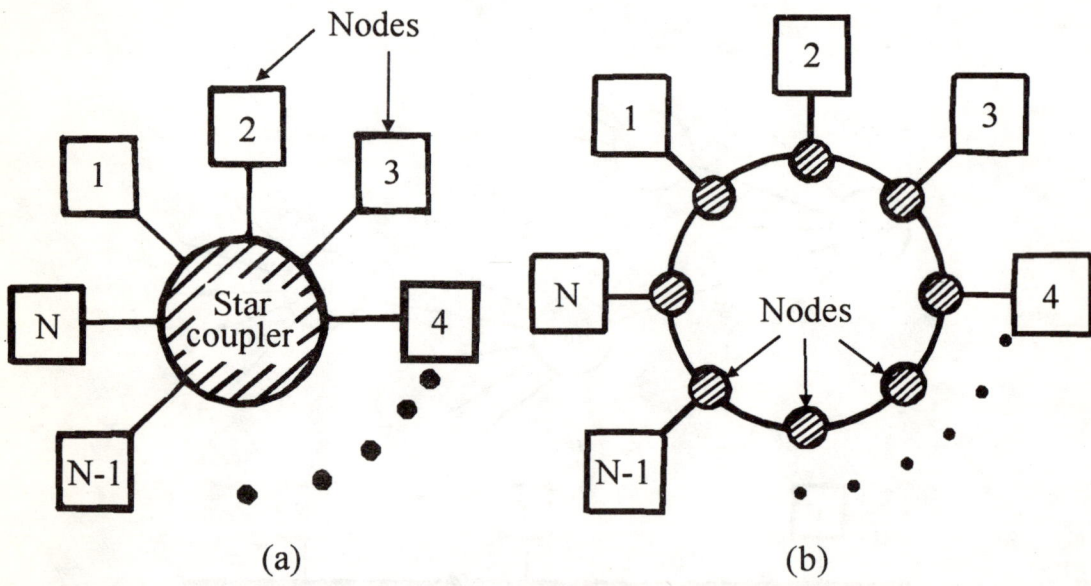

Fig. 8.7: (a) Star and (b) ring topologies for local area networks

In the star topology, all nodes are connected through point-to-point links to a central node called a hub or simply a star (Fig. 8.7a). This star may be either passive or active. In passive star, power distribution among the output lines takes place in the optical domain, whereas in active star it is in the electrical domain i.e., all incoming optical signals are converted into electrical signals through optical receivers. The composite electrical signal is then distributed to drive the individual node transmitters. In this topology, signal loss in dB increases logarithmically with the number of nodes. It is this feature that attracts star topology for LAN applications.

In the ring topology, consecutive nodes are connected by point-to-point links to form a single closed ring as shown in Fig. 8.7b. Each node can transmit and receive the message by using a transmitter-receiver pair, which also acts as repeater. Each station receives a message from its upstream neighbour and if it is not the addressee, repeat the message to its downstream station. The disadvantage in this is that damage to any station on the ring will disrupt the whole network unless bypass switches or redundancies in the network are built-in.

Most of the technical considerations of point-to-point links are also applicable to multiple-access networks. However, there are a few key considerations which must be taken care of in the latter networks. For example, in a point-to-point network, available system gain (allowable loss between transmitter and receiver) must be allocated to transmission loss, connector loss and margin. The available system gain in a multiple-access configuration with passive access couplers must also be partially allocated to these couplers. Further, one must also take into account the dynamic range and synchronization problems which are commonly (but not always) present in multiple-access networks. In LANs, it may be possible that one of the transmitters accessing the network generates a short message which must be received and processed by all the receivers on the networks. Each receiver must quickly adjust itself to accommodate the level of incoming signal and must synchronize its clock (if it uses a clocked generator) to the phase of incoming signal. These are important in point-to-point links also, but may be even more important in LANs due to the presence of multiple transmitters.

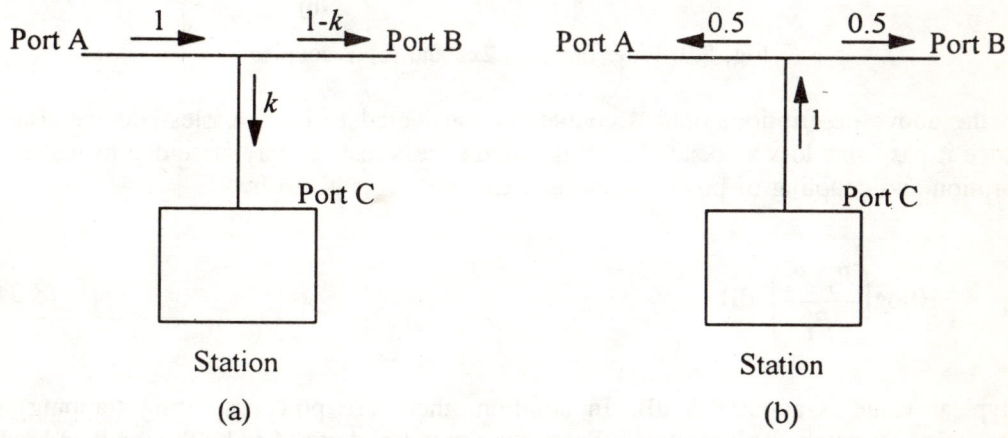

Fig. 8.8: Ideal tap on a bus in the (a) receiving and (b) transmitting mode

The primary issue involved in optical fiber networks is the implementation of tap (Fig. 8.8 above). When the signal is to be tapped off from the optical fiber bus, a tap as shown in Fig.8.8a is required. Incoming signal from port A is transmitted to port B with a small percent k, being tapped off for port C. In the same way, signal from port B will get through to port A. For transmitting the signal, a tap as shown in Fig. 8.8b is required. In this tap, signal transmitted by port C must enter the bus without significant reflections and should equally split towards ports A and B. These coupling mechanism are practically implemented using a 2×2 fiber optic coupler shown in Fig. 8.9a. At times, one of the fiber is removed to give it the appearance of a Y-coupler (Fig. 8.9b). The loss matrix of this coupler is [240]

$$
\begin{bmatrix} P_4 \\ P_3 \end{bmatrix} = \begin{bmatrix} 1-k & k \\ k & 1-k \end{bmatrix} \begin{bmatrix} P_1 \\ P_2 \end{bmatrix}
\tag{8.33}
$$

With k varying from 0.5 to as low as 0.01. A value of $k=0.5$ implies that input power P_1 or P_2 will get split equally between ports P_4 and P_3 and hence it will be a 3 dB coupler. If $k=0.01$, then 99% of power launched in port 1 will go to port 4 and remaining 1% to port 3. However, it will be other way round for the power at port 2 i.e., 99% of the launched power will go to port 3 and remaining 1% to port 4.

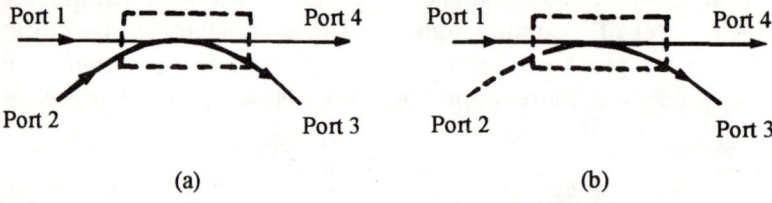

Fig. 8.9: Fiber optic (a) 2x2 and (b) Y-coupler

In the above description, optical coupler is considered to be a lossless device. But, in practice it has some loss associated with it called excess loss. It may arise due to scattering, absorption and coupling of power to the isolated port. It is given by

$$
L_{ex} = -10\log\left(\frac{P_3+P_4}{P_1}\right) \text{ dB}
\tag{8.34}
$$

Its typical value is about 0.5 dB. In addition, there are power splitting (tapping) and transmission (throughout) losses also. Presuming port 1 and port 4 to be the input and output ports respectively, the power splitting loss L_{ps} and the transmission loss L_{tl} are defined as:

$$
L_{ps} = -10\log(k) \text{ dB}
\tag{8.35}
$$

It specifies the loss between input port 1 and the port to be isolated (port 3).

$$
L_{tl} = -10\log(1-k) \text{ dB}
\tag{8.36}
$$

It specifies the loss between the input port 1 and the output port 4 [205].

It is well-known that propagation of light through the optical fiber is unidirectional. It implies that no part of the optical power coupled at port 2 in Fig. 8.9a will travel in the bus in the direction of port 1. It forces the network designer to use either more than one coupler or use dual bus for carrying signals in both the directions.

Optical networks based on the bus, star and ring topologies are discussed in the following Sections.

8.5.1   BUS TOPOLOGY

Bus topology was initially considered in FOLAN. Three commonly used structures are described below.

(i) SINGLE FIBER BUS

This is a direct fiber optic adaptation of an electrical bus configuration. Each node consists of bus and station couplers as shown in Fig. 8.10.

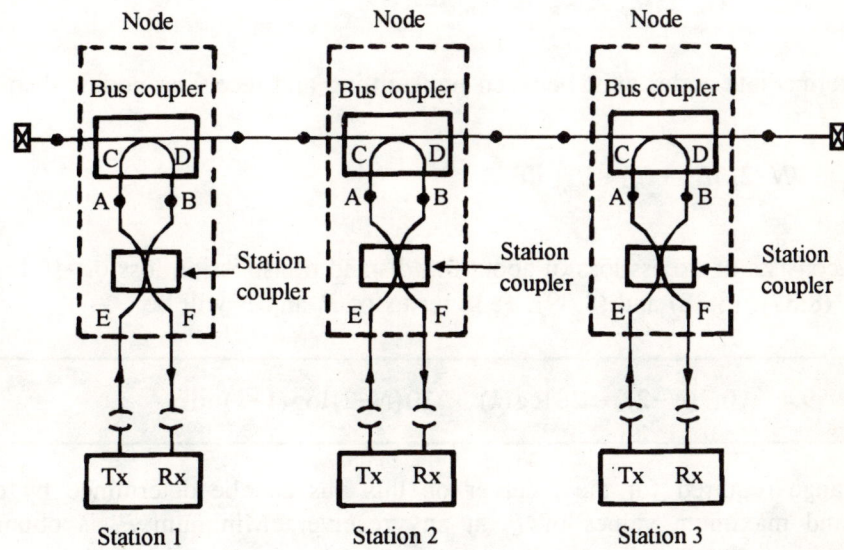

Fig. 8.10: Single fiber bus implementation using bus and station couplers

Split ratio of the bus coupler can be varied resulting into different amount of power tapped from/launched onto the bus by a station. For any given number of nodes, there is an optimum value of split ratio which will maximize the available power budget/system gain [240]. Station coupler is a 3 dB coupler which splits the transmitter power equally in both the output ports A and B. These ports couple power onto the bus in both the directions through the bus coupler. Similarly, an incoming signal from either direction of the bus is coupled to either port A or B and is then passed onto the receiver through the station coupler.

Since the power division in ports A and B is for power launching in opposite directions on the bus, it is natural to have a 3 dB station coupler.

Let P_t be the transmitted power by a node in dBm, the power launched on the fiber bus in dBm will be

$$P_{b1} = P_t - \left(L_{cn} + L_{sc} + L_{sp} + L_{bc} + L_{sp}\right) \text{ dB} \tag{8.37}$$

Here, L_{cn} and L_{sp} are the connector and splice losses typically in the order of 0.5 dB and 0.1 dB respectively; L_{sc} the station coupler loss consisting of excess loss of about 0.5 dB and power splitting loss of 3 dB and L_{bc} the bus coupler loss which consists of about 0.5 dB excess loss and $-10 \log(k)$ dB loss due to power splitting.

Let P_{b2} be the power available on the bus in dBm. The power received P_r at the receiver in dBm will be

$$P_r = P_{b2} - \left(L_{sp} + L_{bc} + L_{sp} + L_{sc} + L_{cn}\right) \text{ dB} \tag{8.38}$$

If $(N-2)$ intermediate nodes exist between transmitting and receiving nodes, then

$$P_{b2} = P_{b1} - (N-2) \left(L_{sp} + \tilde{L}_{bc} + L_{sp}\right) \text{ dB} \tag{8.39}$$

where $\tilde{L}_{bc}$ consists of excess loss of about 0.5 dB and transmission loss of $-10 \log(1-k)$ dB. Using Eqs. (8.37), (8.38) and (8.39), P_r in terms of P_t and N will be

$$P_r = P_t - \left[9.4 + 0.7(N-2) - 20\log(k) - 10(N-2)\log(1-k)\right] \text{ dB} \tag{8.40}$$

Dynamic range required for the receiver on this bus can be determined by computing minimum and maximum values of P_r at any receiver. Minimum P_r is obtained when transmitter and receiver are located at the extreme ends i.e., N in Eq. (8.40) is equal to the maximum number of nodes supported by the bus. Maximum P_r occurs when the transmitting node is an immediate neighbour of the receiver and can be obtained from Eq. (8.40) by taking $N = 2$.

(ii) DUAL FIBER BUS

In dual fiber bus structure, all transmitters are connected to one segment of the bus and receivers to the other. These segments are interconnected at one end as shown in Fig. 8.11. This is also known as folded bus [179]. In this structure, station couplers are basically Y-couplers.

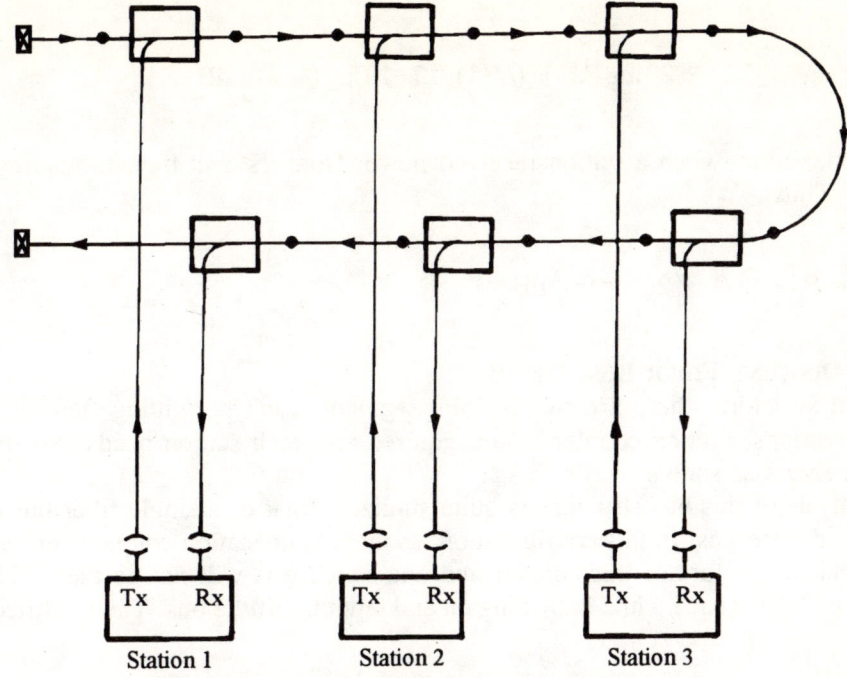

Fig. 8.11: Folded bus implementation using Y-couplers

For a bus of N stations, the received power level P_r will be

$$P_r = P_t - \left[2(L_{cn} + L_{sp} + L_{yc}) + f(N)(L_{sp} + \tilde{L}_{yc} + L_{sp})\right] \text{ dB} \tag{8.41}$$

where $\tilde{L}_{yc}$ is the loss associated with the Y-coupler, which consists of power transmission loss $-10 \log(1-k)$ dB and excess loss (≈ 0.5 dB). L_{yc} is the loss of the Y-coupler, but it consists of split loss $-10 \log(k)$ dB and excess loss (≈ 0.5 dB). The term $f(N)$ represents the number of intermediate couplers between receiver and transmitter [240]. Assuming the same values of L_{cn} and L_{sp} losses as in single fiber bus, above equation gives

$$P_r = P_t - \left[2.2 - 20\log(k) + f(N)(0.7 - 10\log(1-k))\right] \text{ dB} \tag{8.42}$$

Dynamic range required by the optical receiver can be determined by computing the minimum and maximum values of P_r received by the extreme right station. Minimum P_r is received when the transmitter is on the extreme left and is given by

$$P_r(\text{min}) = P_t - \left[2.2 - 20\log(k) + (N-1)\big(0.7 - 10\log(1-k)\big)\right] \text{dB} \qquad (8.43)$$

P_r will be maximum when a station received power from its own transmitter i.e., $N=0$ in Eq. (8.42). Thus

$$P_r(\text{max}) = P_t - \left[2.2 - 20\log(k)\right] \text{dB} \qquad (8.44)$$

(iii) DUAL DISJOINT FIBER BUS, DDFB

In DDFB structure, there are two disjoint segments each permitting flow of signals in opposite directions. Station coupler is not required and each station needs two transmitters and two receivers as shown in Fig. 8.12.

Loss analysis of this bus structure is quite similar to that of a single fiber bus except for a few minor differences. In this configuration, as there is no station coupler, terms involving split loss and excess loss of this coupler and one splice loss will not be there. This results in a saving of 3.6 dB each while launching on and tapping off the bus. The modified equation will be

$$P_r = P_t - \left[2.2 + 0.7(N-2) - 20\log(k) - 10(N-2)\log(1-k)\right] \text{dB} \qquad (8.45)$$

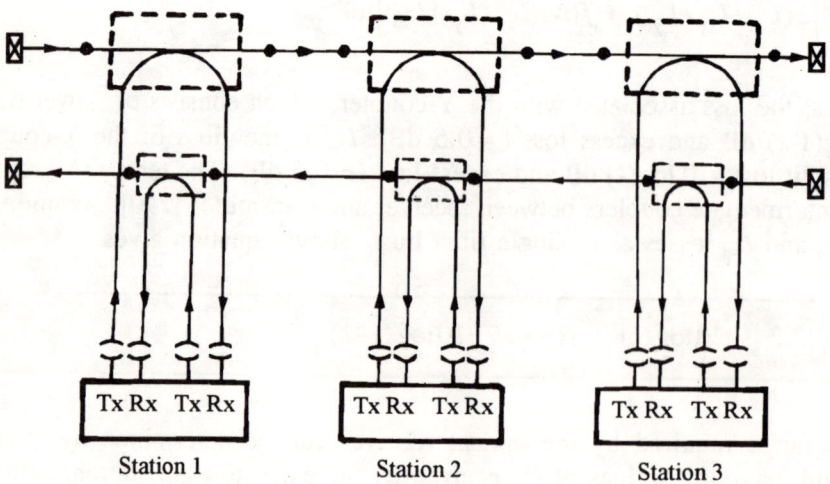

Fig. 8.12: Dual disjoint fiber bus implementation using station couplers

The passive linear bus structures discussed above can support only a few stations with the available sources and detectors. Assuming a power margin of 40 dB, single fiber bus and folded fiber bus can support only 8 and 6 stations respectively and dual disjoint fiber bus supports around 12 stations [240].

8.5.2   STAR TOPOLOGY

Star topology has been implemented in several FOLANs. The star coupler could be passive, active or directional. Sometimes several passive stars are interconnected to accommodate large number of users or to reduce the fiber clusters at the coupler. It will lead to a multistar topology. Various star topologies are discussed in the following Subsections.

(i) PASSIVE STAR

In this topology (Fig. 8.13), transmitter and receiver of each station are connected to the opposite ends of the star coupler. It appears that a star is a logical bus where all the taps are concentrated at one point. Whenever any station transmits, all the stations listen and any two stations transmitting simultaneously results in a collision.

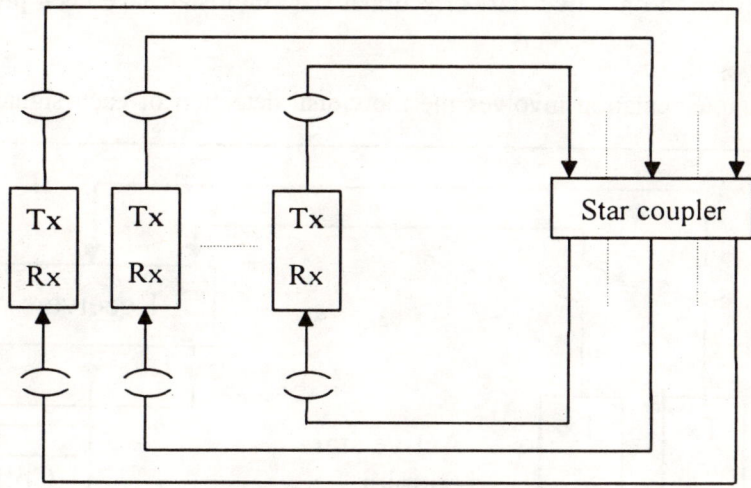

Fig. 8.13: Passive star LAN

In the star coupler, power received in any output port P_r in dBm is given by

$$P_r = P_t - L_{st} \tag{8.46}$$

where P_t is the transmitted power in dBm and L_{st} is loss due to star coupler in dB. It consists of power split loss $10 \log(N)$ dB for N×N star coupler and excess loss which may be about 2 dB. Therefore,

$$P_r = P_t - [10\log(N) + 2\,\text{dB}]\tag{8.47}$$

It is seen from the above equation that the received power level in dBm varies as $\log(N)$, whereas in bus topology it varies linearly with N (Eqs. (8.40), (8.42) and (8.45)). It means that the number of nodes N places more stringent requirement on the received power level in the bus topology than in the star topology. Another very important advantage of the star topology is that all the receivers get the same amount of power and hence the requirement of dynamic range is minimal in this topology. Of course, there may be some variations due to imperfect star coupler, difference in the path lengths or connector losses etc.

Disadvantages of star topology are: (i) no scope of direct expendability i.e., $N \times N$ star can support at most N nodes. Some expendability is possible using multistars, but these have their own limitations, (ii) clustering and presence of long lengths of cables from the star to each node and (iii) unreliable collision detection. The nonideality of the star coupler, difference in the path lengths or source power variations may make the two signals differ from each other by more than 5-6 dB. In that case, stronger signal would mask the weaker signal and the collision would not be reliably detected. In order to avoid these problems, several networks based on active star, directional star, multistar have been proposed.

(ii) ACTIVE STAR

Active star implementation involves the individual detection of each signal (Fig. 8.14).

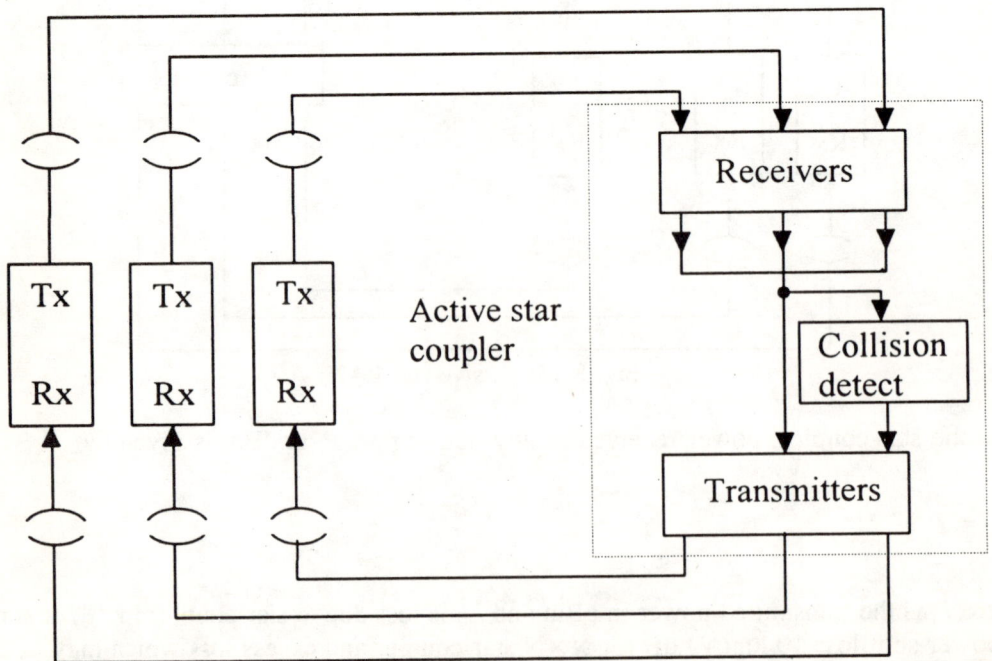

Fig. 8.14: Active star LAN

When two or more signals are present at the receiver outputs, a collision is detected. These signals are then combined and the composite signal is transmitted to all the stations. Since the optical signal is regenerated at the star coupler and a transmitter is dedicated to one or a few outgoing lines, there is no power splitting as in the case of passive star coupler. This improves the available power margin and as a consequence more number of nodes can be supported by the active star coupler. This, of course, is achieved at the cost of reliability.

(iii) DIRECTIONAL STAR

In directional star, several small couplers are joined together to perform the function of power splitting and combining. It is so constructed that any node can receive transmission of all other nodes except its own. It means that the reception of any signal by a transmitting station is a collision condition. A LAN can be implemented by cascading these directional couplers (Fig. 8.15). However, the high excess loss of directional couplers puts a severe limitation on the number of nodes that can be interconnected which in turn limits the size of network.

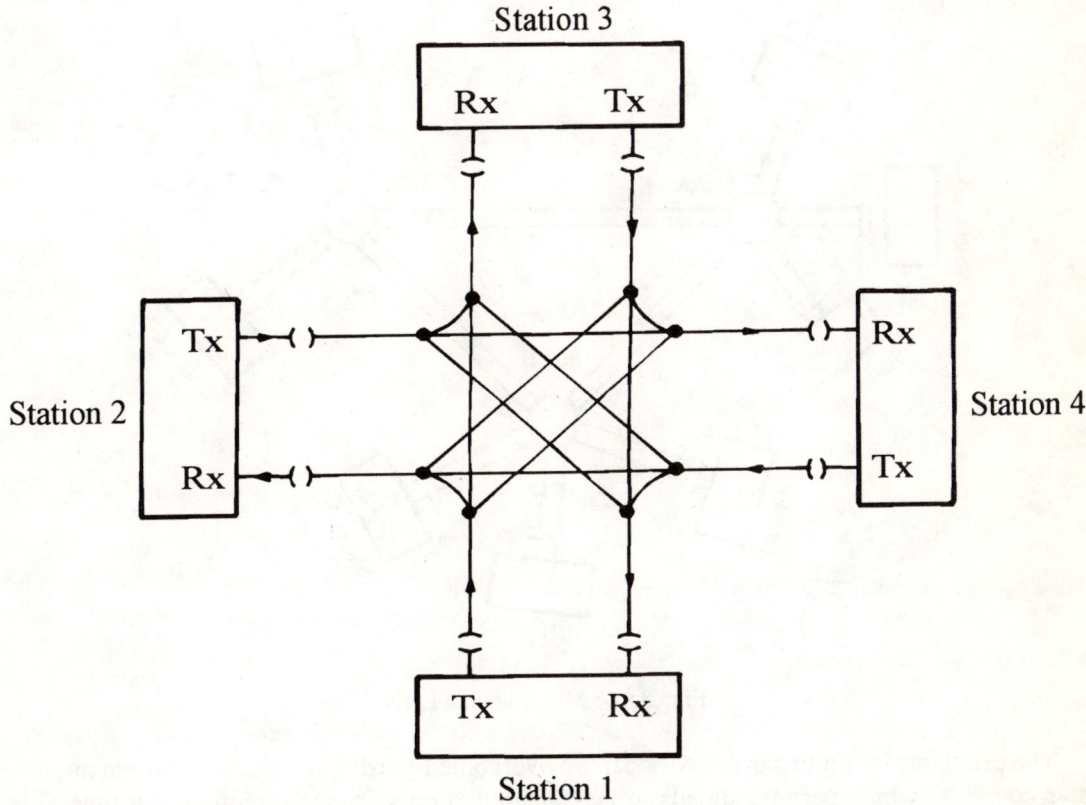

Fig. 8.15: Directional star LAN

(iv) MULTISTAR

As stated earlier, in multistar topology several passive star couplers are interconnected to accommodate larger number of users or to reduce the fiber clusters at the coupler. One way of connecting them is by merely joining each star coupler by another (Fig. 8.16). This is referred as local star network. With multistar, any signal transmitted by a station will be received by all the stations on the bus, but the received power level will depend on whether the transmitting and receiving stations are connected on the same star coupler or on different couplers. Depending on the size of the network and other constraints like path difference, power splitting at the star coupler etc., difference in the received power levels may be quite high. Another problem associated with the multistar is that a station may receive the same signal two or more times via different couplers. This along with the dynamic range problem complicates the receiver design in the network. It has not been found feasible to connect more than 4 or 5 stars of size 8×8 (or 16×16 with higher receiver complexity) if multiple loop backs are not to create serious problems [240].

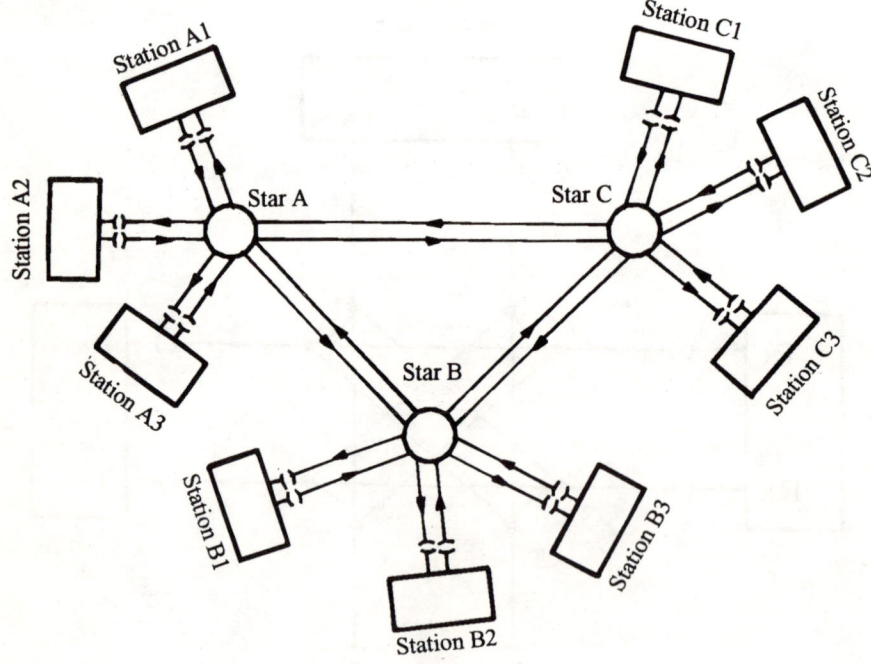

Fig. 8.16: Multiple star LAN

The problems in multistar network can be overcome by using a repeater between any two star couplers, which permits signals to be transmitted only in one direction at a time (Fig. 8.17). These selectively unidirectional repeaters regenerate the signals as well as perform the echo suppression. As a consequence, signals received at stations connected to any star coupler have the same order of magnitudes irrespective of the transmitter location. Similarly, echoes never reach any of the transmitting transceiver avoiding the problem of loop backs. Price paid for the advantages is in terms of reduction in reliability due to the presence of active components in the repeaters.

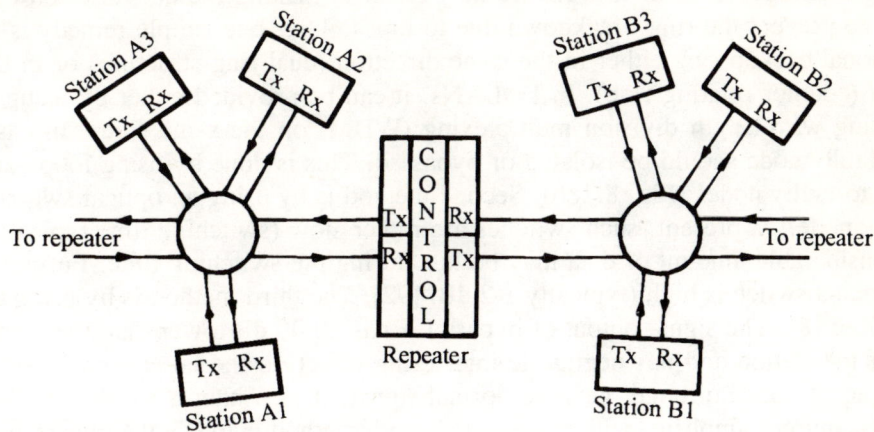

Fig. 8.17: Multiple star LAN with unidirectional repeaters

8.5.3 RING TOPOLOGY

This topology is best suited for FOLAN as it requires point-to-point fiber optic links. In this, a station is connected to adjacent stations through in- and outbound links (Fig. 8.18).

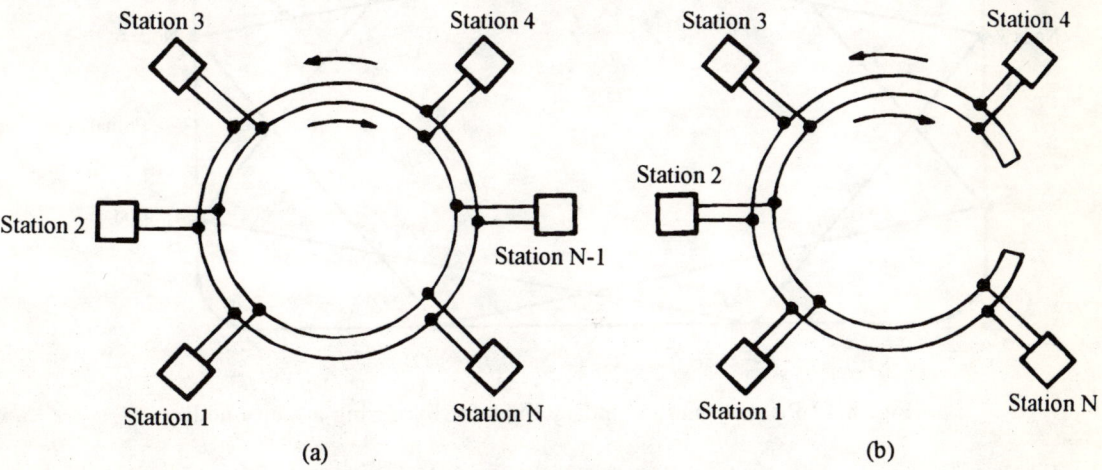

Fig. 8.18: (a) Two counter-propagating rings in FDDI and (b) formation of single ring when both the rings fail at some point

The interface at each node is an active device. It converts the incoming optical signal to an electrical signal. If the signal is destined to the station, it is removed otherwise it is remodulated onto the outgoing optical carrier using an electrical to optical converter. In this process, each node introduces some delay.

In ring topology, node or link failure may result in making the network non-operational. In order to prevent the ring breakdown due to link failure, one simple remedy is to provide an additional backup ring either in the same direction (dual ring structure) or in the reverse direction (counter rotating ring). In FOLANs, it can be provided either by using dual fiber or by using wavelength division multiplexing (WDM) on the same fiber. In case of node failure, faulty node should be isolated or bypassed. This is done by using loop back at node adjacent to faulty nodes (Fig. 8.18b). Second method is by using an optical switch to bypass the faulty node. At present, such switches are rather slow (switching time typically 5-10 ns) and a considerable amount of data may be lost during the switching time. Further, insertion loss of such a switch is high (typically 1-2 dB) [92]. The third method is by using a high loss bypass fiber [8]. The signal output of bypass fiber is 20-30 dB lower than the normal output signal of the station and has negligible interfering effect on the latter signal. When a node fails, adjacent node no longer gets the normal signal. It then haunts for the low level signal which is acquired, amplified and processed. Fourth method is to use the bypass links. These bypass links connect alternate nodes (Fig. 8.19). A node normally locks onto the incoming primary (normal) links. When this signal is absent, it switches to the secondary (bypass) links. Each node transmits data on both primary and secondary links. This method requires extra fiber and some routing decisions.

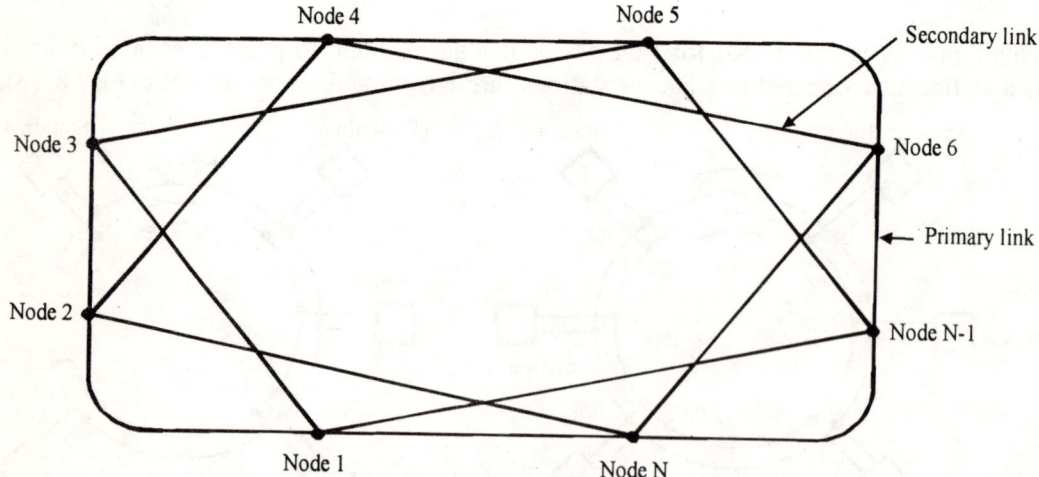

Fig. 8.19 Primary and secondary links for bypassing a faulty node

Fiber optic ring networks have been operating at data rates varying from 4 Mbit/s to 200 Gbit/s [92]. The American National Standard Institute (ANSI) has produced a new FOLAN standard at 100 Mbit/s. This standard is known as Fiber Distributed Data Interface

(FDDI) [30, 226]. The FDDI uses ring topology and is an extension of IEEE 802.5 (token passing ring) protocol. A variant of FDDI, called FDDI-II has been proposed for switched traffic such as private branch exchange (PBX).

The other fiber optic network that has been proposed by ANSI is Synchronous Optical NETwork (SONET). A synchronous network has a master clock that controls the timing of events all through the network. SONET is a proposed wide area network (WAN) operating at a base rate of 155 Mbit/s (with expansion capability to achieve data rates of several Gbit/s). The SONET is also being considered by the CCITT (International Consultative Committee on Telegraphy and Telephony) as an international standard [16]. The international version is called Synchronous Digital Hierarchy (SDH). The extension of SONET to become an international standard SDH holds the promise of breaking down an incompatibility between North American, European and Japanese data rates [217].

9 OPTICAL SPACE COMMUNICATIONS

Optical free-space communication systems for intersatellite links and deep space missions show significant advantages in comparison to alternate microwave systems. This Chapter describes the various aspects of the optical space communications. The Chapter is organized as follows. Section 9.1 gives a brief introduction and Section 9.2 discusses the applications of optical space communication systems. Their comparison with microwave communication systems is made in Section 9.3. Characteristic features of coherent optical space communication systems are explained in Section 9.4. Drawbacks and realization problems of such systems vis-a-vis coherent optical fiber communication systems are given in Section 9.5. The system description and design aspects viz., free-space propagation formula, optical transmitter, background noise, pointing, acquisition and tracking and receiver are discussed in Section 9.6.

9.1   INTRODUCTION

The complexity of space communication systems is growing rapidly with the increase in the transmission requirements. Most of these systems are exclusively based on microwave links. The optical communication systems become more and more attractive as the interest in high-capacity and long-distance space links grows. Advances in laser communication system architectures and optical components technology make such high capacity links feasible. Nevertheless, the choice of an optical source (semiconductor laser diode, solid-state laser or gas laser) as well as the choice of a proper modulation/demodulation technique are still an open question.

The two basically different receiver techniques for optical space communication systems are: direct detection and coherent (heterodyne or homodyne) detection. In direct-detection systems, only light intensity modulation (IM) is practical, whereas in heterodyne and homodyne systems many modulation schemes well-known from radio frequency (RF) transmission can be used; examples include ASK, FSK, PSK, DPSK etc. As is well-known, these schemes have significant advantages in regard to the receiver sensitivity, bit rate and transmission distance which, for example, can reach millions of km in space.

A compromise in sensitivity gain and realization demands is represented by the DPSK heterodyne and phase-diversity receivers. The sensitivity gain of both these receivers is only 3 dB less than the maximum sensitivity gain that can be achieved using a coherent optical homodyne receiver. In contrast to an optical PSK homodyne receiver, these receivers do not require an optical phase-locked loop (OPLL) which is an advantage in system realization.

9.2   APPLICATIONS OF OPTICAL SPACE COMMUNICATIONS

The potential use of optical space communication systems can be divided into three main applications (Fig. 9.1):

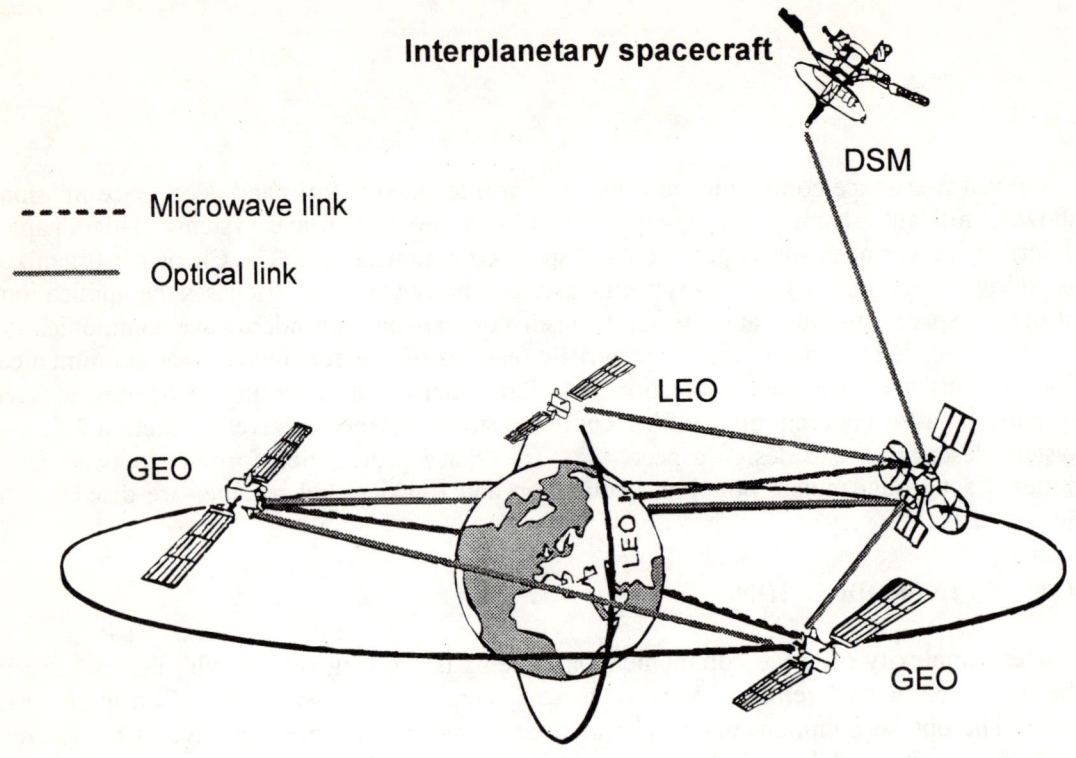

Fig. 9.1: Applications of optical space communications

(i) INTERORBIT LINKS (IOLs):

These links are of great interest for data transmission from a low Earth orbit (LEO) vehicle, such as Earth resource satellites, manned spaced stations, polar platforms of Hermes to a geostationary Earth orbit (GEO) vehicle, usually a data relay satellite (DRS). The return link (LEO-GEO) requires high data rates in the order of 500 Mbit/s, while the forward link (GEO-LEO) only demands telemetry data rates of about 25 Mbit/s.

(ii) INTERSATELLITE LINKS (ISLs):

A second important application, especially for commercial voice, television and data transmission is an intersatellite link between two geostationary telecommunication satellites. This link prevents double hopping arising from the use of an additional third Earth station in establishing a link between two participants living on opposite sides of the Earth. Moreover, interferences with terrestrial microwave links will be avoided.

(iii) DEEP SPACE MISSIONS (DSMs):

A third application of prime importance is the high capacity data transmission from planets such as Mars (78 million km from Earth) or Saturn (1278 million km from Earth) to a geostationary Earth satellite.

9.3   COMPARISON WITH MICROWAVE SYSTEMS

The promise of an optical communication system is mainly based on the strongly improved gain and reduction in the beam width of the antenna. The antenna gain in dB for a diffraction limited beam at short optical wavelengths is given by

$$G_a = 20 \log\left(\frac{\pi D_a}{\lambda}\right) \qquad (9.1)$$

Here, λ represents the wavelength and D_a the antenna (telescope) diameter. The half-power transmitter beam width full angle in radian is given by [218]

$$\psi_a = 1.03\left(\frac{\lambda}{D_a}\right) \qquad (9.2)$$

It may be noted that Eqs. (9.1) and (9.2) are valid for microwave as well as optical antennas.

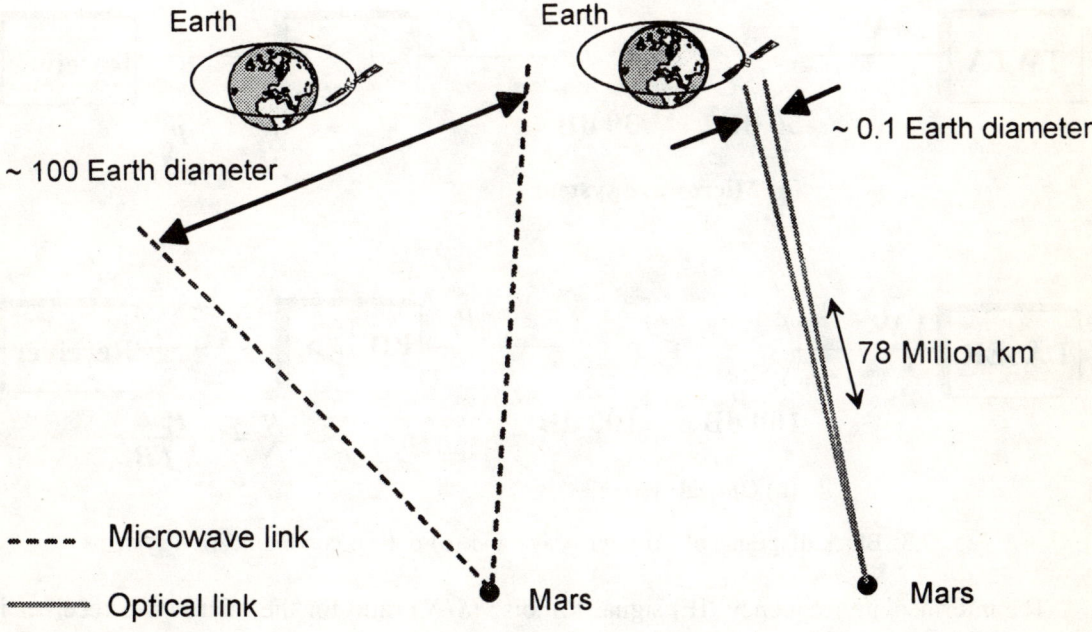

Fig. 9.2: Spot size at the Earth for microwave beam and optical beam transmitted from Mars

Most of the differences between optical and microwave space links are due to the difference in wavelengths. The much shorter optical wavelength results in a very small beam width and a very small spot size. As shown in Fig. 9.2, spot size of an optical beam transmitted from planet Mars is only about 10 percent of the Earth's diameter, whereas in case of microwave beam it is about 100 times. The small spot size of an optical beam results in substantial increase in the received optical power and therefore improvement in the receiver performance.

The much higher antenna gain achieved at the optical frequency allows to

- decrease the antenna (telescope) diameter (for example $D = 10$ cm),

- increase the data rate and

- increase the transmission distance which can, for example, reach millions of km in space.

Block diagrams of microwave and optical communication systems with typical values of transmitting power and antenna gains are shown in Fig. 9.3.

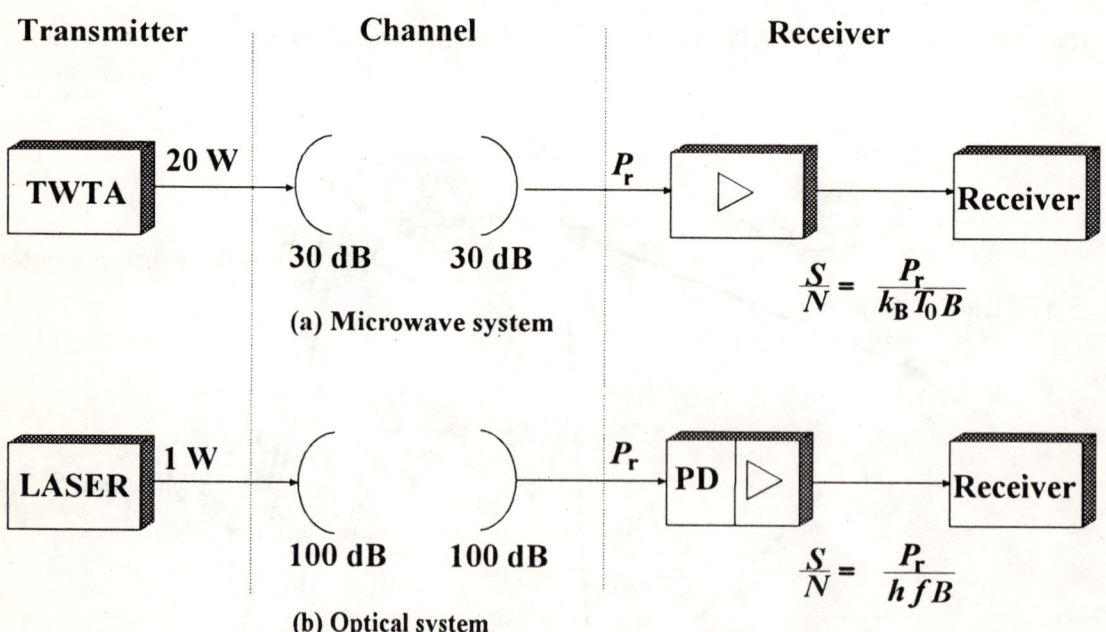

Fig. 9.3: Block diagrams of (a) microwave and (b) optical communication systems

The intermediate frequency (IF) signal-to-noise (S/N) ratio for the microwave receiver is given by

$$\frac{S}{N} = \frac{P_r}{k_B T_0 B} \qquad (9.3)$$

where P_r represents the received signal power, k_B the Boltzmann's constant, T_0 the temperature in degree Kelvin and $B = 1/T$ (T is the bit duration) the Nyquist bandwidth of the IF signal. The corresponding expression for the coherent (heterodyne) optical receiver is given by [202]

$$\frac{S}{N} = \frac{P_r}{hfB} \qquad (9.4)$$

When practical bandwidth is taken to be two times of the Nyquist bandwidth, denominators in Eqs. (9.3) and (9.4) will get multiplied by a factor of 2.

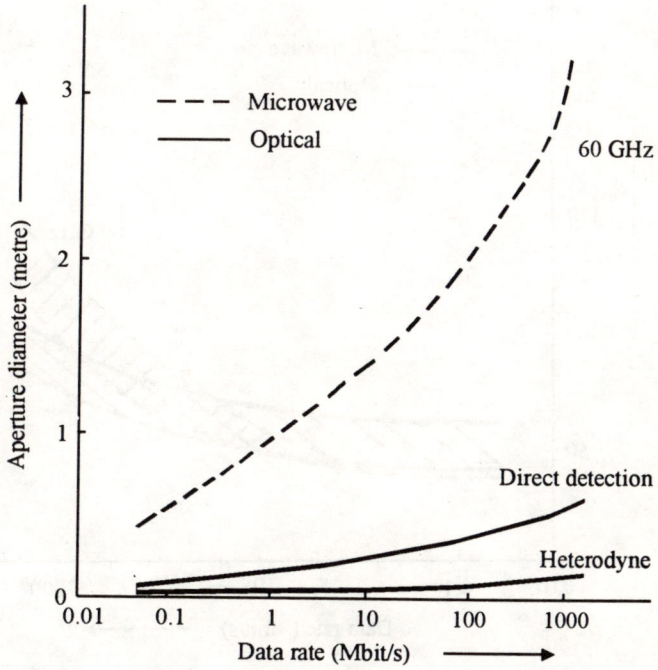

Fig. 9.4: Variations of antenna aperture diameter with data rate for microwave and optical systems

A comparison of Eqs. (9.3) with (9.4) reveals that the thermal noise ($k_B T_0$) in microwave systems corresponds to the quantum noise hf in optical systems. This analogy is not surprising. It is well-known from thermodynamics that $k_B T_0$ and hf are the two limiting forms of the noise psd of an oscillator at frequency f and in thermal equilibrium at temperature T_0.

At the microwave frequencies, $hf \ll k_B T_0$. As a consequence, thermal noise becomes a limiting factor, but it may not be at the optical frequencies [120, 202, 255]. In optical space communications, thermal and other sources of noise like background noise, optical preamplifier noise etc. may become comparable with the quantum noise. In that case, an additional parameter F_i called noise figure (NF) of the receiver will come in the denominator of Eq. (9.4). It varies from 1 to ∞. Lesser is the NF, better is the receiver performance. Variations of antenna aperture diameter and terminal weight with data rate for microwave and optical systems are shown in Figs. 9.4 and 9.5 respectively [68].

It is infer from these figures that the use of optical technology in space communications offers a tremendous reduction in antenna diameter and weight as compared to microwave technology. This is quite promising for satellite communications particularly when the data rate is high. The high data rates is essentially required to obtain a better resolution and a large frame rate of image sequences. For Earth observation satellite data rate of 1 Gbit/s is desired, whereas from the planet Mars more than 10 Mbit/s is the goal (compared to a few kbit/s with the existing RF systems).

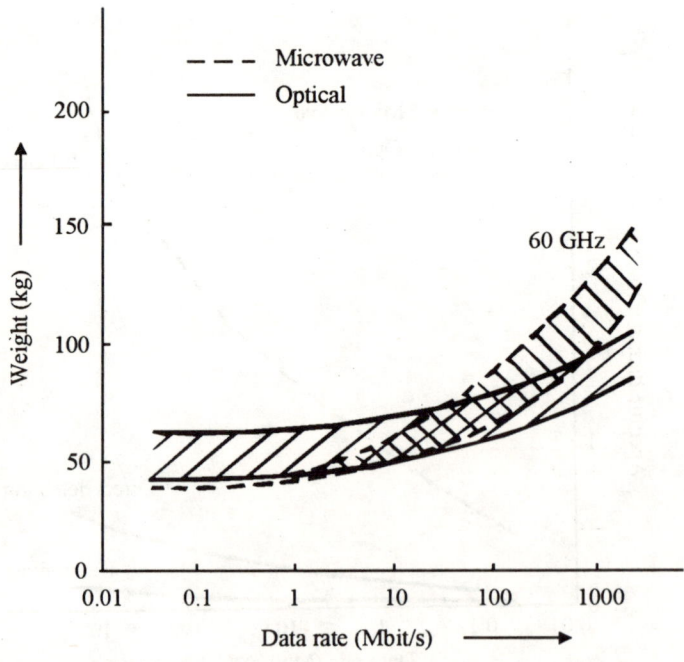

Fig. 9.5: Variations of weight with data rate for microwave and optical communication systems

9.4   COHERENT OPTICAL SPACE COMMUNICATIONS

The main advantage of using a coherent optical space communication system lies in the receiver sensitivity gain of more than 20 dB. It may be mentioned that a sensitivity improvement of 6 dB doubles the transmission distance (for example 30 million km with a direct detection system will become more than 60 million km with a coherent system). This substantial improvement in transmission distance is particularly attractive for the long-range space links in DSMs. For the same transmission distance, highly sensitive coherent optical receivers require much less transmitter power than an optical system with a direct detection receiver. And, therefore, the power consumption of satellite becomes very efficient. Moreover, the influence of background noise (for example from the Sun, Moon, Earth, Venus etc.) is reduced in coherent systems [150]. The main advantages of using coherent communication systems in space are the following:

- High capacity long-range optical space links are practicable,

- ASK, FSK, DPSK and PSK modulation techniques (binary and multilevel) with and without coding scheme are feasible,

- less sensitivity to background noise and

- better frequency selectivity permits the deployment of high density multichannel systems.

9.5 DRAWBACKS AND PROBLEMS OF REALIZATION

Unfortunately, the small beam width as the main advantage of optical space communication systems is also responsible for the main disadvantage. As shown in Fig. 9.6, small satellite vibrations may result in a complete link failure.

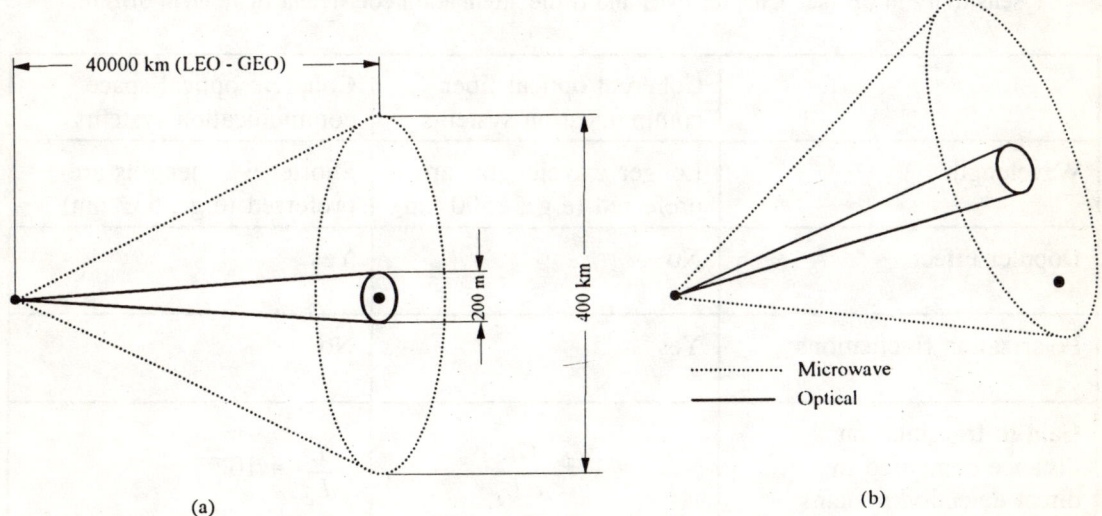

Fig. 9.6: (a) Main advantage and (b) drawback of optical space communication systems

The very small beam width complicates pointing, acquisition and tracking (PAT). Therefore, in addition to the communication subsystem, high-accuracy and speed PAT subsystems are required to reduce the influence of satellite vibrations. The main problem in the realization of optical space communication systems is the nonavailability of optical components at the wavelengths used in space. In optical fiber communication systems, higher wavelengths are preferred (for example 1500 nm), whereas in optical space systems lower wavelengths (for example 532 nm or visible) are preferred. It is worth mentioning that the fiber attenuation loss (due to material effects) decreases at longer wavelengths, whereas the free-space beam spread loss due to geometrical effects decreases at shorter wavelengths (refer to Eq. (9.8) derived latter).

Optical components viz., optical phase modulators, optical isolators, couplers etc. are available at wavelengths typically used in optical fiber applications such as 850 nm, 1300 nm or 1500 nm. In contrast to this, no such components are available at the Nd:YAG laser wavelength of 1064 nm or at its double the frequency i.e., 532 nm wavelength (except high-cost single productions). The Doppler frequency shift ($\approx \pm 10$ GHz) as a result of the relative motion of two linked satellites is another problem in realizing a coherent optical space communication system. This problem can be tackled by including a tuneable local laser with a tuneable frequency span of 20 GHz and an automatic frequency control (AFC) in the receiver circuit. In fiber optics, Doppler effect is non-existent.

In the following table, main differences between coherent optical fiber and coherent optical space communications have been summarized.

Table 9.1: Comparison of coherent optical fiber and space communication systems
* L_d is the distance in km for direct detection system, GR the improvement in receiver sensitivity in dB (see Chapter two) and α the attenuation coefficient of fiber in dB/km.

	Coherent optical fiber communication systems	Coherent optical space communication systems
Wavelength	Longer wavelengths are preferred (e.g., 1500 nm)	Shorter wavelengths are preferred (e.g., 532 nm)
Doppler effect	No	Yes
Polarization fluctuations	Yes	No
Gain in transmission distance compared to direct detection systems	$\dfrac{L}{L_d} = 1 + \dfrac{GR}{\alpha L_d}$	$\dfrac{L}{L_d} = 10^{\frac{GR}{70 \text{ dB}}}$
Special problems of realization	Adjustment of polarization	High-accuracy and speed PAT subsystems and reduction of Doppler effect

9.6 SYSTEM DESCRIPTION AND DESIGN

A physical model of laser communication system is shown in Fig. 9.7. The relationship between the transmitted and received optical power is described by the range equation. This equation provides a characterization of propagation in the communication channel, free-space propagation loss, transmitting and receiving antenna gains.

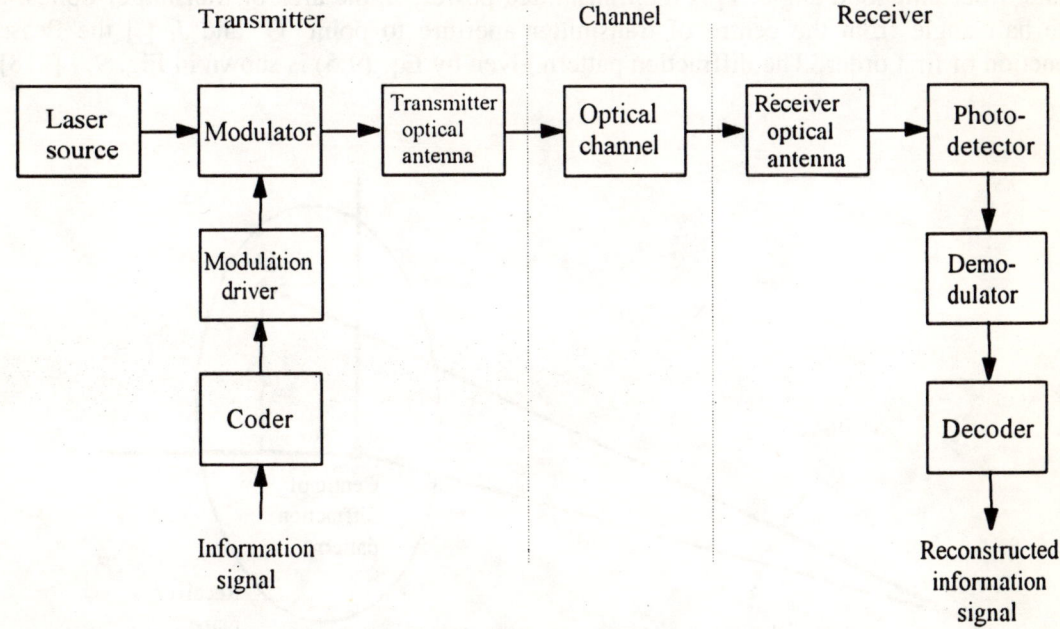

Fig. 9.7: Physical model of optical communication system

9.6.1 FREE-SPACE PROPAGATION FORMULA

When the source radiating diameter D_s is too small, size of its image in the receiver plane is determined by diffraction at the transmission aperture. The diffraction pattern produced by a uniformly illuminated circular aperture of diameter D_t consists of a set of concentric rings. The intensity per unit solid angle centred at point 'Q' in the receiver plane (Fig. 9.8) can be expressed as [218].

$$I(\alpha) = \left[\frac{2J_1(\pi D_t \alpha / \lambda)}{\pi D_t \alpha / \lambda} \right]^2 I(0) \tag{9.5a}$$

where

$$I(0) = \pi D_t^2 \frac{P_t}{4\lambda^2} \tag{9.5b}$$

The parameter $I(0)$ (also equals to $P_t A_t / \lambda^2$) represents the intensity at the centre of diffraction pattern per unit solid angle. P_t is the transmitted power, A_t the area of transmitter optics, α the half angle from the centre of transmitter aperture to point 'Q' and J_1 [.] the Bessel function of first order. The diffraction pattern given by Eq. (9.5) is shown in Fig. 9.9 [218].

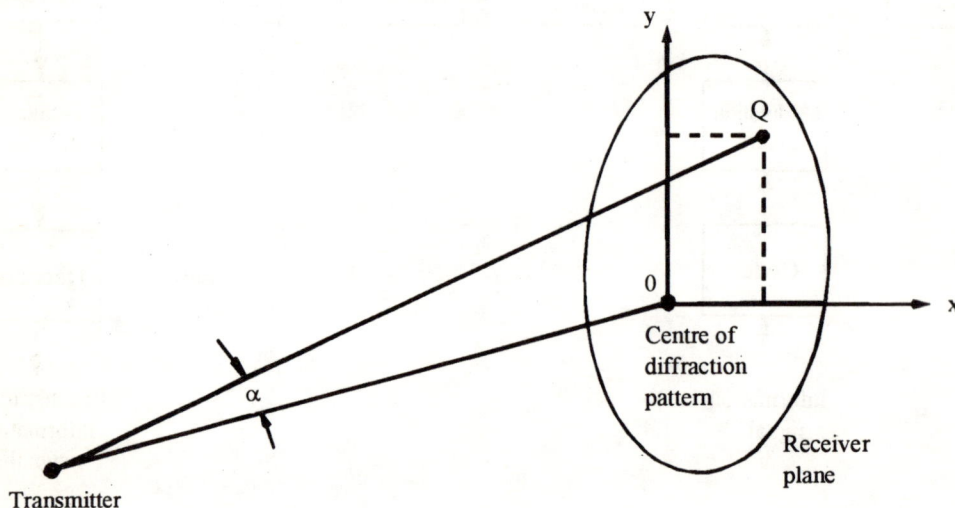

Fig. 9.8: Geometry of transmitter antenna diffraction pattern in the receiver plane

The power incident upon the receiver plane, $P(\theta)$ about the centre of diffraction pattern assuming medium transmissivity factor to be unity is given by

$$P(\theta) = I(0) \int_0^{2\pi} \int_0^{\theta} \left[\frac{2J_1(\pi D_t \alpha / \lambda)}{\pi D_t \alpha / \lambda} \right]^2 \alpha d\alpha d\phi \tag{9.6}$$

If the receiver optical antenna of diameter D_r is located at a distance L from the transmitter and centred on the optic axis, diffraction angle θ is approximately $D_r/2L$. For a large L, power density in the receiver plane is nearly constant at a maximum value of $I(0)$. Therefore, the maximum received power level from Eq. (9.6) is given by

$$P_r = I(0) \int_0^{2\pi} \int_0^{D_r/2L} \alpha \, d\alpha \cdot d\phi = \frac{\pi^2 P_t D_t^2 D_r^2}{16 L^2 \lambda^2} \tag{9.7}$$

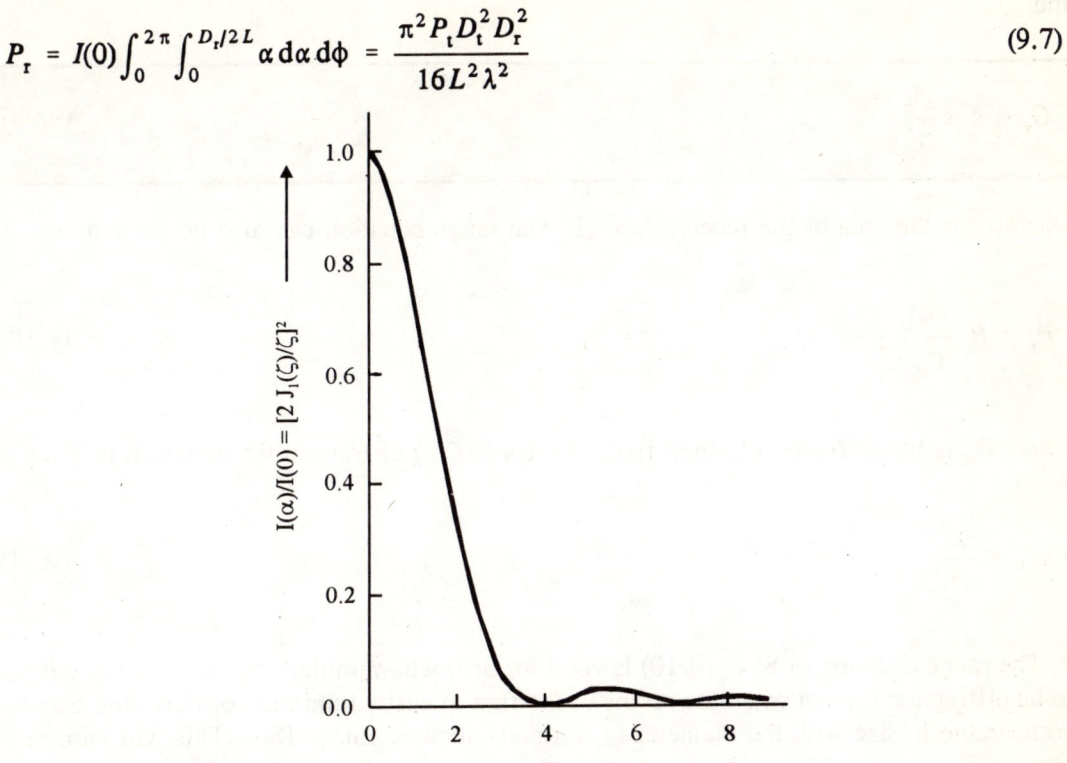

Fig. 9.9: Diffraction pattern of uniformly illuminated circular aperture

The equation above can be put in the form

$$P_r = P_t G_t \left(\frac{\lambda}{4\pi L} \right)^2 G_r \tag{9.8}$$

It is referred as range equation. Here, $(\lambda/4\pi L)^2$ represents the free-space loss, G_t and G_r are the gains of the transmitting and receiving antenna/optics respectively. These are given by

$$G_t = 4\pi \frac{A_t}{\lambda^2} \tag{9.9a}$$

and

$$G_r = 4\pi \frac{A_r}{\lambda^2} \tag{9.9b}$$

where A_r is the area of the receiver optics. The range equation can also be written as

$$P_r = P_t \frac{A_r}{\Omega_{dt} L^2} \tag{9.10}$$

where Ω_{dt} is the diffraction limited field of view (FOV) of transmitter optics. It is given by

$$\Omega_{dt} = \frac{\lambda^2}{A_t} \tag{9.11}$$

The range equation (9.8) or (9.10) is valid for diffraction limited sources. A source is said to be diffraction limited when the radius of the first intensity minimum or dark ring becomes comparable in size with the diameter of normally focused image [86]. This will happen if

$$D_s < \frac{1.22 f_c \lambda}{D_t} \tag{9.12}$$

where f_c is the focal length of the transmitter optics. When a source satisfies Eq. (9.12), P_r can be determined by using either Eq. (9.8) or Eq. (9.10) [86]. Otherwise P_r will be reduced by a factor Ω_{dt}/Ω_t where Ω_t is the transmitter optics FOV [32].

The P_r may also get reduced due to obscuration caused by the secondary mirrors or field stops of the telescopes (see Figs. 9.11 and 9.16), poor quality of the optics etc. And also due to inaccuracy in the alignment of the optical transmitter and receiver i.e., the pointing error. For a fixed pointing error $\pm\psi_e$, optical beam width has to be increased to compensate the pointing error. It will cause the reduction of power by a factor of $(1+2\psi_e/\psi_a)^2$ [70]. Generally, ψ_e is a random variable. It is considered to be Gaussian distributed with zero mean and variance, σ_e^2. Setting $6\sigma_e \leq \psi_a$ gives the probability of $\psi_e > \psi_a$ less than 0.01 [218]. When σ_e^2 is known, one can find out the required ψ_a and hence D_a.

As shown in Fig. 9.7, space optical communication systems have four components: optical transmitter, channel, receiver and PAT subsystem involving both transmitter and receiver (not shown explicitly). These components have been briefly discussed in the following Sections.

9.6.2   OPTICAL TRANSMITTER

The optical transmitter consists of coder, modulator, optical source and transmitting antenna. Depending upon the type of source, modulation may be impressed during the generation of light or after the light has been generated. The transmitting antenna is a lens system that focuses the optical beam for transmission through the optical channel.

(i) OPERATING WAVELENGTHS AND SOURCES

Use of shorter wavelengths in optical space communications give rise to smaller spot size. It produces improvement in the receiver performance/increase in the transmission distance. While selecting a wavelength, in addition to above, many other factors like availability of source, output power level, life time etc. are also to be taken into account.

Three basic laser technologies are existing for space communications. These are based on: (i) CO_2 gas laser operating in the infrared region at 10 μm, (ii) Laser diode pumped Nd:YAG laser operating at 1.064 μm and can be frequency doubled and (iii) Semiconductor laser diode operating in the region 0.78 μm-1.55 μm. During the last decade, CO_2 laser communication links appeared most attractive because of their higher electrical-to-optical conversion efficiency and longer life time. The problems with this laser are its long wavelength and less reliability by virtue of a gas laser. The rapid development of laser diode technology since 1980 has considerably changed this scenario. Now the major light sources are semiconductor laser devices (lasers and laser arrays) and Nd:YAG lasers.

Semiconductor lasers because of their high reliability, efficiency, small size and weight are excellent light sources in optical space communications. Their major drawback is that a single device typically cannot radiate the required power (few hundred milliwatts to a watt) in a stable single-mode operation. This problem can be overcome by combining the power of several semiconductors lasers. The objective is to increase the source brightness (define as the power divided by product of source area and solid angle into which light is emitted), rather than the power. The reason for this is that higher source brightness and not just more source power, translates into higher intensity at the receiver [32].

The source brightness can be increased by combining the power of several semiconductor lasers either noncoherently or coherently. The noncoherent methods employ lasers that emit light at different polarizations or different wavelengths (a combination of two is also possible). The power of lasers is combined by using a polarization beam combiner or wavelength selective elements like filters or wavelengths dispersive elements like gratings. The non-coherent method of power combining is conceptually a simpler approach, but the entire transmitter head becomes a sophisticated optical system that includes, in addition to the laser themselves, many other optical components. Further, increase in the brightness level is not much.

The coherent methods are usually much more complicated to implement. The potential payoff is mainly an increased source brightness due to phase-locking of the lasers (similar to microwave phased antenna arrays). The three basic methods of coherent power combining are: mutual coupling, injection locking and coherent amplification. In mutual coupling method, no laser in the array has a privileged status. There is some coupling among the lasers, which under certain conditions results in their synchronization. The amount of coupling depends on the ratio of separation ($\Delta\lambda_i$) between the intrinsic wavelengths of oscillation of the interacting lasers and the centre wavelength (λ) of the laser transition linewidth [32]. It becomes more difficult to phase-lock the array as the number of elements increases [125]. In injection locking, there is single master laser oscillator. Portions of its emitted radiation are coupled simultaneously into all the other lasers in the array forcing them to oscillate at its frequency. In coherent amplifications, light generated in the master laser is split and fed simultaneously into gain elements, where a travelling wave amplification is employed. If the output of the master laser is coherent over its near-field pattern, outputs of the amplifiers are automatically phase-locked. It has the advantage of avoiding the generally stringent conditions required for establishing phase-locking [32].

In both the injection locking and coherent amplification, failure of master oscillator leads to the failure of the entire source. It may not be so critical in the mutual phase-locking. However, failure of a single device may alter the modal shape of the entire array and make it unusable. The mutual phase-locking is more amenable than the other methods to monolithic implementation.

In the optical space communications, when the background noise or receiver noise or both are high, receiver performance is severely degraded. This can often be improved by using pulse laser i.e., higher amplitude and narrower pulses, while still maintaining the same average power level. The minimum pulse duration is determined by the detector bandwidth and the maximum power level by the largest peak power that the laser source can generate. Due to peak power limitations, semiconductor lasers in low duty cycle operation will emit much less average power than their continuous wave (CW) rating. This problem can be overcome by employing a different type of source. Among all the possible alternatives, Nd:YAG laser seems to be the most attractive. It is a solid-state laser and can be operated in both CW and pulse modes. The most efficient pumping band for Nd:YAG laser is $0.81\ \mu$m. It is relatively easy to fabricate AlGaAs laser sources which emit at the pump wavelength. Even when Nd:YAG laser operates in a pulse mode (employing Q-switching, cavity dumping or mode locking), pump is continuously operated. Thus, this laser may be considered as a highly efficient DC-to-pulse converter to the light that is (efficiently) generated by the semiconductor laser arrays. It may be mentioned that beam quality requirements from the semiconductor laser arrays are less stringent when they are used as pump sources and not directly as the light sources for communications. Pulse position modulation (PPM) is one of the modulation schemes which requires the use of a pulse laser. In this modulation scheme, laser pulse is delayed into one of the "m" possible time slots during each frame interval. Therefore, correct decision of a single pulse conveys $\log_2 m$ bits of information.

Theoretically, spectral linewidth of a semiconductor diode pumped Nd:YAG laser is less than 1 Hz. For a typical semiconductor laser diode, it is in the order of 1 MHz. Further, in Nd:YAG laser, excess intensity noise is negligibly small. Therefore, it provides improvement in the receiver performance for single-detector configuration [141]. Its main drawback is extremely low (below 0.5 percent) electrical-to-optical conversion efficiency. This drawback has now been overcome to a great extent. These lasers with an efficiency of 8.5% to 17% have been developed and demonstrated [153].

(ii) TRANSMITTING ANTENNA

The transmitting antenna should always be designed as close to the diffraction limit as possible since this results in the smallest size. As mentioned earlier, narrowness of beamwidth is limited by the pointing error. The above two requirements put a limitation on the antenna diameter to be 10 centimetres to several metres depending on the wavelength of operation.

The transmitting antenna can be based either on refractive optics or reflective optics. Both types of antennas are shown in Fig. 9.10 and Fig. 9.11 respectively [218]. For small size apertures, lenses are practical for forming the transmitter antenna. When the aperture size is several centimetres in diameter, reflective type is preferred because of lower cost and weight.

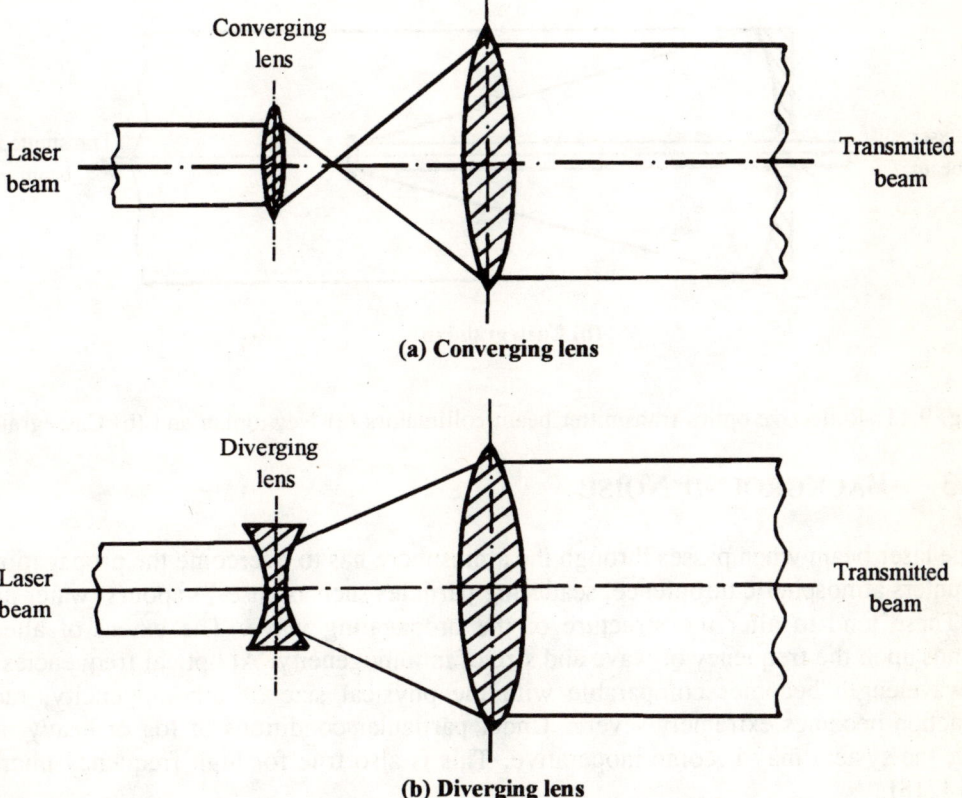

(a) Converging lens

(b) Diverging lens

Fig. 9.10: Refractive optics transmitter beam collimators with (a) converging and (b) diverging lens

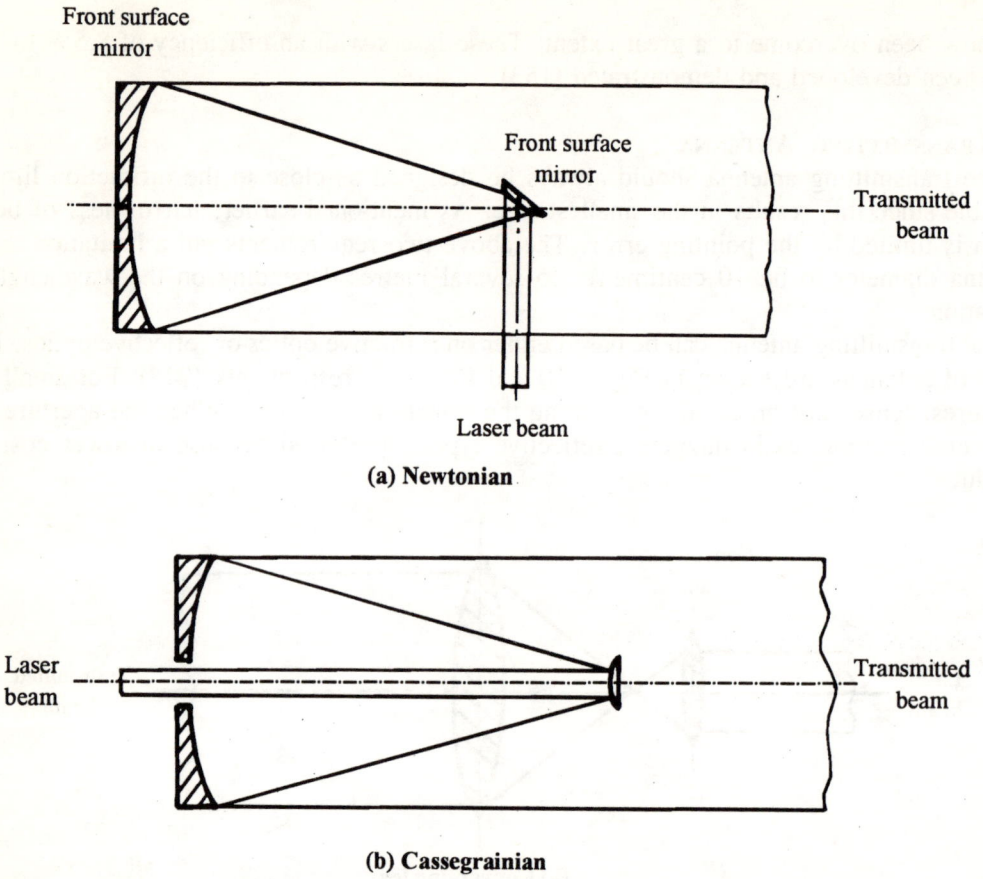

Fig. 9.11: Reflective optics transmitter beam collimators (a) Newtonian and (b) Cassegrainian

9.6.3   BACKGROUND NOISE

The laser beam when passes through the atmosphere has to overcome the propagation loss, encounters atmospheric turbulence, scattering particles such as gases, vapours, water droplets etc. These tend to alter the structure of the propagating wave. The extent of alterations depends upon the frequency of wave and size of inhomogeneity. At optical frequencies where the wavelength becomes comparable with the physical size of inhomogeneity, radiation interaction becomes extremely severe. Under particular conditions of fog or heavy rains or snow, the system may become inoperative. This is also true for high frequency microwave links [218].

When the propagation medium is free-space as in IOLs, ISLs or DSMs, principal channel effect is the propagation loss. The transmission is free of distortion and the objective of system design is to overcome loss by the appropriate choice of source, transmitter and receiver antennas or by increasing the receiver sensitivity or their combinations. Of course, optical channel will introduce background noise in the information signal. This background noise when present in large proportion, can severely degrade the receiver performance and may lead to link failure. The Sun is the strongest background noise source (power spectral density per mode at $\lambda = 1.064$ μm has been reported to be $1.9 \cdot 10^{-20}$ Watt/Hz·mode) [150]. If the receiver FOV is quite large and the communication satellites and the Sun are in line, interruption of communication due to background noise is quite likely [294]. In the following, background noise has been discussed.

The background noise power P_b pick up by the receiver in a bandwidth B around the frequency f is given by [70]

$$P_b = \begin{cases} N(f)BA_r\Omega_r & \text{if} \quad \Omega_r \in \Omega_s \\ N(f)BA_r\Omega_s & \text{if} \quad \Omega_s \in \Omega_r \end{cases} \tag{9.13}$$

where $N(f)$ is the spectral radiance function for the single polarization state. It represents the power radiated at frequency f in a unit bandwidth into a unit solid angle per unit source area. The solid angle Ω_s is the angle subtended by the source when viewed from the detector. It is given by

$$\Omega_s = \frac{A_s}{L^2} \tag{9.14}$$

Here, L is the distance between source and receiver and A_s the source area. The parameter, Ω_r represents the FOV of the receiver. It is the solid angle within which all arriving plane waves project their diffraction pattern onto the detector. By standard geometric analysis in Fig. 9.12, it is given by

$$\Omega_r \approx \frac{A_d}{f_c^2} \tag{9.15}$$

Here, A_d is the detector area and f_c the focal length of the lens at the receiver. Eq. (9.13) for $\Omega_r \in \Omega_s$ is used when the angular extent of source is larger than the receiver FOV. Such sources include sky, sunlit Earth etc. When the angular extent of the source is smaller than the receiver FOV, noise power is determined by using Eq. (9.13) for $\Omega_s \in \Omega_r$. It may be the case with sources like stars, planets, Sun etc.

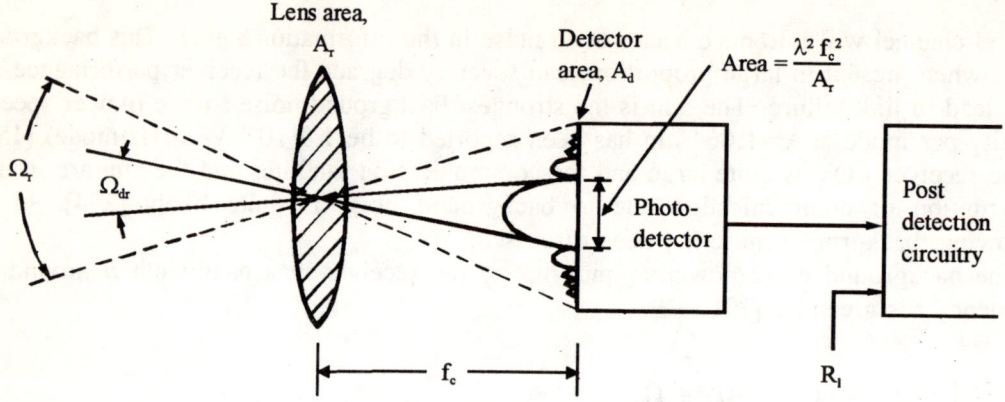

Fig. 9.12: Schematic diagram for relationship between Ω_r and Ω_{dr}

When $\Omega_r \in \Omega_s$, number of background spatial modes received by the receiver is

$$m_t = \frac{\Omega_r}{\Omega_{dr}} \tag{9.16a}$$

Otherwise (i.e., when $\Omega_s \in \Omega_r$), number of modes is given by

$$m_t = \frac{\Omega_s}{\Omega_{dr}} \tag{9.16b}$$

Here, Ω_{dr} represents the diffraction limited FOV of the receiver. It is the minimum FOV consisting of set of arrival angles whose pattern superimpose on the detector surface and are indistinguishable in terms of direction of arrival. As shown in Fig. 9.12, it is given by

$$\Omega_{dr} \sim \frac{(\lambda f_c)^2 / A_r}{f_c^2} = \frac{\lambda^2}{A_r} \tag{9.17}$$

Using Eqs. (9.13) and (9.16), background noise psd $S_n = P_b / (B \cdot m_t)$ per spatial mode for both the types of sources (i.e., when $\Omega_r \in \Omega_s$ or $\Omega_s \in \Omega_r$) becomes

$$S_n = N(f) \lambda^2 \tag{9.18}$$

This psd will get multiplied by a factor of two if the other polarization states of the background noise are considered [150]. The Eq. (9.18) is used to determine S_n at wavelengths greater than 3 μm. For shorter wavelengths, radiation due to Sun reflectance has also to be considered along with the self-emission. To determine S_n in these instances, one may resort to spectral radiance, $H(f) = N(f)\Omega_s$ having the units of power per unit bandwidth per unit area [70]. This function can be easily measured without the explicit knowledge of Ω_s. Data on $H(f)$ are readily available in the literature [218]. The background noise power at the receiver is given by

$$
P_b = \begin{cases} H(f)BA_r\Omega_r/\Omega_s & \text{if} \quad \Omega_r \in \Omega_s \\ H(f)BA_r & \text{if} \quad \Omega_s \in \Omega_r \end{cases} \tag{9.19}
$$

As before, noise psd for spatial mode in both the cases will be same and is given by

$$
S_n = H(f)\frac{\lambda^2}{\Omega_s} \tag{9.20}
$$

Using the relationship

$$
H(f) = H(\lambda)\frac{\lambda^2}{c} \tag{9.21}
$$

Eq. (9.20) can be written as

$$
S_n = \frac{H(\lambda)\lambda^4}{c\Omega_s} \tag{9.22}
$$

Eqs. (9.18) and (9.22) are used to determine S_n at different wavelengths and for various sources. Some of these values are given in the following table [150].

Table 9.2: Power spectral density per mode (in W/Hz) for various celestial bodies at selected wavelengths

	0.85 μm	1.06 μm	1.3 μm	10.6 μm
Earth	$2.6 \cdot 10^{-26}$	$4.0 \cdot 10^{-26}$	$5.5 \cdot 10^{-26}$	$2.0 \cdot 10^{-22}$
Moon	$1.8 \cdot 10^{-26}$	$2.7 \cdot 10^{-26}$	$3.6 \cdot 10^{-26}$	$6.5 \cdot 10^{-22}$
Sun	$1.3 \cdot 10^{-20}$	$1.9 \cdot 10^{-20}$	$2.6 \cdot 10^{-20}$	$7.0 \cdot 10^{-20}$
Venus	$2.1 \cdot 10^{-25}$	$3.1 \cdot 10^{-25}$	$4.1 \cdot 10^{-25}$	$3.1 \cdot 10^{-22}$

For determining the receiver performance, time modes are also to be considered along with the spatial modes. For the description of time modes, let the wave velocity is c and there is no dispersion. Then in time t, receiver measures the field which had previously occupied a path length $L_p = ct$. This field can be described by expanding it into a series of orthogonal modes using spatial Fourier series. For this expansion, the q th mode varies with distance and time according to the exponential factor

$$\exp\left[jq\frac{2\pi}{L_p}(z-ct)\right] \tag{9.23}$$

where q is an integer. It is seen from the above equation that the frequency of the q th mode is qc/L_p. The condition for orthogonality of the modes is that different values of q differ by an integer. This means that the frequency separation between adjacent modes is $\Delta f = c/L_p$. In a receiver of bandwidth B, there will be $B/\Delta f = BL_p/c$ orthogonal modes. Since L_p is equal to ct, receiver measures the state of excitation of Bt such modes. Therefore, the rate of arrival of independent spatial modes at the receiver is B [85].

9.6.4   POINTING, ACQUISITION AND TRACKING

In order to establish an optical space link, transmitted field in addition to having to overcome the path loss must also be properly aimed towards the receiver. Likewise, the receiver must determine the direction of arrival of the transmitted beam. The operation of aiming a transmitter in the proper direction is referred as pointing. The receiver operation of determining the direction of arrival of an impinging beam is called spatial acquisition. The subsequent operation of maintaining the pointing and acquisition throughout the communication time period is called spatial tracking. Thus, the PAT function is to keep aligned the opposite terminals before and throughout the communication. Further, PAT has to compensate the motion of opposite terminals during the round trip time of light by using a point-ahead angle. In a coherent receiver, it has to take care of Doppler frequency shift or perform the frequency pointing, acquisition and tracking also.

The pointing function of PAT subsystem is basically an open-loop system. The inherent errors of the open-loop pointing are typically orders of magnitude higher than the divergence of transmitter beam, so that the tracking function cannot yet lock onto the received beam. The acquisition system performs the function of reducing the pointing error to allow the tracking system to take over. The required accuracy makes a closed-loop operation mandatory. The tracking function operates throughout the communication to keep the antenna aligned within the required sub-microradian accuracies. The tracking is a closed-loop process. Generally, the communication beam of opposite terminal is used by the tracking subsystem to obtain the pointing information for its own antenna.

To understand the pointing, acquisition and tracking operations, consider a link to be established between station 1 and station 2. An important step in establishing the link involves the process of spatial acquisition i.e., aiming the receiving antenna in direction of the arriving field. The acquisition subsystem can use either an expanded transmitter laser beam as a beacon or a separate beacon [26]. In the latter case, optical beacon is an unmodulated laser beam at a wavelength different from the transmitter. It is located at or near the receiver. Alternatively, in a duplex system, each transmitter can serve as a beacon to the other terminal.

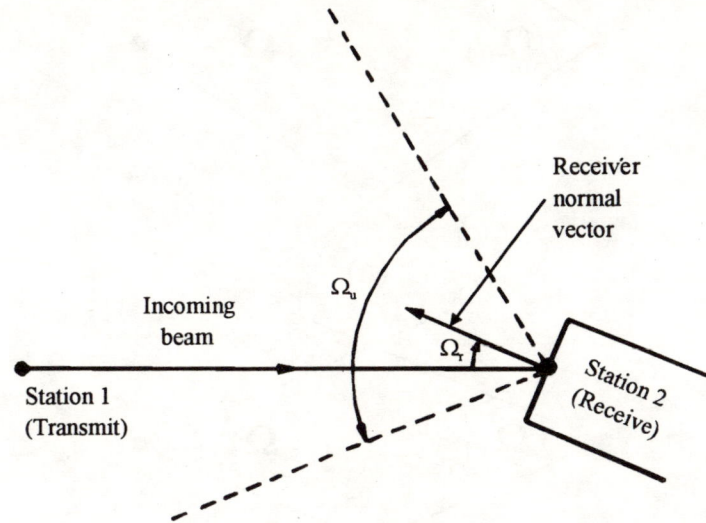

Figure 9.13: One-way beam acquisition sequence

The acquisition can be accomplished by one-way or two-way procedures. Let us consider expanded laser beam as a beacon. As shown in Fig. 9.13, a single transmitter or solar-illuminated planets as in deep space communication at one point transmit to a single receiver located at another point. If the pointing is satisfactory, receiver knows the transmitter direction within some uncertainly solid angle Ω_u defined from the receiver location. The receiver will then align the antenna normal to the direction of arriving field within some pre-assigned resolution solid angle Ω_r. In two-way acquisition procedure (refer to Fig. 9.14), station 1 transmits a beam wide enough (solid angle Ω_{u1}) to cover its pointing error. The station 2 will search its uncertainly FOV Ω_{u2} to acquire. After successful acquisition, station 2 transmits to the station 1 with beam width Ω_{r2} using the arrival direction obtained from the acquisition. The station 1 can now acquire the return beam with its desired resolution Ω_{r1}. For still narrower beams, the procedure can be repeated again. The link is now established with the desired resolution and the communication can begin.

When a separate beacon is used for acquisition (refer to Fig. 9.15), station 2 after the acquisition of transmitted beam sends a beacon signal towards the station 1. The station 1 first receives the beacon and then transmits its modulated laser beam towards the beacon direction of arrival. When the beams are extremely narrow, the station 2 may have moved out of the station 1 beam width during the round trip transmission time. In that case, station 1 must point-ahead its transmission from the measured beacon arrival direction.

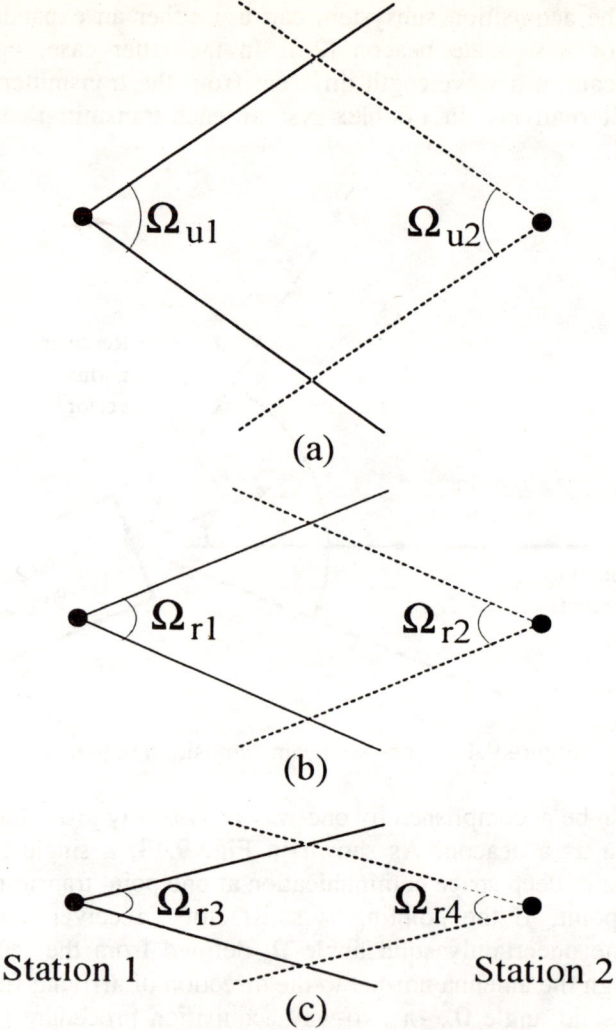

Station 1   Station 2
(c)

Fig. 9.14: Two-way beam acquisition sequence

For two-way acquisition, both the stations 1 and 2 must contain a transmitting laser and an optical receiver. The modulated laser beam serves as a beacon for the return direction. Usually, separate wavelengths are used for the optical beam in each direction. For point-ahead correction, command control must adjust transmit direction relative to receive direction. Such an adjustment requires accurate satellite altitude control [71].

Although acquisition and tracking systems are distinct from the communication system, the two are interrelated in defining system performance. As clear from above, there are basically three link types: acquisition, tracking and communication which operate at the same time. Performance criteria for the above three links are quite different. For acquisition, criteria are typically the acquisition time, false alarm rate (FAR), probability of detection (p_d) and if a multiple detections scheme is used, number of detections needed. For the tracking link, key consideration is the angle error induced by the receiver circuitry. This angle error is commonly referred as noise effective angle (NEA). For the communication link, key parameters are: required data rate, probability of error (p_e)/bit error rate (BER) and proba-bility of burst error (PBE) [21]. The bit error occurs due to the presence of various internal and external sources of noise in the optical receiver. It is normally determined by the average/peak signal power and rms noise power during a bit interval. The burst error occurs when the signal irradiance at the distant receiver falls below the level required to maintain average BER. When considered from the pointing and tracking point of view, burst error occurs when the instantaneous mispoint loss exceeds the value included in the link power budget.

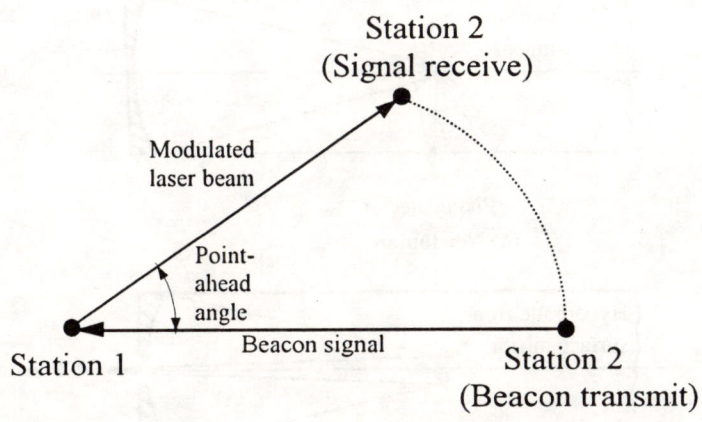

Fig. 9.15: Point-ahead model for optical pointing

9.6.5   OPTICAL RECEIVER

The optical receiver basically consists of a receiving optics (antenna) followed by a demodulator/detector. The latter may be based on noncoherent or coherent techniques depending upon the type of modulation scheme used. The various receiver components are described below.

(i) RECEIVING ANTENNA

The receiving antenna can be a reflective or refractive type collimator. It will gather the optical energy over an aperture of diameter D_r and produce a collimated beam of diameter somewhat smaller than the photodetector area to allow for pointing inaccuracy of the receiver. The photodetector will be placed at the focal point of antenna (Fig. 9.16) [218].

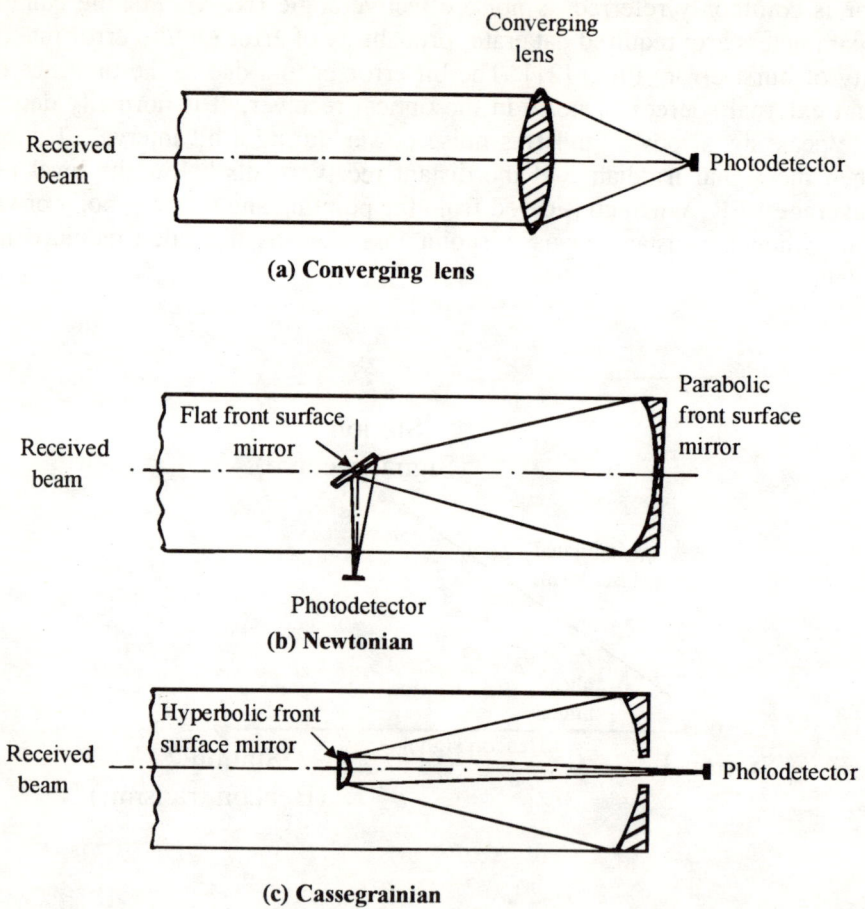

Fig. 9.16: Receiver optical antennas (a) converging lens, (b) Newtonian and (c) Cassegrainian

In a direct detection receiver, diameter of the receiving antenna should be as large as possible to gather the maximum amount of signal energy. Increasing the antenna diameter not only increases the average signal level, but also decreases the ripple in it. In a coherent receiver (both homodyne and heterodyne), there are different constraints. As the aperture diameter increases, effect of intensity fluctuations decreases, but contribution to the variance from phase fluctuations increases. Thus, there is some compromise value of receiver antenna diameter which minimizes the signal variance [218].

Irrespective of whether it is a direct detection or a coherent receiver, amount of optical power focused onto the detector is equal to the amount of power incident over the focusing lens. Thus, the photodetected power can be equivalently computed by determining received power over the lens area instead of the focused power over the detector area. The former is obtained directly from the range equation. It may be mentioned that it does not explicitly involve the actual photodetecting area [71].

A photodetector will respond to all radiation focused on its photoemissive surface. Thus, the receiver FOV given by Eq. (9.15) will depend on detector area A_d and focal length f_c rather than on size A_r of receiving lens. This is in contrast to a RF system wherein receiver FOV depends on λ^2/A_r when diffraction limited. Thus, the optical system FOV can be adjusted independently of A_r. Typically, the focal length of receiver lens is approximately taken as square root of A_r (i.e., $f_c \simeq \sqrt{A_r}$). With this, the receiver FOV becomes

$$\Omega_r \cong \frac{A_d}{f_c^2} \approx \frac{A_d}{A_r} = \left(\frac{A_d}{\lambda^2}\right)\left(\frac{\lambda^2}{A_r}\right) \tag{9.24}$$

At the optical frequency, A_d is in the order of millimetres and λ in the order of microns. As a consequence, (A_d/λ^2) is greater than unity and Ω_r is many times larger than the diffraction limited FOV (λ^2/A_r) for the same A_r. This is quite important in assessing optical visibility and optical receiver noise [71].

(ii) OPTICAL DETECTORS AND MINIMUM REQUIRED POWER EVALUATION

The photodetectors used in optical space communications include PIN-photodiode, avalanche photodiode (APD), photomultiplier tube (PMT) and photon counter. These detectors have their own merits and demerits. As a consequence, an optical receiver with a particular detector may perform better than with other detectors under certain conditions.

For direct detection intersatellite links based on Nd:YAG laser, detector used is either a PIN-photodiode or an APD. For applications requiring gain, silicon APD is generally preferred because of its internal gain. The gain of an APD is limited to about 50-300 because it is a smaller and lighter device. Its excess noise factor increases with the increase in the average gain, M and is usually in the range of 2-5 for the above gain values. When the gain is not required, a PIN-photodiode would be preferred because of its higher quantum efficiency ($\eta \simeq 0.8$). At double the frequency of Nd:YAG laser (i.e., $\lambda = 532$ nm), PMT is generally preferred. Its main advantages are very high gain and low excess noise factor. Typical value of gain lies between 10^4-10^6 and excess noise factor between 1 and 2. However, aside from the physical limitation of size and reliability of high voltage components (5 kV or higher), PMTs have lower quantum efficiencies. This is the result of limited materials and deposition techniques for the photoactive surfaces and the inability to collect efficiently all generated photoelectrons. In addition to this, each photoelectron follows a slightly different trajectory as it passes from photocathode to dynodes and finally to anode. Thus, there is a dispersion of arrival times for the photoelectrons. It leads to reduction in the receiver bandwidth. Usually, higher the gain, greater is the reduction in the bandwidth [126]. In coherent communication systems, PIN-photodiode is generally used.

Irrespective of whether the receiver is of direct detection (PIN/APD/PMT) or coherent detection (homodyne/heterodyne) type, background noise sources like Sun, sunlit Earth, Venus etc. may enter in its FOV. Since photodetector in the optical receiver acts as a mixer for the incident optical fields, it will produce signal-background, background-background beat noise components along with the desired signal component. In a coherent receiver, in addition to above noise components, there will be a LO-background beat noise component too. These components degrade the signal-to-noise (S/N) ratio [150].

In space communications, strength of the received signal may become considerably low due to increase in the transmission distance. In such cases, use of a semiconductor optical amplifier (OA) as preamplifier in the receiver may be of great help. When the amplifier output is detected by a photodetector, it will produce signal-ASE and ASE-ASE beat noise components. In a heterodyne receiver, there will be an additional LO-ASE beat noise component. In a TWA, ASE-ASE beat noise component is spread over a wide continuum instead of being concentrated in discrete modes [143].

In the following, performance evaluation of commonly used coherent and noncoherent receivers in the background noise environment has been discussed. The receiver structures considered are: (a) coherent PSK homodyne receiver, (b) coherent FSK heterodyne receiver, (c) direct detection (PIN+OA) receiver for OOK, (d) direct detection (APD) receiver for OOK, (e) direct detection (PMT) receiver for OOK, (f) direct detection (APD) receiver for m-PPM and (g) direct Detection (PMT) receiver for m-PPM. The general expressions for the output S/N ratio in terms of received optical power level P_r and other receiver parameters are given. These expressions are used to obtain the variations of p_e with P_r in the graphical form for a given data rate R. The minimum required P_r for achieving a given p_e at a given R can easily be determined with the help of these curves. This minimum P_r is subsequently used in the preparation of link power budget.

(a) COHERENT PSK HOMODYNE RECEIVER

A general schematic diagram of an optical communication receiver (valid for different configurations after the deletion of some blocks) is shown in Fig. 9.17.

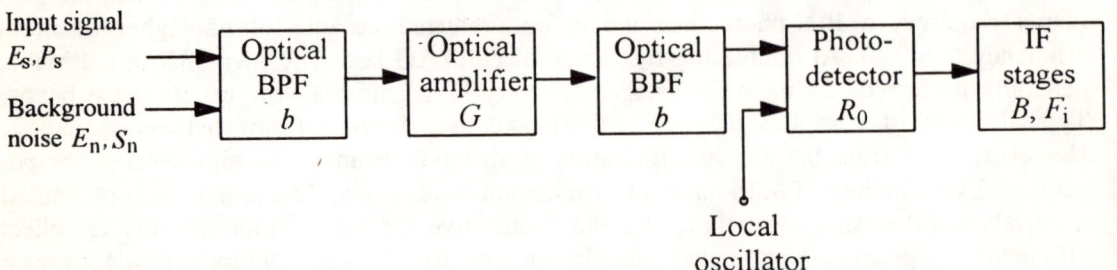

Fig. 9.17: Schematic diagram of optical communication receiver

For PSK homodyne receiver, optical amplifier, following optical BPF and IF stages blocks will not be there. The photodetector current

$$i \propto |E_r(t) + E_n(t) + E_l(t)|^2 \tag{9.25}$$

where E_r, E_n and E_l are the electric fields due to signal, background noise and local oscillator (LO) respectively. Following [115, 150, 154], various signal and noise components from Eq. (9.25) under the condition that the optical filter bandwidth b is much greater than $B = R = 1/T$ (where T is the bit duration), are given by

Signal Current: $\quad i_s(t) = R_0 2 \sqrt{P_r P_l} \; \cos(\theta(t))$ \hfill (9.26a)

DC Current: $\quad I_{ch} = R_0[P_r + P_l + m_t S_n b]$ \hfill (9.26b)

RMS noise current: $\overline{i_{nh}^2} = eI_{ch}2B + R_0^2 S_n[2P_r + 2P_l + m_t S_n b]2B$

$$+ \; \frac{2k_B T_0}{R_l} F_i 2B \tag{9.26c}$$

where $\theta(t) = \theta_l(t) - \theta_r(t)$. The parameters P_r and $\theta_r(t)$ represent the power and phase of received optical signal respectively. Similarly, P_l and $\theta_l(t)$ represent the above quantities for LO signal. In binary PSK signalling scheme, transmitter is considered to send a light pulse with a phase of either zero or π radians. The bandwidth occupied by PSK signal is taken as $B = 1/T$. The $R_0 = e\eta/hf$ represents the responsivity of the photodetector, where η is the quantum efficiency, e the electron charge, f the optical frequency of received signal ($f = \omega/2\pi$) and h the Planck's constant. The R_l represents the photodetector load resistance and F_i the NF of the following stages.

In the derivation of Eq. (9.26), it has been considered that the received data field is in single mode as it originates from a distant point source. At the optical frequency, receiver FOV Ω_r is much greater than Ω_{dr} (refer to Fig. 9.12). When a fiber telescope is used due to size and weight constraints, Ω_{dr} as given by Eq. (9.17) will increase considerably and it may become comparable with Ω_r. In that case, m_t will become nearly unity [154]. Because of this, m_t has been taken as unity in the analysis of this receiver as well as others. For other values of m_t, analyses can be carried out in the same way.

In Eq. (9.26b), first and second terms arise due to received and LO signal respectively. The third term represents the DC component of the current arising due to background noise. In Eq. (9.26c), first term represents the shot noise component. The second, third and fourth terms represent the contribution due to signal-background, LO-background and background-background beat noise respectively. The last term represents the thermal noise contribution of the photodetector load resistance and following stages.

The peak output S/N ratio for binary PSK signalling scheme from Eq. (9.26) is given by

$$\frac{S}{N} = \frac{\overline{i_s^2(t)}}{\overline{i_{nh}^2}} \tag{9.27}$$

Substituting $i_s(t)$ and $\overline{i_{nh}^2}$ from Eq. (9.26a) and Eq. (9.26c) respectively and after further simplification, above equation becomes

$$\frac{S}{N} = \frac{2P_r}{hfBF_h} \tag{9.28}$$

where

$$F_h = \frac{1}{\eta}\left[1 + \frac{P_r}{P_l} + \frac{S_n b}{P_l}\right] + 2\frac{S_n}{hf}\left[1 + \frac{P_r}{2P_l} + \frac{S_n b}{2P_l}\right] + \frac{k_B T_0}{hf}\frac{F_i}{L_m} \tag{9.29a}$$

and

$$L_m = \frac{1}{2} R_0^2 P_l R_l \tag{9.29b}$$

In a balanced homodyne receiver, signal-background and background-background components will be cancelled out [291]. Therefore, F_h from Eq. (9.29a) becomes

$$F_h = \frac{1}{\eta}\left[1 + \frac{P_r}{P_l} + \frac{S_n b}{P_l}\right] + 2\frac{S_n}{hf} + \frac{k_B T_0}{hf}\frac{F_i}{L_m} \tag{9.30}$$

The p_e is, therefore, given by

$$p_e = \frac{1}{2}\mathrm{erfc}\left(\sqrt{\frac{S}{N}}\right) \tag{9.31}$$

Variations of p_e with P_r computed from the above equation for $\eta=0.7$, $\lambda=1064$ nm, $b=5$ nm, $n_{sp}=1.0$, $F_i=2$ dB, $T_0=300$ K, $R_l=100$ Ω, $S_n=1.9\cdot10^{-20}$ Watt/Hz·mode, $R=560$ Mbit/s and $P_l=10$ dBm are shown in Fig. 9.18.

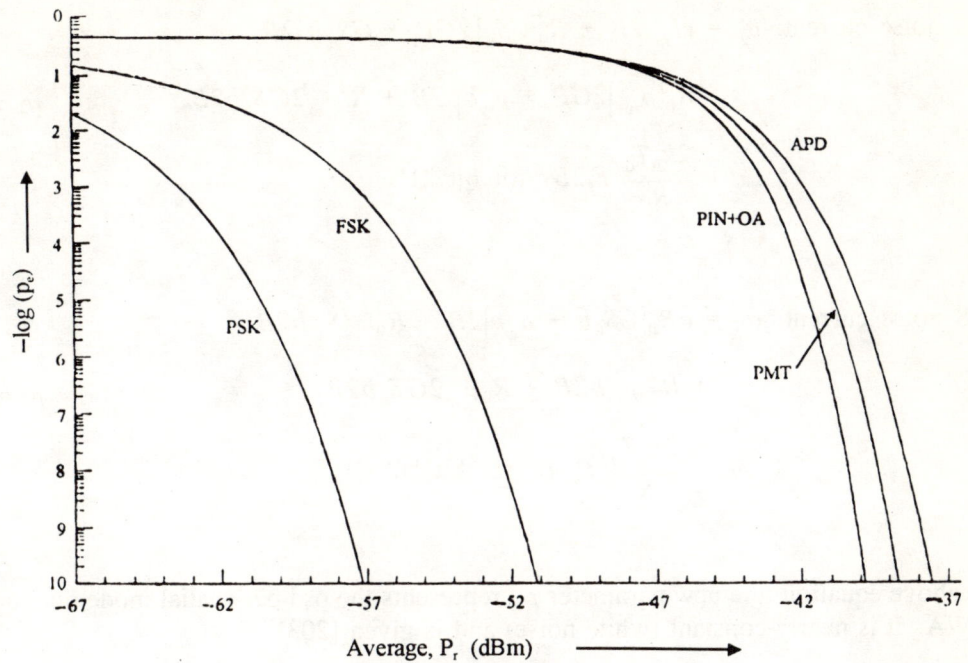

Fig. 9.18: Variations of $\log(p_e)$ with average P_r for different receiver configurations

(b) COHERENT FSK HETERODYNE RECEIVER

Analysis of this receiver structure can be carried out in the same way as for PSK homodyne receiver. The error curve i.e., the variations of p_e with P_r for binary orthogonal FSK signalling scheme remain same as in the earlier case except that it will shift towards right by 6 dB (refer to Fig. 9.18).

(c) DIRECT DETECTION (PIN+OA) RECEIVER FOR OOK

Block diagram of this receiver structure remains same as in Fig. 9.17 except that LO and IF stages blocks will not be there. In this case, various signal and noise components of the photodetector current under the same condition as in homodyne PSK receiver i.e., $b \gg B$ are given by

Signal current: $I_{sp} = R_0 G P_r$ (9.32a)

DC current: $I_{cp} = R_0 \left[G P_r' + G S_n b + \rho_{sp} b \right]$ (9.32b)

RMS noise current: $\sigma_1^2 = eI_{cp} 2B + R_0^2 GS_n \left[2GP_r + GS_n b\right]2B$

$$+ R_0^2 \rho_{sp} \left[2GP_r + \rho_{sp}b\right] 2B + R_0^2 \rho_{sp} 2GS_n b 2B \tag{9.32c}$$

$$+ \frac{2k_B T_0}{R_1} F_i 2B \quad \text{for bit "1"}$$

RMS noise current: $\sigma_0^2 = eR_0\left[GS_n b + \rho_{sp}b\right]2B + R_0^2 G^2 S_n^2 b 2B$

$$+ R_0^2 \rho_{sp}^2 b 2B + R_0^2 \rho_{sp} 2GS_n b 2B \tag{9.32d}$$

$$+ \frac{2k_B T_0}{R_1} F_i 2B \quad \text{for bit "0"}$$

In the above equation, the new parameter ρ_{sp} represents the psd per spatial mode due to ASE for TWA. It is nearly constant (white noise) and is given [203]

$$\rho_{sp}(f) = (G - 1)n_{sp}hf \tag{9.33}$$

where n_{sp} and G are the population inversion parameter and gain of the optical amplifier respectively.

For this signalling scheme, probability of errors for bit "1" and "0" are given by

$$P_{e1} = \frac{1}{2}\text{erfc}\left[\frac{R_0 GP_r - D}{\sqrt{2}\,\sigma_1}\right] \tag{9.34}$$

and

$$P_{e0} = \frac{1}{2}\text{erfc}\left[\frac{D}{\sqrt{2}\,\sigma_0}\right] \tag{9.35}$$

where D is the threshold level. If an optimum threshold level which equalizes p_{e1} and p_{e0} is used, then average p_e from Eqs. (9.34) and (9.35) will be

$$p_e = p_{e1} = p_{e0} = \frac{1}{2} \text{erfc}\left(\sqrt{\frac{S}{2N}}\right) \tag{9.36}$$

Here, S/N represents the signal-to-noise ratio and is given by

$$\frac{S}{N} = \frac{R_0^2 \, G^2 \, P_r^2}{(\sigma_1 + \sigma_0)^2} \tag{9.37}$$

The Eq. (9.32c) after some mathematical simplifications can be written as

$$\frac{\sigma_1^2}{2 \, \eta \, e \, R_0 G^2 P_r B} = F_{p1} \tag{9.38a}$$

where

$$F_{p1} = \frac{1}{\eta \, G} \left[1 + \frac{S_n b}{P_r} + \frac{\tilde{n}_{sp} h f b}{P_r} \right] + 2 \left[\tilde{n}_{sp} + \frac{S_n}{hf} \right] \cdot \left[1 + \frac{S_n b}{2P_r} + \frac{\tilde{n}_{sp} h f b}{2P_r} \right]$$

$$+ \frac{k_B T_0}{hf} \frac{F_i}{G^2 \tilde{L}_m} \tag{9.38b}$$

$$\tilde{n}_{sp} = (1 - 1/G) n_{sp} \tag{9.38c}$$

and

$$\tilde{L}_m = \frac{1}{2} R_0^2 \, P_r \, R_1 \tag{9.38d}$$

Here, the sign " ~ " is used to identify direct detection. Similarly, Eq. (9.32d) can also be expressed in the following form:

$$\frac{\sigma_0^2}{2\eta \, e \, R_0 G^2 P_r B} = F_{p0} \tag{9.39a}$$

where

$$
F_{p0} = \frac{1}{\eta G} \left[\frac{S_n b}{P_r} + \frac{\tilde{n}_{sp} hfb}{P_r} \right] + 2 \left[\tilde{n}_{sp} + \frac{S_n}{hf} \right]
$$

$$
\cdot \left[\frac{S_n b}{2P_r} + \frac{\tilde{n}_{sp} hfb}{2P_r} \right] + \frac{k_B T_0}{hf} \frac{F_i}{G^2 \tilde{L}_m}
$$

(9.39b)

From Eqs. (9.38a) and (9.39a)

$$
(\sigma_1 + \sigma_0)^2 = 2\eta e R_0 G^2 P_r B \left(\sqrt{F_{p1}} + \sqrt{F_{p0}} \right)^2
$$

(9.40)

Substitution of Eq. (9.40) in Eq. (9.37) and further simplification gives

$$
\frac{S}{N} = \frac{P_r}{2hfB \left(\sqrt{F_{p1}} + \sqrt{F_{p0}} \right)^2}
$$

(9.41)

Variations of p_e with P_r obtained from Eqs. (9.36) and (9.41) at $\lambda = 850$ nm and for the same values of the other parameters as in PSK homodyne receiver are shown in Fig. 9.18.

(d) DIRECT DETECTION (APD) RECEIVER FOR OOK

Schematic diagram of this receiver configuration will remain same as in Fig. 9.17 except that optical amplifier, following BPF, LO and IF stages blocks will not be there. Further, PIN-photodiode will be replaced by APD. In this case, various signal and noise components under the condition $b \gg B$ are given by

Signal current: $I_{sa} = R_0 M P_r$ (9.42a)

DC current: $I_{ca} = M^2 F R_0 \left[P_r + S_n b \right]$ (9.42b)

RMS noise current: $\sigma_1^2 = e I_{ca} 2B + M^2 R_0^2 S_n \left[2P_r + S_n b \right] 2B$

$$
+ \frac{2k_B T_0}{R_l} F_i 2B \quad \text{for bit "1"}
$$

(9.42c)

RMS noise current: $\sigma_0^2 = eM^2FR_0S_nb2B + M^2R_0^2S_n^2b\,2B$

$$+ \frac{2k_BT_0}{R_1}F_i\,2B \quad \text{for bit } "0" \tag{9.42d}$$

where M and $F=M^x$ are the multiplication and excess noise factors of the APD respectively. The optimum value of M which maximizes S/N ratio is given by [66]

$$M_{\text{opt}}^{x+2} = \frac{4k_BT_0F_i/R_1}{xeR_0(P_r + S_nb)} \tag{9.43}$$

Following the same approach as in direct detection (PIN+OA) receiver for OOK, p_e can be obtained by using Eqs. (9.36) and (9.41) with F_{p1} and F_{p0} replaced by F_{a1} and F_{a0} respectively. The F_{a1} and F_{a0} are given by

$$F_{a1} = \frac{F}{\eta}\left[1 + \frac{S_nb}{P_r}\right] + 2\frac{S_n}{hf}\left[1 + \frac{S_nb}{2P_r}\right] + \frac{k_BT_0}{hf}\frac{F_i}{M_{\text{opt}}^2\tilde{L}_m} \tag{9.44a}$$

and

$$F_{a0} = \frac{F}{\eta}\frac{S_nb}{P_r} + 2\frac{S_n}{hf}\frac{S_nb}{2P_r} + \frac{k_BT_0}{hf}\frac{F_i}{M_{\text{opt}}^2\tilde{L}_m} \tag{9.44b}$$

Variations of p_e with P_r for the same parameters as in direct detection (PIN+OA) receiver for OOK and $x=0.5$ are shown in Fig. 9.18.

(e) DIRECT DETECTION (PMT) RECEIVER FOR OOK

In contrast to APD receiver, M_{opt} and F are almost independent of the received optical signal power level in PMT receiver. Further, M_{opt} is very high ($\approx 10^5$) and F is quite low (≈ 1.5). Therefore, the variations of p_e with P_r for this receiver can be obtained as for the APD receiver for OOK by taking $M = 10^5$, $F = 1.5$, $\lambda = 532$ nm, $\eta = 0.3$ and keeping the other parameters same. It is shown in Fig. 9.18.

(f) DIRECT DETECTION (APD) RECEIVER FOR m-PPM

In the m-PPM signalling scheme, let each word convey n bits of information. It means

$$m = 2^n \tag{9.45}$$

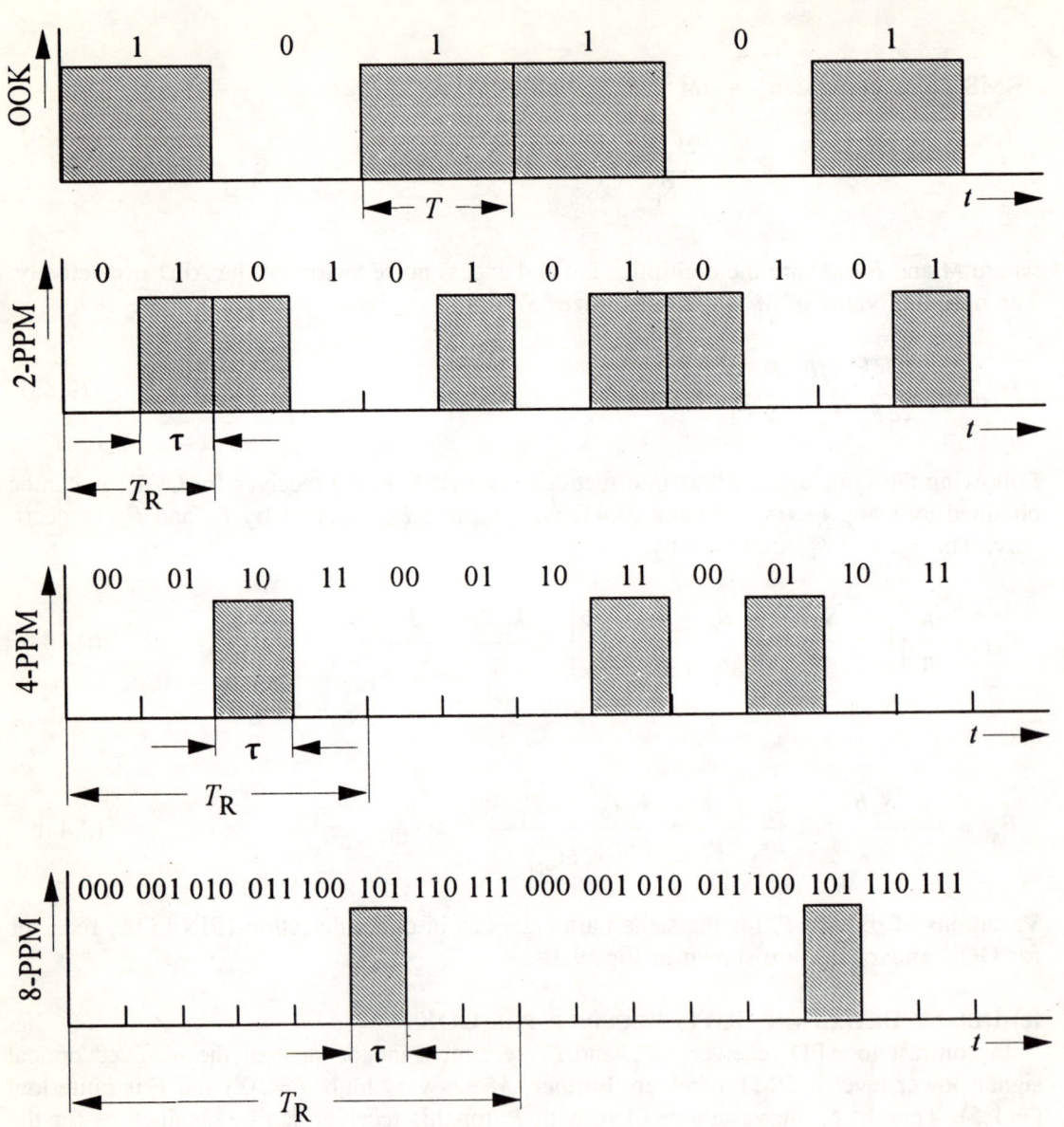

Fig. 9.19: OOK and m-PPM signals ($m=2$, 4 and 8)

For the same R as in OOK (also in binary PSK and FSK), the frame interval T_R (Fig. 9.19) is given by

$$T_R = n \cdot T \qquad (9.46)$$

The slot interval τ from Eq. (9.46) will be

$$\tau = \frac{n \cdot T}{m} \tag{9.47}$$

Therefore, the bandwidth $B_P = 1/\tau$ occupied by an m-PPM signal is given by

$$B_P = \frac{m}{n} B \tag{9.48}$$

In a 2-PPM system, signal input d to the decision device (similar to dual filter FSK) is given by

$$d = \pm R_0 M P_r + I_n \tag{9.49}$$

where I_n is a Gaussian distributed random variable with zero mean and the following variance

$$\sigma_n^2 = \sigma_1^2 + \sigma_0^2 \tag{9.50}$$

Here, σ_1^2 represents the variance of random variable arising due to noise current I_1, when both signal and noise are present in a time slot. Similarly, σ_0^2 represents the variance of the random variable due to noise current I_0, when only noise is present in a time slot. These can be determined by using Eqs. (9.42c) and (9.42d) and by substituting B_P in place of B. Using Eq. (9.49), p_e for bit "0" and "1" are given by

$$P_e = P_{e1} = P_{e0} = \frac{1}{2} \operatorname{erfc} \left[\sqrt{\frac{R_0^2 M^2 P_r^2}{2 \left(\sigma_1^2 + \sigma_0^2 \right)}} \right] \tag{9.51}$$

This can be written in the same form as Eq. (9.36) with the following S/N ratio

$$\frac{S}{N} = \frac{R_0^2 M^2 P_r^2}{\left(\sigma_1^2 + \sigma_0^2 \right)} \tag{9.52}$$

Substitution of σ_1^2 and σ_0^2 from Eq. (9.42c) and Eq. (9.42d) respectively and after further simplification, above equation can be written as

$$\frac{S}{N} = \frac{P_r}{2 h f B_P \left(F_{a1} + F_{a0} \right)} \tag{9.53}$$

In an m-PPM system, decision regarding the presence of pulse in a time slot is made on the basis of (m-1) comparisons. Following the same approach as for an orthogonal m-FSK signalling scheme, probability p_{ew} of wrongly decoding a word is given by

$$P_{ew} = 1 - \int_{-\infty}^{\infty} \left[\int_{-\infty}^{R_0 M P_r + I_1} p(I_0)\, dI_0 \right]^{m-1} p(I_1)\, dI_1 \tag{9.54}$$

where $p(I_1)$ and $p(I_0)$ are the probability density functions of I_1 and I_0 respectively. These are considered to be Gaussian distributed with variance σ_1^2 and σ_0^2 respectively. Substitution of $p(I_1)$ and $p(I_0)$ in the above equation and further simplification gives

$$P_{ew} = 1 - \frac{1}{\sqrt{2\pi\,\sigma_1^2}} \int_{-\infty}^{\infty} \left[1 - \frac{1}{2}\mathrm{erfc}\left(\frac{R_0 M P_r + I_1}{\sqrt{2}\,\sigma_0} \right) \right]^{m-1} \cdot \exp\left[-\frac{I_1^2}{2\sigma_1^2} \right] dI_1 \tag{9.55}$$

This p_{ew} can be used to obtain the upper bound on the probability of error p_e. Following [264], it is given by

$$P_e \leq \frac{m/2}{(m-1)} P_{ew} \tag{9.56}$$

If the quantity inside the second square bracket in Eq. (9.55) is very small ($\ll 1$), then Eq. (9.55) can be approximated as [114]

$$P_{ew} \approx \frac{(m-1)}{2} \cdot \mathrm{erfc}\left[\sqrt{\frac{R_0^2 M^2 P_r^2}{2\left(\sigma_1^2 + \sigma_0^2\right)}} \right] \tag{9.57}$$

For a 2-PPM system, Eq. (9.57) reduces to Eq. (9.51) which is an exact expression for p_e. For an m-PPM system (m $>$ 2) and $p_e < 10^{-4}$, Eq. (9.57) is a good approximation of p_{ew} given in Eq. (9.55). Variations of p_e with average P_r for different m at $\lambda = 1064$ nm and for the same values of the other parameters as in direct detection (PIN+OA) receiver and $x = 0.5$ obtained from Eq. (9.57) are shown in Fig. 9.20.

(f) DIRECT DETECTION (PMT) RECEIVER FOR m-PPM

The error curves for this receiver can be obtained as for direct detection (APD) receiver for m-PPM by taking $\lambda = 532$ nm, $M = 10^5$, $F = 1.5$, $\eta = 0.3$ and keeping the other receiver parameters same. It is shown in Fig. 9.21.

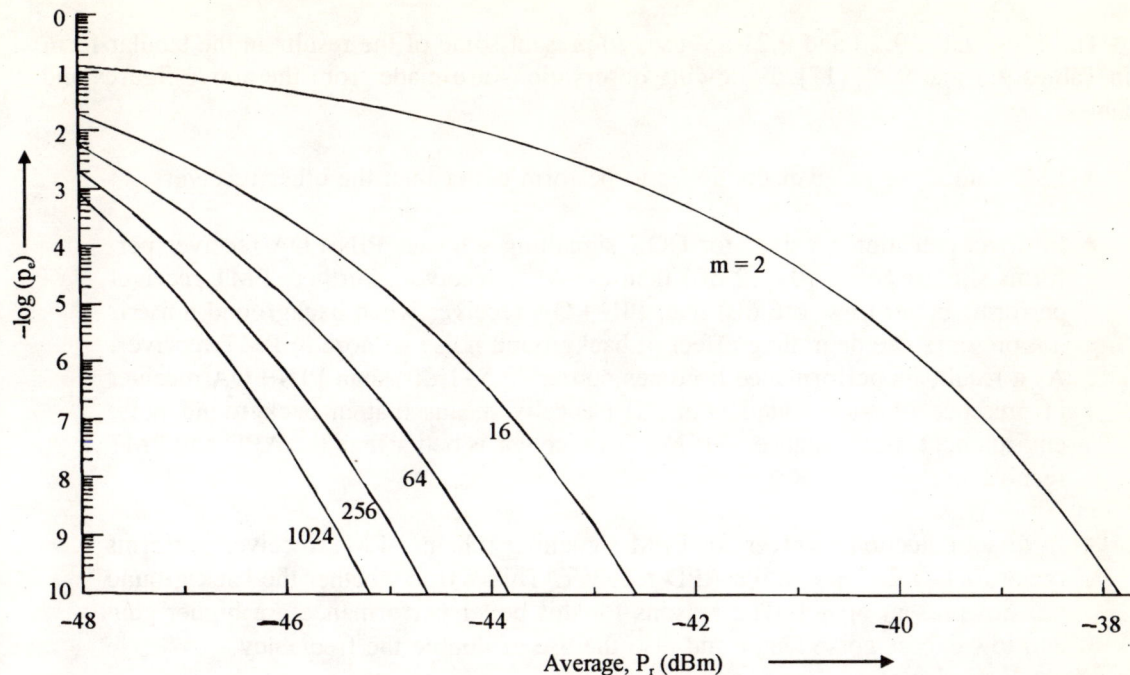

Fig. 9.20: Variations of $\log(p_e)$ with average P_r for direct detection (APD) receiver for m-PPM

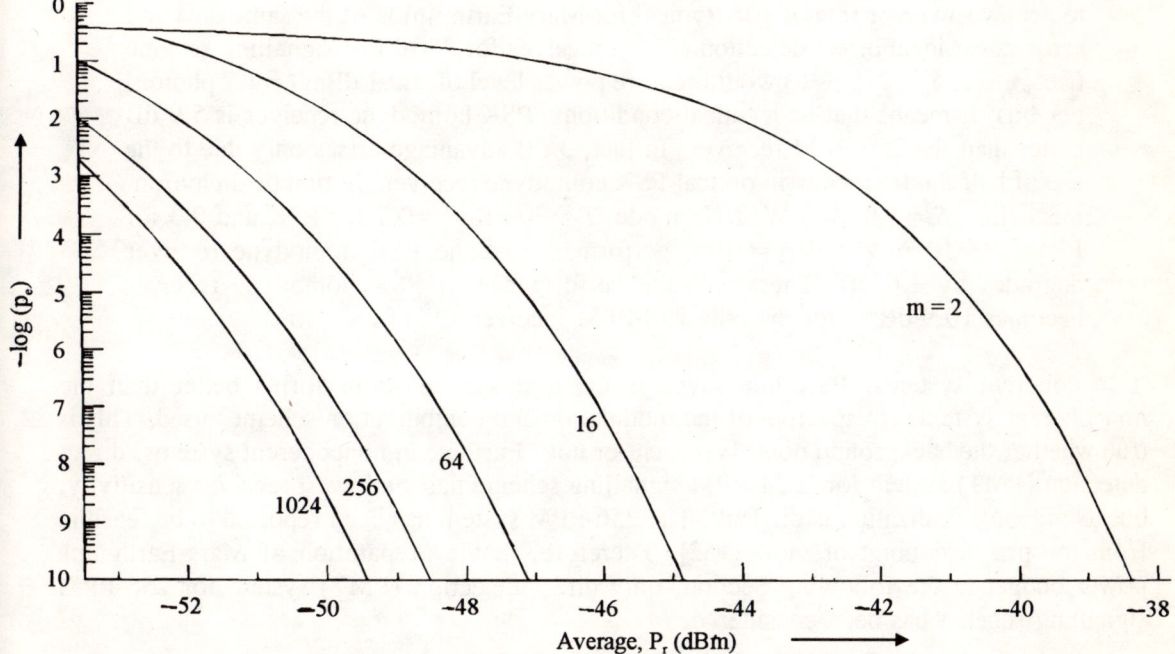

Fig. 9.21: Variations of $\log(p_e)$ with average P_r for direct detection (PMT) receiver for m-PPM

The Figs. 9.18, 9.20 and 9.21 are used to present some of the results in the tabular form in Tables 9.3 and 9.4 [117]. Following observations are made from the above figures and tables.

- PSK homodyne receiver continues to perform better than the other receivers.

- In direct detection receiver for OOK signalling scheme, PIN+OA receiver performs slightly better (0-2.2 dB) than the APD receiver. Further, PMT receiver performs better (3.4-6.6 dB) than PIN+OA receiver when background noise is not present. The degrading effect of background noise is more in PMT receiver. As a result, its performance becomes poorer (0.5-1 dB) than PIN+OA receiver in presence of background noise. It basically means that in background noise environment, performance of PIN+OA receiver is better than the APD and PMT receivers.

- In direct detection receiver for PPM signalling scheme, PMT receiver performs better (0.1-5.2 dB) than the APD receiver. This is true whether the back-ground noise is present or not. The reasons for this better performance are higher gain and low excess noise factor and also the use of double the frequency.

- Ideal optical PSK homodyne receiver (i.e., $P_1 \to \infty$, $\eta \to 1$, $S_n \to 0$, $T_0 \to 0$) at a data rate of 560 Mbit/s would require -66.0 dBm optical power ($=2.4$ photons per bit) to achieve an error rate of 10^{-3} (typical for Mars-Earth link). At the same data and error rates, ideal direct detection (PMT) receiver for 256-PPM signalling scheme (i.e., $\eta \to 1$, $S_n \to 0$, $T_0 \to 0$) would require power level of -60.1 dBm ($=4.7$ photons per bit). It means that under ideal conditions, PSK homodyne receiver is 5.9 dB better than the 256-PPM receiver. In fact, 3 dB advantage arises only due to the use of half the frequency in optical PSK homodyne receiver. In practical environment (i.e., $S_n = 1.9 \cdot 10^{-20}$ Watt/Hz·mode, $T_0 = 300$ K, $\eta = 0.7$ for PSK and 0.3 for PPM, $P_1 = 10$ mW and $p_e = 10^{-3}$), performance of the PSK homodyne receiver degrades by 2.6 dB. Therefore, the performance of PSK homodyne receiver becomes 10.9 dB better than the 256-PPM receiver.

In coherent systems, PSK homodyne is the best system. It performs better than the noncoherent systems irrespective of the modulation and demodulation schemes used. This is true whether the background noise is present or not. Further, in noncoherent systems, direct detection (PMT) system for 1024-PPM signalling scheme has the lowest receiver sensitivity, but its practical realization is difficult. The 256-PPM system has been reported to be feasible from the practical point of view [152]. Therefore, in the preparation of Mars-Earth link power budget in the following Section, only direct detection (PMT) system for 256-PPM signalling scheme has been considered.

Table 9.3: Minimum required average power P_r in dBm at a data rate of 560 Mbit/s with and without background noise
     * It will be slightly different at different λ, but its order of magnitudes remains same.
     £ For FSK heterodyne, required P_r will be 6 dB higher than in PSK homodyne.
     $ Computation of P_r is a good approximation of exact expression for $p_e < 10^{-4}$.

p_e	S_n^* Watt/ Hz·mode	PSK Homodyne£ $\lambda=1064$nm $\eta=0.7$	PIN+OA for OOK $\lambda=850$nm $\eta=0.7$	APD for OOK $\lambda=850$nm $\eta=0.7$	PMT for OOK $\lambda=532$nm $\eta=0.3$	APD for 256-PPM$ $\lambda=1064$nm $\eta=0.7$	PMT for 256-PPM$ $\lambda=532$nm $\eta=0.3$
10^{-9}	0.0	-58.1	-40.2	-38.6	-43.6	-45.2	-49.1
	$1.9\cdot10^{-20}$	-57.6	-40.0	-37.8	-39.0	-44.8	-48.8
10^{-6}	0.0	-60.1	-41.4	-40.3	-46.2	-46.6	-50.8
	$1.9\cdot10^{-20}$	-59.6	-41.1	-39.3	-40.2	-46.0	-50.2
10^{-3}	0.0	-63.9	-43.4	-43.4	-50.0	-48.6	-53.1
	$1.9\cdot10^{-20}$	-63.4	-43.1	-41.8	-42.6	-47.7	-52.5

Table 9.4: Minimum required power P_r in dBm and photons count at a data rate of 560 Mbit/s and probability of error $p_e = 10^{-3}$.
     * At 1064 nm wavelength.
     $ At 532 nm wavelength.

System	Optical power P_r in dBm	Number of photons per bit
PSK* (Practical)	-63.4	4.3
256-PPM$ (Practical)	-52.5	27.1
PSK* (Ideal)	-66.0	2.4
256-PPM$ (Ideal)	-60.1	4.7

9.6.6   LINK BUDGET

As an example, the following table presents the power budget for Mars-Earth link in the tabular form.

Table 9.5: Power budget for Mars-Earth link at data rates of 10 Mbit/s and 1 Mbit/s for $p_e = 10^{-3}$

System parameter	Data rate = 10 Mbit/s		Data rate = 1 Mbit/s	
	PSK homodyne $\lambda = 1064$ nm	256-PPM direct detection $\lambda = 532$ nm	PSK homodyne $\lambda = 1064$ nm	256-PPM direct detection $\lambda = 532$ nm
Minimum P_r	-80.9 dBm	-65.0 dBm	-90.9 dBm	-70.5 dBm
G_r	149.4 dB	155.4 dB	149.4 dBm	155.4 dB
Space loss	360.7 dB	354.7 dB	360.7 dB	354.7 dB
G_t	117.4 dB	123.4 dB	117.4 dB	123.4 dB
Misalignment loss	6.1 dB	6.1 dB	6.1 dB	6.1 dB
Atmospheric loss	10.0 dB	10.0 dB	10.0 dB	10.0 dB
Safety margin	5.0 dB	5.0 dB	5.0 dB	5.0 dB
Required average P_t	34.1 dBm	32 dBm	24.1 dBm	26.5 dBm
	2.6 W	1.6 W	0.3 W	0.4 W
peak P_t	2.6 W	405.7 W	0.3 W	114.4 W

In the above table, diameter of transmitting antenna on Mars is presumed to be 25 cm. Similarly, let the diameter of receiving antenna on Earth be 10 m. Using Eqs. (9.9a) and (9.9b), gains of the transmitting and receiving antennas at 532 nm are 123.4 dB and

155.4 dB respectively. At this wavelength, free-space loss i.e., $(\lambda/4\pi L)^2$ considering Mars-Earth distance to be 78 million km is 254.7 dB. Substitution of these values of G_t, free-space loss and G_r in Eq. (9.8) gives

$$P_r(\text{dBm}) = P_t(\text{dBm}) - 75.9 \text{ dB} \qquad (9.58)$$

If misalignment loss of 6.1 dB, atmospheric loss of 10.0 dB and safety margin of 5.0 dB are taken into account, then the required P_t at 532 nm from Eq. (9.58) will be

$$P_t(\text{dBm}) = P_r(\text{dBm}) + 97.0 \text{ dB} \qquad (9.59)$$

At 1064 nm wavelength and for the same diameters of the transmitting and receiving antennas, gain of each antenna will be reduced by 6 dB as compared to 532 nm wavelength. At this new wavelength, space loss will increase by 6 dB. The required transmitted power in this case, considering the same additional losses and safety margin, is given by the equation

$$P_t(\text{dBm}) = P_r(\text{dBm}) + 115.0 \text{ dB} \qquad (9.60)$$

For PSK homodyne receiver, minimum required average P_r at $\lambda = 1064$ nm, $R = 1$ Mbit/s and $p_e = 10^{-3}$ from Eq. (9.31) is -90.9 dBm. Using Eq. (9.60), required average P_t will be 24.1 dBm. It means an average/peak power level of 0.3 W. For 256-PPM direct detection receiver at 532 nm and for the same data and error rates, minimum required average P_r from Eqs. (9.56) and (9.57) is -70.5 dBm. Therefore, required average P_t from Eq. (9.59) is 26.5 dBm. It corresponds to an average power level of approximately 0.4 W and a peak power level of 114.4 W. With the current source technology, it is possible to generate this power level in a pulse mode. Thus, it is feasible to establish the above link using either of the system.

At $R = 10$ Mbit/s and for the same p_e and λ, required peak power level for PSK homodyne system is 2.6 W. It is difficult to generate this peak power level in a continuous mode. The 256-PPM direct detection system at the above data rate will require a peak power level of 405.7 W. Though this power level is much higher than the power level required by the PSK homodyne system, it is practically possible to generate such a peak power level in a pulse mode.

10 REFERENCES

[1]   Abramowitz, M.; Stegun, I. A.: Handbook of Mathematical Functions. Dover Publications Inc., 1965.

[2]   Adams, M. J.; Payne, D. N.; Ragdale, C. M.: Birefringence in optical fibers with elliptical cross-section. Electron. Lett. 15(1979)10, 298-299.

[3]   Adams, M. J.; Westlake, H. J.; O'Mahony, M. J.; Henning, I. D.: A comparison of active and passive bistability in semiconductors. IEEE J. QE-21(1985)9, 1498-1504.

[4]   Agrawal, G. P.: Amplifier induced crosstalk in multichannel coherent lightwave systems. Electron. Lett. 23(1987)17, 1175-1177.

[5]   Agrawal, G. P.: Nonlinear Fiber Optics. Academic Press Inc., 1989.

[6]   Agrawal, G. P.: Fiber Optic Communication System. John Wiley & Sons Inc., 1992.

[7]   Akiba, S. et. al.: Recent progress of transoceanic optical amplifier transmission systems. ECOC, Vol. 2(1992), 719-726.

[8]   Alabanese, A.: Fall safe nodes for light guide digital networks. Bell System Tech. J. 61(1982)2, 247-257.

[9]   Amann, M.-C.: Tunable semiconductor lasers. ECOC (1994), 1011-1018.

[10]   Andrekson, P. A.: Ultrahigh speed transmission and multiplexing. ECOC (1994), 109-116.

[11]   Atkins, C. G.; Cotter D.; Smith, D. W.; Wyatt, R.: Application of Brillouin amplification in coherent optical transmission. Electron. Lett. 22(1986)10, 556-558.

[12]   Auracher, F.; Schicketanz, D.; Zeitler, K.-H.: High-speed $\Delta\beta$-reversal directional coupler modulator with low insertion loss for 1.3 μm in LiNbO$_3$. J. Opt. Commun. 5(1984)1, 7-9.

[13]   Baack, C.; Bachus, E.-J.; Strebel, B.: Zukünftige Lichtträgerfrequenztechnik in Glasfasernetzen. ntz 35(1982)11, 686-689.

[14]   Baack, C.: Optical Wideband Transmission Systems. CRC Press Inc., 1986.

[15]   Bachus, E.-J.; Böhnke, F.; Braun, R.-P.; Eutin, W.; Foisel, H.; Heimes, K.; Strebel, B.: Two-channel heterodyne-type transmission experiment. Electron. Lett. 21(1985)1, 35-36.

[16]   Ballart, R.; Ching, Y.: Sonet-Now it's the standard optical network. IEEE Commun. Magazine 26(1989)3, 8-15.

[17]   Barlow, A. J.; Payne, D. N.: Polarisation maintenance in circularly birefringent fibres. Electron. Lett. 17(1981)11, 388-389.

[18]   Barry, J. R.; Lee, E. A.: Performance of coherent optical receivers. Proc. IEEE 78(1990)8, 1369-1394.

[19]   Basch, E. E.; Brown, T. G.: Introduction of coherent optical fiber transmission. IEEE Commun. Magazine 23(1985)5, 23-30.

[20]   Becker, P. C.: Erbium-doped fiber makes promising amplifiers. Laser Focus World, 1970, 197-203.

[21] Begley, D. L.; Kobylinski, R. A.; Ross, M.: Solid-state laser cross-link systems and technology. International J. Satellite Commun. (1988)6, 94-105.

[22] Best, R.: Theorie und Anwendungen des Phase-locked Loops. Fachschriftenverband Aargauer Tagblatt AG, Aarau/Schweiz 1976.

[23] Bonek, E.; Leeb, W. R.; Scholtz, A. L.; Philipp, H. K.: Optical PLLs see the light. Microwaves & RF (1983), 65-70.

[24] Booth, R. C.: Integrated optic devices for coherent transmission. IOOC-ECOC (1985), 89-96.

[25] Börner, M.; Müller, R.: Silizium für die Integrierte Optoelektronik? ntz 41(1988)2, 64-75.

[26] Boutemy, J.: Use of CCD arrays for optical link acquisition and tracking. Proc. SPIE on Optical Systems for Space Applications, Vol. 810(1987), 215-222.

[27] Braun, R.-P.; Ludwig, R.; Molt, R.: Ten-channel optic fibre transmission using an optical travelling wave amplifier. IOOC-ECOC (1986), 29-32.

[28] Bronstein, I. N.; Semendjajew, K. A.: Taschenbuch der Mathematik, 19. Auflage. Harri Deutsch Verlag, Thun und Frankfurt/Main, 1980.

[29] Burns, W. K.; Moeller, R. P.: Measurement of polarization mode dispersion in high-birefringence fibers. Optics Lett. 8(1983)3, 195-197.

[30] Burr, W.: The FDDI optical data link. IEEE Commun. Magazine 24(1986)5, 8-23.

[31] Chikama T. et. al.: Modulation and demodulation techniques in optical heterodyne PSK transmission systems. IEEE J. LT-8(1990)3, 309-321.

[32] Chinlon, L.: Optoelectronics Technology and Lightwave Communication Systems. Van Nostrand Reinhold Company Inc., 1989.

[33] Cochrane, P.: Future directions in long haul fiber optic systems. British Telecom. Tech. J. 8(1990)1, 5-17.

[34] Costa, B.: Towards a future all-optical telecommunication network. Lecture notes in third college on physics and technology of lasers and optical fibers held at I.C.T.P. Trieste, Italy, 27 January-21 February, 1992.

[35] Cotter, D.: Observation of stimulated Brillouin scattering in low-loss silica fiber at 1.3 μm. Electron. Lett. 18(1982), 495-496.

[36] Cramér, H.: Mathematical Methods of Statistics. 10. Aufl. Princeton University Press, Princeton 1963.

[37] Cygan, D.: Berechnung der Wahrscheinlichkeitsdichtefunktion am Ausgang eines Filters bei beliebiger Eingangsverteilung und beliebiger Autokorrelationsfunktion. Diplomarbeit, TU München, Lehrstuhl für Nachrichtentechnik, 1986.

[38] Cygan, D.; Franz, J.; Söder, G.: Einfluß eines Filters auf nicht-gaußverteilte Zufallsprozesse. AEÜ 40(1986)6, 377-384.

[39] Daino, B.; Spano, P.; Tamburrini, M.; Piazolla, S.: Phase noise and spectral line shape in semiconductor laser. IEEE J. QE-19(1983)3, 266-270.

[40] Darcie, T. E.; Jopson, R. M.; Tkach, R. W.: Intermodulation distortion in optical amplifier from carrier-density modulation. Electron. Lett. 23(1987)25, 1392-1394.

[41] Davis, A. W.; Pettitt, M. J.; King, J. P.; Wright, S.: Phase diversity techniques for coherent optical receivers. IEEE J. LT-5(1987)4, 561-572.

[42] De Lange, O. E.: Wide-band optical communication systems: Part II - Frequency-division multiplexing. Proc. IEEE 58(1970)10, 1683-1690.

[43] Derr, F.: Comparison of electrical and optical BPSK and QPSK systems. J. Opt. Commun. 10(1989)4, 127-131.

[44] Derr, F.: Optical QPSK homodyne transmission of 280 Mbit/s. Electron. Lett. 26 (1990)6, 401-403.

[45] Derr, F.: System performance of an optical QPSK homodyne transmission system with Costas loop. J. Opt. Commun. 14(1985)2, 42-44.

[46] Draper, N. R.; Tierney, D. E.: Exact formulas for additional terms in some important series expansions. Communications in Statistics 1(1973), 495-524.

[47] Dyott, R. B.; Cozens, J. R.; Morris, D. G.: Preservation of polarisation in optical-fibre waveguides with elliptical cores. Electron. Lett. 15(1979)13, 380-382.

[48] Elion, G. R.; Elion, E. A.: Fiber Optics in Communication Systems, Electro-optics Series/2, Marcel Dekker Inc., 1978.

[49] Favre, F.; Jeunhomme, L.; Joindot, I.; Monerie, M.; Simon, J. C.: Progress towards heterodyne-type single-mode fiber communication systems. IEEE J. QE-17(1981)6, 897-906.

[50] Favre, F.; Le Guen, D.: Emission frequency stability in single-mode-fibre optical feedback controlled semiconductor lasers. Electron. Lett. 19(1983)17, 663-665.

[51] Fischer, E.: Auswirkungen eines ZF-Filters auf phasenverrauschte Eingangssignale. Diplomarbeit, Tech. Univ. München, 1989.

[52] Fischer, G.: The Faraday optical isolator. J. Opt. Commun. 8(1987)1, 18-21.

[53] Fisher, R. A.; Cornish, E. A.: The percentile points of distributions having known comulants. Technometrics 2(1960), 209-225.

[54] Fleischmann, M.: Berechnung, Optimierung und Vergleich verschiedener optischer Übertragungssysteme mit Überlagerungsempfang. Diplomarbeit, TU München, Lehrstuhl für Nachrichtentechnik, 1987.

[55] Fleischmann, M.; Franz, J.: Optimization of coherent optical homodyne systems. J. Opt. Commun. 9(1988)2, 72-77.

[56] Fleming, M. W.; Mooradian, A.: Spectral characteristics of external-cavity controlled semiconductor lasers. IEEE J. QE 17(1981)1, 44-59.

[57] Fleming, M. W.; Mooradian, A.: Fundamental line broadening of single-mode (GaAl)As diode lasers. Appl. Phys. Lett. 38(1981)7, 511-513.

[58] Foschini, G. J.; Greenstein, L. J.,; Vannucci, G.: Noncoherent detection of coherent lightwave signals corrupted by phase noise. IEEE COM-36(1988)3, 306-314.

[59] France, P.W.: Optical Fiber Lasers and Amplifiers. CRC Press Inc., 1991.

[60] Franz, J.: Grundzüge des kohärent optischen Heterodynempfanges. TU München, Lehrstuhl für Nachrichtentechnik, Nachr.-techn. Ber. Band 14, 1984.

[61] Franz, J.: Evaluation of the probability density function and bit error rate in coherent optical transmission systems including laser phase noise and additive Gaussian noise. J. Opt. Commun. 6(1985)2, 51-57.

[62] Franz, J.; Rapp, C.; Söder, G.: Influence of baseband filtering on laser phase noise in coherent optical transmission systems. J. Opt. Commun. 7(1986)1, 15-20.

[63]   Franz, J.; Helnerus, U.: Calculation of bit error rate in ASK heterodyne systems with envelope detection influenced by laser phase noise. Electron. Lett. 22(1986)20, 1072-1073.

[64]   Franz, J.: Receiver analysis of incoherent optical heterodyne systems. J. Opt. Commun. 8(1987)2, 57-66.

[65]   Franz, J.: Berechnung, Optimierung und Vergleich optischer Übertragungssysteme mit Überlagerungsempfang. Dissertation, TU München, Lehrstuhl für Nachrichten-technik, 1987.

[66]   Franz, J.: Optische Übertragungssysteme mit Überlagerungsempfang. Springer-Verlag, 1988.

[67]   Franz, J.; Schaller, N.: Limits of optical heterodyne DPSK transmission systems with envelope detection. J. Opt. Commun. 10(1989)1, 28-32.

[68]   Franz, J.: Optical space communication. Lecture notes at IIT Delhi, 1993.

[69]   Franz, J.; Correll, C.; Dolainsky, F.; Schweikert, R.; Wandernoth, B.: Error correcting coding in optical transmission systems with direct detection and heterodyne detection. J. Opt. Commun. 14(1994)5, 194-199.

[70]   Gagliardi, R. M.; Karp, S.: Optical Communications. John Wiley & Sons Inc., 1976.

[71]   Gagliardi, R. M.: Satellite Communications. Van Nostrand Reinhold Inc., 1984.

[72]   Gardner, F. M.: Phaselock Techniques. John Wiley & Sons Inc., 1979.

[73]   Garrett, I.; Jacobsen, G.: Influence of (semiconductor) laser linewidth on the error-rate floor in dual-filter optical FSK receivers. Electron. Lett. 21(1985)7, 280-282.

[74]   Garrett, I.; Jacobsen, G.: Statistics of laser frequency fluctuations in coherent optical receivers. Electron. Lett. 22(1986)3, 168-170.

[75]   Garrett, I.; Jacobsen, G.: Theoretical analysis of heterodyne optical receivers for transmission systems using (semiconductor) lasers with nonneglible linewidth. IEEE J. LT-4(1986)3, 323-334.

[76]   Garrett, I.; Jacobsen, G.: Phase noise in weakly coherent systems. IEE Proc. Pt. J, (1989)3, 159-165.

[77]   Garrett, I. et. al.: Weakly coherent optical heterodyne ASK receivers; impact of phase noise. IEEE J. LT-8(1990)3, 329-337.

[78]   Giles, C.R.; Desurvire, E.: Modelling erbium-doped fiber amplifiers. IEEE J. LT-9(1991)2, 271-283.

[79]   Glance, B.: Polarisation independent coherent optical receiver. IEEE J. LT-5(1987)2, 274-276.

[80]   Glance, B. et. al.: Multichannel FSK heterodyne optical communication system entirely controlled by computer. IEEE PTL(1991)1, 83-85.

[81]   Gnauck, A. H. et. al.: 4-Gb/s heterodyne transmission experiments using tunable DBR laser as signal source for DPSK and ASK operation and DFB laser source for FSK. IEEE PTL(1990)12, 908-910.

[82]   Golanbari, M. et. al.: Channel coding for coherent lightwave FSK communication deteriorated by phase noise. IEEE PTL(1992)6, 651-654.

[83]   Goldberg, L.; Taylor, H. F.; Dandrige, A.; Weller, J. F.; Miles, R. O.: Spectral characteristics of semiconductor lasers with optical feedback. IEEE J. QE-18(1982)4, 555-564.

[84]   Goodwin, F. E.: A 3.39-micron infrared optical heterodyne communication system. IEEE J. QE-3(1967)11, 524-531.

[85]   Gordon, J. P.: Quantum effects in communication systems. IRE Proc., 50(1962)9, 1898-1908.

[86]   Gower, J.: Optical Communication Systems. Second Edition, Prentice-Hall, 1993.

[87]   Grau, G.; Freude, W.: Optische Nachrichtentechnik. Springer-Verlag, 1991.

[88]   Gross, R.; Meissner, P.; Patzak, E.: Theoretical investigation of local oscillator intensity noise in optical homodyne systems. IEEE J. LT-6(1988)4, 521-530.

[89]   Gross, R. et. al.: Multichannel coherent FSK lightwave system experiments using subcarrier multiplexing. IEEE J. LT-8(1990)3, 406-415.

[90]   Gross, R. et. al.: Fiber-optic 60-channel video system using coherent subcarrier multiplexing. IEEE PTL(1990)4, 288-290.

[91]   Grosskopf, R; Ludwig, W.; Weber, H. G.: Cross-talk in optical amplifiers for two channel transmission. Electron. Lett. 22(1986)17, 900-902.

[92]   Gupta, H. M.: Fiber Optic Local Area Networks. Technical Report No. CS-TR-2130, Department of Computer Science, University of Maryland, USA, September 1988.

[93]   Haber, F.: An Introduction to Information and Communication Theory. Addison-Wesley, 1974.

[94]   Hasegawa, A.; Tappert, F.: Transmission of stationary nonlinear optical pulses in dispersive dielectric fibers: Anomalous dispersion. Appl. Phys. Letter, 23(1973)3, 142-146.

[95]   Helnerus, U.: Der Einfluß des ZF-Filters auf das Laserphasenrauschen im optischen ASK-Heterodynempfänger mit Hüllkurvendemodulation. Diplomarbeit, TU München, Lehrstuhl für Nachrichtentechnik, 1986.

[96]   Henry, C.: Theory of the linewidth of semiconductor lasers. IEEE J. QE-18(1982)2, 259-264.

[97]   Henry, C.: Theory of the phase noise and power spectrum of a single mode injection laser. IEEE J. QE-19(1983)9, 1391-1397.

[98]   Henry, C.: Phase noise in semiconductor lasers. IEEE J. LT-4(1986)3, 298-311.

[99]   Hill, P.; Olshansky, R.: Multigigabit subcarrier multiplexed coherent lightwave systems. IEEE J. LT-10(1992)11, 1656-1664.

[100]   Hodara, H.: Statistics of thermal and laser radiation. Proc. IEEE, 53(1965)7, 696-704.

[101]   Hodgkinson, T. G.: Phase-locked-loop analysis for pilot carrier coherent optical receivers. Electron. Lett. 21(1985)25/26, 1202-1203.

[102]   Hodgkinson, T. G.: Costas loop analysis for coherent optical receivers. Electron. Lett. 22(1986)22, 394-396.

[103]   Hodgkinson, T. G.: Receiver analysis for synchronous coherent optical fibre transmission systems. IEEE J. LT-5(1987)4, 573-586.

[104] Hodgkinson, T. G.; Harmon, R. A.; Smith, D. W.: Polarisation-insensitive hetero-dyne detection using polarisation scrambling. Electron. Lett. 23(1987)10, 513-514.

[105] Hodgkinson, T. G.; Harmon, R. A.; Smith, D. W.: Performance comparison of ASK polarisation diversity and standard coherent optical heterodyne receivers. Electron. Lett. 24(1988)1, 58-59.

[106] Hosaka, T.; Okamoto, K.; Sasaki, Y.; Edahiro, T.: Single mode fibres with asymmetrical refractive index pits on both sides of core. Electron. Lett. 17(1981)5, 191-193.

[107] Hosaka, T.; Okamoto, K.; Miya, T.; Sasaki, Y.; Edahiro, T.: Low-loss single polarization fibres with asymmetrical strain birefringence. Electron. Lett. 17(1981)15, 530-531.

[108] Imoto, N.; Ikeda, M.: Polarization dispersion measurement in long single-mode fibers with zero dispersion wavelength at 1,5 μm. IEEE QE-17(1981)4, 542-545.

[109] Inoue, K.: Observation of crosstalk due to four-wave mixing in a laser amplifier for WDM transmission. Electron. Lett. 23(1987)24, 1293-1295.

[110] Jacobsen, G.; Garrett, I.: Error-rate floor in optical ASK heterodyne systems caused by nonzero (semiconductor) laser linewidth. Electron. Lett. 21(1985)7, 268-270.

[111] Jacobsen, G.; Garrett, I.: The effect of laser linewidth on coherent optical receivers with asynchronous demodulation. IOOC-ECOC (1986), 61-66.

[112] Jain, V. K.; Gupta, H. M.: Optical fiber communication link design. J. Opt. Commun. 6(1985)2, 58-66.

[113] Jain, V. K.: An approach to optical fiber link design. Students' Journal IETE, 29 (1988)2, 35-41.

[114] Jain, V. K.: Study on Space Optical Communication System. Technical Report, DLR Institute for Communication Technology, Germany, 1991.

[115] Jain, V. K.: Optical preamplifier in noncoherent space communications. J. Opt. Commun. 14(1993)2, 75-88.

[116] Jain, V. K.: Study of heterodyne and direct detection receivers with an optical pre-amplifier in space communications. J. Opt. Commun. 14(1993)5, 189-193.

[117] Jain, V. K.: Effect of background noise in space optical communication systems. AEÜ 47(1993)2, 98-107.

[118] Jeunhomme, L.; Monerie, M.: Polarisation-maintaining single-mode fibres cable design. Electron. Lett. 16(1980)24, 921-922.

[119] Jodoin, R.; Mandel, L.: Information rate in an optical communication channel. J. of Optical Society of America, 61(1971)2, 191-198.

[120] Joindet, M.: Heterodyne receiving techniques in microwave and optics: comparison of some concepts. Workshop on Digital Communication. Tirrenia, 1985, 179-191.

[121] Kaminov, I. P.; Ramaswamy, V.: Single-polarisation optical fibres: Slab model. Appl. Phys. Lett. 34(1979)4, 268-270.

[122] Kaminov, I. P.: Polarisation in optical fibres. IEEE J. QE-17(1981)1, 15-22.

[123] Kasper, B. L.; Burrus, C. A.; Telman, J. R.; Hall, K. L.: Balanced dual-detector receiver for optical heterodyne communication at Gbit/s rates. Electron. Lett. 22 (1986)8, 413-414.

[124] Katsuyama, T.; Matsumura, H.; Suganuma, T.: Low-loss single-polarisation fibers. Electron. Lett. 17(1981)13, 473-474.

[125] Katz, J.: Phase Locking of Semiconductor Injection Lasers. TDA Progress Report 42-66, Jet Propulsion Laboratory, Pasadena, CA, 1981, 101-114.

[126] Katzman, M.: Laser Satellite Communication. Prentice-Hall, 1987.

[127] Kazovsky, L. G.: Optical heterodyning versus optical homodyning: a comparison. J. Opt. Commun. 6(1985)1, 18-24.

[128] Kazovsky, L. G.: Decision-driven phase-locked loop for optical homodyne receivers: performance analysis and laser linewidth requirements. IEEE J. LT-3(1985)6, 1238-1247.

[129] Kazovsky, L. G.: Balanced phase-locked loops for optical homodyne receivers: performance analysis, design considerations, and laser linewidth requirements. IEEE J. LT-4(1986)2, 182-195.

[130] Kazovsky, L. G.: Performance analysis and laser linewidth requirements for optical PSK heterodyne communication systems. IEEE J. LT-4(1986)4, 415-425.

[131] Kazovsky, L. G.: Impact of phase noise on optical heterodyne communication systems. J. Opt. Commun. 7(1986)2, 66-78.

[132] Kazovsky, L. G.; Meissner, P.; Patzak, E.: ASK multiport optical homodyne receivers. IOOC-ECOC (1986), 395-398.

[133] Kazovsky, L. G.: Recent progress in phase and polarization diversity coherent optical techniques. ECOC (1987) Vol. I, 83-90.

[134] Keiser, G.: Optical Fiber Communications. Second Edition, McGraw-Hill, 1991.

[135] Kersey, A. D.; Yurek, A. M.; Dandrige, A.; Weller, J. F.: New polarisation-insensitive detection technique for coherent optical fibre heterodyne communications. Electron. Lett. 23(1987)18, 924-926.

[136] Kidoh, Y.; Suematsu, Y.; Furuya, K.: Polarization control on output of single-mode optical fibres. IEEE J. QE-17(1981), 991-994.

[137] Kikuchi, K.; Okoshi, T.; Nagamatsu, M.; Henmi, N.: Bit-error rate of PSK heterodyne optical communication system and its degradation due to spectral spread of transmitter and local oscillator. Electron. Lett. 19(1983)11, 417-418.

[138] Kikuchi, K.; Okoshi, T.; Nagamatsu, M.; Henmi, N.: Degradation of bit-error rate in coherent optical communications due to spectral spread of the transmitter and the local oscillator. IEEE J. LT-2(1984), 1024-1033.

[139] Kimura, T.: Coherent optical fibre transmission. IEEE J. LT-5(1987)4, 414-428.

[140] Kleekamp, C.; Metcalf, B.: Designer's Guide to Fiber Optics. Cahners, 1978.

[141] Kobylinski, R. A.: Coherent Nd:YAG laser communications. Proc. SPIE on Optical Technologies for Space Communication Systems, Vol.756(1987), 117-121.

[142] Kobayashi, S.; Yamamoto, Y.; Ito, M.; Kimura, T.: Direct frequency modulation in AlGaAs semiconductor lasers. IEEE J. QE-18(1982)4, 582-595.

[143] Kobayashi, S.: Semiconductor optical amplifiers. IEEE Spectrum, 21(1984)5, 26-32.

[144] Kreit, D.; Youngquist, R. C.: Polarisation-insensitive optical heterodyne receiver for coherent FSK communications. Electron. Lett. 23(1987)4, 168-169.

[145] Kubota, M.; Oohara, T.; Furuya, K.; Suematsu, Y.: Electro-optical polarisation control on single-mode optical fibres. Electron. Lett. 16(1980)15, 573.

[146] Künzel, T.: Simulation und Analyse optischer Überlagerungsempfänger unter Berücksichtigung des Laserphasenrauschens. Diplomarbeit, TU München, Lehrstuhl für Nachrichtentechnik, 1986.

[147] Kuwahara, H.; Chikama, T.; Ohsawa, C.; Kiyonaga, T.: New receiver design for practical coherent lightwave transmission system. IOOC-ECOC (1986), 407-410.

[148] Lax, M.: Classical noise V. Noise in self-sustained oscillators. Phys. Rev. 160 (1967) 290.

[149] Lee, T. P.: Linewidth of single-frequency semiconductor lasers for coherent lightwave communications. IOOC-ECOC (1985), 189-196.

[150] Leeb, W. R.: Degradation of signal-to-noise ratio in optical space data links due to background illuminations. Applied optics, 28(1989)15, 3443-3449.

[151] Lefevre, H. C.: Single-mode fibre fractional wave devices and polarisation controllers. Electron. Lett. 16(1980)20, 778-780.

[152] Lesh, J. R.: Optical communication research to demonstrate 2.5 bits/detected photon. Commun. Magazine, 1982, 35-37.

[153] Lesh, J. R.; Deutsch L. J.; Weber, W. J.: Plan for the development and demonstration of optical communications for deep space. Proc. SPIE on Optical Space Communication II, Vol.1552 (1991), 27-35.

[154] Letterer, R.; Krichbaumer, W.; Wallmeroth, K.: Signal-to-noise ratio considerations for an analog direct detection receiver with integrated optical amplifier. ESA Contract Number 7660/88/JS, 1989.

[155] Lightwave Electronics: Data sheet of Nd:YAG Q-switched laser. Series 110, California, USA, 1990.

[156] Linke, R. A.; Gnauck, A. H.: High-capacity coherent lightwave systems. IEEE J. LT-6(1988)11, 1750-1769.

[157] Love, J. D.; Sammut, R. A.; Snyder, A. W.: Birefringence in elliptically deformed optical fibres. Electron. Lett. 15(1979)20, 615-616.

[158] Lutz, E.; Söder, G.; Tröndle, K.: Generation of discrete stochastic processes with given probability density and autocorrelation on a digital computer. 4. Seminar, Akademie der Wissenschaften der CSSR, Prag (1979), 308-329.

[159] Machida, S.; Sakai, J.; Kimura, T.: Polarisation conservation in single-mode fibres. Electron. Lett. 17(1981)14, 494-495.

[160] Mahon, C. J.; Khoe, G. D.: Compensational deformation; new endless polarisation matching control schemes for optical homodyne or heterodyne receivers which require no mechanical drivers. IOOC-ECOC (1986), 267-270.

[161] Mahon, C. J.; Khoe, G. D.: Endless polarisation state matching control experiment using two controllers of finite control range. Electron. Lett. 23(1987)23, 1234-1235.

[162] Mahr, H.: Ein Plädoyer für den Begriff Frequenzmultiplex in der optischen Nachrichtentechnik. Frequenz 39(1985)12, 314-319.

[163] Malyon, D. J.; Hodgkinson, T. G.; Smith, D. W.; Booth, R. C.; Daymond-John, B. E.: PSK homodyne receiver sensitivity measurement at 1.5 μm. Electron. Lett. 19(1983)4, 144-146.

[164] Mandel, L.: Fundamental limits on information capacity of an optical communication channel. Kinam, Vol.5, Series C (1983), 213-232.

[165] Marko, H.: Methoden der Systemtheorie. Springer-Verlag, 1977.

[166] Mellis, J.: Optical phase modulation using semiconductor laser amplifiers. IOOC ´89 Kobe, Japan (1989), Paper 21B4-3.

[167] Minoli, D.; Keinath, R.: Distributed Multimedia Through Broadband Communications Services. Artech House, 1993.

[168] Mochizuki, K.; Namihira, Y.; Wakabayashi, H.: Polarization mode dispersion measurements in long single mode fibres. Electron. Lett. 17(1981)4, 153-154.

[169] Mollenauer, L. F.; Smith, K.: Demonstration of soliton transmission over more than 400 km in fiber with loss periodically compensated by Raman gain. Opt. Lett. 13 (1988)8, 675-680.

[170] Monerie, M.: Polarisation-maintaining single-mode fibre cables: influence of joints. Appl. Optics 20(1980), 712-713.

[171] Monerie, M.; Jeunhomme, L.: Polarization mode coupling in long single-mode fibres. Optical and Quantum Electronics 12(1980), 449-461.

[172] Monerie, M.; Lamouler, P.; Jeunhomme, L.: Polarization mode dispersion measurements in long single mode fibres. Electron. Lett. 16(1980)24, 970-980.

[173] Monerie, M.; Lamouler, P.: Birefringence measurement in twisted single-mode fibres. Electron. Lett. 17(1981)7, 252-253.

[174] Murakami, M. et. al.: 10 Gbit/s, 6000 km transmission experiment using erbium-doped fibre in-line amplifiers. Electron. Lett. 28(1992)24, 2254-2255.

[175] Naito, T.; Chikama, T.; Ishikawa, G.; Kuwahara, H.: 4 Gbit/s, 233-km optical fibre transmission experiment using newly proposed direct-modulation PSK. Electron. Lett. 26(1990)20, 1734-1736.

[176] Nakano, H. et. al.: 10 Gbit/s, 100 km nonrepeatered fibre transmission experiment using a high-sensitivity semiconductor optical preamplifier. Electron. Lett. 26(1990)17, 1364-1366.

[177] Namihira, Y.; Ryu, S.; Mochizuki, K.; Furusawa, K.; Iwamoto, Y.: Polarisation fluctuation in optical-fibre submarine cable under 8000 m deep sea environmental conditions. Electron. Lett. 23(1987)3, 100-101.

[178] Namihira, Y.; Horiuchi, Y.; Wakabayashi, H.: Dynamic polarisation fluctuation characteristics of optical-fibre submarine cable coupling under periodic variable tension. Electron. Lett. 23(1987)22, 1201-1202.

[179] Nassehi, M. M.; Tobagi, F. A.: Fiber optic configurations for local area networks. IEEE J. SAC-3(1985)11, 941-949.

[180] Neidlinger, S.: DPSK polarisation diversity receiver with novel switching-demodulators and maximal-ratio combining. Electron. Lett. 26(1990)14, 1070-1071.

[181] Nicholsen, G.: ASK homodyne system receiver using a 6-port fiber coupler. J. Opt. Commun. 9(1988)1, 13-16.

[182] Noé, R.: Endless polarization control in coherent optical communications. Electron. Lett. 22(1986)15, 772-773.

[183] Noé, R.: Endless polarizations control experiment with three elements of limited birefringence range. Electron. Lett. 22(1986)25, 1341-1343.

[184] Noé, R.: Sensitivity comparison of coherent optical heterodyne, phase diversity, and polarization diversity receivers. J. Opt. Commun. 10(1989)1, 11-18.

[185] Noé, R.; Rodler, H. J.; Ebberg, A.; Gaukel, G.; Noll, B.; Wittmann, J.; Auracher, F.: Comparison of polarisation handling methods in coherent optical systems. IEEE J. LT-9(1991)10, 1353-1365.

[186] Noé, R. et. al.: Direct modulation 565 Mb/s DPSK experiment with 62.3 dB loss span and endless polarization control. IEEE PTL(1992)10, 1151-1154.

[187] Nussmeier, T. A.; Goodwin, F. E.; Zavin, J. E.: A 10.6-μm terrestrial communication link. IEEE J. QE-10(1974)2, 230-235.

[188] Okamoto, K.; Sasaki, Y.; Miya, T.; Kawachi, M.; Edahiro, T.: Polarization characteristics in long length v.a.d. single-mode fibres. Electron. Lett. 16(1980)25, 768-769.

[189] Okamoto, K.; Hosaka, T.; Edahiro, T.: Stress analysis of single polarization fibers. Review of the Electrical Communication Laboratories, Vol.31(1983)3, 381-392.

[190] Okoshi, T.; Kikuchi, K.; Nakayama, A.: Novel method for high resolution measurement of laser output spectrum. Electron. Lett. 16(1980)16, 630-631.

[191] Okoshi, T.; Oyamada, K.: Single-polarisation single-mode optical fibre with refractive index-pits on both sides of core. Electron. Lett. 16(1980)18, 712-713.

[192] Okoshi, T.: Single-polarization single-mode optical fibers. IEEE J. QE-17(1981)6, 879-884.

[193] Okoshi, T.; Emura, K.; Kikuchi, K.; Kersten, R. Th.: Computation of bit-error rate of various heterodyne- and coherent-type optical communication schemes. J. Opt. Commun. 2(1981)3, 89-96.

[194] Okoshi, T.; Kikuchi, K.: Heterodyne-type optical fiber communications. J. Opt. Commun. 2(1981)3, 82-88.

[195] Okoshi, T.; Ryu, S.; Emura, K.: Measurement of polarization parameters of a single-mode optical fiber. J. Opt. Commun. 2(1981)4, 134-141.

[196] Okoshi, T.: Review of polarization-maintaining single-mode fiber. ECOC (1983), 57-59.

[197] Okoshi, T.; Ryu, S.; Kikuchi, K.: Polarisation-diversity receiver for heterodyne/coherent optical fiber communications. IOOC, Tokyo (1983), 386-387.

[198] Okoshi, T.: Ultimate performance of heterodyne/coherent optical fiber communications. IEEE J. LT-4(1986)10, 1556-1562.

[199] Okoshi, T.: Recent advances in coherent optical fiber communication systems. IEEE J. LT-5(1987)1, 44-52.

[200] Okoshi, T.; Cheng, Y. H.: Four-port homodyne receiver for optical fibre communication comprising phase and polarisation diversities. Electron. Lett. 23(1987)8, 377-378.

[201] Okoshi, T.; Kikuchi, K.: Coherent Optical Fiber Communications. Kluwer Academic Publisher, 1988.

[202] Oliver, B. M.: Thermal and quantum noise. Proc. IEEE, 53(1965)5, 436-454.

[203] Olsson, N. A.: Lightwave systems with optical amplifiers. IEEE J. LT-7(1989)7, 1071-1082.

[204] O'Mahony, M. J.: Semiconductor laser optical amplifiers for use in future fiber systems. IEEE J. LT-6(1988)4, 531-544.

[205] Palais, J. C.: Fiber Optic Communications. Second Edition, Prentice-Hall, 1988.

[206] Papoulis, A.: Probability, Random Variables and Stochastic Processes. McGraw-Hill, 1985.

[207] Park, Y. K.; Mizuhara, O.; Tzeug, L. D.: Multigigabit repeaterless optical fiber transmission systems. ECOC (1994), 607-617.

[208] Payne, D. N.; Barlow, A. J.; Ramskow Hansen, J. J.: Development of low-and high birefringence optical fibers. IEEE J. QE-18(1982)4, 477-488.

[209] Personick, S. D.: Optical Fiber Transmission System. Plenum Press, 1981.

[210] Pettitt, M. J.; Remedios, D.; Davis, A. W.; Hadjifotiou, A.; Wright, S.: Optical FSK transmission system using a phase-diversity receiver. Electron. Lett. 23(1987)20, 1075-1076.

[211] Piazzolla, S.; Spano, P.; Tamburrini, M.: Characterization of phase noise in semiconductor lasers. Appl. Phys. Lett. 41(1982)8, 695-696.

[212] Piazzolla, S.; Spano, P.: Analytical evaluation of the line shape of single-mode semiconductor lasers. Optics Commun. 51(1984)4, 278-280.

[213] Pietzsch, J.: Der Einfluß des Phasenrauschen auf die Fehlerquote bei Übertragung von frequenzmodulierten Signalen. AEÜ 42(1988)2, 132-138.

[214] Poole, C. D.; Bergano, N. S.; Schulte, H. J.; Wagner, R. E.; Nathu, V. P.; Amon, J. M.; Rosenberg, R. L.: Polarisation fluctuations in a 147 km undersea lightwave cable during installation. Electron. Lett. 23(1987)21, 1113-1115.

[215] Poole, C. D.; Bergano, N. S.; Wagner, R. E.; Schulte, H. J.: Polarisation dispersion in a 147 km undersea lightwave cable. ECOC (1987), 321-324.

[216] Poole, C. D.: Polarization dispersion and principal states in a 147 km undersea lightwave cable. IEEE J. LT-6(1988), 1185-1190.

[217] Powers, J. P.: An Introduction to Fiber Optic System. Aksen Associates Inc., 1993.

[218] Pratt, W. K.: Laser Communication Systems. John Wiley & Sons Inc., 1969.

[219] Purcell, E. M.: Elektrizität und Magnetismus. Berkeley Physik Kurs Band 2. Vieweg-Verlag, 1979.

[220] Ramachandran, G. N.; Ramaseshan, S.: Crystal optics. Handbuch der Physik Band 25/1 (S. Flügge), Springer-Verlag, 1962.

[221] Ramaswamy, V.; French, W. G.; Standley, R. D.: Polarization characteristics on noncircular core single-mode fibres. Appl. Optics 17(1978)18, 3014-3017.

[222] Ramaswamy, V.; Kaminov, I. P.; Kaiser, P.: Single polarization optical fibers: exposed cladding technique. Appl. Phys. Lett. 33(1978)9, 814-816.

[223] Ramaswamy, V.; Standley, R. D.; Sze, D.; French, W. G.: Polarization effects in short length, single mode fibers. Bell Systems Tech. J. 57(1978), 635-651.

[224] Rashleigh, S. C.; Stolen, R. H.: Preservation of polarization in single-mode fibers. Fiber-optic Tech. (1983)5, 155-161.

[225] Rocks, M.: Optischer Überlagerungsempfang: Die Technik der übernächsten Generation glasfasergebundener optischer Nachrichtensysteme. Der Fernmelde-Ingenieur 3 (1985) 2.

[226] Ross, F. E.: FDDI-A tutorial. IEEE Commun. Magazine. 24(1986)5, 10-17.

[227] Ryu, S. et. al.: Polarization diversity techniques for use in coherent optical fiber submarine cable sytems. IEEE J. LT-9(1991)5, 675-682.

[228] Saito S.; Yamamoto, Y.; Kimura, T.: Optical heterodyne detection of directly frequency modulated semiconductor laser signals. Electron. Lett. 16(1980)22, 826-827.

[229] Saito S.; Yamamoto, Y.: Direct observation of Lorentzian lineshape of semiconductor laser and linewidth reduction with external grating feedback. Electron. Lett. 17 (1981)9, 325-327.

[230] Saito, S.; Nilsson, O.; Yamamoto, Y.: Oscillation center frequency tuning, quantum FM noise, and direct frequency modulation characteristics in external grating loaded semiconductor lasers. IEEE J. QE-18(1982)6, 961-970.

[231] Saito, S.; Yamamoto, Y.; Kimura, T.: S/N and error rate evaluation for an optical FSK heterodyne detection system using semiconductor lasers. IEEE J. QE-19(1983)2, 180-193.

[232] Saitoh, T.; Mukai, T.: 1.5 μm GaInAsP travelling-wave semiconductor laser amplifier. IEEE J. QE-23(1987)6, 1010-1020.

[233] Saitoh, T.; Mukai, T.: Recent progress in semiconductor laser amplifiers. IEEE J. LT-6(1988)11, 1656-1664.

[234] Sakai, J.-I.; Kimura, T.: Birefringence and polarization characteristics of single-mode optical fibers under elastic deformations. IEEE J. QE-17(1981)6, 1041-1051.

[235] Sakai, J.-I.; Machida, S.; Kimura, T.: Existence of eigen polarization modes in anisotropic single-mode optical fibers. Optics Lett. 6(1981)10, 496-498.

[236] Sakai, J.-I.; Kimura, T.: Polarization behavior in multiple perturbed single-mode fibers. IEEE J. QE-18(1982)1, 59-65.

[237] Sakai, J.-I.; Machida, S.; Kimura, T.: Degree of polarization in anisotropic single-mode optical fibers: theory. IEEE J. QE-18(1982)4, 488-495.

[238] Sakai, J.-I.; Machida, S.; Kimura, T.: Twisted single-mode optical fiber as polarization-maintaining fiber. Review of Electrical Communication Laboratories Vol.31 (1983)3, 372-380.

[239] Salz, J.: Coherent lightwave communications. AT&T Tech. J. 64(1985)10, 2153-2209.

[240] Sanghi, R. K.; Jhunjhunwala, A.: Implementation considerations in fiber optic bus networks. IETE Technical Review, 6(1989)2, 97-1.10.

[241] Saruwatori, M.: All-optical signal processing in ultrahigh-speed optical transmission. IEEE Commun. Magazine 32(1994)9, 98-105.

[242] Sasaki, Y.; Shibata, N.; Hosaka, T.: Fabrication of polarization-maintaining and absorption-reducing optical fibers. Review of Electrical Communication Laboratories Vol. 31(1983)3, 400-409.

[243] Schaller, H. N.: Berechnung optischer DPSK-Überlagerungssysteme unter Berücksichtigung des Laserphasenrauschens. Diplomarbeit, TU München, Lehrstuhl für Nachrichtentechnik, 1987.

[244] Schneider, R.; Pietzsch, J.: Coherent 565 Mbit/s DPSK transmission experiment with a phase diversity receiver. ECOC (1987), Vol. III, 5-8.

[245] Scholz, A.; Leeb, W. R.; Philipp, H. K.: Detection homodyne pour systemes de communication laser. IOOC-ECOC (1982), 541-546.

[246] Scholz, A.; Philipp, H. K.; Leeb, W. R.: Receiver concepts for data transmission at 10 microns. ESA SP-202 (1984), 107-114.

[247] Senior, J. M.: Optical Fiber Communication-Principles and Practice. Second Edition, Prentice-Hall, 1992.

[248] Sharma, A. B.; Halme, S. J.; Butosov, M. M.: Optical Fiber Systems and Their Components. Springer-Verlag, 1981.

[249] Shibata, N.; Okamoto, K.; Sasaki, Y.: Structure design for polarization-maintaining and absorption-reducing optical fibers. Review of Electrical Communication Laboratories Vol. 31(1983)3, 393-399.

[250] Shikada, M. et. al.: 100-Mbit/s ASK heterodyne detection experiment using $1.3\mu m$ DFB lasaer diodes; OFC-84, Paper No. TUK 6(1984).

[251] Shinda, S: Optical amplifiers for optical communications systems, IEIC Trans. E74(1991)1, 65-74.

[252] Simon, J. C.: GaInAsP semiconductor laser amplifiers for single-mode fiber communications. IEEE J. LT-5(1987)9, 1286-1295.

[253] Singh, N.; Jain, V. K.; Gupta, H. M.: Performance of heterodyne coherent optical fiber communication systems in presence of shot noise, LO excess noise, laser phase noise and time jitter. J. Opt. Commun. 10(1989)2, 48-53.

[254] Smith, D. W.; Stanley, I. W.: The worldwide status of coherent optical fibre transmission systems. IOOC-ECOC (1983), 263-266.

[255] Sobel, H.: The application of microwave techniques in lightwave systems. IEEE J. LT-5(1987)3, 293-299.

[256] Söder, G.; Tröndle, K.: Digitale Übertragungssysteme. Springer-Verlag, 1984.

[257] Spälti, A.: Der Einfluß des thermischen Widerstandsrauschens und des Schroteffektes auf die Störmodulation von Oszillatoren. Bulletin des schweizerischen Elektrotechnischen Vereins, 39(1948)13, 419-426.

[258] Spano, P.; Piazzolla, S.; Tamburrini, M.: Phase noise in semiconductor lasers: A theoretical approach. IEEE J. QE-19(1983)7, 1195-1199.

[259] Spirit, D. M.; Blank, L. C.: Raman-assisted long-distance optical time domain reflectometry. Electron. Lett. 25(1989), 1687-1688.

[260] Stein, S.; Jones, J. J.: Modern Communication Principles. McGraw-Hill, 1967.

[261] Stolen, R. H.; Ramaswamy, V.; Kaiser, P.; Pleibel, W.: Linear polarization in birefringent single-mode fibers. Appl. Phys. Lett. 33(1978)8, 699-701.

[262] Tamburrini, M.; Spano, P.; Piazzolla, S.: Influence of semiconductor-laser phase noise on coherent optical communication systems. Optics. Lett. 8(1983)3, 174-176.

[263] Tatam, R. P.; Pannell, C. N.; Jones, J. D. C.; Jackson, D. A.: Full polarisation state control utilizing linearly birefringent monomode optical fiber. IEEE J. LT-5(1987)7, 980-985.

[264] Taub, H.; Schilling, D. L.: Principles of Communication Systems. Second Edition, McGraw-Hill, 1986.

[265] Tjaden, D. L. A.: Birefringence in single-mode optical fibres due to core ellipticity. Phillips J. Res. 33(1978)5/6, 254-263.

[266] Tonguz, O. K. et. al.: Equivalence between preamplified direct detection and heterodyne receivers. IEEE PTL(1991)9, 835-837.

[267] Tradowsky, K.: Laser kurz und bündig. Vogel-Verlag, 1977.

[268] Treiber, H.: Laser Technik. Frech-Verlag, 1982.

[269] Tröndle, K.; Söder, G.: Optimization of Digital Transmission Systems. Artech House, 1987.

[270] Tsushima, H. et. al.: 1.244 Gbit/s 32 channel 121 km transmission experiment using shelf-mounted continuous-phase FSK optical heterodyne system. Electron. Lett. 27(1991)25, 2336-2337.

[271] Tzeng, L. D.; Emkey, W. L.; Jack, C. A.; Burrus, C. A.: Polarisation-insensitive coherent receiver using a double balanced optical hybrid system. Electron. Lett. 23(1987)22, 1195-1196.

[272] Ulrich, R.: Polarization stabilization on single-mode fiber. Appl. Phys. Lett. 35 (1979), 840-842.

[273] Ulrich, R.; Simon, A.: Polarization optics of twisted single-mode fibers. Appl. Optics 18 (1979)13, 2241-2251.

[274] Unger, H.-G.: Optische Nachrichtentechnik. Hüthig-Verlag 1990.

[275] Unger, H.-G.: Optische Nachrichtentechnik Teil II. Hüthing-Verlag, 1985.

[276] Urquhart, P.: Review of rare-earth doped fiber lasers and amplifiers. IEE Proc. Pt. J, 135(1988)6, 385-407.

[277] Vahala, K.; Yariv, A.: Semiclassical theory of noise in semiconductor lasers-Part I. IEEE J. QE-19(1983)6, 1096-1101.

[278] Vahala, K.; Yariv, A.: Semiclassical theory of noise in semiconductor lasers-Part II. IEEE J. QE-19(1983)6, 1102-1109.

[279] Vodhanel, R. S. et. al.: Performance of directly modulated DFB lasers in 10-Gb/s ASK, FSK, and DPSK lightwave systems. IEEE J. LT-8(1990)9, 1379-1386.

[280] Wandernoth, B.: 1064 nm, 565 Mbit/s PSK transmission experiment with homodyne receiver using synchronization bits. Electron. Lett., 27(1991)19, 1692-1693.

[281] Wandernoth, B.: 5 photon/bit low complexity 2 Mbit/s PSK transmission breadboard experiment with homodyne receiver applying synchronization bits and convolutional coding. ECOC (1994), 59-62.

[282] Wagner, R. E.: Coherent optical systems technology. ECOC (1986), 71-78.

[283] Wagner, R. E.; Linke, R. A.: Heterodyne lightwave systems: Moving towards commercial use. IEEE LCS, Nov. 1990, 28-35.

[284] William, B.; Jones, J.: Optical Fiber Communication Systems. Holt Rinehart and Winston Inc., 1988.

[285] Wu, Jingshown et. al.: Coding to relax laser linewidth requirements and improve receiver sensitivity for coherent optical binary phase-shift keying (BPSK) communications. IEEE J. LT-8(1990)4, 545-553.

[286] Yamamoto, Y.: Noise and error rate performance of semi-conductor laser amplifiers in PCM-IM optical transmission system. IEEE J. QE-16(1980)10, 1073-1081.

[287] Yamamoto, Y.: Receiver performance evaluation of various digital optical modulation-demodulation systems in the 0.5-10 μm wavelength region. IEEE J. QE-16(1980) 11, 1251-1259.

[288] Yamamoto, Y.; Kimura, T.: Coherent optical transmission systems. IEEE J. QE-17 (1981)6, 919-934.

[289] Yamamoto, Y.: AM and FM quantum noise in semiconductor lasers-Part I: Theoretical analysis. IEEE J. QE-19(1983)1, 34-46.

[290] Yamamoto, Y.; Saito, S.; Mukai, T.: AM and FM quantum noise in semiconductor lasers-Part II: Comparison of theoretical and experimental results for AlGaAs lasers. IEEE J. QE-19(1983)1, 47-58.

[291] Yamashita, S.; Okoshi, T.: Analysis of Optical Fiber Communication System with Optical Amplifiers. Tech. Rep. IEICE Japan, No.OCS-90-20, 1990.

[292] Yamazaki, S. et. al.: Optical CATV distribution system based on coherent FDM techniques. IEEE J. LT-8(1990)3, 396-405.

[293] Yoneda, E.; Suto, K.-I.; et. al.: Erbium-doped fiber amplifiers for all fiber video distribution (AFVD) systems. IEICE Trans. on Commun., E-75B(1992)9, 850-861.

[294] Zhu, Z.: Principles and design of bidirectional optical ISL using GaAlAs laser at λ=0.85 μm and direct detection PPM. International J. Satellite Commun. (1988)6, 81-90.

INDEX